T0260871

American Geophysical Union

ANTARCTIC
RESEARCH
SERIES

Antarctic Research Series Volumes

THE ANTARCTIC RESEARCH SERIES

The Antarctic Research Series, published since 1963 by the American Geophysical Union, now comprises more than 70 volumes of authoritative original results of scientific work in the high latitudes of the southern hemisphere. Series volumes are typically thematic, concentrating on a particular topic or region, and may contain maps and lengthy papers with large volumes of data in tabular or digital format. Antarctic studies are often interdisciplinary or international, and build upon earlier observations to address issues of natural variability and global change. The standards of scientific excellence expected for the Series are maintained by editors following review criteria established for the AGU publications program. Priorities for publication are set by the Board of Associate Editors. Inquiries about published volumes, work in progress or new proposals may be sent to Antarctic Research Series, AGU, 2000 Florida Avenue NW, Washington, DC 20009 (http://www.agu.org), or to a member of the Board.

Mount Discovery, McMurdo Sound, moraine deposits which contain the most important record, preserved in glacial erratics, of Paleogene life and paleoenvironments of this interval from East Antarctica. The rich suite of fossiliferous erratics and various lithofacies recovered provides the groundwork for reconstructing Eocene high-latitude nearshore environments and planktic/benthic communities preserved in the erratics. The implications for East Antarctic paleoclimate and paleoceanography prior to the onset of significant glaciation in Antarctica, are far reaching and are of major interest to the global scientific community. Photograph taken by Dr. Jeffrey D. Stilwell, January 3, 1993.

Volume 76

ANTARCTIC
RESEARCH
SERIES

Paleobiology and Paleoenvironments of Eocene Rocks, McMurdo Sound, East Antarctica

Jeffrey D. Stilwell and Rodney M. Feldmann
Editors

American Geophysical Union
Washington, D.C.
2000

PALEOBIOLOGY AND PALEOENVIRONMENTS OF EOCENE ROCKS, MCMURDO SOUND, EAST ANTARCTICA
Jeffrey D. Stilwell and Rodney M. Feldmann, Editors

Published under the aegis of the Board of Associate Directors, Antarctic Research Series

Library of Congress Cataloging-in-Publication Data
Paleobiology and paleoenvironments of Eocene rocks, McMurdo Sound, East Antarctica
/ Jeffrey D. Stilwell and Rodney M. Feldmann, editors
            p.cm -- (Antarctic research series ; v. 76)
        Includes bibliographical references.
        ISBN 0-87590-947-7
         1. Paleontology--Eocene. 2. Fossils--Antarctica--McMurdo Sound Region. 3.
        Boulders--Antarctica--McMurdo Sound Region. 4. Geology, Stratigraphic--Eocene. 5.
        Geology--Antarctica--McMurdo Sound Region. I. Stilwell, Jeffrey D. II. Feldmann,
        Rodney M. III. Series

QE737.P35 2000
560'.1784--dc21

00-20726
CIP

ISBN 0-87590-947-7
ISSN 0066-4634

Cover
Coastal glacial moraine deposits at Mount Discovery, McMurdo Sound, where a wealth of fossiliferous erratics of predominantly Eocene age have been recovered. Many important rock and fossil specimens were discovered in the deposits shown here. Note Ice Pinnacles and Black Island in background. Photograph taken by Jeffrey D. Stilwell, November 11, 1995.

Published by
American Geophysical Union
2000 Florida Avenue, N.W.
Washington, D.C. 20009
With the aid of grant OPP-9414962
from the National Science Foundation

Printed in the United States of America.

CONTENTS

PREFACE

Michael K. Brett-Surman, George Washington University, observed that, "being a paleontologist is like being a coroner except all the witnesses are dead and all the evidence has been left out in the rain for 65 million years." In the study of paleontology in Antarctica it could also be added that, if not left out in the rain, most of the evidence remains buried beneath several thousand feet of ice. Elucidating the geologic history of the Antarctic continent will always be plagued with this problem. Nonetheless, numerous clever means have been used to extract as much information as is possible, and as presented in this volume.

In this light, one of the most intriguing time intervals in Antarctic history is the Eocene Epoch. During this time, the climatic conditions deteriorated rapidly from the so-called "Greenhouse" conditions that dominated Earth's conditions from mid-Mesozoic time through the early Cenozoic to the "Icehouse" conditions that have dominated the climate since that time. Unfortunately, the record of Eocene rocks on the continent is sparse. On the Antarctic Peninsula, specifically on Seymour Island, a robust record of Eocene rocks and fossils has provided virtually all the information we possess about this time interval. Thus the discovery and description of Eocene erratic boulders in morainal deposits in the McMurdo Sound region provides only the second site on the entire continent where we can study the paleontology of this time interval. In all likelihood, the description of erratics containing fossils from any other place in the world would warrant little study and would attract even less attention. However, when most of the vast area of Antarctica lies beneath ice and when clues to the nature of the crust of that part of the continent can be extracted only from study of erratics, the discovery carries with it some excitement.

The study of fossiliferous erratics grows more significant when they are found to contain a diverse array of vertebrate, invertebrate, and plant material, which makes it possible to interpret the Eocene climatic and paleoceanographic setting of the McMurdo Sound region in some detail. Furthermore, the work reinforces conclusions drawn about the climate of the continent from research conducted on Seymour Island. As a result, the study of these erratics has strengthened our understanding of the conditions that prevailed in the high southern latitudes just prior to final separation of Antarctica from the Australian continent and establishment of the isolation of the southernmost continent.

Because of the wide range of fossils known from the erratics, numerous specialists were recruited to study the specimens. The result is a collection of highly authoritative articles providing a benchmark for further work in the area. As with many such studies, this work may be regarded as preliminary, with a next step undoubtedly requiring the serendipitous discovery of a new site.

We thank all the contributors to the volume for their efforts in bringing the work to completion. In addition to the authors, a large cadre of reviewers read the contributions and provided valuable suggestions. The content of the papers is the responsibility of the authors and editors; the quality of the final product was much enhanced by the reviewers and we thank them. Those who chose to be identified are acknowledged in the individual articles. Finally, the transformation from our idea of "camera-ready" copy to the finished product was the task of Karen Smith, Department of Geology, Kent State University, who did a magnificent job. Financial support for the work was largely through the National Science Foundation. As with all projects of this type, the volume of work necessary to complete the task is always underestimated. We beg your indulgence. However, as with fine wine, ...

Jeffrey D. Stilwell
James Cook University
Townsville, Queensland, Australia

Rodney M. Feldmann
Kent State University
Kent, Ohio

Editors

THE McMURDO ERRATICS: INTRODUCTION AND OVERVIEW

David M. Harwood and Richard H. Levy

Department of Geosciences, University of Nebraska-Lincoln, Lincoln, Nebraska 68588-0340

This volume presents paleontological, lithological and paleoenvironmental information derived from a suite of fossiliferous erratics from coastal moraines of the southern McMurdo Sound area. These "McMurdo Erratics" provide a record of conditions in East Antarctica when global "Greenhouse-Earth" conditions prevailed prior to the development of Cenozoic continental-scale ice sheets, and conditions after the transition to Oligocene-Recent "Icehouse-Earth." Microfossil biostratigraphy indicates an age of middle Eocene to late Eocene for most of the erratics. The Eocene fossiliferous erratics preserve a history of marine deposition in a fertile, coastal setting with abundant life. The fossils reflect Eocene paleobiogeography in the southern high latitudes at a time when Antarctica was becoming isolated from Gondwana and marine seaways developed across and around Antarctica in response to the rifting and fragmentation of West Antarctic basins. Fossil wood, leaves and pollen in the Eocene fossiliferous erratics suggest a cool temperate climate in coastal areas, adjacent to the uplifting Transantarctic Mountains. Vertebrate fossils include shark, fish, bird, and crocodile remains. Three sedimentary facies are identified in the Eocene erratics: sandstone, sandy-mudstone, and conglomerate. These are further divided into twelve sub-facies. Erratics of Oligocene and Miocene age include lithologies of diamictite and mudstone with dropstones that originated from strata deposited at the margin of a glaciated Antarctic continent. Fossiliferous clasts of the upper Pliocene Scallop Hill Formation are the youngest recognized in the McMurdo Erratics; they provide age constraint on emplacement of the McMurdo Erratics. During the late Pliocene-early Quaternary an expanded East Antarctic ice sheet excavated Eocene to Pliocene sedimentary rocks from the "Discovery Deep" basin and transported the erratics to McMurdo Sound. This event also transported basement igneous, metamorphic and sedimentary (Beacon Supergroup) clasts to the southern McMurdo Sound moraines. The source strata of the McMurdo Erratics are not known to crop out in East Antarctica, but are assumed to lie beneath the Ross Ice Shelf near the confluence of the Byrd, Skelton and Mulock glaciers in the "Discovery Deep" basin. In addition to providing significant paleoenvironmental data for the Eocene, the restricted distribution of the fossiliferous erratics provides important constraint on the drainage and extent of Plio-Pleistocene ice sheets in southern McMurdo Sound.

APPROACH TO UNCOVER ANTARCTICA'S HIDDEN GEOLOGY AND PALEONTOLOGY

Much of Antarctica's Cenozoic geological record is covered by the Antarctic ice sheet. Glacial erratics eroded and transported from subglacial basins, at times when the Antarctic ice sheet was expanded, provide a means to obtain information about stratigraphic units hidden beneath the ice. Fossiliferous erratic boulders of Eocene-Pliocene age, the "McMurdo Erratics", are present in coastal moraines of southern McMurdo Sound [Wilson, this volume]. They contain important paleontological and sedimentological information that documents the paleoclimate and paleoenvironment of poorly known periods of Antarctic geologic history.

The McMurdo Erratics are thought to have been derived from a broad, deep basin (>1000 mbsf) on the western side of the Ross Embayment, 'Discovery Deep' [Rowe, 1974; Stilwell et al., 1997], in front of the Byrd, Mulock and Skelton glaciers (Figure 1). 'Discovery Deep' was likely carved throughout the Late Cenozoic when outlet glaciers of an expanded, polythermal East Antarctic ice sheet advanced across the continental shelf of the western Ross Embayment. Some erratics may also originate from erosion in the Transantarctic Mountains at the margins of these large outlet glaciers. The presence of erratics of the Pliocene Scallop Hill Formation [Speden, 1962; Vella, 1969; Eggers, 1979; Leckie and Webb, 1979; Webb and Andreasen, 1986; Jonkers, 1998] in the coastal moraines indicate that the most recent interval of scouring of 'Discovery Deep' occurred during a latest Pliocene to early Quaternary advance of the southern McMurdo Sound ice sheet [Wilson, this volume]. This advance sampled a diversity of rocks [Levy and Harwood, this volume a] from stratigraphic sections that included parts of the middle Eocene to upper Miocene (an interval of more than 35 million years). The boulder size of many erratics provided sufficient material to enable paleoecological community and lithofacies analysis, paleoenvironmental reconstruction, and recovery of rare vertebrate teeth and bones.

It was understood at the outset of this project that reconstructing the stratigraphic record from these erratics would be a difficult 'jig-saw puzzle'. But, it was worthy of considerable effort, as Eocene strata do not crop out in East Antarctica and Oligocene-Miocene strata are known from only a few drillholes. Webb (1990) noted, "We must concentrate on ways in which to uncover the 98% of Cenozoic geology we have still not encountered!"

Each erratic represents a piece of strata removed from its stratigraphical context. Our task, through 'reconstructive biostratigraphy', was to develop a temporal and spatial matrix indicated by paleontological and sedimentological characteristics of the erratics, and then position individual erratics within this matrix. Approaches to reconstruct the environment at the time of deposition of the fossils in each erratic are developed and applied in Levy [1998].

PRE-GLACIAL TO GLACIAL TRANSITION IN ANTARCTICA

Antarctica has held a polar position for the last 100 million years [Lawver, et al., 1992]. This region experienced significant steps of climate cooling that divide Cenozoic glacial history into distinct intervals [Moriwaki

et al., 1992; Barrett, 1996]: (1) Cretaceous-Eocene period of global "Greenhouse-Earth" conditions when ice was restricted to inland areas and montane settings; (2) the post-Eocene "Icehouse-Earth" period, characterized by growth and retreat of multiple, wet-based/polythermal ice sheets; and finally (3) the period including the shift to the present stable(?) cold-polar ice sheet in East Antarctica. The last period began between the middle Miocene to late Pliocene, and is a matter of current debate [Webb and Harwood, 1991; Sugden et al., 1993; van der Wateren and Hindmarsh, 1995; Wilson, 1995; Quilty, 1996; Miller and Mabin, 1998; Harwood and Webb, 1998].

The Eocene McMurdo Erratics date from an important time in Antarctic geological evolution: (1) near the end of global "Greenhouse-Earth" conditions; (2)shortly before the postulated onset and continental expansion of Cenozoic temperate Antarctic ice sheets [Wise et al., 1991]; (3) just after separation of the last two Gondwana continents, Australia and Antarctica; (4) during an early stage of uplift of the Transantarctic Mountains [Fitzgerald, 1992]; and (5) rifting of West Antarctic basins [Davey, 1987]. Cenozoic evolution of both terrestrial and marine flora and fauna reflect climate cooling, continental isolation and changes in marine connections across and around Antarctica. Future comparison of the paleobiota and paleoenvironments of East Antarctica with coeval assemblages from the Antarctic Peninsula and around the Antarctic periphery should document trans-Antarctic and circum-Antarctic marine and terrestrial migration routes.

"The paleontological database (of Antarctica) has undergone a radical improvement over the past two decades and now contributes to the solution of a wide range of biostratigraphic, paleontologic, biogeographic, paleoceanographic and evolutionary studies within and beyond the south polar basins" [Webb, 1991]. This reflects our increased knowledge from (1) drilling by DSDP/ODP drilling on the Antarctic continental shelf and in the Southern Ocean [Hayes et al., 1975; Kennett et al., 1975; Hollister, 1976; Barker et al., 1977, 1990; Ludwig et al., 1983; Webb, 1990; Barron et al., 1991; Ciesielski, 1991; Kennett and Barron, 1992; Wise et al., 1992]; (2) other drilling efforts on the Antarctic margin [Barrett, 1986, 1989; McKelvey et al., 1991; Hambrey and Barrett, 1993; Hambrey and Wise, 1998; Cape Roberts Science Team, 1998, 1999]; and (3) land-based Cenozoic research in Antarctica [Feldmann and Woodburne, 1988; Webb, 1991; Birkenmajer, 1991; Moriwaki et al, 1992; Prentice et al., 1993; Quilty, 1993; Sugden et al., 1993; Webb et al., 1996; Jonkers, 1998; Ashworth et al., 1997].

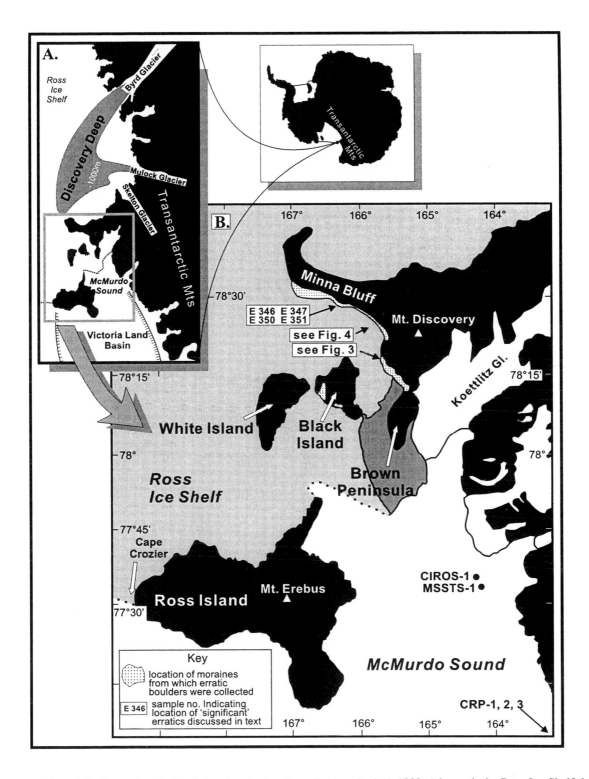

Figure 1 A. Map of Southern Victoria Land showing the location of a deep basin (<1000m) beneath the Ross Ice Shelf, known informally as 'Discovery Deep'. This basin is the likely source for the McMurdo Erratics. B. Map of McMurdo Sound showing the locations of erratic-bearing moraines and the locations of areas (Figs. 3, 4) where the majority of the erratics discussed in this volume were collected. Location of key fossiliferous erratics is also indicated.

TABLE 1. Summary list of the McMurdo Erratics reported in this volume. Samples in bold reflect key samples documented in the papers in this volume; their locations are indicated on Figures 1, 3, or 4. "Lithofacies" include: Sm - massive sandstone, Smc - massive sandstone with intraclasts, Ss - stratified sandstone, Sst - stratified sandstone with trough cross-strata, Ssg - stratified sandstone with grading, Sw - weakly stratified sandstone, Mmb - massive bioturbated sandy mudstone, Ms-d - stratified sandy mudstone with dropstones, Mwb -weakly stratified bioturbated sandy mudstone, Mm-d - massive mudstone with dropstones, Cmc - massive clast supported conglomerate, Cmm - massive matrix supported conglomerate, Cscg - stratified clast supported conglomerate, Dm - massive diamictite, Dw - weakly stratified diamictite. "Fossils Recovered" include: MP - marine palynomorphs, TP - terrestrial palynomorphs, S - siliceous microfossils, F - foraminifera, L - leaves, W - wood, M - molluscs, B - brachiopods, MV - marine vertebrates, TV - terrestrial vertebrate, C - corals, A - arthopods (Ab - barnacle plates, Ac - crustaceans, Ao - ophiomorpha). N/A in the 'Age' column, indicates that no biostratigraphic data is available for the sample. 'Key Fossils' are those utilized for biostratigraphic age control and include: Dn - dinoflagellate cysts, Si - siliceous microfossils, and M - molluscs.

Sample	Collected From:	Lithofacies	Fossils Recovered	Stratigraphic level	Key Fossil
MTD 1	Mt. Discovery	Sm	MP, TP, L	middle to upper Eocene	Dn
MTD 42	Mt. Discovery	Cmm	MP, TP	middle to upper Eocene	Dn
MTD 46	Mt. Discovery	Mmb	S	middle Miocene	Si
MTD 56	Mt. Discovery	Sm	MP, TP	?lower Oligocene	Dn
MTD 95	Mt. Discovery	Mwb	S	middle Eocene	Si
MTD 148	Mt. Discovery	Cmm		N/A	
MTD 153(1)	Mt. Discovery	Smc (clast)	MP, TP	middle to upper Eocene	Dn
MTD 153(2)	Mt. Discovery	Smc (matrix)	MP, TP	middle to upper Eocene	Dn
MTD 154	Mt. Discovery	Sm	MP, TP, M	middle to upper Eocene	Dn
MTD 166	Mt. Discovery	Sm/Quartzite	M	middle to upper Eocene	M
MTD 174A	Mt. Discovery	Sm/Quartzite	TP	?Paleozoic/Mesozoic	
MTD 189	Mt. Discovery	Cmm	MP, TP	middle to upper Eocene	Dn
MTD 190	Mt. Discovery	Sm	MP, TP, M	middle to upper Eocene	Dn
MTD 193A	Mt. Discovery	Sm		N/A	
MTD 197	Mt. Discovery	Ms-d		N/A	
MTD 203	Mt. Discovery	Sm		N/A	
MTD 211A(1)	Mt. Discovery	Ms-d (matrix)	MP, TP,	post-Eocene	Dn
MTD 211A(2)	Mt. Discovery	Quartzite (clast)		?Paleozoic/Mesozoic	
MTD 211B	Mt. Discovery	Sm/Quartzite		N/A	
MB 80	Minna Bluff	Ss	MP, TP,	middle to upper Eocene	Dn
MB 103	Minna Bluff	Sst/Cmm	MP, TP,	middle to upper Eocene	Dn
MB 109(1)	Minna Bluff	Smc (clast)	MP, TP,	middle to upper Eocene	Dn
MB 109(2)	Minna Bluff	Smc (matrix)	MP, TP,	middle to upper Eocene	Dn
MB 172	Minna Bluff	Mm-d	F	N/A	
MB 181	Minna Bluff	Ssg/Csgc	MP, TP, S, M, C	middle to upper Eocene	Dn, Si
MB 187A	Minna Bluff	Sm		N/A	
MB 188B	Minna Bluff	Sm	MP, TP	middle to upper Eocene	Dn
MB 188F	Minna Bluff	Cmm		N/A	
MB 188G	Minna Bluff	Cmm	MP, TP	middle to upper Eocene	Dn
MB 191A	Minna Bluff	Dm		N/A	
MB 202	Minna Bluff	tuff	TP	???	
MB 210	Minna Bluff	Ss/volcanoclastic		N/A	
MB 212I	Minna Bluff	Mm-d		N/A	
MB 212K	Minna Bluff	Mm-d	MP, TP, F	post-Eocene	Dn
MB 213C	Minna Bluff	Dm		N/A	
MB 217A	Minna Bluff	Mm-d	MP, TP, F	post-Eocene	Dn
MB 220	Minna Bluff	Sm	L	N/A	
MB 223F	Minna Bluff	Mm-d		N/A	
MB 224	Minna Bluff	Sm	M	N/A	
MB 235A	Minna Bluff	Dm	MP, TP, S	Miocene	Si
MB 235C	Minna Bluff	Cmc	MP, TP	middle to upper Eocene	Dn
MB 244C	Minna Bluff	Mm-d	S, F	upper Miocene	Si
MB 245	Minna Bluff	Mmb	MP, TP, S	middle to upper Eocene	Dn, Si
MB 249	Minna Bluff	Sm	M	middle to upper Eocene	M
MB 285	Minna Bluff	Sm		N/A	
MB 288B	Minna Bluff	Mm-d		N/A	
MB 290G	Minna Bluff	Mm-d	MP	post-Eocene	Dn
MB 292C	Minna Bluff	Dm		N/A	
MB 292D	Minna Bluff	Sm		N/A	
MB 299	Minna Bluff	Dm	MP	post-Eocene	Dn
MB 301	Minna Bluff	Sm/Sw	M	middle to upper Eocene	M
E 100	Mt. Discovery	Sm	MP, TP, M, Ao	middle to upper Eocene	Dn
E 115	Minna Bluff	Ms-d	MP	post-Eocene	Dn
E 145	Mt. Discovery	Sm	MP, TP, M, MV, L, W	middle to upper Eocene	Dn

TABLE 1 (continued). Summary list of the McMurdo Erratics reported in this volume

Sample	Collected From:	Lithofacies	Fossils Recovered	Age	Key Fossil
E 151	Mt. Discovery	Sm	Ab	N/A	
E 153	Minna Bluff	Sm	MP, TP, M, B, L, W	middle to upper Eocene	Dn
E 155	Mt. Discovery	Sm	MP, TP, M, Ab	middle to upper Eocene	Dn
E 163	Mt. Discovery	Sm	MP, TP, M	middle to upper Eocene	Dn
E 165	Mt. Discovery	Sm	MP, Ac	middle to upper Eocene	Dn
E 168	Mt. Discovery	Sm	MP, TP, M	middle to upper Eocene	Dn
E 169	Mt. Discovery	Sm	MP, TP, M, B	middle to upper Eocene	Dn
E 171	Mt. Discovery	Sm	MP, TP, M	middle to upper Eocene	Dn
E 181	Mt. Discovery	Sm	MP, M	middle to upper Eocene	M
E 183	Mt. Discovery	Sm	M	middle to upper Eocene	M
E 184	Mt. Discovery	Smc	MP, TP, M, L	middle to upper Eocene	Dn
E 185	Mt. Discovery	Sm	M	middle to upper Eocene	M
E 189	Mt. Discovery	Sm	MP, TP, Ab	middle to upper Eocene	Dn
E 191	Mt. Discovery	Sm	MP, TP,	middle to upper Eocene	Dn
E 192	Mt. Discovery	Sm	MP, TP, M	middle to upper Eocene	M
E 194	Mt. Discovery	Sm	MP, TP, M	middle to upper Eocene	Dn
E 200	Mt. Discovery	Sm	MP, TP, M	middle to upper Eocene	M
E 202	Mt. Discovery	Sm	MP, TP, M	middle to upper Eocene	M
E 203	Mt. Discovery	Sm	MP, M	middle to upper Eocene	M
E 207	Mt. Discovery	Sm	MP, TP, M	middle to upper Eocene	M
E 208	Mt. Discovery	Sm	MP, TP	middle to upper Eocene	Dn
E 214	Mt. Discovery	Ms-d	MP, TP, M	middle to upper Eocene	Dn
E 215	Mt. Discovery	Sw	MP, TP, L	middle to upper Eocene	Dn
E 216	Mt. Discovery	Ms-d	MP, TP, F, M	post-Eocene	Dn
E 219	Minna Bluff	Mmb	MP, TP, M, L, W	middle to upper Eocene	Dn
E 240	Minna Bluff	Ms-d	M	???	
E 242D	Minna Bluff	Dm	MP, TP	post-Eocene	Dn
E 243	Minna Bluff	Dm/Dw	MP, TP	post-Eocene	Dn
E 244		Mm-d	TP, F	???	
E 303(1)	Mt. Discovery	Sm (matrix)	MP, TP, M, TV	middle to upper Eocene	Dn
E 303(2)	Mt. Discovery	Mmb (clast)	MP, TP	middle to upper Eocene	Dn
E 308	Mt. Discovery	Sm	Ab	N/A	
E 313	Mt. Discovery	Smc	MP, TP, M	middle to upper Eocene	M
E 317	Mt. Discovery	Sm	MP, TP	middle to upper Eocene	Dn
E 323	Mt. Discovery	meta	MP, TP	???	
E 331	Mt. Discovery	Sm	M	middle to upper Eocene	M
E 339	Minna Bluff	Mm-d		N/A	
E 344(1)	Mt. Discovery	Mmb (clast)	M	N/A	
E 344(2)	Mt. Discovery	Sm (matrix)		N/A	
E 345	Mt. Discovery	Sm	MP, TP, S, M	middle to upper Eocene	Dn, Si
E 346	Minna Bluff	Dm	TP, S	lower to middle Miocene	Si
E 347	Minna Bluff	Dm	MP, TP, S	upper Oligocene to lower Miocene	Si
E 350	Minna Bluff	Mmb	MP, TP, S, M	upper Eocene	Dn, Si
E 351	Minna Bluff	Dm	TP, S	Miocene	Si
E 355	Mt. Discovery	?metased	MP, TP	???	Dn
E 356	Mt. Discovery	Sm	MP, TP	?lower Oligocene	Dn
E 357	Mt. Discovery	Smc	MP, TP, L	middle to upper Eocene	Dn
E 360	Mt. Discovery	Mm-d		???	
E 363	Mt. Discovery	Mm-d	MP	post-Eocene	Dn
E 364	Mt. Discovery	Mw	MP, TP, S	middle to upper Eocene	Dn, Si
E 365(1)	Mt. Discovery	Mmb	MP, TP	middle to upper Eocene	Dn
E 365(2)	Mt. Discovery	Sm	MP, TP	middle to upper Eocene	Dn
E 372	Mt. Discovery	Sm	M	middle to upper Eocene	M
E 380	Mt. Discovery	Sm	Ab	N/A	
E 381	Mt. Discovery	Sm	MP, TP	middle to upper Eocene	Dn
SV 3	Salmon Valley	Sm/quartzite	TP	?Paleozoic/Mesozoic	
SV 12	Salmon Valley	Cmc	TP	???	
SV 14	Salmon Valley	Ss/quartzite		N/A	
SIM 1	Sea Ice Moraine	Cmm	TP	???	
SIM 5	Sea Ice Moraine	Ss		???	
SIM 6	Sea Ice Moraine	Sm		N/A	
SIM 9	Sea Ice Moraine	Cmm		N/A	
SIM 11	Sea Ice Moraine	Sm	MP, TP	middle to upper Eocene	Dn
BG 1	Blue Glacier	Cmm		???	
D1	Mt. Discovery	Mwb	MP, TP, S	middle to upper Eocene	Dn, Si

However, most intervals of pre-Oligocene time in Antarctica remain poorly sampled and understood (Webb, 1990). A significant time period that required additional focus was the Paleocene-Eocene pre-glacial interval and the transition period when the Antarctic paleoenvironment cooled, resulting in the inception of glaciation in Antarctica. Stratigraphic records of this time interval are known from only 5 other locations on the Antarctic continental shelf and islands: La Meseta Formation on Seymour and Cockburn islands in the Antarctic Peninsula [Feldmann and Woodburne, 1988; Stilwell and Zinsmeister, 1992]; strata on King George Island [Birkenmajer, 1991, 1996]; the upper Eocene interval of the CIROS-1 drillcore [Barrett, 1989; Wilson et al., 1998]; the poorly dated, but inferred, middle Eocene(?) glacial deposits cored at Site 742 in Prydz Bay [Barron et al., 1991; Ehrmann et al., 1992]; and in a core from Prydz Bay, East Antarctica [Quilty et al., 1999].

As a result of the limited geological information from the Antarctic continent, "?the Paleogene glacial history of Antarctica has been largely inferred from indirect evidence of glaciation gathered from the oceans beyond that remote, ice-shrouded, and inhospitable continent." "There is little agreement, however, among investigators as to whether an ice sheet was present at any time during the Eocene, particularly during early-middle Eocene times" [Wise et al., 1991].

Oxygen isotope studies indicate that the long-term cooling trend of the Cenozoic began at ~52 Ma [Shackleton, 1986; Ehrmann et al., 1992]. Sedimentology, paleontology and isotope geochemical records from Kerguelen Plateau suggest that Antarctic ice, if present, did not reach sea-level during the Paleocene to early Eocene [Wise et al., 1991; 1992]. Sea surface temperatures at Kerguelen Plateau cooled from between 10° to 14°C at 52 Ma to between 5° to 9°C at 40 Ma. Extrapolation of these isotope temperatures southward suggests sea surface temperatures of 6°C at the Antarctic coastline between 45 to 40 million years ago [Wise et al., 1991, 1992; Ehrmann, et al., 1992].

A shift in deep-sea oxygen isotope records is inferred to have resulted from the rapid growth of ice sheets during the early Oligocene [Zachos et al, 1992; Salamy and Zachos, 1999]. Stratigraphic evidence in Antarctica for this event is the first record of grounded ice on the continental shelf during the Oligocene [Barrett et al., 1989; Anderson and Bartek, 1992; Hambrey and Barrett, 1993; Wilson et al., 1998]. However, terrestrial climate was still humid and temperate enough to support woody vegetation in coastal regions during the Oligocene [Mildenhall, 1989; Hill, 1989]. Lower Oligocene-Lower Miocene palynoflo-

ras recovered from the Cape Roberts Project drillcores reflect a landscape of herb-moss tundra, with summer temperatures similar to that of islands near the modern Antarctic Convergence, with woody vegetation growing in warmer locations and times during the Oligocene-early Miocene [Jiang and Harwood, 1992; Raine, 1998; Cape Roberts Science Team, 1999].

The Eocene fossils and lithologies from the McMurdo Erratics documented in this volume provide the first clear view of East Antarctic paleoclimate and coastal paleoenvironments before the development of Cenozoic ice sheets. The Eocene fossiliferous erratics also reflect a time when shallow marine pathways existed between Antarctica and Australia across the South Tasman Rise and Tasmania [Lawver et al., 1992]. The shallow seas enabled the connection of marine communities, and probably also many terrestrial biotic elements between these two continents [Zinsmeister and Camacho, 1980; Woodburne and Zinsmeister, 1984]. As Australia moved northward, fauna and flora on each continent began to adapt to different climatic conditions. Knowledge of the fossils in the erratics provides a useful starting point to monitor subsequent evolutionary changes in the Antarctic biota that result from isolation, cooling, and the effects of advance and retreat of multiple ice sheets during the Late Paleogene and Neogene.

THE McMURDO ERRATICS

The McMurdo Erratics represent sediments that were deposited in a diverse range of marine and terrestrial environments. They comprise a wide range of sedimentary lithologies [Levy and Harwood, this volume a] and ages (Table 1). We presently consider the McMurdo Erratics to comprise any Tertiary sedimentary rocks present in coastal moraines of the McMurdo Sound region. It is likely that erratics of Cretaceous and Paleocene age will eventually be recovered from this suite of displaced sedimentary rocks. This volume focuses on fossils and lithologies recovered from the Eocene fossiliferous erratics, although Oligocene and Miocene erratics are treated in some papers.

Lithofacies

The McMurdo Erratics are classified into five lithofacies: (1) sandstone, (2) sandy-mudstone, (3) conglomerate, (4) mudstone and (5) diamictite. These are further divided into fourteen sub-facies [Table 2 in Levy and Harwood, this volume a]. The sandstone, sandy-mudstone and conglomerate facies are inferred to have been deposited within coastal marine settings adjacent to the rising Transantarctic Mountains during the middle to late Eocene. Most of these erratics are cemented by calcium carbonate. Marine macrofauna are almost exclusively recovered from the arenaceous lithologies. Microfossils were most abundant in the sandy mudstone and mudstone lithofacies; interpreted to have been deposited in deeper-water. The fifth lithofacies, diamictite, is observed to be restricted to the Oligocene and Miocene erratics and marks the influence of glaciers at sea-level during these times.

Other fine-grained, less indurated facies were probably also transported to southern McMurdo Sound with the suite of calcareous-cemented erratics described herein, but these lithologies did not survive weathering in the coastal moraines. If this were the case, our sampling is biased towards the coarse-grained, calcareous-cemented erratics.

Paleontology

These McMurdo Erratics have yielded a wealth of Early Tertiary, particularly Eocene, marine and terrestrial fossil remains and enabled the reconstruction of paleoclimate and paleoenvironments before the onset of significant glaciation of Antarctica. Fossil marine phytoplankton, including dinoflagellate cysts [Levy and Harwood, this volume b], ebridians, silicoflagellates, chrysophyte cysts and endoskeletal dinoflagellates [Bohaty and Harwood, this volume], and diatoms [Harwood and Bohaty, this volume] provide the best age control for the McMurdo Erratics. Most of the erratics are of middle to late Eocene age, based on dinoflagellate biostratigraphy. In erratics where ebridian biostratigraphy can also be applied, this age range is shortened to middle to early late Eocene. Younger erratics of Oligocene to late Miocene age are also present [Harwood and Bohaty, this volume].

The abundant marine phytoplankton provided a fertile base for higher marine organisms including: molluscs [Stilwell, this volume], bryozoans [Hara, this volume], decapod and cirriped crustaceans [Schweitzer and Feldmann, this volume; Buckeridge, this volume], brachiopods [Lee and Stilwell, this volume], a crocodile [Willis and Stilwell, this volume], sharks and teleost fish [Long and Stilwell, this volume] and a marine false-toothed bird (Pseudodontorn) [Jones, this volume]. Terrestrial communities are represented by leaves [Pole et al., this volume], wood [Francis, this volume], and palynomorphs [Askin, this volume]. Stilwell and Zinsmeister [this volume] discuss the evolutionary and paleobiogeographic processes that shaped the Eocene Southern Hemisphere biota and coastal to pelagic marine paleo-communities and paleoenvironments from fossil fauna and flora recovered in the McMurdo Erratics.

FIELDWORK

This volume presents the results from studies of McMurdo Erratics collected during four Antarctic field seasons by scientists from the University of Nebraska-Lincoln.

During 1991-92 approximately 250 erratics were collected from southern McMurdo Sound during a 3-week reconnaissance of the coastal moraines. Areas visited include northern coasts of White and Black islands, Brown Peninsula, Minna Bluff and the northeastern coast of Mt. Discovery (Figure 1). Preliminary studies on these samples revealed the presence of diatoms, dinoflagellates and abundant macrofossils in the McMurdo Erratics.

In 1992-93, approximately 800 erratics, many of which contained diatoms, dinoflagellates, foraminifera and pollen were collected from Minna Bluff and Mt. Discovery. Several large meter-scale fossiliferous sandstone boulders containing many macrofossils and fossil wood were also discovered [see Figure 2d in Stilwell et al., 1993; and Plate 1, Figure 9 and Plate 2 in Levy and Harwood, this volume a]. These boulders provide short stratigraphic sequences of Eocene strata. Smaller sandstone erratics contained leaf (*Nothofagus*) remains, sponge spicules, serpulid worms, burrows and a decapod [Stilwell et al., 1997]. Other lithofacies collected include diamictite and mudstone of Oligocene and Miocene age. A suite of more than 150 erratics of the Pliocene Scallop Hill Formation was also collected for future study, but are not treated in this volume. Areas visited during reconnaissance include the western coast of southern McMurdo Sound and Koettlitz Glacier margin: Cape Chocolate, Garwood Valley, Miers Valley, The Pyramid, The Bulwark and the western margin of Brown Peninsula. Eocene McMurdo Erratics were not encountered at these sites, but erratics of the Scallop Hill Formation were noted across this region.

In 1993-94 and 1995-96 fieldwork concentrated on the collection of new, previously undocumented lithologies, and additional fossiliferous remains. More than 300 erratics, including many of fine-grained, deeper-water facies, were collected for micropaleontological investigations. The distribution of the fossiliferous erratics and their host moraines was mapped on the ground and by aerial photography [Wilson, this volume]. Moraine composition data were also collected. Significant field discoveries in the 1995-96 field seasons included the discovery by J. Kaser of the first fossil bird bone from East Antarctica, shark teeth, many additional invertebrates and the recogni-

Figure 2. Oblique air-photograph of Mt. Discovery and Minna Bluff. Fossiliferous erratics are most abundant on the outer moraines along the northeastern coast of Mt. Discovery and along the northern edge of Minna Bluff.

tion of callianassid decapods preserved in their burrows [Stilwell et al., 1997].

LOCATION OF THE McMURDO ERRATICS

Fossiliferous erratics were collected from coastal moraines in the southern McMurdo Sound Region, at Minna Bluff, Mount Discovery, Black Island and Brown Peninsula (Figure 1). The McMurdo Erratics are not uniformly distributed across all of the coastal moraines in southern McMurdo Sound. Instead, they are concentrated in moraines that contain abundant light-colored erratics derived from basement granites and metamorphic lithologies. The McMurdo Erratics are less common in dark moraines that are mostly comprised of material of the Erebus Volcanic Province, McMurdo Volcanic Group [Kyle, 1981, 1990]. All of the fossiliferous Eocene erratics are restricted to moraines east of the Koettlitz Glacier [Wilson, this volume], and they are most abundant in the coastal moraines northeast of Mount Discovery (Figures 2 and 3). Erratics collected from Minna Bluff, including several of the largest erratics (1 - 2m), were concentrated in the most seaward moraine (Figure 4). Erratics were also present in moraines on the ice shelf north of Minna Bluff (Figure 4). Fossiliferous erratics are less common on Black Island, but they occur in moraines in the vicinity of Scallop Hill and along the northern coast facing Ross Island (Figure 1). A fossiliferous erratic was reported from Cape Crozier, Ross Island [Hertlein, 1969].

LOCATION AND INDEX OF ERRATIC SAMPLES

The McMurdo Erratics documented in the various chapters of this volume are listed in Table 1. The field locations of these specimens (identified in bold in Table 1) are also identified on Figures 1, 3 and 4. See also Figure 4 in Wilson [this volume] for a detailed view of the moraines along Minna Bluff. Most of the erratics that contained significant and unique macrofossil assemblages were also examined for dinoflagellates and siliceous microfossils, in order to define their age. Many fine-grained erratics contained several microfossil groups. These were of value for comparison to existing biochronological schemes. These specimens (Table 1) and others are currently curated within the Department of Geosciences at the University of Nebraska-Lincoln.

HISTORY OF RESEARCH ON THE McMURDO ERRATICS

The value of the McMurdo Erratics as a potential source of information on the Paleogene biota of Antarctica has been recognized for more than forty years. Cranwell et al. [1960] were the first to report pre-Quaternary fossils in East Antarctica. The promise for recovery of Late Cretaceous and other Paleogene marine fossil material in the western Ross Sea has been advanced by the report of widespread recycled Paleogene dinoflagellate cysts in Holocene sediment [Wilson, 1968; Truswell, 1983] and in

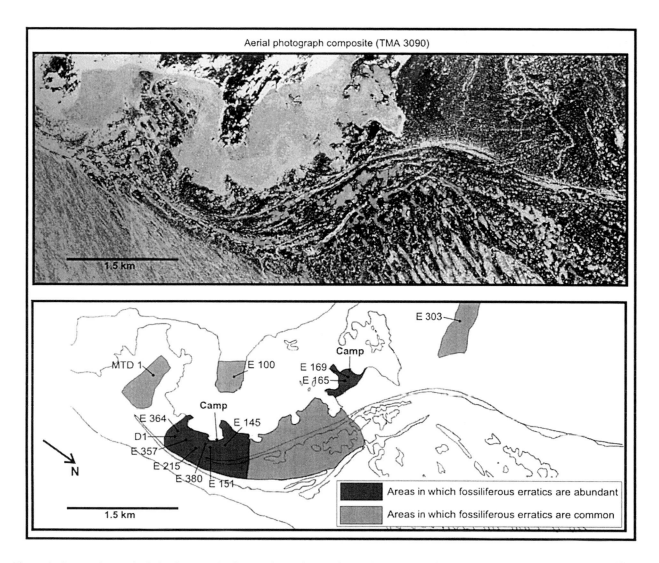

Figure 3. Composite vertical air photograph of coastal moraine on the northeast coast of Mt. Discovery, between the ice shelf and a meltwater lake. Location of several key fossiliferous erratics is indicated. See Figure 1 for location of this vertical frame.

upper Miocene sediment [Wrenn and Beckman, 1982; Harwood et al., 1989]. Upper Cretaceous and Lower Tertiary foraminifera were reported in stream deposits in Taylor Valley [Webb and Neall, 1972] and in glacial marine sediments in DSDP Hole 270 [Leckie and Webb, 1983] and in the CIROS-1 drillhole [Webb, 1989]. Below we review the history of collection, interpretation, and publication of results on the McMurdo Erratics, prior to the present study.

H. J. Harrington Collection - New Zealand Geological Survey

In 1959 H. J. Harrington collected samples of gray, white-weathering mudstone or vitric tuff from Minna Bluff and White Island. These yielded the first pre-Quaternary fossils from East Antarctica. From these erratics, two samples were sent to L. M. Cranwell, who identified abundant dinoflagellate cysts, micro-foraminifera, scolecodonts, (?)alcyonarian spicules, tracheids and a minor amount of pollen and spores [Cranwell et al., 1960; Cranwell, 1964]. An Early Tertiary age was suggested for this assemblage, but later revised to late Eocene by Cranwell [1969]. Deposition was inferred to have been in normal marine waters of 50 to 150 meters paleodepth, and terrestrial temperatures warm enough to support *Nothofagus* woody vegetation on the adjacent land. They assumed the source for these erratics was beneath the Ross Sea.

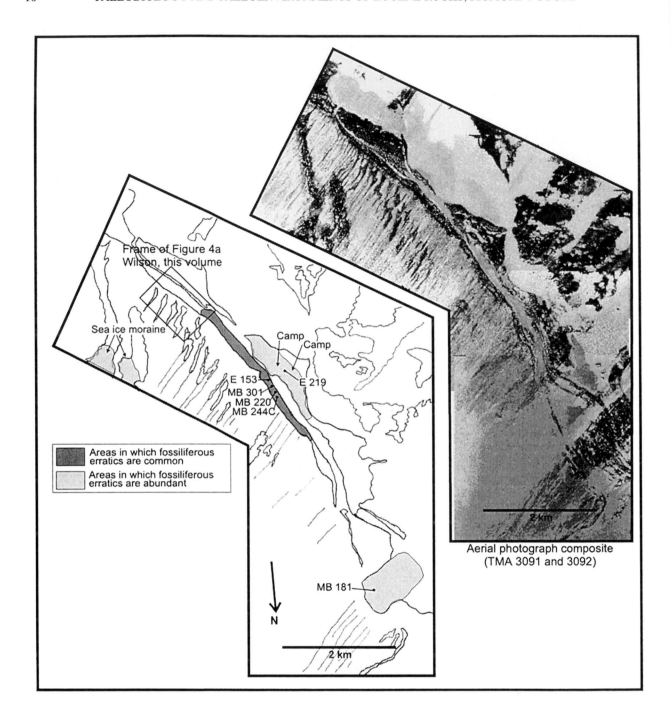

Figure 4. Composite vertical air photograph of coastal moraine on the northern coast of Minna Bluff, between the ice shelf, a melt-water lake and the slopes of Minna Bluff. Location of several key fossiliferous erratics is indicated. See Figure 1 for location of this vertical frame. For both an oblique aerial view of this location and a detailed view of the area framed by the rectangle see Figure 4 in Wilson [this volume].

In 1969 H.J. Harrington and R. Korsch traversed more than 60km along Minna Bluff, and noted the 'erratic' distribution of McMurdo Erratics into distinctive groups, describing "scattered patches and streamlines of Tertiary mudstones". They collected several hundred fragments of mudstone, calcareous sandstone and conglomerate or diamictite from which, "the microfossils in the sediments will be processed in 1969 by L.M. Cranwell and others" [Harrington, 1969]. No results appear to have been subsequently published. These erratics are now part of the collection of P.-N. Webb (see below).

P. Vella Collection- Victoria University of Wellington

In 1964-1965, during Victoria University of Wellington Antarctic Expedition (V.U.W.A.E.) 9 led by P. Vella, several samples were collected from fossil-bearing erratic boulders (up to 45 cm in diameter) from benches described by Vella [1969] as the "fossil-bearing moraine", associated with bench N on Black Island. McIntyre and Wilson [1966] described several dinoflagellates from five of these erratics collected from Black Island and Minna Bluff. These authors assigned a probable Eocene age to the assemblage. They also noted the presence of *Nothofagus*-dominated vegetation that reflected a cool- to moderately-warm temperate climate. Wilson [1967] later described 14 dinoflagellate species from the five Black Island erratics (Vella collection) and one erratic from Minna Bluff (Harrington collection) and revised the dinoflagellate identifications of McIntyre and Wilson [1966]. Hotchkiss and Fell [1972] described scutelline echinoid fragments in one of the calcareous sandstone erratics collected from Black Island.

Rowe [1974] examined the above five fossiliferous erratics collected from Black Island. He provided detailed petrographic descriptions and reported many macrofossils (identifications aided by A. Beu), plant fossils, and possible bone remains. Rowe [1974] suggested that the area south of Minna Bluff was the most likely source area for the erratics. Rowe believed the erratics were deposited during a transgressive cycle when rising sea-level reworked, mobilized and winnowed immature sediments previously deposited by fluvial processes.

L. G. Hertlein Report

One fossiliferous erratic boulder collected from the vicinity of Cape Crozier, Ross Island by R.C. Wood (1968-69 field season) contained a layer of gastropods identified as *Struthiolarella* cf. *S. variabilis* [Hertlein, 1969]. Petrographic description of this erratic as a fossilif-

erous subfeldsarenite with numerous rounded quartz grains with abraded quartz overgrowths was reported by Landis [1974]. Landis noted that the source of the texturally mature sands were likely derived from the Devonian-Triassic Beacon Supergroup of the Transantarctic Mountians.

P.-N. Webb Collection ? The Ohio State University

During the 1979 austral summer, a field party led by P.-N. Webb collected a large number of fossiliferous erratics from the northeastern coast of Mount Discovery. A study on 45 erratics from this collection and 15 erratics from the N.Z.G.S. Harrington collection [Harrington, 1969] was conducted by L. D. Stott [1982]. Stott et al. [1983] concentrated on the sedimentary petrology, with which he divided these erratics into three lithologic groups: (1) metamorphic arkose lithology, (2) granitic arkose lithology, and (3) quartzitic lithology. Five erratics were processed for siliceous and organic microfossils. Diatoms and palynomorphs were found in at least four of these, and one contained dinoflagellates. Pliocene and Miocene diatoms reported by Stott et al. [1983] are considered to be surface contaminants incorporated into these rocks as they sat at the sea-floor, or, during transport within Late Neogene, diatom-bearing diamicton [Harwood and Bohaty, this volume].

W. J. Zinsmeister Collection - Purdue University

In 1982, W.J. Zinsmeister and R. D. Powell visited sites in McMurdo Sound from which erratics were collected previously. From this search the most abundant fossil bearing erratics were found along the northeastern coast of Mount Discovery. R.M. Feldmann and W.J. Zinsmeister report the occurrence of a fossil decapod crustacean, identified as *Callianassa symmetrica* Feldmann and Zinsmeister [1984] from an erratic boulder collected by R.D. Powell on the coast of Mount Discovery. A moderately abundant pollen assemblage was reported from this erratic by R.A. Askin, who suggested an Eocene age.

RECONSTRUCTION OF THE EOCENE-OLIGOCENE COASTLINE, PALEOENVIRONMENT AND PALEOCLIMATE OF EAST ANTARCTICA

The Eocene McMurdo Erratics record sediments that were deposited in coastal-terrestrial and nearshore (inner shelf) marine environments along a steep coastline that formed as the paleo-Transantarctic Mountains rose rapidly along the western margin of the Victoria Land Basin [see Fitzgerald, 1992]. This coastline was likely dissected by steep walled estuaries (rias) and embayments [e.g. Webb, 1994]. We infer that many of the source-beds for the Eocene erratics were deposited in fan deltas that formed along this rugged coastline based on (1) sedimentary facies dominated by coarse-grained clastics [see plates in Levy and Harwood, this volume a], many of which exhibit sedimentary structures formed in high-energy environments; (2) abundant terrestrial organic material; (3) low diversity dinoflagellate cyst assemblages (4); presence of brackish water dinoflagellate cyst and acritarch species; and (5) presence of parautochthonous molluscan assemblages.

Beach facies consist of massive fine to coarse sands, while tidal channel facies comprise channel lag conglomerate and shell lags. Sandy fluvial channel deposits contain sparse pollen assemblages and lack marine dinoflagellate cysts. The molluscan fauna [Stilwell, this volume; Stilwell and Zinsmeister, this volume] and decapods [Schweitzer et al., this volume; Stilwell et al., 1997] indicate a shallow shelf environment, probably above wave-base and perhaps shallow sub-tidal.

The rich fossil phytoplankton assemblages (dinoflagellate cysts, ebridians, silicoflagellates and diatoms) reflect a highly fertile environment that would support higher organisms. The recovery of a 'false-toothed' bird (Pseudodontorn) [Jones, this volume], commonly associated with eutrophic marine settings of upwelling, strengthens the interpretation of a fertile coastal environment.

The terrestrial margins of this coastline were forested by *Nothofagus* (southern beech) and associated *Araucaria* [Pole et al., this volume; Francis, this volume]. This is an association typical of vegetation in southern South America today, while beech forests are typical of the moist, cool-temperate climate in New Zealand and Tasmania. Similar conditions are inferred for coastal regions of East Antarctica during the middle to late Eocene. Eocene marine faunas indicate a warm temperate climate at the Ross Sea margin of East Antarctica.

A study of the clay mineralogy [Holmes, this volume] notes the conspicuous absence of abundant kaolinite, even in the lithic- and fespathic-arenites. Holmes concludes that during the middle to late Eocene, climates were not notably warm or wet, otherwise weathering would have produced more kaolinite.

It is significant that no Eocene glacial facies were identified, suggesting the absence of ice at sea-level [Levy and Harwood, this volume a]. This suggests that the McMurdo Erratics of middle to early late Eocene age likely preceded the growth of large ice sheets in Antarctica.

However, at higher elevations and in the Antarctic interior, away from the coast, montane glaciers and ice caps could have been present. Future studies on clast shape and grain surface micromorphology on the McMurdo Erratics may identify the presence of glacially-derived features, but at the present time, there is no unequivocal evidence in the McMurdo Erratics that indicates an influence of glaciation at sea-level during the middle to early-late Eocene.

Conditions of "Icehouse Earth", following the growth of large ice sheets that discharged into the Ross Embayment are documented by the recovery of diamictite lithofacies of Oligocene and Miocene age in the McMurdo Erratics [Harwood and Bohaty, this volume]. The stratigraphic interval marking the onset of glacially-influenced sedimentation has not been recovered in drillcore records, though this has been a major objective. The CIROS-1 drillcore recorded relatively warm conditions with glacial activity of marine sedimentation from icebergs during the late Eocene, and a shift by the late early Oligocene to large scale glaciation where ice was grounded on the continental shelf [Barrett et al., 1989; Wilson et al., 1998].

Several current initiatives, like the Cape Roberts Project in the Western Ross Sea [Barrett and Davey, 1992; Webb and Wilson, 1995] and ODP Leg 188 to Prydz Bay, are positioned to recover more complete, and older records through this important time interval by drilling and recovering the pre-late Eocene through Oligocene transition on the Antarctic shelf. The McMurdo Erratics will provide a base from which future drilling projects will refine the history of this important interval of Antarctic and Earth history.

McMURDO ERRATICS AS MARKERS OF EOCENE-QUATERNARY GLACIAL HISTORY OF McMURDO SOUND

The distribution of fossiliferous erratics in the coastal moraines of McMurdo Sound is not random. Their occurrence is restricted to individual moraines east of Brown Peninsula (Figure 1). Wilson [this volume] outlines the distribution of the erratics in coastal moraines, proposes a glacial history for this region to explain this distribution, and describes processes that are active along the coastal zone. Advance of the southern McMurdo Sound ice sheet into the study area during the late Pliocene to early Pleistocene is inferred from the distribution of the McMurdo Erratics. During eustatic lowstands of the last million years the ice shelf has been interactive with relict moraines on the sea-floor through grounding; glacial and other sediments were incorporated by freezing into the basal ice. Ablation of the surface of the ice shelf by strong winds resulted in the vertical advection of basal sediments, including the McMurdo Erratics, toward the surface, where they are now exposed in moraine bands.

CONCLUSIONS

This volume represents an initial characterization of the paleontological and lithological information preserved within the McMurdo Erratics. This remote approach to reconstructing Antarctic history by studying glacial erratics has produced important new paleoenvironmental and paleontological information. Many erratics require further paleontological sampling for vertebrate and other fossil remains. The fossil assemblages documented herein record a rich middle to late Eocene Antarctic paleocommunity, from a time after separation of Australia, and near the end of "Greenhouse-Earth" conditions. The coastal environment on the Ross Sea margin of East Antarctica included a cool temperate climate, similar to the South Island of New Zealand and southern Chile today. It is significant that no Eocene erratics of diamictite or mudstone with dropstones were identified in the McMurdo Erratics. Middle to early late Eocene climate was apparently too warm for continental ice sheets to reach sea-level in this region of the Ross Embayment. Diamictite facies were recovered in erratics of Oligocene and Miocene age, which document the shift to increased glaciation in Antarctica and the start of glacial influence on marine sedimentation on the Antarctic continental shelf.

Interpreting the glacial geologic history of the McMurdo Sound region is aided by knowledge of the distribution of the McMurdo Erratics. Advance of the southern McMurdo Sound ice sheet during the Pliocene and early Quaternary(?) eroded and transported sedimentary rocks of Eocene, Oligocene, Miocene and Pliocene age from 'Discovery Deep' and deposited them into moraines in southern McMurdo Sound. Processes involving ice shelf grounding and freeze-on of basal sediments, followed by subsequent wind-enhanced ablation, are proposed to explain the present distribution of the erratics on the surface of coastal moraines. A better understanding of these processes will aid future prospecting for McMurdo Erratics in this and other coastal areas within and beyond the Ross Embayment.

Similar scenarios involving regional sampling of subglacial strata by ice likely occurred in other areas of Antarctica. Field studies similar to that employed in this study in other coastal areas of the Antarctic margin may provide a means to conduct regional geological reconnaissance of areas covered by ice and further advance the paleontological record of Antarctica. The approach of 'reconstructive biostratigraphy' outlined in this volume demonstrates the potential to obtain an excellent and previously undocumented record of middle to late Eocene paleoenvironments and paleocommunities on the Ross Sea margin of East Antarctica. Stratigraphic drilling and reconstructive biostratigraphic studies of reworked material present the greatest potential for new information on the Cenozoic geologic history of Antarctica.

Acknowledgements. We acknowledge the able assistance of S. Bohaty, J. Francis, R. Graham, X. Jiang, J. Kaser, M. Pole, A. Srivastav, J. Stilwell, D. Watkins, G. Wilson, and D. Winter in collection of erratics during the four field seasons between 1991-1995. Antarctic Support Associates (ASA) and the US Navy helicopter squadron VXE-6 provided logistical support. J. Mullins of USGS provided aerial photography. M. Reed and S. Williams of ASA provided technical skill in 'dissecting' several large erratics to aid recovery of fossil specimens. We thank P. Webb for introducing the potential of this study and for general advice on matters related to the McMurdo Erratics. S. Bohaty suggested the term 'reconstructive biostratigraphy' to describe the general process of geological investigation employed in the papers in this volume. J. Stilwell and R. Feldmann are acknowledged for dedicated effort and patience as editors of this volume. This paper benefited from careful review by R. Askin, S. Bohaty, R. Powell and G. Wilson. Field and laboratory research was supported by National Science Foundation grants OPP-9158078 and OPP-9317901 to D. Harwood and by generous donations of the alumni of the Department of Geosciences, University of Nebraska-Lincoln.

REFERENCES

Anderson, J.A. and L.R. Bartek
1992   Cenozoic glacial history of the Ross Sea revealed by intermediate resolution seismic reflection data combined with drill site information, in *The Antarctic paleoenvironment: a perspective on global change, part one*, edited by J.P. Kennett and D.A. Warnke, Antarctic Research Series, *56*: 231-263, American Geophysical Union, Washington, D.C.

Ashworth, A.C., D.M. Harwood, P.-N. Webb and M.C.G. Mabin
1997   A weevil from the heart of Antarctica, in, Studies in Quaternary entomology - an inordinate fondness for insects, *Quaternary Proceedings No. 5*, Chichester: John Wiley & Sons Ltd., 15-22.

Askin, R.A.
This volume   Spores and pollen from the McMurdo Sound Erratics, Antarctica, in *Paleobiology and Paleoenvironments of Eocene Rocks, McMurdo Sound, East Antarctica*, edited by J.D. Stilwell and R.M. Feldmann, Antarctic Research Series, American Geophysical Union

Barker, P.F., et al.
1977   Leg 36, *Initial Report of the Deep Sea Drilling Project, 36*
1990   Leg 113, *Proceedings of the Ocean Drilling Program Science Results, 113,* 1033 pp.

Barrett, P.J., ed.
1986   Antarctic Cenozoic history from the MSSTS-1 drill-hole, McMurdo Sound, Bulletin in the Miscellaneous Series of the New Zealand Department of Scientific and Industrial Research, *237,* 174 pp.
1989   Antarctic Cenozoic history from the CIROS-1 drill-hole, McMurdo Sound. Bulletin in the Miscellaneous Series of the New Zealand Department of Scientific and Industrial Research, *245*: 254 pp.

Barrett, P.J.
1996   Antarctic paleoenvironment through Cenozoic times - A review, *Terra Antarctica, 3*: 103-119.

Barrett, P.J., M.J. Hambrey, D.M. Harwood, A.R. Pyne, and P.-N. Webb
1989   General Synthesis, in: *Antarctic Cenozoic History of the CIROS-1 drillhole*, edited by P.J. Barrett, Bulletin in the Miscellaneous Series of the New Zealand Department of Scientific and Industrial Research, *245*: 241-251.

Barrett, P.J. and F.J. Davey, eds.
1992   Antarctic stratigraphic drilling, Cape Roberts Project Workshop Report, The Royal Society of New Zealand Miscellaneous Series *23,* Wellington, New Zealand, 38 pp.

Barron, J.A., B. Larsen and J.G. Baldauf
1991   Evidence for late Eocene to early Oligocene Antarctic glaciation and observations on late Neogene glacial history of Antarctica: results from Leg 119, *Proceedings of the Ocean Drilling Program Scientific Results, 119*: 869-891.

Barron, J.A. et al.
1991   Leg 119, *Proceedings of the Ocean Drilling Program Scientific Results, 119.*

Birkenmajer, K.
1991   Tertiary glaciation in the South Shetland Islands, West Antarctica: evaluation of data, in *Geological Evolution of Antarctica,* edited by M.R.A Thomson, J.A. Crame and J.W. Thomson, pp. 629-632, Cambridge University Press, New York.

Birkenmajer, K.
1996   Polish geological research on King George Island, West Antarctica (1977-1996), *Polish Polar Research,* 17(3-4): 125-141.

Bohaty, S.M. and D.M. Harwood
This volume Ebridian and silicoflagellate biostratigraphy from Eocene McMurdo Erratics and the Southern Ocean, in *Paleobiology and Paleoenvironments of Eocene Rocks, McMurdo Sound, East Antarctica,* edited by J.D. Stilwell and R.M. Feldmann, Antarctic Research Series, American Geophysical Union.

Buckeridge, J., St. J.S.
This volume A new species of *Austrobalanus* (Cirripedia, Thoracica) from Eocene erratics, Mount Discovery, McMurdo Sound, East Antarctica, in *Paleobiology and Paleoenvironments of Eocene Rocks, McMurdo Sound, East Antarctica,* edited by J.D. Stilwell and R.M. Feldmann, Antarctic Research Series, American Geophysical Union.

Cape Roberts Science Team
1998   Initial Report on CRP-1, Cape Roberts Project, Antarctica, *Terra Antarctica, 5,* 187 pp.
1999   Studies from the Cape Roberts Project, Ross Sea, Antarctica, Initial Report on CRP-2/2A, *Terra Antarctica,* 6, 173 pp.

Ciesielski, P.F. et al.
1991   Leg 114, *Proceedings of the Ocean Drilling Program, 114*: 826 pp.

Cranwell, L.M.
1964   Hystrichospheres as an aid to Antarctic dating with special reference to the recovery *Cordosphaeridium* in erratics at McMurdo Sound, *Grana Palynologica, 5*(3): 397-405.
1969   Antarctic and circum-Antarctic palynological contributions, *Antarctic Journal of the U.S., 4*: 197-198,

Cranwell, L.M., H.J.Harrington and I.G. Speden
1960   Lower Tertiary microfossils from McMurdo Sound, Antarctica, *Nature, 186*(4726): 700-702.

Davey, F.J.
1987   Geology and structure of the Ross Sea region, in *The Antarctic Continental Margin: Geology and Geophysics of the western Ross Sea,* edited by A.K. Cooper and F. J. Davey, Circum-Pacific Council for Energy and Mineral Resources, Houston, Texas, Earth Science Series, *5B*: 1-15.

Eggers, A.J.
1979   Scallop Hill Formation, Brown Peninsula, McMurdo Sound, Antarctica, *New Zealand Journal of Geology and Geophysics, 22*: 353-361.

Ehrmann, W.U., M.J. Hambrey, J.G. Baldauf, J.A. Barron, B. Larsen, A. Mackensen, S.W. Wise, Jr., and J.C. Zachos
1992   History of Antarctic Glaciation: An Indian Ocean Perspective, in *Synthesis of Results from Scientific Drilling in the Indian Ocean,* edited by R.A. Duncan, D.K. Rea, R.B. Kidd, U. von Rad and J.K. Weissel, Geophysical Monograph *70,* American Geophysical Union, pp. 423-446.

Feldmann, R.M. and M.O. Woodburne, eds.
1988   Geology and Paleontology of Seymour Island, Antarctic Peninsula, *The Geological Society of America, Memoir 169,* Boulder, Colorado, 566 pp.

Feldmann, R.M. and W.J. Zinsmeister
1984   First occurrence of fossil decapod crustaceans (Callianassidae) from the McMurdo Sound region, Antarctica, *Journal of Paleontology, 58*(4): 1041-1045.

Fitzgerald, P.G.
1992   The Transantarctic Mountains of Southern Victoria Land: the application of apatite fission track analysis to a rift shoulder uplift, *Tectonics, 11*(3): 634-662.

Francis, J.E.
This volume   Fossil wood from Eocene high latitude forests, McMurdo Sound, Antarctica, in *Paleobiology and Paleoenvironments of Eocene Rocks, McMurdo Sound, East Antarctica,* edited by J.D. Stilwell and R.M. Feldmann, Antarctic Research Series, American Geophysical Union, Washington, D.C.

Hambrey, M.J. and P.J. Barrett
1993   Cenozoic sedimentary and climatic record, Ross Sea region, Antarctica, in *The Antarctic Paleoenvironment: a Perspective on Global Change, part two,* edited by J. P. Kennett and D.A. Warnke, Antarctic Research Series, *60*: 67-74, American Geophysical Union, Washington, D.C.

Hambrey, M.J. and Wise, S.W., Jr.
1998   Scientific Results of the Cape Roberts Drilling Project, CRP-1, *Terra Antarctica, 5*(3).

Hara, U.
This volume   Bryzoan remains from Eocene glacial erratics of McMurdo Sound, East Antarctica, in *Paleobiology and Paleoenvironments of Eocene Rocks, McMurdo Sound, East Antarctica,* edited by J.D. Stilwell and R.M. Feldmann, Antarctic Research Series, American Geophysical Union, Washington, D.C.

Harrington, H.J.
1969   Fossiliferous rocks in moraines at Minna Bluff, McMurdo Sound, *Antarctic Journal of the United States, 4*(4): 134-135.

Harwood, D.M., R.P. Scherer and P.-N. Webb
1989   Multiple Miocene marine productivity events in West Antarctica as recorded in upper Miocene sediments beneath the Ross Ice Shelf (Site J-9). *Marine Micropaleontology, 15*: 91-115.

Harwood, D.M. and P.-N. Webb
        Glacial transport of diatoms in the Antarctic Sirius Group: Pliocene refrigerator, *GSA Today, 8*(4): 1-8.

Harwood, D.M. and S.M. Bohaty
This volume   Marine diatom assemblages from Eocene and Younger Erratics, McMurdo Sound, Antarctica, in *Paleobiology and Paleoenvironments of Eocene Rocks, McMurdo Sound, East Antarctica,* edited by J.D. Stilwell and R.M. Feldmann, Antarctic Research Series, American Geophysical Union, Washington, D.C.

Hayes, D.E. et al.
1975   Leg 28, *Initial Reports of the Deep Sea Drilling Project, 28.*

Hertlein, L.G.
1969   Fossiliferous boulder of Early Tertiary Age from Ross Island, Antarctica, *Antarctic Journal of the U.S., 4*(4): 199-201.

Hill, R.S.
1989   Fossil leaf, in Antarctic Cenozoic History of the CIROS-1 drillhole, edited by P.J. Barrett, *Bulletin in the Miscellaneous Series of the New Zealand Department of Scientific and Industrial Research, 245*: 143-144.

Hollister, C.D. et al.
1976   Leg 35, *Initial Reports of the Deep Sea Drilling Project, 35.*

Holmes, M.A.
This volume   Clay mineral composition of glacial erratics, McMurdo Sound, in *Paleobiology and Paleoenvironments of Eocene Rocks, McMurdo Sound, East Antarctica,* edited by J.D. Stilwell and R.M. Feldmann, Antarctic Research Series, American Geophysical Union, Washington, D.C.

Hotchkiss, F.M.C. and B.H. Fell
1972   Zoogeographical implications of a Paleogene echinoid from East Antarctica, *Journal of Geophysical Research, 78*: 3448-3468.

Jiang, X. and D.M. Harwood
1992   Antarctic palynology from RISP diatomite: a glimpse of early Miocene south polar forests. *Antarctic Journal of the U.S., 27*: 3-6.

Jones, C.M.
This volume   First fossil bird from East Antarctica, in *Paleobiology and Paleoenvironments of Eocene Rocks, McMurdo Sound, East Antarctica,* edited by J.D. Stilwell and R.M. Feldmann, Antarctic Research Series, American Geophysical Union, Washington, D.C.

Jonkers, H.A.
1998   Stratigraphy of Antarctic late Cenozoic pectinid-bearing deposits, *Antarctic Science, 10*(2): 161-170.

Kennett, J.P. et al.
1975   Leg 29, *Initial Reports of the Deep Sea Drilling Project, 29.*

Kennett, J.P. and J.A. Barron
1992   Introduction, in *The Antarctic paleoenvironment: a perspective on global change, part one,* edited by J.P. Kennett and D.A. Warnke, Antarctic Research Series, *56,* American Geophysical Union, Washington, D.C., 1-6.

Kyle, P.R.
1981   Glacial history of the McMurdo Sound area as indicated by the distribution and nature of McMurdo Volcanic Group rocks, in *Dry Valley*

Drilling Project, edited by L.D. McGinnis, Antarctic Research Series, *33,* American Geophysical Union, Washington, D.C., 403-412.

1990   Summary, A.III. Erebus Volcanic Province, in *Volcanoes of the Antarctic Plate and Southern Oceans,* edited by W.E. LeMasurier and J.W. Thomson, Antarctic Research Series, *48,* American Geophysical Union, Washington, D.C., 81-88.

Landis, C.A.
1974   Petrography of an early Tertiary fossiliferous sandstone from the Ross Sea area, Antarctica, *New Zealand Journal of Geology and Geophysics, 17*(3): 715-718.

Lawver, L.A., L.M. Gahagan and M.F. Coffin
1992   The development of paleoseaways around Antarctica, in *The Antarctic paleoenvironment: a perspective on global change, part one,* edited by J.P. Kennett and D.A. Warnke, Antarctic Research Series, *56,* American Geophysical Union, Washington, D.C., 7-30.

Leckie, R.M. and P.-N. Webb
1979   Scallop Hill Formation and associated Pliocene marine deposits of southern McMurdo Sound, *Antarctic Journal of the U.S., 14*(5): 54-56.

1983   Late Oligocene-early Miocene glacial record of the Ross Sea, *Geology, 11*: 578-582.

Lee, D.E. and J.D. Stilwell
This volume   Rhynchonellide Brachiopods from late Eocene erratics in the McMurdo Sound region, Antarctica, in *Paleobiology and Paleoenvironments of Eocene Rocks, McMurdo Sound, East Antarctica,* edited by J.D. Stilwell and R.M. Feldmann, Antarctic Research Series, American Geophysical Union, Washington, D.C.

Levy, R.H.
1998   Middle to late Eocene coastal paleoenvironments of Southern Victoria Land, East Antarctica: a palynological and sedimentological study of glacial erratics, Ph.D. dissertation, University of Nebraska-Lincoln, Lincoln, Nebraska, 211 pp.

Levy, R.H. and D.M. Harwood
This volume a   Sedimentary lithofacies of the McMurdo Sound erratics, in *Paleobiology and Paleoenvironments of Eocene Rocks, McMurdo Sound, East Antarctica,* edited by J.D. Stilwell and R.M. Feldmann, Antarctic Research Series, Amer. Geophysical Union, Washington, D.C.

This volume b   Tertiary marine palynomorphs from the McMurdo Sound erratics, in *Paleobiology and Paleoenvironments of Eocene Rocks, McMurdo Sound, East Antarctica,* edited by J.D. Stilwell and R.M. Feldmann, Antarctic Research Series, Amer. Geophysical Union, Washington, D.C.

Long, D.H. and J.D. Stilwell
This volume   Fish remains from the Eocene of Mount Discovery, East Antarctica, in *Paleobiology and Paleoenvironments of Eocene Rocks, McMurdo Sound, East Antarctica,* edited by J.D. Stilwell and R.M. Feldmann, Antarctic Research Series, American Geophysical Union, Washington, D.C.

Ludwig, W.J. et al.
1983   Leg 71, *Initial Report of the Deep Sea Drilling Project, 71.*

McIntyre, D.J. and G.J. Wilson
1966   Preliminary palynology of some Antarctic Tertiary erratics, *New Zealand Journal of Botany,* 4(3): 315-321.

McKelvey, B.C.
1991   The Cainozoic record in south Victoria Land — a geological evaluation of the McMurdo Sound Drilling Projects, in *Geological History of Antarctica,* edited by R. J. Tingey, Oxford University Press, pp. 434-454.

Mildenhall, D.C.
1989   Terrestrial Palynology, in: Antarctic Cenozoic History of the CIROS-1 drillhole, edited by P.J. Barrett, Bulletin in the Miscellaneous Series of the New Zealand Department of Scientific and Industrial Research, *245*: 119-127.

Miller, M. and M.C.G. Mabin
1998   Antarctic Neogene landscapes ? in the refrigerator or in the deep freeze?, *GSA Today, 8*(4): 1-8.

Moriwaki, K., Y. Yoshida and D.M. Harwood
1992   Glacial history of Antarctica - a correlative synthesis, in *Progress in Antarctic Earth Science,* edited by Y. Yoshida, K. Kaminuma, K. Shiraishi, Proceedings of the Sixth International Symposium on Antarctic Earth Science, NIPR, Terrapub, Tokyo, p. 773-780.

Pole, M., R. Hill, and D.M. Harwood
This volume   Eocene plant macrofossils from erratics, McMurdo Sound, in *Paleobiology and Paleoenvironments of Eocene Rocks, McMurdo Sound, East Antarctica,* edited by J.D. Stilwell and R.M. Feldmann, Antarctic Research Series, American Geophysical Union, Washington, D.C.

Prentice, M.L., J.G. Bockheim, S.C. Wilson, L.H. Burckle, D.A. Hodell, C. Schlichter, and D.E. Kellogg
1993   Late Neogene Antarctic glacial history: evidence from central Wright Valley, in *The Antarctic paleoenvironment: a perspective on global change, part two,* edited by J.P. Kennett and D.A. Warnke, Antarctic Research Series, *60*: 207-250, American Geophysical Union, Washington, D.C.

Quilty, P.G.
1993   Coastal East Antarctic Neogene sections and their contribution to the ice sheet evolution debate, in *The Antarctic paleoenvironment: a perspective on*

global change, part two, edited by J.P. Kennett and D.A. Warnke, Antarctic Research Series, *60:* 251-264, American Geophysical Union, Washington, D.C.

1996   The Pliocene environment of Antarctica, *Papers and Proceedings of the Royal Society of Tasmania, 130*(2): 1-8.

Quilty, P.G., E.M. Truswell, P.E. O'Brien and F. Taylor

1999   Paleocene-Eocene biostratigraphy and paleoenvironments: new data from Mac Robertson Shelf and western parts of Prydz Bay, East Antarctica. AGSO *Journal of Australian Geology and Geophysics, 17*: 133-143.

Raine, J.I.

1998   Terrestrial palynomorphs from Cape Roberts Project drillhole CRP-1, Ross Sea, Antarctica, *Terra Antarctica, 5*(3): 539-548.

Rowe, G.H.

1974   A petrographic and paleontologic study of Lower Tertiary erratics from Quaternary moraines, Black Island, Antarctica, B.Sc.(Honors) thesis, Victoria University of Wellington, New Zealand, 91 pp.

Salamy, K.A. and J.C. Zachos

1999   Latest Eocene-Early Oligocene climate change and Southern Ocean fertility: inferences from sediment accumulation and stable isotope data, *Palaeogeography, Palaeoclimatology, Palaeoecology, 145*: 61-77.

Shackleton, N.J.

1986   Paleogene stable isotope events, *Palaeogeography, Palaeoclimatology, Palaeoecology, 57*: 91-102.

Schweitzer, C.E. and R.M. Feldmann

This volume   *Callichirus? symmetricus* (Decapoda: Thalassinoidea) and associated burrows, Eocene, Antarctica, in *Paleobiology and Paleoenvironments of Eocene Rocks, McMurdo Sound, East Antarctica,* edited by J.D. Stilwell and R.M. Feldmann, Antarctic Research Series, American Geophysical Union, Washington, D.C.

Speden, I.G.

1962   Fossiliferous Quaternary marine deposits in the McMurdo Sound Region, Antarctica, *New Zealand Journal of Geology and Geophysics, 5:* 746-777.

Stilwell, J.D.

This volume   Eocene Mollusca (Bivalvia, Gastropoda and Scaphopoda) from McMurdo Sound: Systematics and paleoecologic significance, in *Paleobiology and Paleoenvironments of Eocene Rocks, McMurdo Sound, East Antarctica,* edited by J.D. Stilwell and R.M. Feldmann, Antarctic Research Series, American Geophysical Union, Washington, D.C.

Stilwell, J.D., R.H. Levy and D.M. Harwood

1993   Preliminary paleontological investigations of Tertiary glacial erratics from the McMurdo Sound region, East Antarctica, *Antarctic Journal of the U.S., 28*: 1-19.

Stilwell, J.D., R.H. Levy, R.M. Feldmann and D.M. Harwood

1997   On the rare occurrence of Eocene Callianassid decapods (Arthropoda) preserved in their burrows, Mount Discovery, East Antarctica, *Journal of Paleontology, 71*(2): 284-287.

Stilwell, J.D. and W.J. Zinsmeister

1992   *Molluscan systematics and biostratigraphy of the lower Tertiary La Meseta Formation, Seymour Island, Antarctic Peninsula,* American Geophysical Union, Antarctic Research Series 55, xii + 192 pp.

This volume   Paleobiogeographic synthesis of the Eocene macrofauna from McMurdo Sound, Antarctica, in *Paleobiology and Paleoenvironments of Eocene Rocks, McMurdo Sound, East Antarctica,* edited by J.D. Stilwell and R.M. Feldmann, Antarctic Research Series, American Geophysical Union, Washington, D.C.

Stott, L.D.

1982   A re-evaluation of the age of the age and nature of Cenozoic erratics from McMurdo Sound. Unpublished B.S. Thesis, The Ohio State University, Columbus, Ohio, 52 pp.

Stott, L.D., B.C. McKelvey, D.M. Harwood and P.-N. Webb

1983   A revision of the ages of Cenozoic erratics at Mount Discovery and Minna Bluff, McMurdo Sound, *Antarctic Journal of United States, 1983* Review: 36-38.

Sugden, D.E., D.R. Marchant and G.H. Denton, editors

1993   The case for a stable East Antarctic ice sheet, *Geografiska Annaler, 75A*: 151-351.

Truswell, E.M.

1983   Recycled Cretaceous and Tertiary pollen and spores in Antarctic marine sediments: a catalogue, *Palaeontographica Abt. B, 186*: 121-174.

van der Wateren, F.M. and R. Hindmarsh

1995   Stabilists strike again, *Nature, 382*: 389-391.

Vella, P.

1969   Surficial geological sequence, Black Island and Brown Peninsula, McMurdo Sound, Antarctica, *New Zealand Journal of Geology and Geophysics, 12*(4): 761-770.

Webb, P.-N.

1989   Benthic foraminifera, in Antarctic Cenozoic history from the CIROS-1 drill-hole, McMurdo Sound, edited by P.J. Barrett, *Bulletin in the Miscellaneous Series of the New Zealand Department of Scientific and Industrial Research, 245*: 99-118.

1990   The Cenozoic history of Antarctica and its global

impact, *Antarctic Science, 2*: 3-21.

1991   A review of the Cenozoic stratigraphy and pale-
ontology of Antarctica, in *Geological Evolution of
Antarctica,* edited by M.R.A. Thomson, J.A.
Crame, and J.W. Thomson, Cambridge University
Press, Cambridge, 599-608.

Paleo-drainage systems of East Antarctica and
sediment supply to West Antarctic rift system,
Terra Antarctica, 1(2): 457-461.

Webb, P.-N. and V.E. Neall

1972   Cretaceous Foraminifera in Quaternary deposits
from Taylor Valley, Victoria Land, in *Antarctic
Geology and Geophysics,* edited by R.J. Adie, pp.
653-657, Oslo, Universitetsforlaget.

Webb, P.-N. and J.E. Andreasen

1986   Potassium-argon dating of volcanic material asso-
ciated with the Pliocene Pecten Conglomerate
(Cockburn Island) and Scallop Hill Formation
(McMurdo Sound), *Antarctic Journal of the
United States, 21*(5): 59.

Webb, P.-N. and G.S. Wilson, eds.

1995   Cape Roberts Project: Antarctic Stratigraphic
Drilling, Proceedings of a meeting to consider the
project science plan and potential contributions by
the U.S. science community, 6-7 March, 1994,
BPRC Report No. *10,* Bryd Polar Research
Center, The Ohio State University, Columbus,
Ohio, 117 pp.

Webb, P.N., D.M. Harwood, M.C.G. Mabin, and B.C. McKelvey

1996   A marine and terrestrial Sirius Group succession,
middle Beardmore Glacier - Queen Alexandra
Range, Transantarctic Mountains, Antarctica,
Marine Micropaleontology, 27: 273-297.

Willis, P.M.A. and J.D. Stilwell

This volume   A possible Piscivorous crocodile from Eocene
deposits of McMurdo Sound, East Antarctica, in
*Paleobiology and Paleoenvironments of Eocene
Rocks, McMurdo Sound, East Antarctica,* edited
by J.D. Stilwell and R.M. Feldmann, Antarctic
Research Series, American Geophysical Union,
Washington, D.C.

Wilson, G.J.

1967   Some new species of Lower Tertiary dinoflagellates from McMurdo Sound, Antarctica, *New
Zealand Journal of Botany, 5*(1): 57-83.

1968   On the occurrence of fossil microspores, pollen
grains, and microplankton in bottom sediments of
the Ross Sea, Antarctica, *New Zealand Journal of
Marine and Freshwater Research, 2*: 381-391.

1995   The Neogene East Antarctic ice sheet: a dynamic
or stable feature? *Quaternary Science Reviews,
14*(2): 101-123.

This volume   Glacial geology and origin of fossiliferous-erratic-bearing moraines, southern McMurdo Sound,

Antarctica ? an alternative ice sheet hypothesis, in
*Paleobiology and Paleoenvironments of Eocene
Rocks, McMurdo Sound, East Antarctica,* edited
by J.D. Stilwell and R.M. Feldmann, Antarctic
Research Series, American Geophysical Union,
Washington, D.C.

Wilson, G.S., A.P. Roberts, K.L. Verosub, F. Florindo and L.
Sagnotti

1998   Magnetobiostratigraphic chronology of the
Eocene-Oligocene transition in the CIROS-1
core, Victoria Land margin, Antarctica: implications for Antarctic glacial history, *Geological
Society of America Bulletin, 110,* 35-47.

Wise, S.W., Jr., J. Breza, D.M. Harwood, and W. Wei

1991   Paleogene glacial history of Antarctica, in
Controversies in Modern Geology, edited by D.W.
Muller, J.A. Mackenzie and H. Weissert, London,
Academic Press, p. 133-171.

Wise, S.W., Jr., J. Breza, D.M. Harwood, W. Wei, and J. Zachos

1992   Paleogene glacial history of Antarctica in light of
ODP Leg 120 drilling results, *Proceedings of the
Ocean Drilling Program, Science Results, 120*:
1001-1030.

Woodburne, M.O. and W.J. Zinsmeister

1984   The first land mammal from Antarctica and its
biogeographical implications, *Journal of
Paleontology, 58*: 913-948.

Wrenn, J.H. and S.W. Beckman

1982   Maceral, total organic carbon, and palynological
analyses of Ross Ice Shelf Project site J9 cores,
Science, 216: 187-189.

Zachos, J.C., D.K. Rea, K. Seto, R. Nomura, N. Niitsuma

1992   Paleogene and Early Neogene deep water paleo-
ceanography of the Indian Ocean as determined
from benthic foraminifer stable carbon and oxy-
gen isotope records, in *Synthesis of Results from
Scientific Drilling in the Indian Ocean,* R.A.
Duncan, D.K. Rea, R.B. Kidd, U. von Rad, and
J.K. Weissel, eds., Geophysical Monograph *70,*
American Geophysical Union, 351-385.

Zinsmeister, W.J. and H.H. Camacho

1980   Late Eocene Struthiolariidae (molluscan:
Gastropoda) from Seymour Island, Antarctic
Peninsula and their significance to the biogeogra-
phy of early Tertiary shallow-water faunas of the
Southern Hemisphere, *Journal of Paleontology,
54*: 1-14.

David M. Harwood and Richard H. Levy, Department of
Geosciences, University of Nebraska-Lincoln, NE 68588-0340
USA

GLACIAL GEOLOGY AND ORIGIN OF FOSSILIFEROUS-ERRATIC-BEARING MORAINES, SOUTHERN MCMURDO SOUND, ANTARCTICA– AN ALTERNATIVE ICE SHEET HYPOTHESIS

Gary S. Wilson

Byrd Polar Research Center, The Ohio State University, Columbus, Ohio 43210

Glacial sediments on the McMurdo Ice Shelf and constituting adjacent coastal moraines record a complex late Cenozoic glacial history of southern McMurdo Sound. The physiography of the moraines and the distribution of erratics on them document at least two ice grounding events in southern McMurdo Sound. The first: grounding, advance, and retreat across southern McMurdo Sound of an ice sheet, the southern McMurdo Sound Ice Sheet (SMS Ice Sheet). This ice sheet distributed glacial sediments including granitic, volcanic, Pliocene fossiliferous volcaniclastic, and Eocene marine fossiliferous erratics into a distinctive pattern of terminal, lateral and retreat moraines across the floor of McMurdo Sound. The SMS Ice Sheet was different in character to a previous model (the Ross Sea Ice Sheet or RSIS). It comprised three coexisting lobes: 1) An expanded and grounded Koettlitz Glacier Lobe (KG Lobe), that enveloped much of Brown Peninsula, 2) A Ross Ice Shelf Lobe (RIS Lobe), that enveloped White Island and flowed north of and partly over Black Island, and 3) a smaller medial lobe (the Minna Bluff Lobe – MB Lobe), that was fed from a grounded ice sheet in the Ross Sea, flowed northwest along Minna Bluff, then turned north between Brown Peninsula and Black Island and terminated against the confluence of the RIS and KG lobes just north of Black Island. The second glacial event; in more recent but less extensive Quaternary glaciations, the McMurdo Ice Shelf (MIS) thickened and/or lowered and incorporated glacial sediment from the sea floor by basal adfreezing. Surface ablation transported these sediments to the MIS surface.

INTRODUCTION

Southern McMurdo Sound generally refers to that area of McMurdo Sound south of Ross Island and to the perennially sea-ice covered water of McMurdo Sound proper (Figures 1 and 2). Basement geology of the sound comprises lava flows, pyroclastic deposits, scoria cones and lava domes of the McMurdo Volcanic Group [Kyle, 1981]. The western margin of the sound is flanked by the Transantarctic Mountains that expose metasediments of the Koettlitz Group, granitoids of the Granite Harbour Intrusives, sandstones of the Beacon Supergroup, the Ferrar Dolerite, and McMurdo Volcanic Group rocks [Gunn and Warren, 1962].

Southern McMurdo Sound is now occupied by the McMurdo Ice Shelf (MIS) [Stuart and Bull, 1963; Kellogg et al., 1990], which is an extension of the Ross Ice Shelf (RIS). The MIS is bounded to the south by Minna Bluff and Mount Discovery, and is pinned by Brown Peninsula and White and Black islands. Brown Peninsula is an island connected at its southernmost tip to Mount Discovery by a debris-covered ice bridge (Figures 1 and 2). The southeastern Dry Valleys of the Royal Society Range mark the western edge of southern McMurdo Sound, but, these are separated from the MIS by the Koettlitz Glacier (Figures 1 and 2).

Glacial moraines and debris sheets are well developed on the ice surface close to Minna Bluff and in the area between Black Island and Brown Peninsula. Although, in other areas the MIS is free of glacial sediment. Along the northern coast of Minna Bluff and flanking eastern Mount Discovery, ice-cored lateral moraines are present landward of the tide crack(s). North of Mount Discovery and east and north of Brown Peninsula, where there are similar moraines, the position of the tide crack is less clear. All these lateral coastal moraines, most of which are ice-cored, contain a variety of glacial erratics including the Eocene fossiliferous erratics, which are the focus of other papers in this volume.

This study of the glacial geology of southern McMurdo Sound is to support the program of glacial erratic collection for paleontologic and sedimentologic analysis led by D.M. Harwood. During three austral summer field seasons between 1992 and 1995, as part of the effort to collect the fossiliferous glacial erratics distributed along the coastal moraines in southern McMurdo Sound [e.g. Harrington, 1969; Stilwell et al., 1993], the author mapped the distribution of the erratics and other glacial features. The main aim was to provide a means of predicting the occurrences of the erratics by understanding their transport and emplacement processes, so as to direct future collecting. The erratics are excellent tracers of glacial flow and extent, and the study of their distribution led to the interpretation of glacial history of southern McMurdo Sound presented here. The present-day physiography of the MIS was also mapped in some detail as it contains many features that demonstrate recent and current processes that have contributed to the distribution of erratics and distribution and morphology of moraines. The present MIS also contains many features that are relict or derived from former grounded ice in southern McMurdo Sound.

Kellogg and Kellogg [1985; 1987; and 1988] and Kellogg et al. [1990] addressed the late Quaternary history of the western MIS and commented on the origin of the debris bands that mantle the surface of the MIS. They inferred that these features are remnants of the Wisconsin RSIS, which according to Stuiver et al. [1981], was an ice sheet that occupied McMurdo Sound and fed by grounded ice in the Ross Sea (Figure 3). The RSIS enveloped most of Black and White Islands, Brown Peninsula and the northern flanks of Minna Bluff and Mount Discovery and flowed westwards into the southern Dry Valleys of the Transantarctic Mountains [Denton et al., 1971]. According to this model, most of the area studied here would have been covered by 300-700 m of ice only ca. 17-21 ka ago. While Kellogg et al. [1990] recognize that erratic material from the Transantarctic Mountains is widespread, they do no discuss its transport into and

mode of deposition in McMurdo Sound. Furthermore, glaciations older than the inferred late Wisconsin RSIS [Stuiver et al., 1981] are not considered. The inferred flow lines of the RSIS proposed by Stuiver et al. [1981] are in conflict with the distribution of Eocene fossiliferous erratics and when the pattern of coastal and MIS supraglacial moraines is also considered, it is clear a new glacial history reconstruction is required.

METHODS

The southern McMurdo Sound area was mapped using a combination of aerial photographic techniques and ground traverses and surveys. Moraines, surface features of the MIS, and coastal lake and tide crack features were plotted onto a 1:50,000 base map prepared by enlarging the USGS 1:250,000 Ross Island and Mount Discovery topographic maps. For the purposes of this study, the USGS flew 3 lines of color vertical aerial photography (Figure 2) and a series of oblique black and white photographs were prepared from low level helicopter overflights. These were used in conjunction with US Navy vertical and oblique black and white aerial photographs (Figure 2). Ground traverses were designed to cross features and areas of interest identified by analysis of the aerial photographs. The following features were noted and described along each traverse: quasi-linear trends and continuity of moraines and other features; height, steepness and dissectedness of local topography; size, location, composition (sedimentary/ fossiliferous, igneous, or volcanic) and distribution of large boulders; general petrographic, textural, and color appearance of moraine material; degree of ice coring; and ice or debris stratification. At selected locations, more rigorous measurements were made to determine and quantify the differences between features mapped by visual means: A 1 m^2 quadrat survey was carried out at the surface of moraines to determine clast petrography and shape. Gross texture was also estimated in a 10-20 cm pit excavated beneath the armoured surface of the moraines. Petrographic proportions, clast morphology and textures were estimated visually using the methods of Krumbein [1988]. Clasts were grouped into six petrographic categories according to rock types of distinctive color as well as provenance to quantify features mapped by aerial photography and visual observation along traverses (Table 1, Figure 2). These included two igneous categories; granite and dolerite, two volcanic categories, basalt and scoria (felsic igneous rocks were also grouped with scoria as they tended to have a pinkish hue), and two sedimentary categories; metasediments and clasts of Beacon Supergroup sediments and fossiliferous erratics.

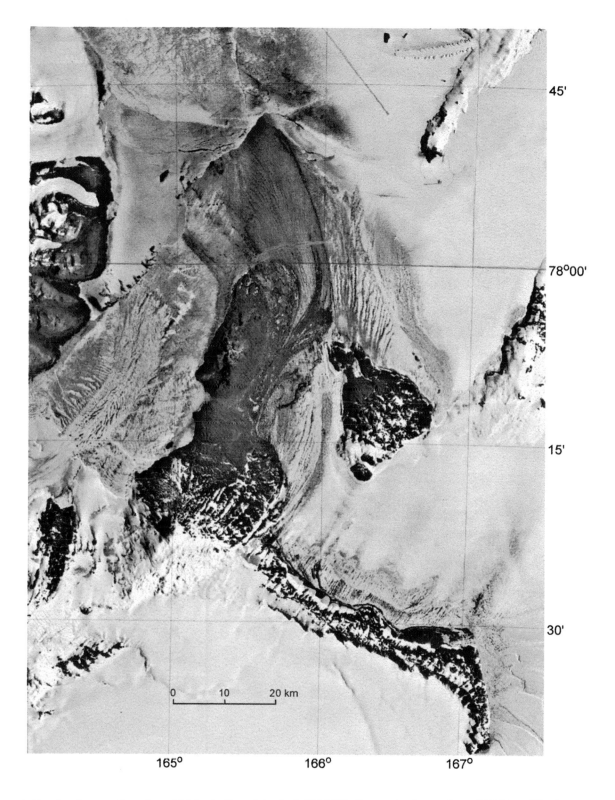

Figure 1. Landsat image of southern McMurdo Sound (scale 1:567,000). Composite image from USGS satellite image maps 77190-SI-250 (Ross Island, 1975) and 78192-SI-250 (Mount Discovery, 1974).

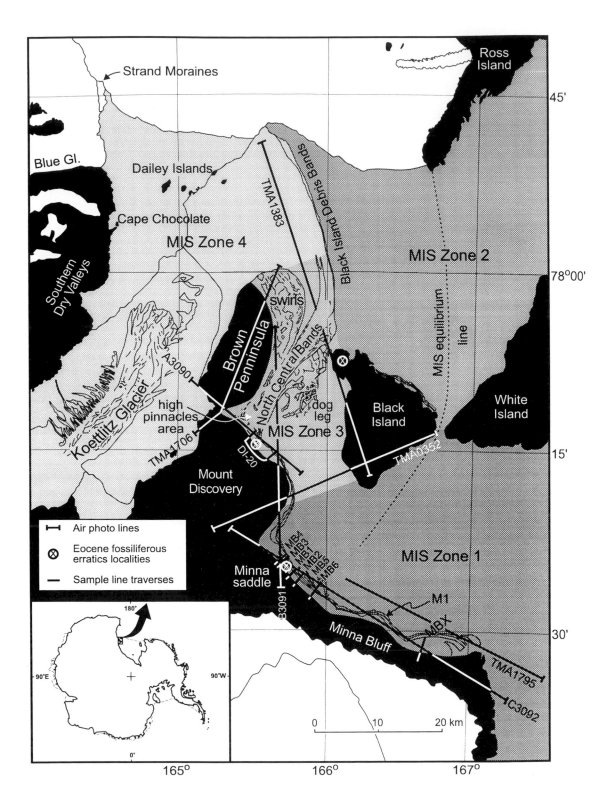

Figure 2. Line drawing of southern McMurdo Sound (scale 1:567,000) illustrating features discussed in text. Shaded areas are physiographic subdivisions of the MIS (discussed in text).

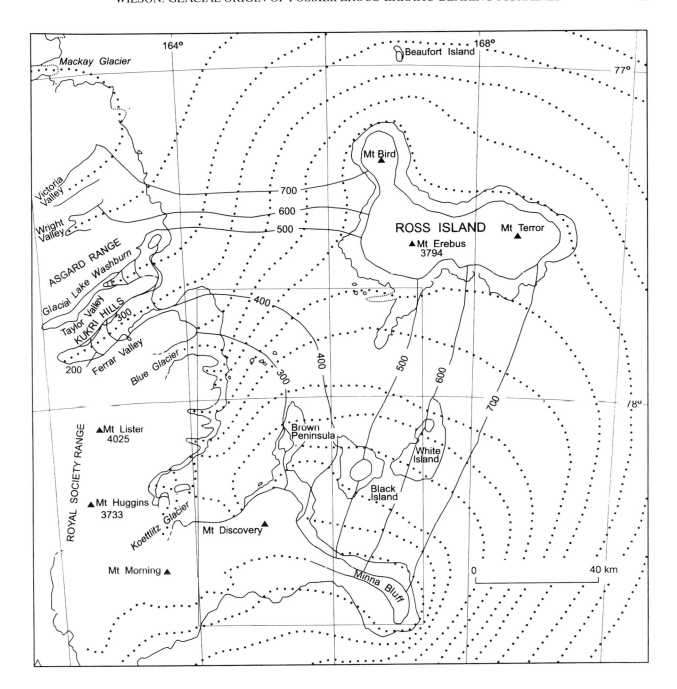

Figure 3. Reconstruction of the Ross Sea Ice Sheet (RSIS) after Stuiver et al. [1981]. Solid lines are proposed ice surface contours based on glacial trim lines of the Ross Sea Drift and dotted lines are flow lines drawn perpendicular to ice surface contours and consistent with striation directions on Ross and Black islands. The box shows the area examined in this study.

RESULTS

Physiography of the McMurdo Ice Shelf (MIS)

The MIS can be subdivided into four physiographic zones (Figure 2): Zone 1) the area to the south and west of White and Black islands, Zone 2) the area to the north of White and Black islands, Zone 3) the area between Black Island and Brown Peninsula, and Zone 4) The area to the north and west of Brown Peninsula. Swithinbank [1970] determined that the MIS is ca. 100 m thick in Zone 1 and

TABLE 1. m^2 quadrat classification of drift/moraine texture and petrography along traverses shown in Figure 2.

Sample number	Matrix at surface	Clast petrography						General moraine texture		
		Igneous		Volcanic		Sedimentary		Boulders/ cobbles	Gravel/ granules	Sand/ silt
		Granitic	Dolerite	Scoria + Felsic	Basalt	Meta-morphic	Fossil-iferous			
MBX-1	33	2	5	5	50	5	-	15	60	25
MBX-2	15	5	5	5	65	5	-	20	70	10
MBX-3	10	10	5	5	60	10	-	50	45	5
MBX-4	4	30	4	2	20	40	-	25	60	15
MBX-5	5	10	10	10	40	25	-	40	50	10
MBX-6	10[α]	20	5	5	35	25	-	30	60	10
MBX-7	5	20	5	5	25	40	-	25	70	5
MB1-1	13	5	-	10	70	2	-	40	55	5
MB1-3	10	15	-	5	65	5	-	50	45	5
MB1-4	10	15	<1	<1	70	5	-	40	45	15
MB1-5	[β]	30	5	-	30	35	-	75	15	10
MB1-6	<1	45	10	10	20	10	5[χ]	90	5	5
MB2-2	5	5	-	15	70	5	-	20	40	40
MB2-3	2	20	10	1	20	45	2[δ]	70	20	10
MB2-4	<1 [β]	5	5	10	65	14	1[ε]	92	8	-
MB3-1	[β]	2	10	25	63	-	-	90	5	5
MB3-2	15	20	10	1	50	2	2[δ]	50	20	30
MB3-3	15	5	-	15	64	1	-	50	30	20
MB3-4	5 [β]	10	5	<1	65	10	5[ε]	55	15	30
MB3-5	[β]	10	30	<1	55	2	3[ε]	80	5	15
MB4-1	[β]	15	25	<1	58	2	-	90	5	5
MB4-2	10	25	10	<1	40	15	-	50	25	25
MB5-1	15	2	5	20	40	10	8[χ]	40	20	40
MB5-2	10 [β]	50	5	5	15	10	5[χ]	50	30	20
MB5-3	[β]	60	<1	2	13	20	5[ε]	40	25	35
MB5-4	25	5	5	5	30	20	10[χ]	40	50	10
MB6-1	5	10	5	5	70	5	-	50	20	30
MB6-2	10	35	10	1	40	4	-	40	30	30
MB6-3	10 [β]	25	4	1	45	5	-	40	40	20
MB6-4	[β]	20	<1	-	80	<1	-	85	7	8
D1-1	20	4	-	74	1	1	-	50	30	20
D1-2	20	50	5	5	10	10	-	25	25	50
D1-3	20	40	3	3	3	31	-	40	25	35
D1-4	15	3	1	10	70	1	-	60	20	20
D1-5	25	30	1	3	6	30	5[χ]	30	10	60
D1-6	40	30	3	<1	2	25	-	25	20	55
D1-7	15	3	-	10	70	2	-	40	30	30
D1-8	5	10	-	75	10	<1	-	70	20	10
D1-9	15	40	5	<1	30	10	<1[χ]	40	30	30
D1-10	5	40	10	<1	30	15	-	60	10	30
D1-11	10	45	1	5	34	5	-	15	50	35
D1-12	20	5	5	15	50	5	-	30	20	50
D1-13	15	40	2	3	20	10	10[χ]	50	30	20
D1-14	10	25	-	5	40	10	10[χ]	55	25	20
D1-15	5	3	-	78	3	10	2[ε]	60	30	10
D1-16	15	10	-	20	50	5	-	50	40	10
D1-17	25	2	-	3	65	5	-	30	50	20
D1-18	5	2	-	60	33	<1	-	70	10	20
D1-19	20	20	-	-	20	40	-	60	10	30
D1-20	23	32	-	-	25	20	-	75	5	20

[α] Surface at this site was completely armored. Immediately beneath surface = 40-50% matrix.
[β] Surface of moraine completely armored.
[χ] Eocene and Scallop Hill Formation fossiliferous erratics.
[δ] Eocene fossiliferous erratics only.
[ε] Scallop Hill Formation fossiliferous erratics only.

moving slowly (< 2 m/yr.) in a northwest direction. In contrast, the MIS in Zone 4 is only 30 m thick on average and is flowing at about 20 m/yr. in a northwest direction. Swithinbank [1970] also identified a broad equilibrium line (where the MIS is neither thickening nor thinning) trending SW from Black Island towards Minna Bluff. To the east in Zones 1 and 2, ice is accumulating and the MIS thicken eastwards. To the north, in Zones 3 and 4, ice is ablating and or melting. The MIS surface has generally low relief. It varies by no more than 20 m in areas of drift covered ice-pinnacles and depressions, and 8 m or less in areas that are free of glacial drift (e.g. between Black Island and Mount Discovery [Heine, 1967; Swithinbank, 1970]).

Ablation is upwarping the MIS at the southwestern edge of Zone 1 and causing stratification in the ice shelf to dip north-northeast at ca 45° (Figure 4a). This dip shallow towards the north-northeast. Intact sponges and other macrofossils frozen into the base of the ice shelf are preserved within discrete stratigraphic horizons of the ice shelf. They are also found in melt pools and hollows on the ice shelf surface, along with coarse sands and gravel deposited by eolian processes [Gow et al., 1965; John Kaser personal communication, 1994]. Two distinct and several less distinct northwest trending moraine bands occur on the surface of the ice shelf, along the northern shores of Minna Bluff (Figure 4a &b). They are relict moraines from beneath the current MIS that have been subsequently elevated to the ice shelf surface by surface ablation and basal adfreezing to its base (a mechanism first suggested by Debenham [1919]). Their present distribution replicates that in existence when the moraines were originally beneath the ice shelf. The moraine bands are cross cut by the upwarping stratified bands in the MIS discussed above (Figure 4a).

Zone 2 of the MIS is mostly free of a sediment mantle. An area of slower moving / stagnant ice in front of Black Island shows wind channelling and has an eolian sediment cover. Flow measurements of Swithinbank [1970] suggest a flow boundary, with ice) of the Ross Ice Shelf that flows northwestwards (i.e. around White Island).

In Zone 3, to the east of Brown Peninsula, extensive moraines and glacial drift [Kellogg et al., 1990] mantle the ice surface (Figure 5). In the southernmost part of Zone 3, north and northeast trending moraine bands (the North Central Bands of Kellogg et al. [1990]) form a curvilinear pattern (Figure 6a). Some of these moraines radiate northwards past Black Island and northwest around the tip of Brown Peninsula, towards the curved Zone 2 / Zone 4 boundary (the "Black Island Debris Bands" of Kellogg et al. [1990] Figure 6b). The sediment cover on these features is mostly greater than 1 m in thickness and the moraine pin-

nacles and ablation depressions form a relief of up to 10 m. Individual moraines are mostly separated by debris-free ice areas 10-500 m in width (Figure 7). Some individual moraines bifurcate. These moraines are relict basin floor moraines, frozen onto the base of the ice shelf and then transported to the surface as a result of surface ablation. Where the basin floor is deeper, moraines are less common on the MIS surface and may represent only the larger relict moraines (e.g. "The Dog Leg" [Kellogg et al., 1990], Figure 6). To the northeast of Brown Peninsula and west of the "Black Island Debris Bands", the MIS is covered by an extensive, thick layer of glacial drift (Figure 5) referred to as the "Swirls" by Kellogg et al. [1990], Figures 1, 2, & 6b). Individual moraine bands are less distinct and debris-free ice, melt channels and ponds are less common. Some of the glacial drift is ice-cored. However, it is difficult to ascertain if it is ice cored everywhere, because the glacial drift, although patchy, is often more than a meter thick and ice cemented below 0.5 m depth. Kellogg et al. [1991a] report very negative $\delta^{18}O$ values (<-35‰) demonstrating that the ice in this area is a relict glacial or ice shelf feature and not the result of basal freezing of sea water. The drift mantle is thick enough to insulate, and so preserve, the underlying stagnant ice. A similar situation exists in the vicinity of Cape Chocolate and the Dailey Islands area where the glacial drift blankets ice islands [Oliver et al., 1978].

To the west of Brown Peninsula (Zone 4, Figures 1 & 2), the MIS comprises mostly the Koettlitz Glacier ice tongue, which is free of surficial debris, except for a few medial moraine bands. The MIS occupies the area beyond the floating terminus of the Koettlitz Glacier. Occasional moraines on the eastern side of Zone 4 (north of Brown Peninsula) cover the ice shelf surface, but otherwise, only patchy occurrences of windblown sand and silt cover the ice shelf surface.

Coastal Moraines

The entire northern coastline of Mount Discovery and Minna Bluff is separated from the MIS by a suite of coastal moraines. The most distinct is an almost continuous ice-cored moraine (referred to here as M1, Figures 1 & 2). It lies landward of the tide-crack and marks the southern edge of the MIS. This moraine closely mimics the coastline, but in places has been deformed by alpine glaciers flowing off Minna Bluff pushing it into convex forms (Figure 8). M1 is the youngest of all the coastal moraines and crosscuts all the other moraines resulting from either the MIS or SMS Ice Sheet. Its topography is steep (at the angle of repose), variable and hummocky, with pinnacles and depressions

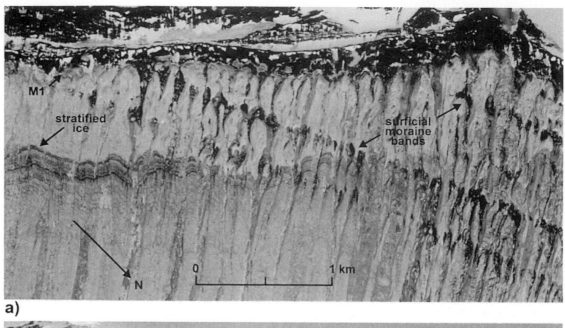

a)

b)

Figure 4. a) USGS vertical aerial photograph of the MIS adjacent to Minna Bluff. Photograph shows stratified ice upwarping from increased surface ablation close to Minna Bluff and oblique debris bands accumulating on the ice shelf surface from basal adfreezing of relict moraines on the McMurdo Sound floor and transport to the ice surface by ablation. M1 moraine (see text) is visible in the top of the photograph. b) U.S. Navy oblique aerial photograph of Minna Bluff, viewed from the west showing location of Figure 4b and the oblique debris bands accumulating on the MIS surface. M1 moraine is labelled (see text). The area of Figure 9 is also shown. Minna Bluff rises more than 700m above the MIS.

Figure 5. Thick ice-cored drift moraine (poorly sorted diamicton) on the MIS surface north of Mount Discovery. Boulders are up to 0.5 m in diameter, well rounded and striated. Hammer (1 m) leans against the ice underlying the drift layer. The ice itself contains no sediment.

resulting in elevation differences of up to 20 m. M1 is separated from the older moraines by a succession of polyhedral ice covered meltwater ponds that infill the intermorainal topographic depressions. The older moraines are less pervasive and more closely mimic local coastal landforms. Embayed and low-lying coastal areas have been infilled by glacial drift by several landward encroachments of the MIS or SMS Ice Sheet. This is best seen on the eastern coastline of Minna Bluff where more than six parallel, ca. 500 m wide moraines form a 3 km wide debris apron between Minna Bluff and the MIS (Figure 9). Topography is subdued (<6 m of elevation variation) and rounded on the older moraines. Sometimes polygonal ground is developed on the older more flat drift in areas with a higher fines content.

Texture of the coastal moraines is variable but does not appear to be related to discrete moraines (Table 1). Matrix content is generally less than 50%, by volume, and is fine sand with occasional coarse silt. Any fines have been winnowed and the moraines were deflated by wind during melt-out. The surfaces of the moraines are well armoured. Boulders, up to 1 m in diameter, are also distributed across

the different moraines. The larger boulders are generally granitic but, occasionally, the fossiliferous Eocene erratics occur as large boulders.

McMurdo Volcanic Group clasts comprise greater than 50% of the coastal moraines (Table 1). Granitic clasts are quite common (3-35 %) and occasionally dominant. The moraines can be defined by their current geomorphology and also differentiated by their variation in clast petrography. However, the variation in clast petrography is not related to the current moraine geomorphology. The current geomorphology crosscuts varying light colored granite clast-rich bands (as much as 50% granite, Table 1) and dark colored granite clast-poor bands (Figure 10). The granite clast-poor bands consist almost exclusively of clasts of the McMurdo Volcanic Group. There is also a covariance of sedimentary clasts (formerly of the Beacon Supergroup) and metamorphosed sediments (formerly of the Skelton Group) with the granite clast-rich bands. The distribution of dolerite clasts is more uniform, with a slight increase in proportion of dolerite clasts within granite clast-rich bands (Table 1). The lighter colored granitic clast-rich bands are concentrated in the area between the Brown Peninsula sad-

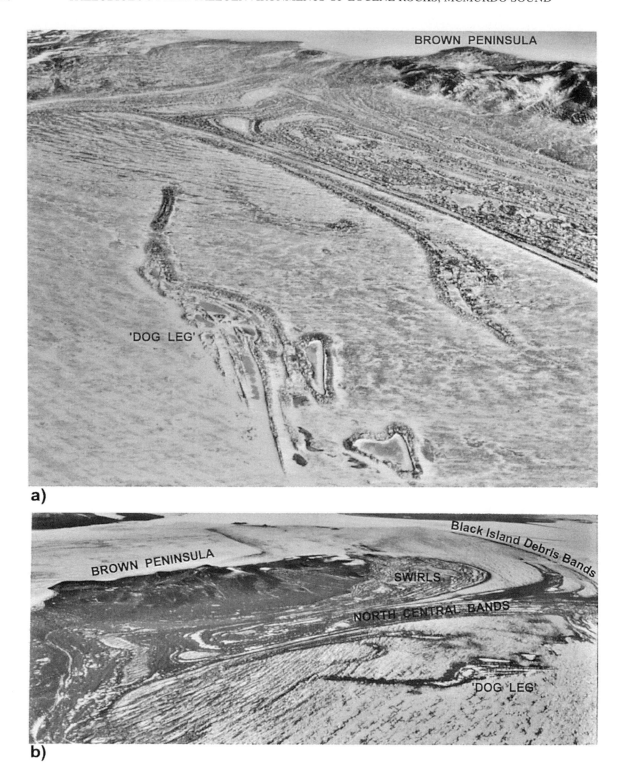

Figure 6. a) U.S. Navy oblique aerial photograph of the MIS east of Brown Peninsula, viewed from the northeast. Brown peninsula is labelled in the top right of the photograph and the northern flanks of Mount Discovery in the top left. The western edge of the MIS is covered by extensive drift, marking the southern end of the "North Central Bands", which is a relict medial moraine (see discussion in text). b) U.S. Navy oblique aerial photograph of the MIS drift between Brown Peninsula and Black Island, viewed from the southeast. Labelled moraine and drift features are discussed in text.

Figure 7. debris-free ice-ridges of the "Central Debris Bands" on the MIS east of Brown Peninsula. Ice ridges have ca. 15 m of relief. Drift is concentrated in low lying areas between ice ridges. Note large melt pool areas on surface of the MIS around drift.

Figure 8. U.S. Navy oblique aerial photograph of alpine glacier moraines on Minna Bluff in front of the saddle between Minna Bluff and Mount Discovery. Snow has accumulated in the troughs between moraines. Alpine moraines cross cut SMS Ice Sheet moraines. View is looking southwest. Location and alternative view is shown on Figure 4b.

Figure 9. U.S. Navy oblique aerial photograph of multiple coastal moraines behind M1 on the eastern end of Minna Bluff. View is looking southeast. Moraine bands are ca. 500 m wide. Minna Bluff is in the background.

dle and a few miles east of the Minna Bluff saddle (Figure 2). The moraine along the eastern and northern coast of Brown Peninsula comprises very few granitic clasts and is a mixture of light brown to black volcanic drift, presumably originating from the McMurdo Volcanic Group. The moraines of the western and northern coastline of Black Island contain a moderate proportion of granitic clasts (≤10%) and also occasional fossiliferous Eocene erratics. The clast-petrographic bands and geomorphic features of the northern Mount Discovery and eastern Brown Peninsula areas are crosscut in several places by the tide crack, demonstrating a common origin for features on the MIS and the coastal moraines.

Several moraine features can be observed in the southern Dry Valleys and along the western flank of the Koettlitz glacier. Towards the heads of the valleys, these originate from through-valley glaciation, but towards the coast they originate from re-entrant glaciation from an expanded marine ice shelf grounded in the Ross Sea [Denton et al., 1970; 1989; Denton and Borns, 1974; Drewry, 1979; Stuiver et al., 1981]. These re-entrant moraines were named the Ross Sea Drift by Denton et al. [1971] and record the youngest glacial expansion dated at ca. 21 ka by Denton et al. [1985], Brook et al. [1995], and Kellogg et al. [1991b]. The Ross Sea Drift is mostly sourced from the McMurdo Volcanic Group, but also contains occasional granite erratics, although these are not as common as the coastal areas of Minna Bluff and Mount Discovery. Volcaniclastic sedimentary erratics (possibly Scallop Hill Formation) are particularly common around the Bullwark and Walcott Bay area. Scallop Hill Formation erratics are also common along the most coastal moraine originating from an expanded ice body grounded east of the southern Dry Valleys.

Elevated Moraines and Glacial Drift

The flanks of Mount Discovery, Minna Bluff, and Brown Peninsula are mantled with moraines and glacial drift sheets. These moraines and glacial drift are generally not ice-cored and exhibit only remnant features of glacial emplacement and ice melt-out. The drift does not exhibit the same topographic features as the coastal and ice shelf moraines described above, but instead blankets the underlying basement topography and exhibits moderate to well-developed polygonal ground. Melt-water streams, fed largely by snowmelt in summer months, break the drift sheets in several places. A subhorizontal moraine ridge marks the highest extent of ice that mantled the area and bounds the upper limit of the drift.

Figure 10. Light and dark bands on the coastal moraines at Mount Discovery. Light colored bands are rich in granitic erratics and matrix, and dark coloured bands are rich in volcanic erratics and matrix. Note that the bands cross the tide crack, which is filled with snow and in the middle ground. Eocene fossiliferous erratics are more common on the lighter bands.

Fossiliferous Erratics

Two suites of marine fossiliferous erratics have been previously described from southern McMurdo Sound: Eocene fossiliferous sandstone and the Pliocene volcaniclastic Scallop Hill Formation. Speden [1962] first described the Scallop Hill Formation. It comprises cemented tuffaceous sandstones, conglomerates and breccias containing the extinct thick-shelled *Zygochlamys*

anderssoni (Hennig). The Eocene fossiliferous erratics are well cemented, stratified, fine to coarse quartz sandstones with gravely facies. They contain micro-flora and faunas, molluscs and decapods of middle Eocene age [Stilwell et al., 1993; 1997]. Their distribution is more limited than erratics of the Scallop Hill Formation and they are only found in areas around Minna Saddle, the northeast coast of Mount Discovery, the northwest coast of Black Island, and at Cape Crozier on Ross Island

(Figure 2). Erratics of the Scallop Hill Formation have a widespread distribution across the coastal moraines and drift sheets examined in this study and are occasionally found on MIS moraines. Sites where the Scallop Hill Formation erratics are concentrated have been reported from the eastern margin of Brown Peninsula, eastern coast of Black Island in the vicinity of Scallop Hill, the northern tip of White Island and the northern tip of Ross Island in the vicinity of Cape Bird [Speden, 1962; Vella, 1969; Cole et al., 1971; Leckie and Webb, 1979; Eggers, 1979; Buckridge, 1989; Stott et al., 1983].

Scallop Hill Formation erratics are ubiquitous around southern McMurdo Sound. They appear to be coeval with McMurdo Volcanic Groups rocks and been incorporated into the same widespread moraines as the McMurdo Volcanic Group rocks. The origin of the Eocene marine fossiliferous erratics is less clear. Drilling in McMurdo Sound indicates that in-situ Eocene strata are still buried deeply beneath the sea floor and that it is unlikely that McMurdo Sound itself is the source. The distribution of the erratics and inferred ice flow directions indicate a source to the south of southern McMurdo Sound.

DISCUSSION

Interrelationship of Coastal Moraines and the MIS Surface Drift

Moraines are not geographically restricted to either the MIS surface or the coastal areas. Individual moraines may cross the tide crack. This demonstrates a common origin for the coastal and MIS moraines. Coastal moraines and drift sheets are commonly ice-cored, but their topographic relief is lower than their ice-shelf counterparts. This is because the coastal moraines are now behaving as melt-out tills. Higher on the flanks of the coastal areas of southern McMurdo Sound, older moraines and drifts from grounded ice in McMurdo Sound (the Southern McMurdo Sound Ice Sheet – SMS Ice Sheet) are no longer ice cored. It is possible to trace moraines from the coastal areas to higher on the flanks of Brown Peninsula and Mount Discovery. This demonstrates a common genetic origin for the coastal moraines and drifts in these regions. But, this study did not find such a link between the coastal moraines and elevated drift on Black Island or Minna Bluff.

Evidence for Former Grounded ice in Southern McMurdo Sound (the Southern McMurdo Sound Ice Sheet)

There are three clear lines of evidence for a former ice sheet grounded on the floor of southern McMurdo Sound in the area that is now occupied by the MIS:

1) The occurrence of drift material on the present MIS surface. Because the MIS does not currently contain any englacial material, nor is there any clear transport path of glacial drift material on the surface of the MIS from areas beyond its margins. This drift must have been brought into southern McMurdo Sound by a former glacier or ice sheet and subsequently incorporated onto the surface of the MIS.

2) The distribution of coastal and present MIS surficial moraines does not reflect MIS flow patterns. Rather, the physical distribution of moraines suggests three distinct former lobes of ice grounded in southern McMurdo Sound (Figure 11). The extent and retreat pattern of these lobes is clearly visible from the distribution of relict terminal, lateral and retreat moraines along the coastal areas and on the surface of the present MIS. These features are clearly visible on aerial photographs and the Landsat image (Figure 1). The western-most lobe (Koettlitz Glacier Lobe – KG Lobe) was an extended former Koettlitz Glacier. It enveloped much of Brown Peninsula and flowed past the Dailey Islands and Cape Chocolate. Cape Chocolate and the Dailey Islands are drift-covered ice pedestals grounded on the floor of McMurdo Sound [Oliver, 1978], which may be relict features of a former expanded Koettlitz Glacier grounded in McMurdo Sound.

The easternmost lobe of the SMS Ice Sheet (Ross Ice Shelf Lobe – RIS Lobe) was a former extension of the Ross Ice Shelf, which was also most likely grounded along the eastern and northern sides of Black Island at this time. The RIS Lobe flowed northwestward past Hut Point Peninsula.

The third lobe of the SMS Ice Sheet (Minna Bluff Lobe – MB Lobe) advanced between the KG and RIS lobes. It was also sourced from the grounded RIS. It flowed northwestward along Minna Bluff and then turned northward in front of Mount Discovery and flowed between the KG Lobe and Black Island. It terminated at about the northernmost extent of Black Island. North of Brown Peninsula and Black Island the KG and RIS lobes coalesced. Distinct suites of lateral moraines mark the various boundaries between the three lobes.

3) A further line of evidence suggesting grounded ice in southern McMurdo Sound is the occurrence and distribution pattern of glacial erratics on the MIS and coastal moraines that are not currently being transported to southern McMurdo Sound by the present MIS. These include igneous and sedimentary rocks from the Skelton and Ferrar groups, and Beacon Supergroup, and more recent Cenozoic sedimentary rocks (the Eocene fossiliferous erratics, that are the subject of this volume, and rocks of

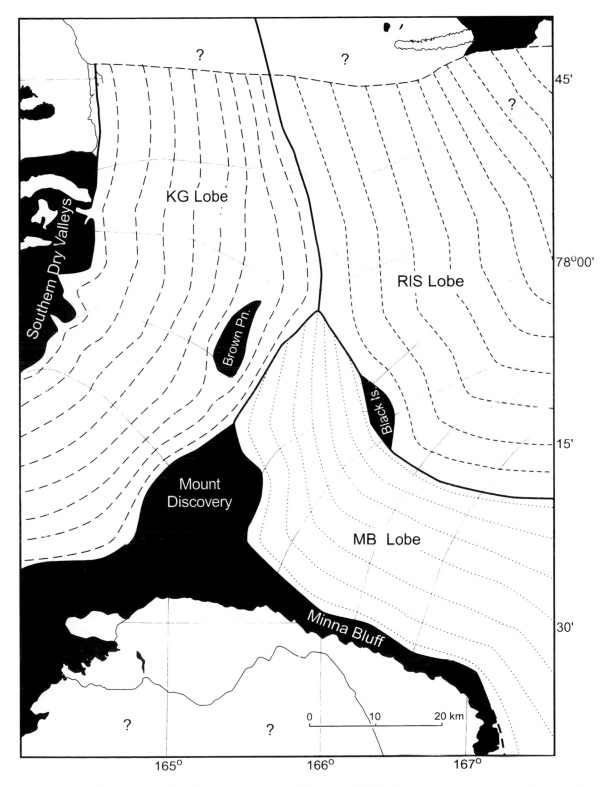

Figure 11. Reconstruction of the SMS Ice Sheet comprising the KG, MB, and RIS Lobes as hypothesis in this study and explained in the text. The reconstruction is qualitative and based on geologic observations outlines in the text. A quantitative model of the SMS Ice Sheet is yet to be undertaken. Solid lines are hypothesized ice surface contours (not to scale) and dotted lines are hypothesized flow lines.

the volcaniclastic Scallop Hill Formation, Table 1, Figure 2). The only Cenozoic rocks currently exposed in southern McMurdo Sound are those of the McMurdo Volcanic Group [Kyle, 1981].

Advance and Retreat of the Southern McMurdo Ice Sheet Lobes

The geographic distribution of the coastal and MIS moraines and the distribution of erratics on those moraines record the flow and retreat patterns of the SMS Ice Sheet lobes. Several moraines mark the lateral extent of the MB Lobe and its narrowing during retreat (Figures 2, 4b & 6b). The northwest extent of the lobe is marked by a relict medial moraine along the eastern side of Brown Peninsula and between Brown Peninsula and just north of the northern tip of Black Island (Figures 1 & 6), where the MB Lobe came into contact with the KG Lobe (Figure 11). In this study of southern McMurdo Sound, Eocene fossiliferous erratics were only recovered around the perimeter of the MB Lobe (the northern flanks of Minna Bluff and Mount Discovery, and the NW tip of Black Island, Figures 2 & 11). Other moraines on the surface of the MIS and the coastal areas in front of Mount Discovery are retreat moraines deposited or accumulated during the retreat of the MB Lobe from southern McMurdo Sound. These terminal and lateral retreat moraines demonstrate that the MB Lobe retreated southeastward at least halfway along Minna Bluff before it began to float.

The KG Lobe of the SMS Ice Sheet was grounded in McMurdo Sound at least as far north as Cape Chocolate and the Dailey Islands. These drift mantled, grounded remnants of the former Koettlitz Glacier demonstrate that the KG Lobe was grounded below sea level before retreating and floating as it does today. Lateral moraines and drift flank the eastern edge of the southern Dry Valleys and show that small lobes flowed into the mouths of the valleys [Denton et al., 1970; 1989; Stuiver et al., 1981; Hall et al., 1993]. Further north, the Strand Moraines are a remnant of a medial moraine between the extended KG Lobe and the Blue Glacier, which was turned northwards by the flow of the KG Lobe. On its eastern side, the KG Lobe bifurcated around Brown Peninsula (Figure 11) and formed a medial moraine made up of the "North Central Bands" against the MB Lobe in the south, and the "Black Island Debris Bands" against the RIS Lobe in the north. A thick ice-cored drift sheet was deposited around Brown Peninsula by the retreat of the KG Lobe. During the retreat, a relatively stagnant area of ice on the NE flank of Brown Peninsula,

the result of an eddy, persisted and produced a winnowed melt-out till that is still ice-cored in "The Swirls" of Kellogg et al. [1990], Figure 6b).

Comparatively few features record the presence of the grounded RIS Lobe of the SMS Ice Sheet. White Island is now mostly covered by ice and snow, preventing the examination of sediment cover. A thin lateral moraine along the northern coast of Black Island demonstrates that the SW edge of the RIS Lobe flowed between White and Black islands and northward toward McMurdo Sound. North of Black Island, the RIS Lobe flowed alongside the KG Lobe forming a medial moraine (the "Black Island Debris Bands"). Ice from the RIS Lobe may have almost completely enveloped White Island.

Advance and Reworking of Moraines by the McMurdo Ice Shelf

A second, but less extensive, glacial advance into southern McMurdo Sound is recorded by the McMurdo Ice Shelf (MIS). After retreat of the SMS Ice Sheet from southern McMurdo Sound, floating ice subsequently refilled the area of McMurdo Sound east of Brown Peninsula. It formed either from a floating ice tongue from the Ross Ice Sheet/Shelf, or by snow and ice build-up on multiple year sea-ice. It did not originate by floating of a grounded ice-sheet because retreat moraines of the SMS Ice Sheet were note eroded by an advancing grounded ice sheet. Once it filled southern McMurdo Sound, the MIS did not continue to flow, probably because it is pinned by Black and White Islands.

Since formation, the MIS has thickened and/or been lowered by a lowering of eustatic sea level at least once, causing it to rest on relict moraines of the SMS Ice Sheet which form bathymetric high points on the floor of southern McMurdo Sound. At times when the MIS rested on the SMS Ice Sheet moraines, it has incorporated sediments by basal adfreezing. Subsequent upwards transport, from surface ablation, resulted in their accumulation on the MIS surface. Original moraine patterns have been maintained due to the very slow flow of the MIS. Currently, the MIS is floating and does not contain any englacial sediment.

Chronology

It is well recognized that ice grounded in McMurdo Sound, dammed the Dry Valleys, and resulted in the formation of Wisconsin age glacial lakes [Denton et al., 1989]. Pliocene Scallop Hill Formation sediments are ubiquitous on the moraines of southern McMurdo Sound. These are the best age constraints available and suggest that all the glacial activity associated with the SMS Ice Sheet and MIS partial grounding is latest Pliocene through Quaternary. If one accepts that the MIS partial grounding occurred at the last glacial maximum and that other glaciations of the last million years were similar in size, then, an obvious conclusion is that the SMS Ice Sheet advance is early Quaternary or latest Pliocene in age. This is not inconsistent with dates afforded by the McMurdo Volcanic Group [Kyle, 1981] or of the Kenyte outcrop on Mt Erebus. However, this argument is still conjecture.

Reconciling These New Observations with the RSIS Hypothesis.

While the ice sheet reconstruction presented here is internally consistent, some discussion of its relationship to the hypothesis of Stuiver et al. [1981], whose reconstruction of the RSIS in McMurdo Sound was based on the Ross Sea drift, is necessary. The two hypotheses have many similarities. Both suggest grounding of ice in McMurdo Sound. The SMS Ice Sheet reconstruction presented here is consistent with Ross Sea Drift trim lines and striation indications of flow directions from Observation Hill on Ross Island and Scallop Hill on Black Island. It is also consistent with grounding of Ross Sea Ice Sheet ice to the north in McMurdo Sound, flowing around Ross Island and damming the northern Dry Valleys. But, key differences between the hypotheses occur in southwestern McMurdo Sound: Stuiver et al. [1981] propose westward and southwestward flow of the RSIS, whereas the reconstruction presented here proposes northerly ice flow. Stuiver et al. [1981] used the distribution of Kenyte erratics on the Ross Sea drift as a key element in reconstructing flow lines of the marine Ross Sea Ice Sheet. However, several factors compromise the robustness of this approach:

1) With very little exposure of bedrock, it may be premature to suggest that the exposure on Ross Island is the only occurrence of Kenyte.

2) Kenyte is not found on the Ross Sea Drift on Ross Island itself.

3) Kenyte occurrences on Black Island [Vella, 1969; Stuiver, 1981] are inconsistent with the reconstruction and flow lines proposed by Stuiver et al. [1981]. Their preferred explanation for this is that earlier glacial events distributed Kenyte erratics around McMurdo Sound, and these were subsequently reworked by the RSIS. By this same argument, it is possible to envisage the SMS Ice

Sheet redistributing Kenyte deposited around McMurdo Sound by previous glacial events.

A further key issue in resolving these differences is the age and timing of these different glacial events. Stuiver et al. [1981] suggest the Ross Sea Drift and hence RSIS glaciation is Wisconsin in age. However, carbon-14 ages presented are from lacustrine strata that superpose the Ross Sea Drift or from material that may have been reworked into parts of the drift by subsequent glaciations [e.g. Denton et al., 1995]. The Ross Sea Drift may considerably predate the formation of glacial lakes in the Dry Valleys.

CONCLUSIONS

At least two major glacial events are recorded in southern McMurdo Sound by the distribution of moraines and erratics on the MIS and along adjacent coastal areas:

1) The grounding, advance and retreat of the SMS Ice Sheet records an extensive phase of glaciation across southern McMurdo Sound. The SMS Ice Sheet comprised three lobes (Figure 11); the KG Lobe to the west, the MB Lobe that occupied the area between Black Island, Brown Peninsula and Minna Bluff, and the RIS Lobe to the east and north of Black Island. The grounded ice pedestals that form the Dailey Islands and Cape Chocolate, on the western margin of McMurdo Sound, demonstrate that the SMS Ice Sheet grounded on the floor of McMurdo Sound at least as far north as 77°45' S. Eocene marine fossiliferous erratics occur only in moraines of the MB Lobe of the SMS Ice Sheet. The only other reported occurrence of Eocene fossiliferous erratics is at Cape Crozier on the eastern edge of Ross Island [Hertlein, 1969]. The source of these erratics was most likely shallow basins south of the Mount Discovery and Minna Bluff area. It is possible that an earlier phase of glaciation delivered these erratics to southern McMurdo Sound and the MB Lobe of the SMS Ice Sheet subsequently redistributed them.

The SMS Ice Sheet was grounded in McMurdo Sound. However, the distribution of terminal and medial moraines and the flow patterns reconstructed from them, along with the distribution of the Eocene fossiliferous erratics does not support an ice sheet as hypothesized by Stuiver et al. [1981]. The reconstruction proposed here (Figure 11), indicates northward flow of the SMS Ice Sheet fed by a grounded proto-Ross Ice Shelf and Koettlitz Glacier, rather than southwestward flow as predicted by Stuiver et al. [1981]. The thickening ice probably reached levels suggested by Stuiver et al. [1981], both in the southern Dry Valley mouths and on Brown Peninsula. But, the ice that re-entered the southern Dry Valleys was from an expanded Koettlitz Glacier (the KG Lobe). This is not inconsistent with an expanded marine ice sheet in the Ross Sea as hypothesised by Stuiver et al. [1981] which most likely dammed the northern Dry Valleys.

2) The thickening and/or lowering of the MIS onto the sea floor, causing it to incorporate and transport, as a result of surface ablation, relict glacial material from the floor of southern McMurdo Sound. Given that late Quaternary glaciations were all of similar size [Martinson et al., 1987], it is likely that grounding of the MIS was repeated, allowing for further incorporation of material from the sea floor. The Wisconsin thickening/lowering of the MIS may have provided a large enough barrier to dam the southern Dry Valleys and allow lakes to fill these valleys as reported by Clayton-Greene et al. [1988].

The SMS Ice Sheet may have grounded in McMurdo Sound as early as latest Pliocene times and the MIS may have thickened several times in the late Quaternary. An exact chronology of these events is not currently available.

Acknowledgements. John Kaser and Richard Levy helped in various aspects of the quadrat and ground survey work. Antarctic Support Associates and US Navy helicopter squadron VXE-6 provided field and logistical support in very difficult weather conditions. Richard Levy took and prepared the low-level oblique black and white photographs. Rosemary Askin, David Harwood and Richard Levy provided discussion on various aspects of this work. This manuscript benefited greatly from reviews by Barrie McKelvey and an anonymous reviewer. Field research was supported by NSF grants OPP9158078 and OPP9317901 to D.M. Harwood. The author acknowledges financial support from the Byrd Fellowship (The Ohio State University).

REFERENCES

Brook, E.J., M.D. Kurz, R.P. Ackert, G. Raisbeck, and F. Yiou
1995   Cosmogenic nuclide exposure ages and glacial history of late Quaternary Ross Sea drift in McMurdo Sound, Antarctica. *Earth Planet. Sci. Lett.*, 131, 41-56.

Buckridge, J.S.
1989   Marine invertebrates from Late Cainozoic deposits in the McMurdo Sound region, Antarctica, *J. Royal Soc. N.Z.*, 19, 333-342.

Clayton-Greene, J.M., C.H. Hendy, and A.G. Hogg
1988   The chronology of a Wisconsin-aged proglacial lake in the Miers Valley, Antarctica, *N.Z. J. Geol. Geophys.*, 31, 353-361.

Cole, J.W., P.R. Kyle, and V.E. Neall
1971   Contributions to Quaternary geology of Cape Crozier, White Island and Hut Point Peninsula, McMurdo Sound, Antarctica, *N.Z. J. Geol. Geophys.*, 14, 528-546.

Debenham, F.
1919   A new mode of transport by ice, *Q. J. Geol. Soc. London*, 75, 51-76.

Denton, G.H., and H.W. Borns, Jr.
1974   Former grounded ice sheets in the Ross Sea, *Antarct. J. U.S.*, 9, 167.

Denton, G.H., R.L. Armstrong, and M. Stuiver
1970   Late Cenozoic glaciation in Antarctica: the record in the McMurdo Sound region, *Antarct. J. U.S.*, 5, 15-21.

Denton, G.H., R.L. Armstrong, and M. Stuiver
1971   The Late Cenozoic glacial history of Antarctica, in *The Late Cenozoic glacial ages*, edited by K.K. Turekian, pp. 267-306, Yale University Press, New Haven Connecticut.

Denton, G.H., J.G. Bockheim, S.C. Wilson, and M. Stuiver
1989   Late Wisconsin and early Holocene glacial history, inner Ross embayment, Antarctica, *Quat. Res.*, 31, 151-182.

Denton, G.H., M. Stuiver, and K.G. Austin
1985   Radiocarbon chronology of the last glaciation in McMurdo Sound, Antarctica, *Antarct. J. U.S.*, 20, 59-61.

Drewry, D.J.
1979   Late Wisconsin reconstruction of the Ross Sea region, Antarctica, *J. Glaciol.*, 24, 231-243.

Eggers, A.J.
1979   Scallop Hill Formation, Brown Peninsula, McMurdo Sound, Antarctica, *N.Z. J. Geol. Geophys.*, 22, 353-361.

Gow, A.J., W.F. Weeks, G. Hendrickson, and R. Rowland
1965   On the mode of uplift of the fish and fossiliferous moraines of the McMurdo Ice Shelf, Antarctica, *CRREL Res. Rep.*, 172, 1-16.

Gunn, B.M., and G. Warren
1962   Geology of Victoria Land between the Mawson and Mulock Glaciers, Antarctica. *N.Z. J. Geol. Surv. Bull.*, 71, 157 p.

Hall, B.L., G.H. Denton, D.R. Lux, and J.G. Bockheim
1993   Late Tertiary Antarctic paleoclimate and ice sheet dynamics inferred from surficial deposits in Wright Valley. *Geograf. Annal.*, 75A, 239-267.

Harrington, H.J.
1969   Fossiliferous rocks in moraines at Minna Bluff, McMurdo Sound, *Antarct. J. U.S.*, 4, 134-135.

Heine, A.J.
1967   The McMurdo Ice Shelf, Antarctica: A preliminary report, *N.Z. J. Geol. Geophys.*, 10, 474-478.

Hertlein, L.G.
1969   Fossiliferous boulder of Early Tertiary age from Ross Island, *Antarct. J. U.S.*, 4, 199-201.

Kellogg, T.B., and D.E. Kellogg
1985   Evidence bearing on the former existence of grounded ice sheets in the Ross Sea, Antarctica, *S. African J. Sci.*, 81, 237-238.

Kellogg, D.E., and T.B. Kellogg
1987   Diatoms of the McMurdo Ice Shelf, Antarctica: implications for sediment and biotic reworking. *Palaeogeogr. Palaeo-climatol. Palaeoecol.*, 60, 77-96.

Kellogg, T.B., and D.E.
1988   Antarctic cryogenic sediments: Biotic and inorganic facies of ice shelf and marine-based ice sheet environments. *Palaeogeogr. Palaeo-climatol. Palaeoecol.*, 67, 51-74.

Kellogg, T.B., D.E. Kellogg, and M. Stuiver
1990   Late Quaternary history of the southwestern Ross Sea: Evidence from debris bands on the McMurdo Ice Shelf, Antarctica, in *Contributions to Antarctic Research I, Antarct. Res. Ser.*, vol. 50, edited by D.H. Elliot, pp. 25-56, AGU Washington D.C..

Kellogg, T.B., D.E. Kellogg, and M. Stuiver
1991a   Oxygen isotope data from the McMurdo Ice Shelf, Antarctica: implications for debris band formation and glacial history, *Antarct. J. U.S.*, 26, 73-76.

Kellogg, T.B., D.E. Kellogg, and M. Stuiver
1991b   Radiocarbon dates from the McMurdo Ice Shelf, Antarctica: implications for debris band formation and glacial history, *Antarct. J. U.S.*, 26, 77-79.

Krumbein, W.C.
1988   Manual of sedimentary petrology, *SEPM Reprint Ser.*, 13, 549 p.

Kyle, P.R.
1981   Glacial history of the McMurdo Sound area as indicated by the distribution and nature of the McMurdo volcanic group rocks, in *Dry Valley Drilling Project, Antarct. Res. Ser.*, vol. 33, edited by L.D. McGinnis, pp. 403-412, AGU Washington, D.C.

Leckie, R.M., and P.N. Webb
1979   Scallop Hill Formation and associated Pliocene marine deposits of southern McMurdo Sound, *Antarct. J. U.S.*, 14, 54-56.

Martinson, D.G., N.G. Pisas, J.D. Hayes, J. Imbrie, T.C. Moore, Jr., and N.J. Shackleton
1987   Age dating and the orbital theory of ice ages: Development of a high resolution 0-300,000-year chronostratigraphy, *Quat. Res.*, 27, 1-29.

Oliver, J.S., E.F. O'Connor, and D.J. Watson
1978   Observations on submerged glacial ice in McMurdo Sound, Antarctica. *J. Glaciol.*, 20, 115-121.

Speden, I.G.
1962   Fossiliferous Quaternary marine deposits in the McMurdo Sound region, Antarctica, *N.Z. J. Geol. Geophys.*, 5, 746-777.

Stilwell, J.D., R.H. Levy, R.M. Feldmann, and D.M. Harwood
1997   On the rare occurrence of Eocene callianassid decapods (Arthropoda) preserved in their burrows, Mount Discovery, East Antarctica, *J. Paleontol.*, 71, 284-287.

Stilwell, J.D., R.H. Levy, and D.M. Harwood
1993   Preliminary paleontological investigation of Tertiary glacial erratics from the McMurdo Sound region, East Antarctica, *Antarct. J. U.S.*, 28, 16-19

Stott, L.D., B.C. McKelvey, D.M. Harwood, P.N. Webb, and others
1983   revision of the ages of Cenozoic erratics at Mount Discovery and Minna Bluff, McMurdo Sound, *Antarct. J. U.S.*, 18.

Stuart, A.W., and C. Bull
1963   Glaciological observations on the Ross Ice Shelf near Scott Base, Antarctica, *J. Glaciol.*, 4, 399-414.

Stuiver, M., G.H. Denton, T.J. Hughes, and J.L. Fastook
1981   History of the marine ice sheet in West Antarctica during the last glaciation: A working hypothesis, in *The Last Great Ice Sheets*, edited by G.H. Denton, and T.J. Hughes, pp. 319-436, Wiley Interscience, New York.

Swithinbank, C.W.M.
1970   Ice movement in the McMurdo Sound area of Antarctica, *IAHS SCAR Publicat.*, 86, 472-487.

Vella, P.
1969   Surficial geological sequence, Black Island and Brown Peninsula, McMurdo Sound, Antarctica, *N.Z. J. Geol. Geophys.*, 12, 761-770.

Gary S. Wilson, Department of Earth Sciences, University of Oxford, Parks Road OX1 3PR UK

SEDIMENTARY LITHOFACIES OF THE MCMURDO SOUND ERRATICS

Richard H. Levy and David M. Harwood

Department of Geosciences, University of Nebraska-Lincoln, Lincoln, NE, U.S.A.

Our knowledge of Cenozoic paleoenvironments in East Antarctica is limited due to the general lack of exposed strata of this age. Fossil-bearing erratic boulders present in coastal moraines in McMurdo Sound, East Antarctica, are derived from strata that are presently covered by ice. These erratics were most likely eroded and transported by ice from the area behind Minna Bluff, in the vicinity of the confluence of the Byrd, Mulock and Skelton Glaciers during the Neogene. The erratics provide a record of Paleogene fossil biotas, climate and paleoenvironments that existed on the coastal margin of the Paleogene Transantarctic Mountains (TAM). In order to increase the paleontologic and sedimentologic database for East Antarctica, more than one thousand erratics were collected from McMurdo Sound between 1992 and 1995. Herein, we describe the following lithofacies: sandstone, sandy-mudstone, conglomerate, mudstone and diamictite, based on the examination of over one-hundred erratic boulders. The facies recorded in these erratics were most-likely present, and perhaps widespread, along the eastern margin of the TAM. Most of the sandstone, sandy-mudstone and conglomerate facies were deposited during the middle to late Eocene within coastal marine settings, proximal to the rising TAM. There is no direct evidence for ice influence in this environment. Sandy mudstone, mudstone and diamictite facies of Oligocene and younger age, were probably deposited in proximal glaciomarine, distal glaciomarine and subglacial environments, respectively. The suite of erratic boulders presented herein, therefore, records a transition from cool-temperate coastal environments of the middle to late Eocene to colder coastal environments of the Oligocene and Miocene, where the influence of glacial processes on sedimentation was strong.

INTRODUCTION

At present, the Tertiary geologic history of East Antarctica is not well understood, as strata of this age are poorly exposed [Webb, 1990; 1991]. Fossiliferous glacial erratics (the McMurdo Erratics) in coastal moraines in McMurdo Sound (Figure 1) provide an accessible record of Paleogene strata that are presently hidden beneath the Antarctic ice sheet. Previous studies highlighted the potential wealth of geologic data contained within the McMurdo Erratics [e.g. Cranwell et al., 1960; Wilson, 1967; Hertlein, 1969; Feldmann and Zinsmeister, 1984]. In an effort to recover more geologic information from these recycled rocks, hundreds of sedimentary erratics

were collected between 1992 and 1995, by a team of scientists from the University of Nebraska-Lincoln. These erratics were recovered from coastal moraines along the shores of Mount Discovery, Brown Peninsula and Minna Bluff, as well as moraine on Black Island and along the floors of Salmon and Miers valleys (Figure 1).

In this paper we describe and illustrate a suite of sedimentary facies recovered during this recent period of collection. We are limited in our ability to develop a detailed understanding of both depositional setting and environmental change represented by these facies, as the erratics are pieces of strata that have been removed from their original temporal and spatial stratigraphic framework. An understanding of these stratigraphic relationships is gener-

Figure 1. Location of sites from which erratics were collected in McMurdo Sound.

ally necessary to identify ancient depositional environments and to note any changes through time [Walker, 1984; Reading, 1986]. In spite of the absence of a sequence of strata, we are able to infer broad depositional environments within which the suite of lithofacies were most likely deposited based on analyses of sediment texture and associated paleontologic control.

Only in Antarctica and other areas where Cenozoic rock exposures are virtually non-existent, would a project such as this be fruitful. Future stratigraphic drilling through these and other sequences will eventually yield a more complete view of the Antarctic environment and paleobiology of the Paleogene. In the meantime, the erratics represent a survey of environments and ages for which we have limited knowledge. Erosion and transport by ice provided us with a means to examine rocks from a wider geographic area than could be covered by several drill holes. Cenozoic studies in Antarctica will benefit from combined data obtained through both glacial paleontology and future stratigraphic drilling.

PREVIOUS WORK

Previous petrographic studies of erratics from McMurdo Sound concentrated on detailed thin section analysis [Rowe, 1974; Landis, 1974], with an emphasis on determining sediment provenance. Stott [1982] examined sixty erratics comprising sandstone, conglomerate and limestone facies, for which he identified the following major 'petrographic groups': metamorphic arkose; granitic arkose; and quartz arenite. Although 'compositional' groups are useful for provenance studies, the goal herein is to identify the most likely depositional environments for the source strata from which the erratics were derived. Textural characteristics are most useful in this regard, and are therefore the main criteria used to identify the lithofacies described here.

METHODS

Lithofacies represented in the McMurdo Erratics were initially identified and characterized in the field. A suite of over one hundred erratics (Table 1) was selected from a large collection housed at the University of Nebraska for the detailed study reported here. Polished rock slabs were prepared for representative erratics of each lithologic type and were examined to determine medium- to large-scale physical characteristics including color, grain shape, grain size, sorting, and sediment structures.

One hundred and four rock thin sections were examined under an Olympus BH-1 transmitted light microscope to determine: (1) fine scale physical characteristics including grain shape, grain size and sorting, and (2) grain composition of the clasts. A minimum of 300 grain counts were obtained for several sandstone erratics and modal proportions were plotted on the ternary diagram of Folk [1968]. Several rock thin sections were stained for calcite following the method outlined by Friedman [1959].

Detailed study of clast composition and surficial features are not undertaken herein, although such studies may prove to be useful for future correlation to stratigraphic sections recovered by future drilling.

RESULTS

Five facies and fourteen sub-facies (Table 2) are identified based on the textural characteristics outlined below: Sandstone Lithofacies (Sm, Smc, Ss, Sst, Ssg, Sw).

The majority of the McMurdo Erratics comprise yellowish-gray to greenish-gray, indurated to friable, well-sorted to poorly-sorted, fine-grained to coarse-grained sandstone lithologies (Plates 1-5). Most of these sandstone

TABLE 1. Sample list and age data. Samples Collected 1992/93: MTD = Mount Discovery; MB = Minna Bluff; SV = Salmon Valley; BG = Blue Glacier; SIM 1 = Sea Ice Moraine. Samples Collected 1993/94 and 95/96: E 100 - E 381 from Minna Bluff and NW flank of Mount Discovery. For an explanation of lithofacies abbreviations see table 1, this paper. Age data are determined from dinoflagellate cyst (Dn) biostratigraphy (Levy and Harwood, this volume) and siliceous microfossil (Si) biostratigraphy (Bohaty and Harwood, this volume; Harwood and Bohaty, this volume). Samples that were not processed for microfossils are indicated with an N/A in the age column.

Location	Sample	Lithofacies	Age	Key Fossil
Minna Bluff	E 153	Sm	middle to upper Eocene	Dn
Minna Bluff	MB 188B	Sm	middle to upper Eocene	Dn
Minna Bluff	MB 187A	Sm	N/A	
Minna Bluff	MB 220	Sm	N/A	
Minna Bluff	MB 224	Sm	N/A	
Minna Bluff	MB 249	Sm	N/A	
Minna Bluff	MB 285	Sm	N/A	
Minna Bluff	MB 292D	Sm	N/A	
Mt. Discovery	E 100	Sm	middle to upper Eocene	Dn
Mt. Discovery	E 145	Sm	middle to upper Eocene	Dn
Mt. Discovery	E 163	Sm	middle to upper Eocene	Dn
Mt. Discovery	E 165	Sm	middle to upper Eocene	Dn
Mt. Discovery	E 168	Sm	middle to upper Eocene	Dn
Mt. Discovery	E 169	Sm	middle to upper Eocene	Dn
Mt. Discovery	E 171	Sm	middle to upper Eocene	Dn
Mt. Discovery	E 189	Sm	middle to upper Eocene	Dn
Mt. Discovery	E 191	Sm	middle to upper Eocene	Dn
Mt. Discovery	E 194	Sm	middle to upper Eocene	Dn
Mt. Discovery	E 208	Sm	middle to upper Eocene	Dn
Mt. Discovery	E 303(1)	Sm	middle to upper Eocene	Dn
Mt. Discovery	E 317	Sm	middle to upper Eocene	Dn
Mt. Discovery	E 345	Sm	middle to upper Eocene	Dn, Si
Mt. Discovery	E 365(2)	Sm	middle to upper Eocene	Dn
Mt. Discovery	E 381	Sm	middle to upper Eocene	Dn
Mt. Discovery	MTD 1	Sm	middle to upper Eocene	Dn
Mt. Discovery	MTD 154	Sm	middle to upper Eocene	Dn
Mt. Discovery	MTD 190	Sm	middle to upper Eocene	Dn
Mt. Discovery	E 155	Sm	upper middle to upper Eocene	Dn
Mt. Discovery	E 200	Sm	?lower Oligocene	Dn
Mt. Discovery	E 202	Sm	?lower Oligocene	Dn
Mt. Discovery	E 203	Sm	?lower Oligocene	Dn
Mt. Discovery	E 356	Sm	?lower Oligocene	Dn
Mt. Discovery	MTD 56	Sm	?lower Oligocene	Dn
Mt. Discovery	MTD 174A	Sm	?Paleozoic/Mesozoic	
Mt. Discovery	E 181	Sm	???	
Mt. Discovery	E 185	Sm	???	
Mt. Discovery	E 192	Sm	???	
Mt. Discovery	E 207	Sm	???	
Mt. Discovery	E 331	Sm	???	
Mt. Discovery	E 372	Sm	???	
Mt. Discovery	E 183	Sm	N/A	
Mt. Discovery	E 344(2)	Sm	N/A	
Mt. Discovery	MTD 193A	Sm	N/A	
Mt. Discovery	MTD 203	Sm	N/A	
Mt. Discovery	MTD 211B	Sm	N/A	

Table 1 (continued)

Location	Sample	Lithofacies	Age	Key Fossil
Sea Ice Moraine	SIM 11	Sm	middle to upper Eocene	Dn
Sea Ice Moraine	SIM 6	Sm	N/A	
Minna Bluff	MB 109	Smc	middle to upper Eocene	Dn
Mt. Discovery	E 184	Smc	middle to upper Eocene	Dn
Mt. Discovery	E 357	Smc	middle to upper Eocene	Dn
Mt. Discovery	MTD 153	Smc	middle to upper Eocene	Dn
Mt. Discovery	E 313	Smc	?lower Oligocene	Dn
Minna Bluff	MB 301	Sm/Sw	N/A	
Mt. Discovery	E 215	Sw	middle to upper Eocene	Dn
Minna Bluff	MB 80	Ss	middle to upper Eocene	Dn
Sea Ice Moraine	SIM 5	Ss	???	
Minna Bluff	MB 181	Ssg/Csgc	middle to upper Eocene	Dn, Si
Minna Bluff	MB 103	Sst/Cmm	middle to upper Eocene	Dn
Minna Bluff	MB 235C	Cmc	upper middle to upper Eocene	Dn
Salmon Valley	SV 12	Cmm	???	
Blue Glacier	BG 1	Cmc	???	
Minna Bluff	MB 188G	Cmm	middle to upper Eocene	Dn
Minna Bluff	MB 188F	Cmm	N/A	
Mt. Discovery	MTD 189	Cmm	middle to upper Eocene	Dn
Mt. Discovery	MTD 42	Cmm	middle to upper Eocene	Dn
Mt. Discovery	MTD 148	Cmm	N/A	
Sea Ice Moraine	SIM 9	Cmm	N/A	
Sea Ice Moraine	SIM 1	Cmm	???	
Mt. Discovery	E 214	Mmb	middle to upper Eocene	Dn
Mt. Discovery	E 303(2)	Mmb	middle to upper Eocene	Dn
Mt. Discovery	E 365(1)	Mmb	upper middle to upper Eocene	Dn
Mt. Discovery	E 344(1)	Mmb	N/A	
Minna Bluff	E 350	Mmb	middle to upper Eocene	Dn, Si
Minna Bluff	MB 245	Mmb	middle to upper Eocene	Dn, Si
Minna Bluff	E 219	Mmb	upper middle to upper Eocene	Dn
Mt. Discovery	D1	Mwb	middle to upper Eocene	Dn, Si
Mt. Discovery	E 364	Mwb	middle to upper Eocene	Dn, Si
Minna Bluff	MB 212K	Mm-d	post-Eocene	Dn
Minna Bluff	MB 217A	Mm-d	post-Eocene	Dn
Minna Bluff	MB 244C	Mm-d	post-Eocene	Si
Minna Bluff	MB 290G	Mm-d	post-Eocene	Dn
Minna Bluff	E 339	Mm-d	N/A	
Minna Bluff	MB 172	Mm-d	N/A	
Minna Bluff	MB 212I	Mm-d	N/A	
Minna Bluff	MB 223F	Mm-d	N/A	
Minna Bluff	MB 288B	Mm-d	N/A	
Minna Bluff	E 244	Mm-d	???	
Mt. Discovery	E 363	Mm-d	post-Eocene	Dn
Mt. Discovery	E 360	Mm-d	???	
Minna Bluff	E 115	Ms-d	post-Eocene	Dn
Minna Bluff	E 240	Ms-d	???	
Mt. Discovery	E 216	Ms-d	post-Eocene	Dn
Mt. Discovery	MTD 211A	Ms-d	post-Eocene	Dn
Mt. Discovery	MTD 197	Ms-d	N/A	

Table 1 (continued)

Location	Sample	Lithofacies	Age	Key Fossil
Minna Bluff	E 242D	Dm	post-Eocene	Dn
Minna Bluff	E 243	Dm/Dw	post-Eocene	Dn
Minna Bluff	MB 299	Dm	post-Eocene	Dn
Minna Bluff	E 347	Dm	Oligocene-Miocene	Si
Minna Bluff	E 346	Dm	Miocene	Si
Minna Bluff	E 351	Dm	Miocene	Si
Minna Bluff	MB 235A	Dm	Miocene	Si
Minna Bluff	MB 191A	Dm	N/A	
Minna Bluff	MB 213C	Dm	N/A	
Minna Bluff	MB 292C	Dm	N/A	
Salmon Valley	SV 3	Sm/quartzite	?Paleozoic/Mesozoic	
Mt. Discovery	MTD 166	Sm/Quartzite	N/A	
Salmon Valley	SV 14	Ss/quartzite	N/A	
Minna Bluff	MB 210	Ss/volcanoclastic	N/A	
Minna Bluff	MB 202	tuff	???	
Mt. Discovery	E 323	?metased	???	
Mt. Discovery	E 355	?metased	???	

lithologies are sublitharenites (Figure 2; Table 3), although lithic arkose lithologies also occur (Figure 2; Plate 1, Figure c; Plate 3, Figures. g and h). The quartz component in these sandstone lithologics is separated into two distinctive groups: (1) rounded to well-rounded grains with common quartz overgrowth (Plate 3, Figure d), and (2) sub-angular to angular grains that lack quartz overgrowth (Plate 3). Well-rounded grains are likely derived from mature sandstone beds from the Devonian to Triassic Beacon Supergroup. These 'inherited' grains affect the textural characteristics of the McMurdo Erratics and must be considered when interpreting sediment maturity of the sandstone facies. Lithic clasts usually include granite, quartzite, dolerite, diorite, and various fine-grained metasediments (Plate 3). The sandstone erratics are cemented with microcrystalline calcite or sparry calcite, which often exhibits poikilotopic texture.

Most of the sandstone lithologies are massive. The majority of the marine macrofossils (molluscs, arthropods, brachiopods, bryozoans) recovered from the McMurdo Erratics are preserved in this lithofacies (Plate 1, Figures d-f; Plate 2; other contributions to this volume). Massive sandstone erratics that contain yellowish-brown to dark gray, fine-grained, angular, pebble-sized rip-up clasts are distinguished from other massive sandstone lithologies (Table 2; Plate 4). These massive sandstone lithologies with rip-up clasts commonly contain abundant marine palynomorphs [Levy and Harwood, this volume].

Stratified sandstone sub-facies are identified based on whether they are well-stratified or weakly-stratified (Table 2; Plate 5). Weakly-stratified sandstone lithologies commonly possess layers of terrestrial organic material (Plate 5, Figures. g and h) and/or shells of molluscs (Plate 2, Figures. c and d). The well-stratified sandstone sub-facies may be trough-cross-stratified (Plate 5, Figure f), or graded (Plate 5, Figures. d and e; Plate 12, Figure b).

Sandy Mudstone Lithofacies (Mmb, Mwb, Ms-d)

Erratics consisting of a yellowish-brown to dark gray, hard to fissile, poorly-sorted sandy mudstone lithology are less common than the above sandstone erratics. Composition of the sandy mudstone lithofacies may comprise up to 50% sand-sized clasts that usually consist of sub-angular to well-rounded quartz (Plate 6, Figures. e-g); the remaining component consists of mud.

The majority of the sandy mudstone lithologies are bioturbated (Plate 6, Figures. a-d, g), but weak stratification is sometimes preserved (Plate 6, Figure c). These rocks are important as they contain relatively rich assemblages of fossil microflora and fauna (dinoflagellate cysts, diatoms, ebridians and silicoflagellates) and reasonably well-preserved terrestrial macroflora (Plate 6, Figure h; Pole et al., this volume; Francis, this volume].

Well-stratified sandy mudstone lithologies commonly contain dispersed pebbles of various composition, including granite, dolerite and metasediments (Plate 7).

Table 2. A summary of McMurdo Erratic lithofacies.

Lithofacies	Abbreviation	Description
Sandstone		
Massive	Sm	Well-sorted to poorly-sorted yellowish gray to greenish gray massive sandstone; scattered pebbles of various lithology may be present; invertebrate fossils are common.
Massive with intraclasts	Smc	Moderately well-sorted to poorly-sorted yellowish gray to grayish brown massive sandstone with intraclasts of dark gray to dark grayish brown fine sandstone or mudstone. Intraclasts from sand to pebble size.
Stratified	Ss	Moderately well-sorted olive brown sandstone with cm-scale stratification; scattered pebbles of various composition shape and size may be present.
Stratified / trough cross-strata	Sst	Moderately well-sorted yellowish gray to olive gray sandstone with well-developed trough cross-stratification.
Stratified / graded	Ssg	Poorly-sorted yellowish gray to dark greenish gray stratified, graded sandstone; grain size grades from basal pebbles to upper sands; beds range in thickness from less than 5mm to 4cm; both complete and fragmented fossil invertebrate shells and terrestrial organic remains may be incorporated in the coarser basal section of the graded beds.
Weakly stratified	Sw	Moderately well-sorted to well-sorted yellowish gray to greenish gray weakly stratified sandstone; stratification is usually indicated by layers of terrestrial organic material (leaves and wood) or marine invertebrate fossils (usually molluscs).
Sandy mudstone		
Massive / bioturbated	Mmb	Poorly-sorted dark gray sandy mudstone; dispersed pebbles and sandy lenses may be present; massive; mottled appearance indicates probable bioturbation; terrestrial macroflora (wood and leaves) and marine invertebrate macrofauna may occur.
Stratified with dropstones	Ms-d	Poorly-sorted light olive gray sandy mudstone with moderately well-developed stratification indicated by diffuse layers of sand and mud; dispersed pebbles (?dropstones) and pelloids may be present.
Weakly stratified	Mwb	Poorly-sorted dark yellowish brown to dark grayish brown weakly stratified sandy mudstone; mudstone pelloids may be present; stratification often masked or destroyed by bioturbation.
Mudstone		
Massive with dropstones	Mm-d	Light olive gray massive mudstone with dispersed pebbles (?dropstones) of various lithology, shape and size; mudstone matrix may contain ostracodes and planktonic foraminifera.
Conglomerate		
Massive - clast supported	Cmc	Unstratified poorly-sorted clast-supported conglomerate; well-rounded to sub-angular clasts range from sand to pebble in size (max 11mm); clast lithologies are varied and may possess circumgranular acicular calcite 'rinds'.
Massive - matrix supported	Cmm	Unstratified poorly-sorted sandy matrix-supported conglomerate; well rounded to subangular clasts range up to cobble size (~90mm); clasts comprise several lithologies.
Stratified / graded - clast supported	Csgc	Poorly-sorted stratified clast-supported conglomerate; subrounded clasts are sand to pebble size and comprise mudstone intraclasts; layers are graded and may be up to 8cm thick; terrestrial organic remains (wood and leaves) and marine invertebrates are usually incorporated within the conglomerate.
Diamictite		
Massive / weakly stratified	Dm / Dw	Olive gray unstratified to ?weakly stratified sandy mudstone with matrix-supported clasts (between 5 and 15%) of various lithology, shape and size.

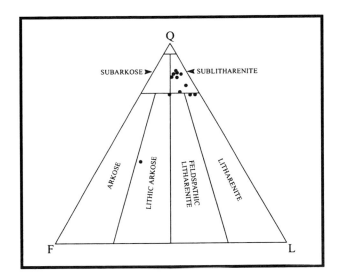

Figure 2. Sandstone classification diagram of Folk (1968) on which modal data from several McMurdo Erratic sandstones are plotted.

TABLE 3. QFL percentage data for sandstone lithofacies.

Erratic	Facies	Q	F	L
MTD 1	Sm	85	4	10
MB 188B	Sm	84	3	13
E 303(1)	Sm	78	4	18
E 345	Sm	74	2	24
E 317	Sm	40	51	9
E 100	Sm	83	5	12
E 155	Sm	83	7	9
E 145	Sm	84	4	11
E 365(2)	Sm	84	6	10
MTD 153	Smc	74	4	22
MB 109	Smc	74	13	12
MB 80	Ss	75	8	17

These lithologies sometimes contain invertebrate macrofauna (Plate 7, Figure g), but do not usually yield microfossils [Bohaty and Harwood, this volume; Levy and Harwood, this volume].

Mudstone Lithofacies (Mm-d)

The mudstone lithofacies is characterized by light olive gray, lithified, massive, mudstone that commonly contain dispersed pebbles of granite, dolerite, metasediments, basalt, and volcanic glass (Plate 8). Foraminiferas, ostracodes, and marine diatoms are commonly present (Plate 8, Figures. c and f) [see Harwood and Bohaty, this volume], whereas dinoflagellate cysts, silicoflagellates, ebridians and marine invertebrates are rare.

Diamictite (Dm/Dw)

Diamictite lithofacies are characterized by lithologies comprising olive gray, lithified, unstratified to weakly-stratified, very poorly sorted, sandy mudstone matrix with pebble clasts of various lithologies, including granite, dolerite, meta-sediments, basalt and vesicular volcanic glass that make up between 5 and 50% of the sediment (Plate 9). Weak stratification is sometimes evident (Plate 9, fig. d). Clasts of basalt and vesicular glass are not present in all of the diamictite erratics (Plate 9, Figures. c, e, f), which suggests that the sediment provenance for these lithologies has varied spatially and/or temporally.

Microfossils are usually rare in the diamictite, although well-preserved lower Oligocene to upper Miocene diatom assemblages [Harwood and Bohaty, this volume] and reworked middle to upper Eocene dinocyst assemblages [Levy and Harwood, this volume] occur in some.

Conglomerate (Cmc, Cmm, Csgc)

Massive, matrix-supported conglomerate lithofacies (Plate 10) are characterized by well-rounded to sub-angular, pebble to cobble-sized clasts of granite, dolerite, and various metasediments, supported by a sandy matrix with textural and compositional characteristics similar to the massive sandstone lithofacies (Plate 10, Figures. d, e, and g). These conglomerate lithologies commonly contain fossil invertebrates (molluscs) and microflora (dinoflagellate cysts and pollen).

Massive, clast-supported conglomerate lithofacies (Plate 11) include Erratics MB 235C and SV 5. These erratics are matrix-free conglomerates that contain sub-angular to well-rounded clasts of various lithologies. Although these erratics have textural characteristics that are broadly similar, they possess significantly different compositional characteristics. Clasts in Erratic MB 235C comprise a variety of lithologies including: biotite-rich granite, hornblende-rich granite, marble, garnet schist, dolerite, mafic-rich volcanoclastic and finely laminated meta-sediments (Plate 11, Figures. d-h). Pore spaces in this erratic are filled with microgranular calcite, columnar calcite, and equant calcite crystals (Plate 11, fig. h). Erratic SV 12 consists of sub-angular to rounded clasts comprising basalt, granite and metasediments (Plate 11, Figures. a-c). Circumgranular, acicular calcite crystals coat each clast and pore spaces are filled with microgranular calcite (Plate 11, Figures. b and c).

Plate 1: Massive sandstone lithofacies (Sm).

Figure a. Polished section: Erratic E 100, a poorly-sorted massive sandstone.

Figure b. Polished section: Erratic E 365(2), a moderately well-sorted, massive sandstone. Note the organic-rich lenses (leaves) dispersed throughout this rock.

Figure c. Polished section: Erratic E 317, a poorly-sorted, massive sandstone comprising abundant feldspar and granite clasts.

Figure d. Abundant bivalves ("Eurhomalea" claudiae Stilwell, this volume) preserved in Erratic E 331.

Figure e. Polished section: Erratic E 382. Note geopetal structures present within the gastropods (Struthiolarella mcmurdoensis Stilwell, this volume).

Figure f. Massive sandstone erratic containing large oysters (Crassostrea antarctogigantea Stilwell, this volume).

Figure g. Erratic E 153, a large sandstone block containing fossil conifer wood (W). Note pen in lower left for scale.

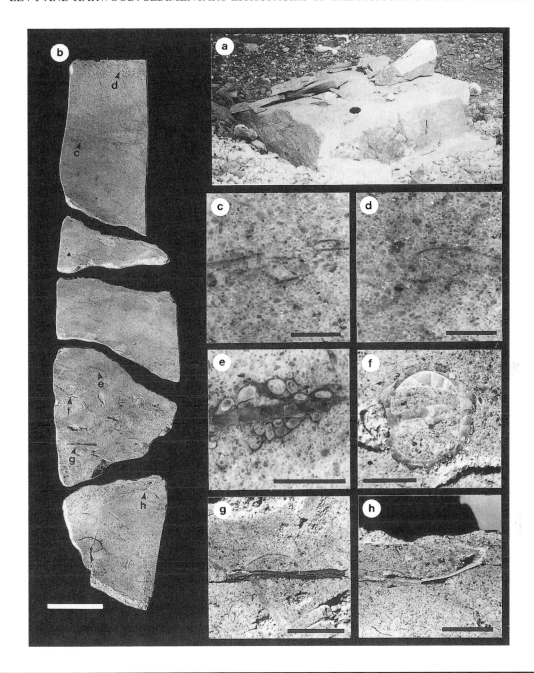

Plate 2: Erratic MB 301, a massive to weakly-stratified sandstone lithology.

Figure a. Erratic MB 301, a large sandstone block (note camera lens in center cap for scale) present in coastal moraine from the western end of Minna Bluff. A thick (~1m) composite section of this erratic (illustrated in figure b.), was cut in the field (note cut surface and rock powder).

Figure b. Composite section (~ 1m thick) composed of multiple pieces from Erratic MB 301. Areas indicated by arrows are shown in detail in figs. c-h, scale bar = 10cm.

Figs. c and d. Layers of shell (bivalves) define stratification, scale bar = 5mm.

Figure e. Transverse section of a bivalve with encrusting serpulid worm tubes, scale bar = 5mm.

Figure f. Transverse section of a gastropod showing geopetal structure, scale bar = 10mm. Note encrusting serpulid worm tubes.

Figure g. Fossil wood, scale bar = 20mm.

Figure h. Articulated bivalve shell, scale bar = 20mm.

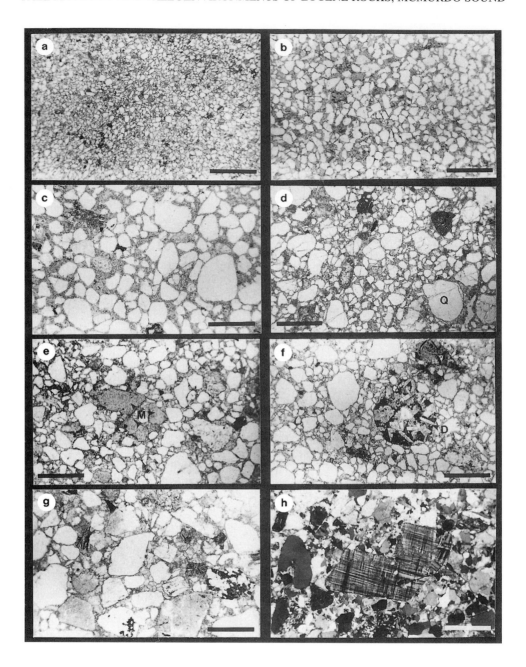

Plate 3: Transmitted light photomicrographs of rock thin sections cut from massive sandstone erratics.
Figs. a-g: plane polarized light; figure h: cross polarized light. Scale bar = 1mm.

Figure a. Well-sorted, fine-grained sublitharenite (Erratic MB 292D).
Figure b. Well-sorted, medium-grained sublitharenite (Erratic E 171).
Figure c. Moderately well-sorted, medium to coarse-grained sublitharenite (Erratic E 183).
Figure d. Poorly sorted sublitharenite (Erratic E 100). Note quartz grain with overgrowth (Q).
Figure e. Poorly sorted sublitharenite (Erratic E 155). Note metasediment clasts (M).
Figure f. Poorly sorted sublitharenite (Erratic SIM 11). Note dolerite clasts (D).
Figs. g and h. Poorly sorted arkose sandstone (Erratic E 317).

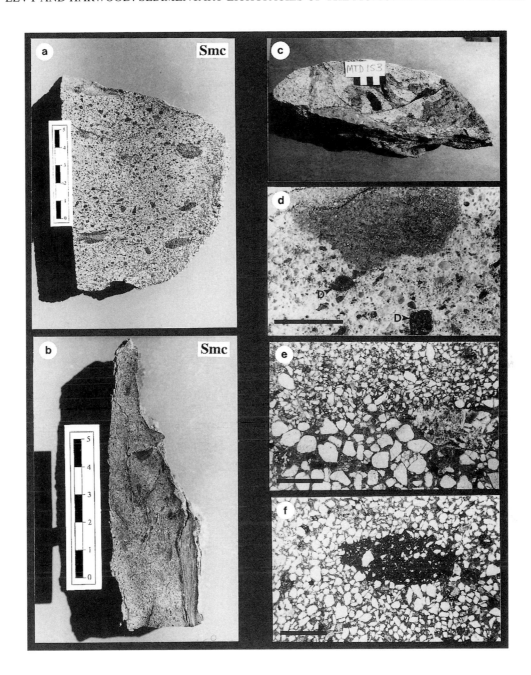

Plate 4: Massive sandstone lithofacies with intraclasts (Smc).

Figs. a. c. and e. Erratic MTD 153, a poorly-sorted sandstone of medium to coarse grain-size, containing abundant fine-grained sandstone intraclasts: (a) polished section; (c) hand specimen showing large (> 10cm) intraclasts, scale bar = 5cm; (e) transmitted light photomicrograph (plane polarized light) showing boundary between medium to coarse-grained sandstone 'matrix' and fine-grained sandstone intraclast, scale bar = 1mm.

Figs. b and f. Erratic MTD 109, a moderately well-sorted sandstone of medium grain-size with abundant sandy-mudstone and mudstone intraclasts: (b) polished section; (f) transmitted light photomicrograph (plane polarized light) showing a sandy-mudstone intraclast contained within a sandstone 'matrix', scale bar = 1mm.

Figure d. Reflected light photomicrograph of Erratic E 313, scale bar = 5mm. Note fine-grained sandstone intraclast and dolerite clasts (D).

Plate 5: Stratified sandstone lithofacies (Ss, Ssg, Sst, Sw)

Figure a. Polished section: Erratic MB 80. Stratified, coarse-grained sandstone with dispersed pebbles comprising various lithologies including granite (G), and metasediment (M).

Figs. b and c. Erratic SV 3, a stratified quartzite erratic likely derived from the Beacon Supergroup: (b) polished section; (c) transmitted light photomicrograph (cross polarized light), scale bar = 1mm.

Figure d. Polished section: Erratic MB 181, showing graded sand and pebble beds.

Figure e. Polished section: Erratic MTD 91, showing graded sand and pebble beds.

Figure f. Polished section: Erratic MB 103, which consists of trough cross-stratified sandstone beds underlain by a massive conglomerate.

Figs. g and h. Erratic E 215, a weakly-stratified sandstone with layers of terrestrial organic material: (g) polished section; (h) fossil leaves (L) define a bedding surface (photograph courtesy of M. Pole).

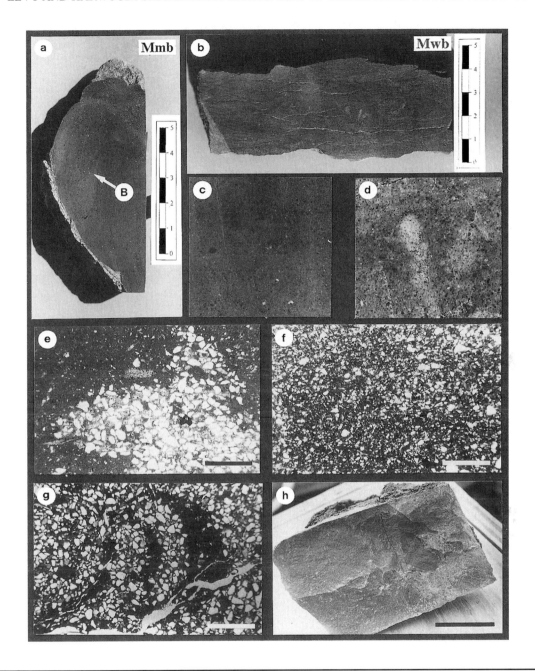

Plate 6: Massive and weakly-stratified, bioturbated, sandy mudstone lithofacies (Mmb and Mwb)

Figs. a. c. and e. Erratic E 365(1), a bioturbated sandy mudstone cobble that was contained within Erratic E 365(2), a massive sandstone lithology (see plate 1, figure b): (a) polished section, note burrow (B); (c) close up of burrow shown in figure a.; (e) transmitted light photomicrograph (plane polarized light) showing sandy lens (?burrow) contained within a mud matrix, scale bar = 1mm.

Figs. b and d. Erratic E 364, a weakly-stratified sandy mudstone lithology: (b) polished section; (d) close up of figure b., showing structure formed through bioturbation.

Figs. f and h. Erratic E 219, a massive sandy mudstone: (f) transmitted light photomicrograph (plane polarized light), scale bar = 1mm; (h) hand specimen showing a fossil leaf, scale bar = 5cm.

Figure g. Erratic D1: transmitted light photomicrograph (plane polarized light) showing meniscus-shaped burrow fill.

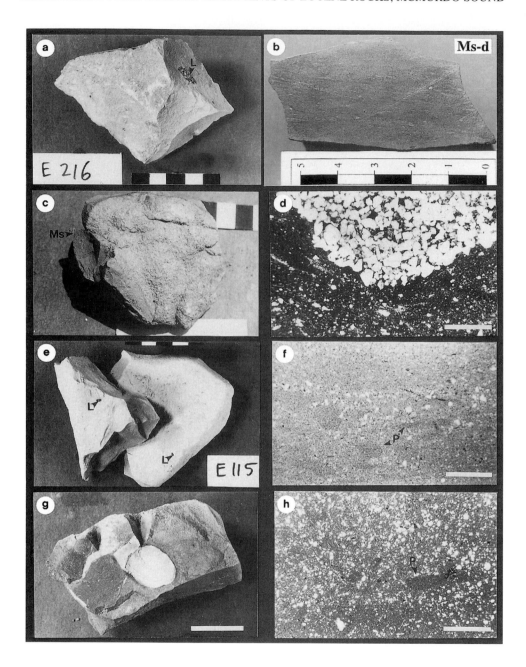

Plate 7: Stratified sandy mudstone lithofacies (Ms-d)

Figs. a and b. Erratic E 216: (a) hand specimen with granite lonestone (L); (b) polished section. Note sand stringers.

Figs. c and d. Erratic MTD 211A: (c) sandy mudstone 'matrix' (Ms) surrounds a quartzite cobble lonestone likely derived from Beacon Supergroup strata; (d) transmitted light photomicrograph (plane polarized light) showing sandy mudstone matrix and quartzite lonestone, scale bar = 1mm. Note deformed bedding beneath the lonestone.

Figs. e and f. Erratic E 115: (e) hand specimen, note small pebble-sized lonestones (L); (f) transmitted light photomicrograph (plane polarized light), showing sandy layers and peloids (P), scale bar = 1mm.

Figs. g and h. Erratic E 240: (g) hand specimen with fossil bivalve, scale bar = 2cm; (h) transmitted light photomicrograph (plane polarized light), showing peloids (P), scale bar = 1mm.

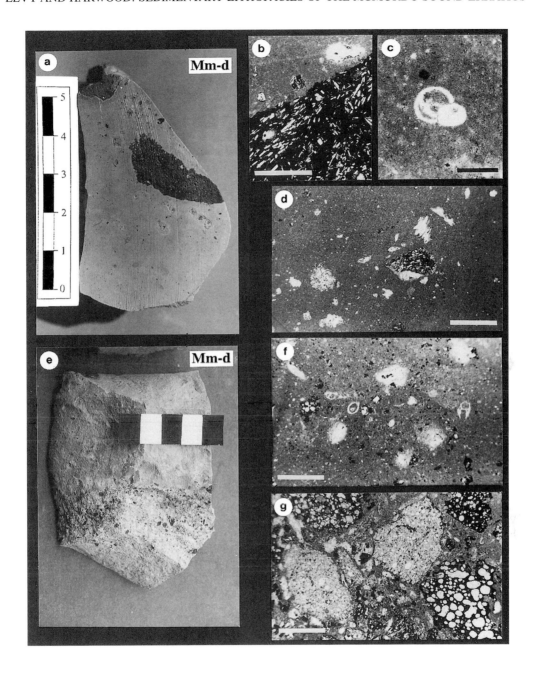

Plate 8: Massive mudstone lithofacies with lonestones (Mm-d)

Figs. a-c. Erratic MB 212K: (a) polished section showing pebble-sized lonestones comprised of vesicular basalt; (b) transmitted light photomicrograph (plane polarized light) showing basalt clast and mudstone 'matrix', scale bar = 1mm; (c) transmitted light photomicrograph (plane polarized light) showing foraminifera, scale bar = 250m.
Figure d. Erratic MB 290G: transmitted light photomicrograph (plane polarized light), scale bar = 1mm. Note coarse sand-sized basalt lonestone.
Figs. e-g. Erratic MB 244C: (e) hand specimen comprised of a layer of massive mudstone with dispersed sand-size basalt clasts and a layer of sand and pebble-size basalt and vesicular volcanic glass clasts; (f) transmitted light photomicrograph (plane polarized light) of massive mudstone layer shown in figure e., scale bar = 1mm; (g) transmitted light photomicrograph (plane polarized light) of sand and pebble layer shown in figure e., scale bar = 1mm. Note clasts of vesicular volcanic glass.

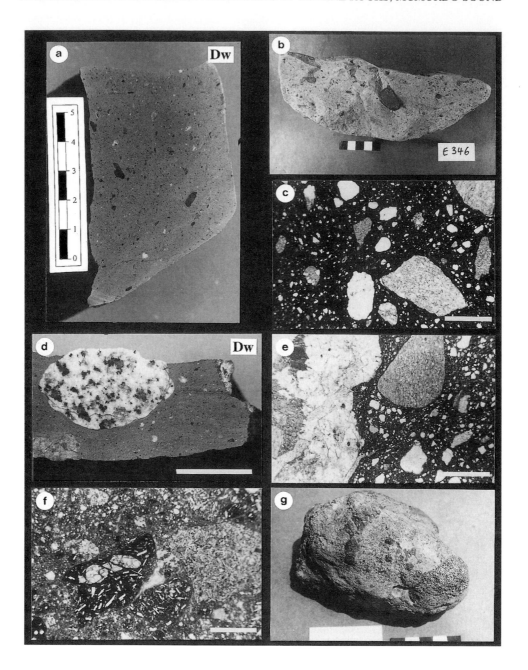

Plate 9: Diamictite lithofacies (Dm/Dw)

Figs. a-c. Erratic E 346: (a) polished section showing weak stratification indicated by light and dark 'layers'; (b) hand specimen, scale bar = 5cm; (c) transmitted light photomicrograph (plane polarized light), scale bar = 1mm.

Figs. d and e. Erratic E 243: (d) polished section, note deformed bedding beneath the pebble-sized granite clast, scale bar = 2cm; (e) transmitted light photomicrograph (plane polarized light), scale bar = 1mm.

Figs. f and g. Erratic MB 299: (f) transmitted light photomicrograph (plane polarized light) showing clasts of various volcanic lithologies, scale bar = 1mm; (g) hand specimen with large pebble-sized vesicular volcanic clasts, scale bar = 5cm.

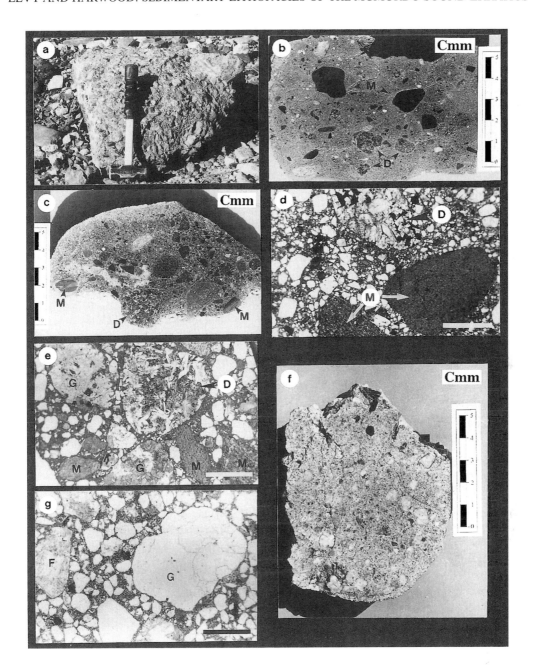

Plate 10: Matrix supported conglomerate lithofacies (Cmm)

Figure a. Large conglomerate boulder in moraine located along the northwestern coast of Mount Discovery.

Figs. b and d. Erratic MTD 189: (b) polished section with abundant clasts of dolerite (D) and various metasediments (M); (d) transmitted light photomicrograph (plane polarized light) showing clasts of dolerite (D) and various metasediments (M), scale bar = 1mm. Note sandy matrix.

Figs. c and e. Erratic MB 188G: (c) polished section with abundant clasts comprising various metasediments (M) and dolerite (D); (e) transmitted light photomicrograph (plane polarized light) showing clasts of dolerite (D), granite (G), and various metasediments (M), scale bar = 1mm.

Figs. f and g. Erratic E 381: (f) polished section showing pebble-sized granite clasts in a sandy matrix; (g) transmitted light photomicrograph (plane polarized light), scale bar = 1mm. Note clasts of feldspar and granite.

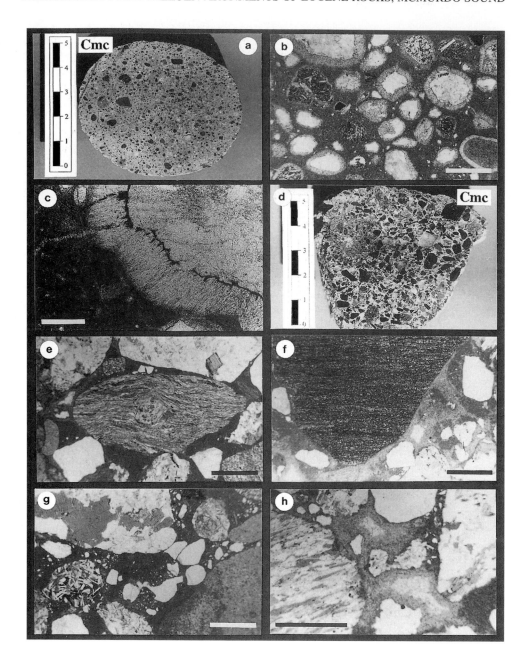

Plate 11: Clast supported conglomerate lithofacies (Cmc)

Figs. a-c. Erratic SV 12: (a) polished section showing calcite 'rinds' that surround the grains; (b) transmitted light photomicrograph (plane polarized light) showing clasts of granite and basalt with circumgranular acicular calcite, scale bar = 1mm; (c) transmitted light photomicrograph (plane polarized light) showing circumgranular acicular calcite crystals and microgranular cement, scale bar = 250m.

Figs. d-h. Erratic MB 235C: (d) polished section showing variety of clast lithologies; (e-g) transmitted light photomicrographs (plane polarized light) showing various clast lithologies, scale bar = 1mm; (h) transmitted light photomicrograph (plane polarized light) showing microgranular, columnar, and equant calcite crystals, scale bar = 1mm.

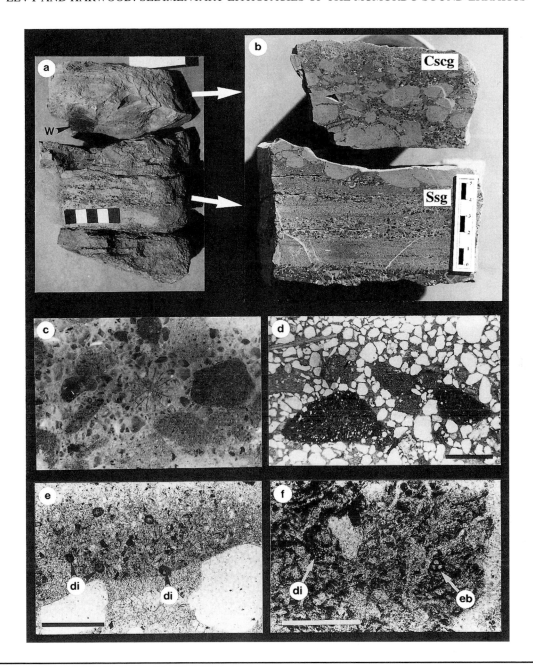

Plate 12: Erratic MB 181, an erratic comprising graded sandstone (Ssg) and matrix supported graded conglomerate lithofacies (Csgc).

Figure a. Hand specimen, scale bar = 5cm. Note fossil wood (W),
Figure b. Polished section showing an upper graded pebble conglomerate composed of mudstone and sandy mudstone intraclasts (Csgc) and lower series of graded sandstone beds (Ssg).
Figure c. Reflected light photomicrograph showing sandy matrix supporting intraclasts of sandy mudstone and mudstone. Note the scleractinian coral in center.
Figure d. Transmitted light photomicrograph (plane polarized light) showing sandy matrix supporting intraclasts of sandy mudstone and mudstone, scale bar = 1mm.
Figure e. Transmitted light photomicrograph (plane polarized light) showing intraclast containing pyritized diatoms (di), scale bar = 200m.
Figure f. Transmitted light photomicrograph (plane polarized light) of intraclast showing pyritized diatom (di) and ebridians (eb), scale bar = 100m.

Stratified conglomerate lithologies are rare but distinctive. Erratic MB 181 is a well-stratified lithology that consists of graded sandstone (lithofacies Ssg) and graded intraclast conglomerate (Plate 12, fig. b). The graded conglomerate consists of pebble-sized, sub-angular to sub-rounded, mudstone intraclasts that commonly contain abundant dinoflagellate cysts and pyritized diatoms, ebridians, silicoflagellates and radiolarians (Plate 12, Figures. e and f). A sandy matrix that infills spaces between the mudstone intraclasts, usually contains shell fragments and terrestrial organic remains (Plate 12, Figures. a-d).

DISCUSSION

The suite of lithofacies represented by the McMurdo Erratics range in age from Paleozoic to Recent, but almost all of the fossiliferous rocks are Paleogene in age [Askin, this volume; Bohaty and Harwood, this volume; Harwood and Bohaty, this volume; Levy and Harwood, this volume; Stilwell, this volume]. The Paleogene erratics are likely derived from strata that occur beneath the Ross Ice Shelf, to the south of Minna Bluff [fig. 3; Wilson, this volume]. The geographic location within which these strata were deposited would have been strongly influenced by: (1) Cenozoic rifting of the Victoria Land Basin and associated uplift of the Transantarctic Mountains [Fitzgerald, 1992]; and (2) Eocene-Oligocene climatic cooling and growth of an ice sheet in East Antarctica [Matthews and Poore, 1980; Miller et al., 1987].

Most of the sandstone, sandy mudstone and conglomerate facies are derived from strata deposited during the middle to upper Eocene [Askin, this volume; Bohaty and Harwood, this volume; Harwood and Bohaty, this volume; Levy and Harwood, this volume]. Most of these lithofacies are texturally sub-mature to immature, which implies relatively rapid sediment deposition and burial in basins proximal to the rising TAM front. Clasts contained within the middle to upper Eocene lithologies were likely derived from strata presently exposed in the Transantarctic Mountains, which include Precambrian to Cambrian metasediments; Early Paleozoic granite; Devonian to Triassic fluvial and shallow marine sediments of the Beacon Supergroup; and Jurassic igneous intrusives and extrusives of the Ferrar Group (fig. 3).

The suite of middle to upper Eocene lithologies were deposited in nearshore to offshore, moderate to high-energy, marine environments. Massive sandstone lithofacies were likely deposited between the backshore and offshore transition (fig. 4) where burrowing fauna disrupted bedding. Rip-up clasts of fine-grained sand-

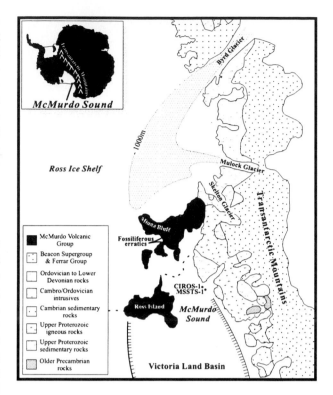

Figure 3. Location of the deep sub-glacial basin from which the McMurdo Erratics are likely derived (behind Minna Bluff). Geologic outcrop map is adapted from Laird (1991), figure 2.4.

stone and sandy mudstone, characteristic of lithofacies Smc, may have been: (a) derived from offshore and reworked shorewards by tidal currents or storm-generated currents, or (b) incorporated into sandy turbidites flowing offshore (fig. 4). Stratified graded sandstone lithologies may represent sandy turbidites deposited in the offshore-transition. Sandy mudstone facies were likely deposited within and beyond the offshore-transition (fig. 4). The presence of terrestrial organic material (wood and leaves) in several of the sandstone and sandy mudstone lithofacies indicates proximity to a fluvial source and suggests that many of these lithofacies may be either estuarine and/or deltaic and pro-deltaic. Conglomerate lithofacies may represent one or more of the following: (a) tidal channel deposits or channel lags; (b) storm-wave induced deposits e.g. storm-scour lags; (c) submarine debris-flow deposits; or (d) proximal submarine conglomerates deposited within fan-deltas that formed along the rising Transantarctic Mountain front. Conglomerate Erratic SV 12 contains no marine fossils and may represent a gravel deposit on an alluvial fan or within a fluvial channel.

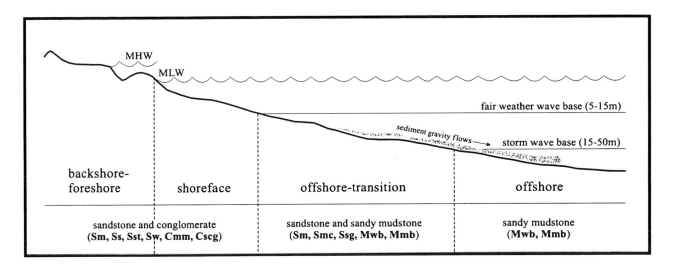

Figure 4. Facies model for middle to upper Eocene erratics, adapted from Elliot (1986), fig. 7.14 and Hambrey and others (1989), fig. 5.

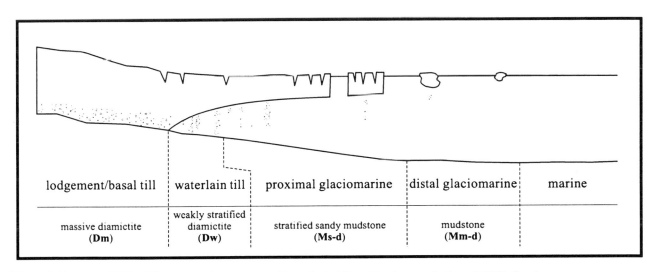

Figure 5. Facies model for Oligocene and younger erratics, adapted from Hambrey and others (1989), fig. 4.

Sandy-mudstone facies (Ms-d), mudstone facies (Mm-d) and diamictite facies (Dm/Dw) are most similar to glacial facies recovered from the CIROS-1 drillcore in McMurdo Sound [Hambrey et al., 1989]. We infer that these lithofacies were deposited in proximal glaciomarine, distal glaciomarine and either sub-glacial or proximal glaciomarine environments respectively (fig. 5). All of the erratics that comprise 'glacial' facies are Oligocene or younger [Askin, this volume; Bohaty and Harwood, this volume; Harwood and Bohaty, this volume; Levy and Harwood, this volume]. The suite of erratics therefore record a change from pre-Oligocene ?ice-free coastal environments to post-Oligocene, glaciated, coastal environments.

Coarse-sand and pebble-sized clasts derived from Cenozoic igneous extrusives (McMurdo Volcanic Group) (fig. 3), occur in many of the 'glacial' facies. Basaltic clasts are also recorded in upper Oligocene and younger strata recovered from the MSSTS-1 and CIROS-1 drillcores in McMurdo Sound [Figure 3; Barrett and McKelvey, 1986; Hambrey et al., 1989]. This suggests that the McMurdo Erratic 'glacial' facies are lithostratigraphic equivalents of the upper Oligocene and younger units recovered in the drillcores.

CONCLUSION

The sedimentary lithologies documented in erratics from coastal moraines in McMurdo Sound are characterized by five facies and fourteen sub-facies (Table 2). Middle to upper Eocene sandstone, sandy mudstone and conglomerate lithofacies were deposited in coastal marine and inner shelf environments, and show no indication of glacial ice. These were proximal to the actively rising and eroding Transantarctic Mountains. A transition to glaciomarine environments in the Oligocene is recorded by the 'glacial' mudstone and diamictite lithofacies. The environmental transition recorded by the McMurdo Erratics, supports the pattern of climate evolution recognized in the CIROS-1 drillcore [Barrett, et al., 1989; Wilson et al., 1998].

Future scientific drilling [e.g. Barrett and Davey, 1992; Webb and Wilson, 1995] will enable a better understanding of the stratigraphic relationships that exist between the facies discussed herein. The McMurdo Erratics provide us with a broad spatial view of facies that may be recovered in these cores. Stratigraphic drilling and reworked sediment and rocks, together, will allow us to better interpret the Cenozoic evolution of Antarctic climate, basin history and paleobiota.

Acknowledgements. The authors thank Brian Bruckno and Justin Spence for their help with thin section preparation. Dr. D.L. Loope provided advice on various aspects and techniques of sedimentology. The authors also thank Dr. Peter Barrett and Dr. Neil Wells for their critical reviews of the manuscript. This work was supported in part by grants from the National Science Foundation, Office of Polar Programs to D.M. Harwood (OPP-9317901 and OPP-9158075), grants from the Geological Society of America to R.H. Levy and generous donations from the geology alumni of the University of Nebraska-Lincoln.

REFERENCES

Askin, R.A
This vol.   Terrestrial Palynomorphs in *Paleobiology and Paleo-environments of Eocene Rocks, McMurdo Sound, East Antarctica,* Antarctic Research Series, edited by J.D. Stilwell and R.M. Feldmann, AGU, Washington, D.C.

Barrett, P.J., and Davey, F.J., (Eds.)
1992   Antarctic Stratigraphic Drilling Cape Roberts Project Workshop Report, *The Royal Society of New Zealand Miscellaneous Series* 23, 38 pp., The Royal Society of New Zealand, Wellington.

Barrett, P.J., and B.C. McKelvey
1986   Stratigraphy, in Antarctic Cenozoic Glacial history from the MSSTS-1 drillhole, McMurdo Sound, *DSIR Bulletin* 237, edited by P.J. Barrett, pp. 9-15.

Barrett, P.J., Hambrey, M.J., Harwood, D.M., Pyne, A.R., and Webb, P.-N.
1989   Synthesis, in Antarctic Cenozoic history from the CIROS-1 drillhole, McMurdo Sound, *DSIR Bulletin* 245, edited by P.J. Barrett, pp. 241-251.

Bohaty, S., and D.M. Harwood
This vol.   Ebridians and Silicoflagellates From McMurdo Sound Glacial Erratics and the Southern Kerguelen Plateau, in *Paleobiology and Paleoenvironments of Eocene Rocks, McMurdo Sound, East Antarctica,* Antarctic Research Series, edited by J.D. Stilwell and R.M. Feldmann, AGU, Washington, D.C.

Cranwell, L.M.
1960   Harrington, H.J., and Speden, I.G., Lower Tertiary microfossils from McMurdo Sound, Antarctica, *Nature,* 184 (4701), 1782-1785.

Elliot, T.
1986   Siliciclastic Shorelines (chapter 7), in *Sedimentary Environments and Facies* (2nd ed.), edited by H.G. Reading, pp. 155-188, Oxford, Blackwell Scientific Publications.

Feldmann, R.M., and W.J. Zinsmeister
1984   First Occurrence of Fossil Decapod Crustaceans (Callianassidae) From the McMurdo Sound Region, Antarctica, *Journal of Paleontology,* 58(4), 1041-1045.

Fitzgerald, P.
1992   The Transantarctic Mountains of Southern Victoria Land: The Application of Fission Track Analysis to a Rift Shoulder Uplift, *Tectonics,* 11(3), 634-662.

Folk, R.L.
1968   *Petrology of Sedimentary Rocks,* 170 pp., Hemphill's Book Store, Austin, Texas.

Francis, J.E.
          Fossil Wood from Eocene High Latitude Forests, McMurdo Sound, Antarctica, in *Paleobiology and Paleoenvironments of Eocene Rocks, McMurdo Sound, East Antarctica,* Antarctic Research Series, edited by J.D. Stilwell and R.M. Feldmann, AGU, Washington, D.C.

Friedman, G.M.
1959   Identification of Carbonate Minerals by Staining Methods, *Journal of Sedimentary Petrology,* 29(1), 87-97.

Hambrey, M.J., Barrett, P.J., and Robinson, P.H.
1989   Stratigraphy, in Antarctic Cenozoic history from the CIROS-1 drillhole, McMurdo Sound, *DSIR Bulletin* 245, edited by P.J. Barrett, pp. 23-48.

Hertlein, L. G.
1969   Fossiliferous Boulder of Early Tertiary Age

from Ross Island, Antarctica, *Antarctic Journal of the United States*, 4(5), 199-201.

Laird, M.G.
1991   The Late Proterozoic-Middle Palaeozoic Rocks of Antarctica (Chapter 2), in *The Geology of Antarctica, Oxford Monographs on Geology and Geophysics,* edited by R.J. Tingey, pp. 74-119, Oxford Scientific Publications.

Landis, C.A.
1974   Petrography of and Early Tertiary Fossiliferous Sandstone from the Ross Sea Area, Antarctica (Note), *New Zealand Journal of Geology and Geophysics*, 17(3), 715-718.

Levy, R.H., and D.M. Harwood
This vol.   Marine Palynomorph Biostratigraphy and Age(s) of the McMurdo Sound Erratics, in *Paleobiology and Paleoenvironments of Eocene Rocks, McMurdo Sound, East Antarctica,* Antarctic Research Series, edited by J.D. Stilwell and R.M. Feldmann, AGU, Washington, D.C.

Matthews, R.K., and R.Z. Poore
1980   Tertiary 18O record and glacio-eustatic sea-level fluctuations, *Geology*, 8, 501-504.

Miller, K.G., Fairbanks, R.G., and Mountain, G.S.
1987   Tertiary Oxygen Isotope Synthesis, Sea Level History, and Continental Margin Erosion, *Paleoceanography*, 2(1), 1-19.

Pole, M., Hill, B., and Harwood, D.M.
This vol.   Eocene Plant Macrofossils from Erratics, McMurdo Sound, in *Paleobiology and Paleoenvironments of Eocene Rocks, McMurdo Sound, East Antarctica,* Antarctic Research Series, edited by J.D. Stilwell and R.M. Feldmann, AGU, Washington, D.C.

Reading, H.G., (Ed.)
1986   *Sedimentary Environments and Facies* (2nd ed.), 615pp., Oxford, Blackwell Scientific Publications.

Rowe, G.H.
1974   A petrographic and paleontological study of lower Tertiary erratics from Quaternary Moraines, Black Island, Antarctica, B.Sc. (Hons.) thesis, 91 pp., Victoria University of Wellington, New Zealand.

Stilwell, J.D., and R.M. Feldmann (Eds.)
This vol.   *Paleobiology and Paleoenvironments of Eocene Rocks, McMurdo Sound, East Antarctica,*

Antarctic Research Series, AGU, Washington, D.C.

Stott, L.D.
1982   A Re-evaluation of the Age and Nature of Cenozoic Erratics from McMurdo Sound, Antarctica, B.S. Thesis, 52pp., The Ohio State University.

Walker, R.G. (Ed.)
1984   *Facies Models* (2nd ed.), 317pp., Geoscience Canada Reprint Series 1.

Webb, P-N.
1990   The Cenozoic history of Antarctica and its global impact, *Antarctic Science*, 2(1), 3-21.

Webb, P-N.
1991   A review of the Cenozoic stratigraphy and paleontology of Antarctica, in *Geological Evolution of Antarctica*, edited by M.R.A. Thomson, J.A. Crame and J.W. Thomson, pp. 599-607, Cambridge University Press.

Webb, P-N., and G.S. Wilson (Eds.)
1995   The Cape Roberts Drilling Project: Antarctic Stratigraphic Drilling, Proceedings of a meeting to consider the project science plan and potential contributions by the U.S. science community, 6-7 March, 1994, *BPRC Report No. 10*, 117 pp., Byrd Polar Research Center, The Ohio State University, Columbus Ohio.

Wilson, G.J.
1967   Some new species of Lower Tertiary dinoflagellates from McMurdo Sound, Antarctica, *New Zealand Journal of Botany*, 5(1), 57-83.

Wilson, G.S., Roberts, A.P., Verosub, K.L., Florindo, F., and Sagnotti, L.
1998   Magnetobiostratigraphic chronology of the Eocene-Oligocene transition in the CIROS-1 core, Victoria Land Margin, Antarctica: Implications for Antarctic Glacial History, *GSA Bulletin*, 110(1), 35-47.

David M. Harwood, Department of Geosciences, 214 Bessey Hall, University of Nebraska - Lincoln, Lincoln, NE 68588-0340, U.S.A.

Richard H. Levy, Department of Geosciences, 214 Bessey Hall, University of Nebraska - Lincoln, Lincoln, NE 68588-0340, U.S.A.

CLAY MINERAL COMPOSITION OF GLACIAL ERRATICS, MCMURDO SOUND

Mary Anne Holmes

Department of Geosciences University of Nebraska-Lincoln, Lincoln, NE

Twenty-two erratics collected from coastal moraines along the shores of Mount Discovery, Brown Peninsula, Minna Bluff, on Black Island, and from the Salmon and Miers valley floors in East Antarctica were examined for their mineral composition in the <2 m fraction by x-ray diffraction to determine their provenance and the climate under which the sediment in the erratics formed. Semi-quantitative results from peak areas were subjected to principal components analysis and indicate that there are two distinct mineral compositions in the erratics (c = 0.05): A) dominant smectite group minerals, minor illite and kaolinite, and no chlorite, and B) dominant illite, subordinate smectite group, and either chlorite and R=1 I/S clay or R=3 I/S clay. Group A erratics include two types: 1) Eocene age siliciclastic sediment and 2) volcaniclastics of unknown age. Group B erratics comprise three types: 1) Eocene age siliciclastic sediment dominated by illite with subordinate smectite, no chlorite, and very low levels of kaolinite and mixed-layer clays; 2) post Eocene age erratics dominated by illite with a major component of chlorite and R=1 I/S clay, minor or no smectite and kaolinite; and 3) post Eocene age erratics dominated by illite and containing R=3 I/S clay. Eocene age sediment occurs in either group and so had two distinct provenances for the clay fraction: a smectite-dominant area and an illite-rich, smectite-poor area. Post Eocene age sediment also had two distinct provenances for the clay fraction and are different from the Eocene sources: a metamorphic + ancient sedimentary terrain that supplied chlorite, illite, and R=1 I/S clay to some of the erratics, and a sedimentary terrain that supplied illite and R=3 I/S clay. Kaolinite levels are low, indicating the absence of intense weathering and/or any significant contribution from the Beacon Supergroup.

INTRODUCTION

A team of University of Nebraska-Lincoln geologists collected over one hundred glacial erratics from coastal moraines along the shores of Mount Discovery, Brown Peninsula and Minna Bluff, as well as from moraine on Black Island and along the floors of Salmon and Miers valleys in Antarctica [Levy and Harwood, this volume; see their Figure 1] during three field seasons (1992-1995) in an effort to learn more about the Paleogene history of East Antarctica. The erratics were analyzed for siliceous microfossil content, palynomorphs, and lithofacies, and the findings are discussed in Levy and Harwood [this volume], and Bohaty and Harwood [this volume]. The clay mineral compositions of 22 selected erratics were examined for this study to provide a representative of each lithotype identified by Levy and Harwood [this volume] in order to determine if the clay composition could shed any light on the climatic conditions that prevailed when the sediment in the erratics was originally deposited.

Provided the clays are detrital, some climatic significance may be gleaned from their occurrence, particularly in sediments formed at high latitudes [Ehrmann et al., 1992; Ehrmann, 1996]. Among the most common clays, chlorite and ordered mixed-layer clays do not survive much weathering and transport by water [e.g., Chamley, 1989]. Their abundant presence in sediments thus indicates a cool and/or dry climate. Smectite forms under more humid conditions, especially when there is a readily soluble precursor, such as volcanic ash or basalt [e.g., Biscaye, 1965; Nadeau and Reynolds, 1981]. Pedogenic kaolinite forms in the wettest and warmest climate among the detrital clays. Illite is a poor climatic indicator. It is apparently principally derived from Paleozoic or older shales, in which diagenesis has formed illite from smectite and nonclay precursors. It can survive temperate and even subtropical weathering conditions.

Abundant chlorite and/or ordered, mixed-layer clays in the <2 m fraction of these erratics would indicate that the sediment initially formed under cool and/or dry conditions, and has not been subjected to significant weathering at any time in its subsequent history. The presence of smectite and/or kaolinite would indicate significant weathering at some time during the erratic's history. Presumably this weathering would have occurred during a period of nonglacial (excluding montane glaciation) conditions in East Antarctica.

The purpose of this study was to determine the clay mineral composition of glacial erratics collected in East Antarctica in an effort to learn the intensity of the weathering regime under which the sediment in the clasts formed, which may assist in the determination of their ages and environments of deposition.

METHODS

Sample Preparation and Diffraction Analysis

Erratics were collected during three field seasons from Minna Bluff, Mount Discovery, and sea ice moraine near McMurdo station [see Levy and Harwood, this volume for collection and locality details]. Twenty-two of these were selected for clay mineral analysis to represent the major lithotypes identified by Levy and Harwood [this volume]. These were soaked in distilled water and dispersed in sodium hexametaphosphate solution, treated with the ultrasound for 3-4 minutes, and centrifuged to remove the <2 m fraction for examination [Jackson, 1975]. This fraction was decanted into a Millipore filtra-

tion apparatus and an oriented mount prepared after the method of Drever [1973]. Oriented mounts were x-rayed using a Scintag PAD V x-ray diffractometer equipped with a graphite monochromator and 0.67° and 0.76° divergence slits on either side of a collimator. Oriented samples were scanned from 2° to 45°2 in the air-dried state, from 2° to 30°2 after ethylene glycol solvation (60°C over ethylene glycol vapor in a desiccator overnight), and from 2° to 15°2 after heating to 350°C for one hour in a muffle furnace. Peak areas and positions were calculated from the glycolated scans using the unweighted split Pearson model in the Profile Fitting package of Scintag's DMS software, v. 3.1.

Mineral Identification and Semi-Quantification

Mineral identification followed standard methods. Kaolinite and chlorite were distinguished by the slow scan method of Biscaye [1964] over the 004 peak of chlorite and the 002 peak of kaolinite. Where these peaks were too small to allow an accurate position, the 002 peak of chlorite and the 001 peak of kaolinite were used. This assumes that a peak position of 7.10Å indicates chlorite and 7.16Å indicates kaolinite [Biscaye, 1964]. Illite and mixed-layer clays were identified from diffractograms of samples treated with ethylene glycol by referring to Reynolds [1980], Moore and Reynolds [1989], and the ° 2 method of Srodon [1980]. In brief, for glycolated samples, any peak at 16-17Å was identified as R=0 I/S clay ("randomly interstratified"); a peak near 12-13Å as R=1 I/S clay ("I/S ordered"), and a low-angle shoulder on the 10Å illite peak near 11Å as R=3 I/S clay ("ISII ordered"). In addition, peak positions for 002/003 I/S peaks between 15°-20°2 were calculated using the profile fitting software. A peak at 5.0Å was used to identify the presence of illite where no 11Å shoulder occurred on the 10Å peak. Had a shoulder appeared, this peak might indicate R=3 I/S clay. Peaks between 5.2 and 5.34Å were identified as R=1 I/S clay, and peaks between 5.4 and 5.6Å were identified as R=0 I/S clay. Where possible, the presence of both 001/002 and 002/003 I/S peak positions were used to obtain a ° 2 value [Srodon, 1980], from which % nonexpandable layers was calculated from values published in Moore and Reynolds [1989]. Non-clay mineral identification also followed standard methods as outlined in *Brown* [1980]. Plagioclase was distinguished from potassium feldspars by the method of Borg and Smith [1969]; opal-CT was identified after Jones and Segnit [1971].

Semi-quantitative results were calculated for clay minerals from profile fit-derived peak areas using the weighted method of Biscaye [1965]. This method was tested on this instrument by the use of a pyrophyllite internal standard [Heiden and Holmes, 1998]. The two methods agree nearly perfectly (confidence level for correlation coefficient = 0.95). The overall error for any particular clay for this method is unknown, but repeated tests indicate it is internally consistent. Other workers estimate an error for xrd-derived results at around 10% [e.g., Moore and Reynolds, 1989].

The semi-quantitative results were subjected to principal components analysis, using Systat v. 5.0, with an Eigenvalue of 1.000 and varimax rotation, in an effort to see if the number of variables might be reduced.

RESULTS

Smectite group minerals occur in all but four of the samples (Tables 1 and 2) and vary from minerals that give large, sharp peaks (Figure 1a), to small, sharp peaks (Figures 1b and 1c), to small, broad peaks (Figures 1d and 1e). Twelve samples contain chlorite (e.g., Figures 1b, 1c,

and 1e), and all but two samples contain illite (e.g., Figures 1a-1c and 1e). Nine samples contain kaolinite (Table 1, Figures 1a and 1d). Eight samples contain R=1 I/S clay (Table 1, Figures 1b and 1c), and only three contain R=3 I/S clay (Table 1, Figure 1e). Most contain quartz (Figures 1a, 1b, 1c and 1e), and about half contain generally small amounts of potassium feldspar and/or plagioclase (Figures 1b-1e). Minor components, identified in only a few samples, include a zeolite at around 9Å (Figure 1b), possibly of the clinoptilolite-heulandite series, amphiboles (sharp peak at around 8 to 8.4Å, Figure 1c), and opal-CT in one sample (Figure 1b).

Semi-quantitative results based on peak areas of clays in individual samples (Table 2) were subjected to principal components analysis. Two factors explain 91.6% of the variance in the data set. Factor loadings indicate that factor 1 causes samples to be enriched either in smectite, or in illite (Table 3). The second factor apparently causes enrichment in chlorite and R=1 I/S clay or, conversely, in R=3 I/S clay. From these results, the sam-

Table 1. Presence/absence data for the <2mm fraction of glacial erratics. For smectite group minerals, the small 's' = small but sharp peak; large 'S' = large, sharp peak; 'b' = small, broad peak. K-spar = potassium feldspar; plag = plagioclase; relative sizes of 'x''s for feldspars indicates relative peak sizes from dif-fractograms.

Sample	smectite	R=1 I/S	R=3 I/S	chlorite	illite	kaolinite	quartz	k-spar	plag	others
E145	X, s	X			X	X	X	x	X	
E219	X, S			X	X		X		X	
E242D	X, b	X		X	X		X	x	X	amphibole
E243		X		X	X		X	x	X	
E317	X, S				X	X	X			
E323		X		X	X					
E347				X	X		X	x	X	amphibole
E360			X	X	X		X	x	X	zeolite
E363	X, s	X		X	X		X	x	X	opal-CT, zeolite
MB80	X, s				X					zeolite
MB181	X, S				X	X		X		
MB202	X, b					X	X	X	X	
MB212K	X, s			X	X		X	x	X	amphibole
MB217A	X, s	X		X	X	X	X	X	X	amphibole, zeolite
MB235A	X, b		X	X	X		X		X	
MTD42	X, S				X	X	X			
MTD153	X, s	X			X	X				
MTD154	X, b				X	X	X			
MTD190	X, b				X	X	X			
SIM5	X, s	X		X	X		X	X	x	
SIM11	X, b		X							
SV12	X, b			X	X		X		X	amphibole

Table 2. Semi-quantitative results on clay minerals of the <2μm fraction of glacial erratics, based on profile fit-derived peak areas and Biscaye's (1965) weight factors.

Sample	smc	chlor	R=1 I/S	R=3 I/S	ill	kao
E145	68	0	6	0	16	10
E219	68	10	0	0	22	0
E242D	36	17	12	0	36	0
E243	0	13	15	0	73	0
E317	92	0	0	0	5	3
E323	0	38	5	0	56	0
E347	0	8	0	0	92	0
E360	0	8	0	21	71	0
E363	10	7	27	0	55	0
MB80	25	0	0	0	75	0
MB181	63	0	0	0	33	4
MB202	91	0	0	0	0	9
MB212K	50	14	0	0	36	0
MB217A	18	6	7	0	67	3
MB235A	19	4	0	34	43	0
MTD42	73	0	0	0	24	3
MTD153	33	0	13	0	52	3
MTD154	44	0	0	0	52	4
MTD190	65	0	0	0	30	5
SIM5	30	12	18	0	40	0
SIM11	90	0	0	10	0	0
SV12	82	10	0	0	8	0

ples appear to fall into two groups which differ in each clay component at the 0.05 confidence level (Table 4). The first, designated 'Group A', has a consistent mineral composition characterized by dominant smectite levels (>60%), higher levels of kaolinite than the other group of samples, and lower levels of the other four clay minerals (Table 4). The second, designated 'Group B', is a more diverse group of samples as indicated by higher standard deviations for all mineral components. The dominant clay in this group is illite. These samples have, on average, higher levels of chlorite and R=1 I/S clays. The differences are statistically significant at the 0.05 confidence level (based on student's t-tests) for all mineral components except the R=3 I/S clays (Table 4).

DISCUSSION

Authigenic vs. Detrital Origin of the Minerals

Clays occurring in sediments may be detrital or authigenic, which may be determined from x-ray diffraction by peak sharpness (generally sharp for authigenic minerals), thin section petrography, and scanning electron microscopy [Wilson and Pittman, 1977]. Authigenic minerals reflect the physico-chemical conditions of diagenesis under which they form. Among the minerals identified in this study, opal-CT in Erratic E363 (Figure 1b) is most probably an alteration product of biogenic opal [Jones and Segnit, 1971; Kastner et al., 1977]. This erratic is a marine mudstone with ostracodes and dropstones [Levy and Harwood, this volume]. In addition, four samples contain sharp smectite group mineral peaks that may indicate an authigenic component: E219, E317, MB181, and MTD42. Photomicrographs of thin sections do not indicate the presence of authigenic clays [Levy and Harwood, this volume]. All four of these erratics were dated using palynomorphs as middle to late Eocene in age [Bohaty and Harwood, this volume]. They represent four different lithofacies: E219 is a massive, sandy, bioturbated mudstone with a leaf fossil; E317 is a poorly sorted, massive sandstone with abundant feldspar and granite clasts; MB181 comprises graded sand and pebbles; and MTD 42 is a massive, matrix-supported conglomerate. Based on petrography, these sharp-peaked smectite group minerals are assumed to be detrital. None of the other clay or non-clay minerals appear to be authigenic, based on petrography [Levy and Harwood, this volume].

Significance of Detrital Clays

Where clays are established to be detrital in origin, they are generally assumed to be derived from the soils in which they formed, and hence, are indicators of the climatic regime from which they were derived [e.g., Chamley, 1989]. However, detrital clays may also be derived from older, exposed and eroding sediment, and the relative contribution of soils and pre-existing sediment to any sedimentary basin remains problematic when

Table 3. Factor score coefficients, factors 1 and 2, from principal components analysis on the results from Table 2, varimax rota-tion. These two factors explain 91.6% of the variance in the data set.

	Factor 1	Factor 2	Factor 3
Illite	0.922	-0.04	0.157
R=1	0.511	0.615	0.123
R=3	0.186	-0.884	0.048
Chlorite	0.137	0.211	0.889
Kaolinite	-0.313	0.236	-0.775
Smectite	-0.922	0.04	-0.348

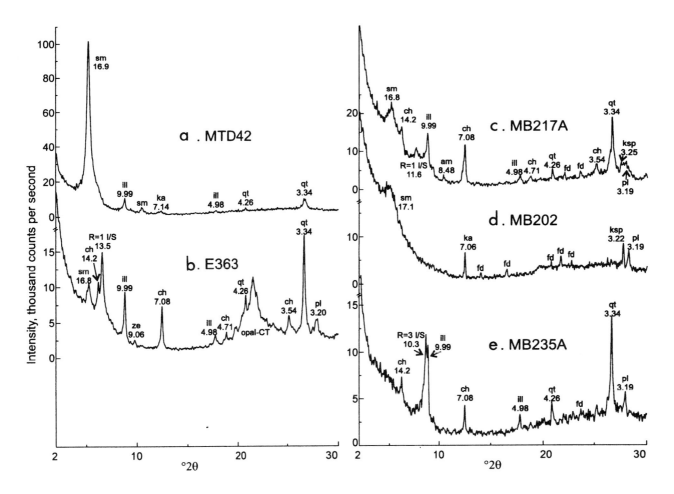

Figure 1. Selected diffractograms of ethylene glycol-treated, <2 m fraction of glacial erratics. The ordinate axis is intensity in thousands of counts per second. Number over peaks are d-spacings in Ångstroms. sm=smectite; ch=chlorite; ill=illite; ka=kaolinite; qt=quartz; ze=zeolite; fd=feldspar; ksp=potassium feldspar; pl=plagioclase feldspar. a. Sample MTD42 is an example of samples that contain a smectite group mineral with a large, sharp peak. Illite is also present (peak at 10Å), as is kaolinite (peak at 7Å) and quartz (peaks at 3.33 and 4.26Å). b. Sample E363 is an example of samples that contain a smectite group mineral with a small but sharp peak. Chlorite is present (peaks at 14, 7 and 4.7Å), as is R=1 I/S (peak near 13Å), illite (peak at 10Å), quartz (peaks at 3.34 and 4.26Å), plagioclase (peak at 3.19Å), and opal-CT (broad area from 20 to 25° 2 , with peak at 4.1Å). A zeolite (clinoptilolite-heulandite) occurs at 9Å. c. Sample MB217A contains a smectite group mineral with a small but sharp peak near 17Å, chlorite with a peak at 14Å, R=1 I/S with a peak near 12Å, a zeolite with a peak near 9Å, and amphibole, with a sharp peak near 8.5Å. d. Sample MB202 is an example of samples that contain a smectite group mineral with a small, broad peak near 17Å. Kaolinite is present (peak at 7Å; no 14 or 4.79Å peak), as is quartz (peaks at 3.33 and 4.26Å), potassium feldspar (peak at 3.22Å) and plagioclase (peak at 3.19Å). e. Sample MB235A contains a smectite group mineral with a small, broad peak near 17Å, chlorite (peak at 14Å), R=3 I/S clay (shoulder on the 10Å peak), quartz and plagioclase.

trying to decipher paleoclimate. However, chlorite generally does not survive cool, moist temperate or more intense weathering regimes, and its presence in sediment as a detrital clay is a good indicator of cool and/or dry climates. Smectite group minerals may indicate monsoonal climates, which, with their alternating wet and dry seasons, promote the formation of this alternately expanded and contracted mineral [Chamley, 1989; Millot, 1970].

Smectite may also indicate the presence of volcaniclastic input such as ash; [Nadeau and Reynolds, 1981] or a basaltic terrain subjected to temperate weathering conditions. Smectite group minerals may concentrate in offshore (outer shelf or farther) marine conditions, due to pericontinental fractionation of clays [Porrenga, 1966; Gibbs, 1977; Holmes, 1987]. Kaolinite may indicate moist and temperate to tropical weathering conditions.

Table 4. Mean and standard deviation of samples grouped ac-cording to the results of principle component analysis. s.d.=standard deviation; C=confidence level.

	Group A		Group B		Student's	C
	mean	s.d.	mean	s.d.	t	
smectite group	76.9	11.9	20.4	17.6	-8.357	0.01
R=1 I/S	0.7	2.0	7.5	8.9	2.246	0.05
R=3 I/S	1.1	3.3	4.2	10.7	0.844	
chlorite	2.2	4.4	9.8	10.1	2.091	0.05
illite	15.3	12.6	57.5	17.1	6.284	0.01
kaolinite	3.8	3.7	0.8	1.5	-2.643	0.05
n	9		13			

This mineral can survive at least moderate weathering, and its presence in sediment may also derive from older sediment. Illite derives from moderate to no weathering of ancient shales. Smectite group minerals decrease in older shales, presumably due to their alteration to illite during diagenesis, and illite levels increase as age of shales increases [Weaver, 1967].

Illite generally forms by a stepwise alteration of smectite group minerals, providing potassium is available, that is observable by x-ray diffraction [e.g., Reynolds, 1980; Pollastro, 1993]. In the first stages of diagenesis, the smectite group 001 peak at 17Å expands less while other peaks shift to larger d-spacings. Such a diffraction pattern indicates a randomly interstratified mixed-layer illite/smectite clay [designated R=0 I/S clay; "R" is for "Reichweite" or ordering; Reynolds, 1980]. In later stages and/or at higher temperatures, the illite interstratification appears to become ordered, which produces a characteristic x-ray pattern with the 17Å peak replaced by a peak at around 13Å. This clay is designated R=1 I/S clay. At still later stages, interstratification appears to be so extensive that only a slight expansion of a 10Å mineral is detected as a shoulder on the 10Å peak or a peak at 11Å. This mineral is designated R=3 I/S clay. The survival of the I/S clays in the weathering environment is not well constrained, and I/S clay that is identified in modern soils may be entirely inherited from the parent material [Wilson and Nadeau, 1985].

The four erratics which contain sharp smectite group minerals, E219, E317, MB181, and MTD42, do not contain volcanic ash and do not appear authigenic [Levy and Harwood, this volume]. These highly crystalline minerals must have formed by chemical weathering in a temperate or monsoonal climate. The weathering environment was too intense to preserve chlorite (0-10% in these samples) but was not intense enough to generate much kaolinite (3-

4% in these samples). These erratics were collected from areas that today have exposed Jurassic age Ferrar Dolerite [Levy and Harwood, this volume]. Mafic rocks weather readily to smectites. The Ferrar Dolerite may have been the source that was moderately weathered to generate these highly crystalline smectite group minerals. Ehrmann [1996] arrived at a similar conclusion to account for abundant smectite in sediment at the base of the CIROS-1 and -2, and MSSTS-1 cores, which were drilled in McMurdo Sound, downstream of the area where these erratics were collected. The depositional environments for the sediment in these erratics ranges from marginal marine to fully marine [Levy and Harwood, this volume].

These samples fall in with the group of samples that contain more than 60% smectite group minerals, based on the results of principal components analysis (Figure 2). This group (designated Group A), has a fairly consistent mineral composition that is dominated by smectite group minerals, followed by minor illite, and has higher kaolinite levels than the erratics not in this group (Table 4). A common mineral composition suggests that these erratics all derive from the same or similar terrains and formed under similar climatic conditions, but at least two of the erratics are distinct from the group: MB202 and SV12. These are a tuff and a volcaniclastic conglomerate, respectively, both barren of fossils [Levy and Harwood, this volume]. Thus high smectite levels in the erratics arise from either demonstrated volcaniclastic input (MB202 and SV12) or from chemical weathering. All nonvolcaniclastic erratics in Group A are middle to upper Eocene in age (Figure 3).

The other group of erratics, designated Group B, has widely variable mineral compositions that almost certainly derive from more than one terrain. Group B erratics tend to be dominated by illite, and contain significantly higher levels of chlorite and R=3 I/S clays. This type of assemblage might be expected in sediment derived by physical weathering, with little or no chemical weathering. As these erratics are sediments, not metamorphic rocks, the presence of chlorite indicates that they have undergone little chemical weathering during their formation. High levels of chlorite in samples E323, E242D, MB212K, E243, and SIM5 suggest that these erratics may be derived from a metamorphic terrain. Six samples with chlorite also contain R=1 I/S clay, indicating that the R=1 I/S supply, probably an ancient (Paleozoic or older) sedimentary terrain, and the chlorite supply were linked, either geographically or by transport mechanism, during deposition of the sediment in these erratics. None of the samples contain both an R=1 I/S and an R=3 I/S clay component, which suggests

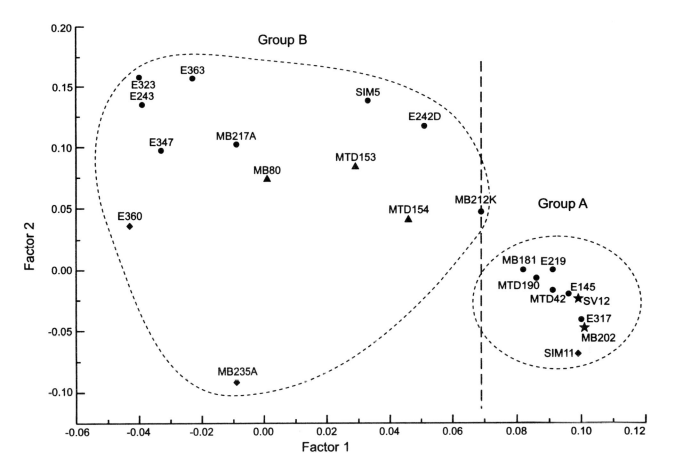

Figure 2. Results of principal components analysis (PCA) on the semi-quantitative results for the clay minerals given in Table 2, indicating how samples plot in Factor 1/Factor 2 space. The dotted line is the "50% smectite" line. Samples with >50% smectite group together ('Group A') and have a significantly different mineral composition from the other samples ('Group B'). The starred samples in Group A are undated volcaniclastic sediment. The triangles in Group B are middle to upper Eocene in age; all other samples are post-Eocene or undated. The diamonds are samples with R=3 I/S clay.

that the terrains supplying these clays are distinct.

All but three of the thirteen samples in this group are post-Eocene age or were not datable because of a lack of fossils [Figure 3; Levy and Harwood, this volume]. Seven of nine Group A samples are Eocene age and two in this group are volcaniclastic and contain no fossils. From these results it appears that chemical weathering declined after the Eocene in east Antarctica. These results agree with those of Ehrmann [1996] from the CIROS-1 and -2 cores, the MSSTS cores drilled in McMurdo Sound, and with results of studies by Robert and Maillot [1990] on clays from cores from the Weddell Sea, that the Eocene-Oligocene climate shift in Antarctica, from a non-glacial to a glacial one, is reflected in the clay mineral composition of marine sediments

as a shift from smectite to chlorite dominance.

Three samples of middle to upper Eocene age occur in Group B, MB80, MTD153, and MTD154. These samples contain large amounts of illite (52 to 75%) and low levels of smectite group minerals (25 to 44%), and also have more species of dinoflagellates (6 to 22) than other members of Group B, which have 0 or 1 species. These three samples contain no chlorite and are all sandstones. They may have been deposited in an area with a dominant illite supply, with smectite limiting, either because it did not occur in abundance in the source area, or because it was bypassing the depositional area. Alternatively, the sediment in these erratics may have been deposited during a cold snap when smectite generation was at an Eocene low.

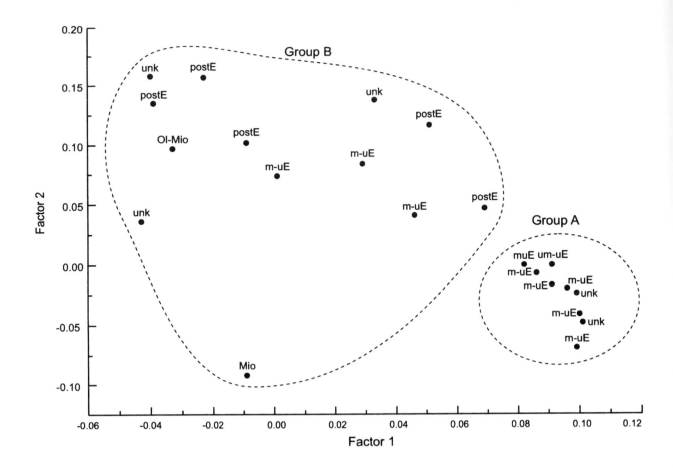

Figure 3. Results of PCA, as in Figure 2, indicating ages of erratics as determined by Levy and Harwood, and Bohaty and Harwood, this volume.

CONCLUSIONS

Varying mineral composition in the <2 m fraction of the erratics divides them into two groups: A) erratics with dominant smectite group minerals, which may or may not be highly crystalline, and B) erratics with widely varying composition, but in general, with dominant illite and low levels (<50%) of smectite group minerals. This latter group includes at least two types of erratics: those with abundant chlorite, often accompanied by R=1 I/S clay, and those erratics with abundant R=3 I/S clay. Most of the erratics in Group B are post Eocene in age, but three of them are Eocene. These three clasts are unusual in that, although formed during a relatively mild climatic period in East Antarctica's history, their mineral composition is dominated by illite rather than smectite. They contain no chlorite. The sediment in these samples may have accumulated nearshore, or may have had a provenance dominated by ancient sedimentary rocks rich in illite.

There are two types of Group A erratics: those that are middle to upper Eocene, and those that are volcaniclastic and barren of fossils. Conspicuously absent from any of the erratics is abundant kaolinite, even in erratics with abundant feldspars and granitic lithofragments. This suggests that even during the Eocene, climates were not particularly warm or wet, and that the Beacon Supergroup, which is kaolinite-rich [Ehrmann et al., 1992], did not contribute significantly to the clay fraction of these sediments.

McMurdo Sound erratics collected for this study that are Eocene age have at least two distinct provenances for the clay fraction: 1) a smectite-rich source area with minor illite, kaolinite, and no surviving chlorite (Group A samples), and 2) an illite-rich source area with subordinate smectite, minor kaolinite, and no surviving chlorite (Group B samples). Post Eocene age erratics have a clay fraction from at least two source areas: 1) a metamorphic + ancient sedimentary terrain that supplied chlorite and

R=1 I/S clay and illite, and 2) an ancient sedimentary terrain that supplied R=3 I/S clay and illite. Smectite-rich volcaniclastics of unknown age comprise a fifth type of erratic.

Acknowledgments. Thanks to Steve Bohaty, David Watkins, Jiang (Mr. Asia), Richard Graham, Aradhna Srivastav, John Kaser, Diane Winter, and David Harwood for collecting the samples, and to David and Richard for 'letting' me work on them as a sort of unfunded mandate. Special thanks and an SPF 30 salute to Ms. Kelly Bergmann, for sample preparation and x-ray analysis.

REFERENCES

Biscaye, P. E.
1964   Distinction between kaolinite and chlorite in recent sediments by X-ray diffraction. *Amer. Miner.*, 49, 1281-1289.

1965   Mineralogy and sedimentation of Recent deep-sea clay in the Atlantic Ocean and adjacent seas and oceans. *Geol. Soc. Amer. Bull.*, 76, 803-832.

Borg, I. Y., and D. K Smith
1969   Calculated x-ray powder patterns for silicate minerals. *Geol. Soc. of Amer. Memoir*, 122. Boulder, CO

Brown, G.
1980   Associated minerals, in *Crystal Structures of Clay Minerals and Their X-ray Identification*, edited by G. W. Brindley and G. Brown, 361-410, Mineral. Soc. Monogr:, 5, London.

Chamley, H.
1989   *Clay Sedimentology.* Berlin, Springer-Verlag, Stuttgart, pp. 117-131.

Drever, J.I.
1973.   The preparation of oriented clay mineral specimens for X-ray diffraction analysis by a filter-membrane peel technique. *Amer. Mineral.*, 58:553-554.

Ehrmann, W. U.
1996   Smectite concentrations and crystallinities: Indications for Eocene age of glaciomarine sediments in the CIROS-1 drill hole, McMurdo Sound, Antarctica. *Ant. Geol. Geophys.*, 1-10.

Ehrmann, W. U., Melles, M., Kuhn, G., and Grobe, H.
1992   Significance of clay mineral assemblages in the Antarctic Ocean. *Marine Geol.*, 107:249-273.

Gibbs, R. J.
1977   Clay mineral segregation in the marine environment. *Jour. Sediment. Petrol.*, 47:237-243.

Heiden, K., and Holmes, M.A.
1998   Grain-size distribution and significance of clay and clay-sized minerals in Eocene to Holocene age sediments from Sites 918 and 919

in the Irminger Basin. *ODP Sci. Results*, 152:39-49.

Holmes, M. A.
1987   Clay mineralogy of the Lower Cretaceous deep-sea fan, deep sea drilling project site 603, lower continental rise of North Carolina. *Init. Repts.*, *DSDP*, 92:1079-1089 (U.S. Govt. Printing Office, Washington).

Jackson, M.L.
1975   *Soil Chemical Analysis-Advanced Course (2nd ed.)*: Madison, WI. (M.L. Jackson), 27-95.

Jones, J. B., and Segnit, E. R.
1971   The nature of opal. I. Nomenclature and constituent phases. *Jour. Geol. Soc. Aust.*, 18:57-68.

Kastner, M., Keene, J. B., and Gieskes, J. M.
1977   Diagenesis of siliceous oozes-I. Chemical controls on the rate of opal-A to opal-CT transformation-an experimental study. *Geochim. et Cosmochim. Acta.*, 41:1041-1059.

Levy, R. H., and Harwood, D. M.
this volume.   Sedimentary lithofacies and inferred depositional environments of the McMurdo Sound erratics.

Millot, G.
1970   *Geology of Clays*, (Farrand, W.R., and Paquet, H., trans.). Springer-Verlag, New York.

Moore, D.M., and Reynolds, R.C., Jr.
1989   *X-ray Diffraction and the Identification and Analysis of Clay Minerals*: Oxford (Oxford University Press).

Nadeau, P. H., and Reynolds, R. C., Jr.
1981   Volcanic components in pelitic sediments. *Nature*, 294:72-74.

Pollastro, R. M.
1993   Considerations and applications of the illite/smectite geothermometer in hydrocarbon-bearing rocks of Miocene to Mississippian age. *Clays Clay Minerals*, 41:119-133.

Porrenga, D. H.
1966   Clay minerals in recent sediments of the Niger delta. *Clays Clay Minerals*, 14th natl. conf., Pergammon, Oxford, New York:221-233.

Reynolds, R. C., Jr.
1980   Interstratified clay minerals. *In* Brindley, G. W., and Brown, G., (eds.), *Crystal Structures of Clay Minerals and Their X-Ray Identification*: Monograph No. 5, Mineralogical Society, London, 249-303.

Robert, C., and Maillot, H.
1990   Paleoenvironments in the Weddell Sea area and Antarctic climates as deduced from clay mineral associations and geochemical data, ODP Leg 113. *In* Barker, P. F., Kennett, J. P., et al. (eds.), *Proc. ODP, Sci. Res.*, 113:51-70. College Station, TX.

Srodon, J.
  1980   Precise identification of illite/smectite interstratifications by X-ray powder diffraction. *Clays Clay Minerals*, 28:401-411.
Weaver, C. E.
  1967   Potassium, illite and the ocean. *Geochim Cosmochim. Acta*, 31:2181-2196.
Wilson, M. D., and E. D. Pittman
  1977   Authigenic clays in sandstones: recognition and influence on reservoir properties and paleoenvironmental analysis. *Jour. Sed. Petrol.*, 47, 3-31.

Wilson, M. J., and Nadeau, P. H.
  1985   Interstratified clay minerals and weathering processes. *In* Drever, J. I. (ed.), *The Chemistry of Weathering*. De. Reidel, Dordrecht, Netherlands, pp. 97-118.

Mary Anne Holmes, 214 Bessey Hall, Department of Geosciences, Univ. of Nebraska-Lincoln, Lincoln, NE 68588-0340

PALEOBIOLOGY AND PALEOENVIRONMENTS OF EOCENE ROCKS, MCMURDO SOUND, EAST ANTARCTICA
ANTARCTIC RESEARCH SERIES VOLUME 76, PAGES 73–98

MARINE DIATOM ASSEMBLAGES FROM EOCENE AND YOUNGER ERRATICS, McMURD0 SOUND, ANTARCTICA

David M. Harwood and Steven M. Bohaty

Department of Geosciences, University of Nebraska-Lincoln, Lincoln, Nebraska 68588-0340

The search for diatoms in erratics from glacial moraines in McMurdo Sound yielded diatom floras in eleven samples. These erratics represent a sampling of Cenozoic strata presently hidden beneath the Antarctic ice sheet or ice shelves. Middle to upper Eocene diatom assemblages extracted from six glacial erratics of mudstone, sandstone and conglomerate are the focus of this biostratigraphic report. In addition, Oligocene and Miocene diatom floras from seven younger erratics of diamictite and mudstone are treated briefly. The middle to late Eocene age, as indicated by dinoflagellate cyst and ebridian biostratigraphy, is supported by diatom biostratigraphy, however, distinct Eocene diatom floras in each of the six erratics suggest that they represent different time intervals within the middle to late Eocene. Further age determination of these erratics will be possible in the future, as local biostratigraphic schemes are developed for the shelf basins of Antarctica. Many diatoms (~70 taxa) are known previously from Southern Ocean and lower latitude sites. The remaining diatoms (~50 taxa) are treated informally. More than 75% of the Eocene assemblage is illustrated herein. Two distinctive diatoms are described as new: *Hemiaulus stilwelli* n. sp. and *Pseudorutilaria levyii* n. sp. Recovery of a diatom assemblage from an erratic of early-late Miocene age that resembles, in composition and structure, the modern sea-ice diatom flora suggests cold marine surface temperatures (-1°C or lower) at this time. In addition to marking the oldest age of sea-ice in Antarctic waters, it demonstrates that this sea-ice diatom community evolved by at least the early-late Miocene. Absence of this sea-ice flora in subsequent times of the early Pliocene and early Pleistocene indicates that the Late Cenozoic history of sea-ice cover in Antarctic waters was episodic.

INTRODUCTION

Although rare outcrops and a few drillcores in Antarctica provide information about paleoenvironments and past fauna and flora, the present mantle of ice hides much of Antarctica's Cenozoic geologic history. Glacial erratics present in coastal moraine of McMurdo Sound (Figure 1) provide a means for studying lithologies and fossils from strata presently covered by the ice sheet. Because

paleontologic information is unavailable for many Cenozoic intervals, the glacially transported erratics provide useful data for biostratigraphy, paleobiogeography and paleoenvironments of this remote, yet significant region. The presence of Eocene sedimentary erratics in coastal moraine in the McMurdo Sound region has been known for many years [Cranwell, 1964; Cranwell et al., 1960; McIntyre and Wilson, 1966; Wilson, 1967; Harrington, 1969; Vella, 1969; Hertlein, 1969; Hotchkiss and Fell, 1972; Rowe, 1974; Stott

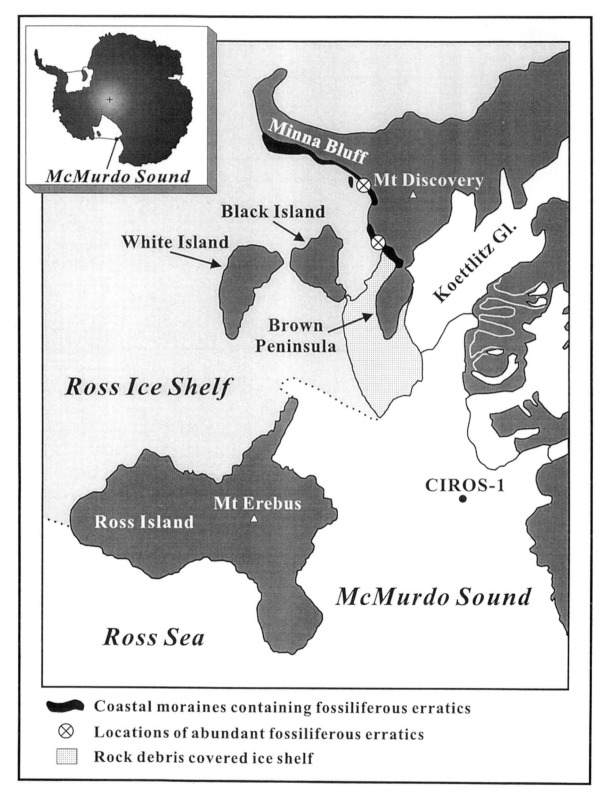

Fig. 1. Map of McMurdo Sound region identifying coastal moraine collection area of Eocene glacial erratics.

et al., 1983; Feldmann and Zinsmeister, 1984] and recognized as a potential source of information on the Paleogene biota of Antarctica.

The source strata of such Eocene erratics are not known to outcrop in the Transantarctic Mountains, but they are thought to be sourced from 'Discovery Deep' at the confluence of the Byrd, Mulock, and Skelton glaciers. Here, a >1000 meter-deep basin was carved by glacial erosion at times when these glaciers grounded and advanced across the shelf of the western Ross Embayment (Wilson, this volume). The erratics currently represent the only record for the middle and early-late Eocene in this region, although the younger erratics do overlap in age with the uppermost Eocene to Oligocene intervals in the CIROS-1 drillcore (366 to 702 mbsf) [Barrett, 1989; Hannah et al., 1997; Wilson et al., 1998].

We recently began an intensive phase of sample collection and description of these erratics and the fossils they contain. The chief goals were to establish the ages and paleoenvironments represented by the Eocene erratics. The papers in the present volume provide the results of this effort. More than 1000 specimens of erratics were collected during the 1992-93, 1993-94, and 1995-96 field seasons from the coastal areas of Mt. Discovery, Minna Bluff and Black Island (Figure 1).

A diversity of environments from terrestrial to marine is represented by the McMurdo erratics. Sediments range from fine siltstone to coarse conglomerate and diamictite [Levy and Harwood, this volume b], with marine lithofacies indicating deposition in an inner shelf environment, most likely within an estuarine system representing a fjord or ria (a long, narrow inlet of the sea formed by submergence of the lower part of a narrow, transverse to the coast, river valley) [Levy, 1998].

Information on Eocene macrobiota has expanded significantly with the documentation of a broad range of fossil groups: marine invertebrates [Stilwell, this volume; Willis and Stilwell, this volume], brachiopoda [Lee and Stilwell, this volume], bryozoa [Hara, this volume], barnacles [Buckeridge, this volume], and decapods [Stilwell et al., 1997; Schweitzer-Hopkins and Feldmann, this volume] and marine and avian vertebrates [Long, this volume; Jones, this volume]. Terrestrial communities are represented by leaves [Pole et al., this volume], wood [Francis, this volume], and palynomorphs [Askin, this volume]. The age of the erratics is best constrained by the diverse microfossil assemblages of dinoflagellate cysts [Levy and Harwood, this volume a], ebridians, silicoflagellates, chrysophycean cysts, and endoskeletal dinoflagellates [Bohaty and Harwood, this volume]and by diatoms [this report].

Existing paleontologic reports of Eocene age from Antarctic shelf sediments are restricted, chiefly, to the fauna and flora from the La Meseta Formation on Seymour Island in the Antarctic Peninsula [Woodburne and Feldmann, eds., 1988] and the lower interval of the CIROS-1 drillcore [Harwood et al., 1989a; Wilson et al., 1998]. Other Eocene records are restricted to core recovered in ODP holes 739 and 742 in Prydz Bay [Barron et al., 1991]. It is anticipated that a stratigraphic section representing similar intervals of the erratics will be recovered during the Cape Roberts Project (CRP) in the western Ross Sea [Barrett and Davey, eds., 1992; Webb and Wilson, 1995].

As each erratic records a portion of geologic time, but is removed from a stratigraphic context, the information obtained can be used to reconstruct Antarctic geologic history in only a broad sense. Nevertheless, this suite of rocks presently bears the only record of many groups of organisms from the Paleogene of Antarctica. Sufficient information is now available to infer the environmental conditions on the marine margin of the Transantarctic Mountains under a cool-temperate climate. As there is no evidence identified in the Eocene erratics to suggest the presence of ice at sea-level, the paleobiota and sedimentary rocks reflected in the erratics likely preceded the growth of large ice sheets in Antarctica.

EOCENE DIATOM BIOSTRATIGRAPHIC REFERENCE SECTIONS

Our knowledge of Eocene diatom assemblages in the southern high latitudes is limited to a few short stratigraphic intervals drilled by the Deep Sea Drilling Project, particularly during Leg 29 [Hajós, 1976], Leg 36 [Gombos, 1977], Leg 71 [Gombos, 1983; Gombos and Ciesielski, 1983]; the CIROS-1 drillcore [Harwood, 1989; Wilson et al., 1998], and at ODP Site 739 [Mahood, et al., 1993]. Detailed descriptions of diatom floras from the upper Eocene Oamaru Diatomite in New Zealand [Desikachary and Sreelatha, 1989; Edwards, 1991] provide another useful comparative reference. Other useful sources of biostratigraphic information are from other areas [Kanaya, 1957; Schrader and Fenner, 1976; Dzinoridze et al., 1978; and Fenner, 1978, 1985, among numerous other reports].

The diatom assemblages recovered from the McMurdo erratics include elements of all the above studies, yet do not match any one assemblage. Several diatoms recognized for their biostratigraphic utility aid in age assignment and relative stratigraphic position to the diatom assemblages, though reference control from local drillcores on the Antarctic shelf would provide better control.

METHODS

Preparations to extract siliceous microfossils were performed on seventy of the fine-grained erratics [Table 1 in Bohaty and Harwood, this volume]. The same preparations and slides were used in this study and the ebridian and silicoflagellate report by Bohaty and Harwood [this volume]. Approximately 50 g of each sample was treated in a hot solution of 50% HCl to remove the calcareous cement. The sample was then rinsed with filtered water through repeated settling. In order to concentrate the siliceous microfossils, the coarser materials were settled out in a 600 ml beaker for 30 seconds and the suspended materials decanted to another beaker, which was allowed to settle overnight. The final residue was concentrated by centrifugation for 5 minutes at 1500 rpm. Strewn slides of this reside were mounted on 22 x 40 mm coverglass and mounted with Norland Optical Adhesive #61. Slides were examined under transmitted light and differential interference contrast with Olympus BH-2 and Leica DMRX microscopes.

If siliceous microfossils were noted in these preparations, further steps were taken to concentrate the assemblages. Residues were sieved through a 10mm polyester mesh sieve and washed with a weak Calgon solution to remove some of the clay-size material. Samples were then rinsed in deionized water, and centrifuged at 1500 rpm for 5 minutes (repeated 3 times). The residues were transferred to a glass vial, shaken, and allowed to settle for 1 minute. The suspended material was strewn on 20 x 40 mm coverglass. Several slides were examined from each sample. The diatom assemblages were routinely examined using a 40x objective, with closer examination of noted specimens with a 100x objective.

DIATOM ASSEMBLAGES

Diatoms were present in samples from 10 of the erratics samples examined, and these are separated into two groups: the Eocene erratics, and the Oligocene and Miocene erratics (Table 1). Abundance and preservation varied between the different samples of all the erratics, likely due to the diagenetic effects of dissolution, which altered and removed them.

Assemblages of diatoms from the Oligocene and Miocene erratics (Table 2) were encountered in the diamictite and some fine-grained lithologies. Lithologic designations of these erratics are from Levy and Harwood

TABLE 1. Collection location, lithology, and age of erratics containing abundant and well-preserved siliceous micro-fossils.

Erratic	Collection Location	Lithology	Age
MB244C	Minna Bluff	mudstone (Mm-d)	late Miocene
MtD46	Mt. Discovery	mudstone (Mmb)	middle Miocene
E351	Minna Bluff	diamictite (Dm)	middle (?) Miocene
E346	Minna Bluff	diamictite (Dm)	early to middle Miocene
E347	Minna Bluff	diamictite (Dm)	late Oligocene to early Miocene
D1	Mt. Discovery	mudstone (Mwb)	middle to early-late Eocene
MtD95	Mt. Discovery	mudstone (Mwb)	middle Eocene
E345	Mt. Discovery	sandstone (Sm)	late middle to late Eocene
E350	Minna Bluff	mudstone (Mwb)	late Eocene
MB181	Minna Bluff	sandstone (Ssg)/ conglomerate (Csgc)	middle to late Eocene

E346	E347	E351	MtD 46	MB 244c
Actinocyclus octonarius	*Actinoptychus senarius*	*Actinocyclus* sp. cf. *octonarius*	*Actinocyclus octonarius*	*Actinocyclus karstenii*
Coscinodiscus sp. A MSSTS-1	*Arachnoidiscus* sp.	*Asteromphalus symmetricus*	*Actinoptychus senarius*	*Actinocyclus octonarius*
Eucampia antarctica	*Cocconeis* cf. *antiqua* v. *tenuistriata*	*Chaetoceros* spp.	*Asterolampra* sp.	*Chaetoceros* cf. *bulbosum*
Liradiscus sp.	*Cocconeis* spp.	*Coscinodiscus oculusiridus*	*Asteromphalus symmetricus*	*Corethron criophilum*
Nitzschia sp. A RISP J-9	*Coscinodiscus* spp.	*Coscinodiscus* sp. A MSSTS-1	*Chaetoceros* spp.	*Coscinodiscus oculusiridus*
Paralia sulcata	*Entopyla* sp. A	*Endictya hungarica*	*Coscinodiscus oculusiridus*	*Coscinodiscus* sp.
Stellarima microtrias	*Grammatophora* sp.	*Eucampia antarctica*	*Coscinodiscus* sp. A MSSTS-1	*Dactyliosolen antarcticus*
Thalassiosira irregulata	*Isthmia* fragments	*Isthmia* fragments	*Dactyliosolen antarcticus*	*Denticulopsis simonsenii*
Thalassiothrix sp.	*Paralia sulcata*	*Liradiscus* sp. RISP	*Nitzschia grosspunctata*	*Denticulopsis lauta* ?
Trinacria excavata	*Pyxilla* fragments	*Nitzschia maleinterpretaria*	*Denticulopsis maccollumii*	*Denticulopsis* sp.
Xanthiopyxis acrolopha	*Rhizosolenia hebetata*	*Paralia sulcata*	*Endictya hungarica*	*Entomoneis* sp.
	Stellarima microtrias	*Rhizosolenia hebetata*	*Entopyla* sp.	*Eucampia antarctica*
	Stephanopyxis cf. *megapora*	*Rouxia* sp.	*Fragilariopsis* sp.	*Fragilariopsis* sp. A
	Stephanopyxis grunowi	*Stellarima microtrias*	*Isthmia* fragments	*Fragilariopsis* spp.
	Stephanopyxis turris	*Stephanopyxis turris*	*Liradiscus* sp. RISP J-9	*Hyalodiscus* spp.
	Trinacria excavata	*Thalassionema* sp.	*Nitzschia maleinterpretaria*	*Pinnularia quadratarea*
		Trinacria excavata	*Paralia sulcata*	*Porosira* spp.
			Rhizosolenia hebetata	*Rouxia* sp.
			Rouxia sp.	*Rouxia californica* ?
			Stellarima microtrias	*Stellarima microtrias*
			Stephanopyxis spinosissima	*Thalassiosira nansenii*
			Stictodiscus hardmanianus	*Thalassiosira* spp.
			Thalassiosira irregulata	*'Tigeria'* sp.
			Trinacria excavata	*Trinacria* spp.
			Trinacria racovitzae	*Distephanus speculum*
			Xanthiopyxis spp.	Radiolarian fragments

TABLE 2. Diatom occurrence in post-Eocene McMurdo Sound erratics.

TABLE 3. Occurrence of Eocene diatom taxa in McMurdo Sound erratics.

MB-181	MTD-95	D-1	E-345	E-350	Samples
				×	Pseudopyxilla stylifera
				×	Pseudorutilaria levyi
		×	×	×	Pterotheca aculeifera
	×		×	×	Pterotheca carnifera
		×	×	×	Pterotheca danica
				×	Pterotheca minor
		×	×		Pterotheca sp. A
	×	×	×	×	Pyxilla reticulata
	×	×	×	×	Pyxilla sp. A
		×	×		Pyrgupyxis eocena
				×	Rhizosolenia spp.
				×	Sceptroneis lingulatus
	×	×			Sheshukovia sp. A
		×			Sheshukovia sp. B
				×	Sphynctolethus cf. pacificus
	×				Sphynctolethus sp. A
		×			Sphynctolethus sp. B
		×			Spiniviculum sp. A
		×			Stellarima sp.
	×	×	×	×	Stephanopyxis grunowii
		×	×		Stephanopyxis megapora
				×	Stephanopyxis cf. oamaruensis
		?			Stephanopyxis subantarctica
				×	Stephanopyxis superba
	×	×	×	×	Stephanopyxis turris
				×	Stephanopyxis sp. A
			×	×	Stephanopyxis? sp. B
		×	×		Stephanopyxis sp. C
		×			Stephanopyxis? sp. D
	×	×			Stephanopyxis sp. E
	×	×			Stephanopyxis sp. F
×					Stephanopyxis spp.
		×	×	×	Stictodiscus californicus var. nitidus
	×	×	×	×	Stictodiscus hardmanianus
			×		Triceratium americanum
		×			Triceratim castellatum var. fractum
			×		Triceratium castelliferum
			×		Triceratium columbi?
	×				Triceratium inconspicuum var. trilobata
		×			Triceratium lineatum Greville var.
	×	×			Triceratium cf. russlandicum
		×			Triceratium unguiculatum
		×		×	Trigonium arcticum
	×	×			Trinacria acutangulum
	×	×			Trinacria cornuta
				×	Trinacria excavata
		×			Trinacria fragilis
			×		Trinacria lingulata
		×			Trinacria sp. A
	×				Trochosira spinosa
		×			Trochosira sp. A
		×	×		Xanthiopyxis acrolophra
				×	Xanthiopyxis diaphana
				×	Xanthiopyxis globosa
	×	×	×		Xanthiopyxis oblonga
		×	×	×	Xanthiopyxis panduraeformis
	×			×	Genus et species indet. A
			×		Genus et species indet. B
		×			Genus et species indet. C
	×	×	×	×	Gn. et sp. indet (e) Schrader & Fenner
	×				Actinocyclus octonarius var. tenellus
	×	×	×		Actinoptychus senarius
	×	×	×		Annalus sp. A
		×	×	×	Arachnoidiscus sp.
		×	×		Aulacodiscus cf. huttonii
	×				Aulacodiscus rattrayii
	×				Auliscus sp. A
				×	Biddulphia elegantula
	×	×	×		Biddulphia rigida
	×	×			Biddulphia tenera
	×		×	×	Biddulphia? sp. A
				×	Biddulphia? sp. B
				×	Biddulphia? sp. C
		×	×		Biddulphia? sp. D
	×				Biddulphia? sp. E
		×			Biddulphia sp. F
	×	×			Biddulphia sp. G
		×			Biddulphia sp. H
		×	×		Briggeria siberica
		×			Briggeria sp.
			×		Chaetoceros didymus
		×			Chaetoceros spp.
×			×	×	Cocconeis costata
×		×	×	×	Cocconeis spp.
	×	×	×		Coscinodiscus radiatus
		×			Craspedodiscus molleri
		×	×		Dicladia sp. A
		×	×		Dicladia sp. B
			×	×	Dicladia sp. 1 Kanaya
		×			Distephanosira architecturalis?
		×			Drepanotheca bivittata
		×			Endictya sp.
				×	Eurossia irregularis var. incurvatus
		×			Glyphodiscus sp. A
		×	×	×	Goniothecium odontella
		×			Helminthopsidella ortha
		×			Hemiaulus altus
				×	Hemiaulus caracteristicus
	×	×			Hemiaulus danicus
		×	×	×	Hemiaulus dissimilis
	×	×			Hemiaulus hostilis var. polaris
		×	×	×	Hemiaulus polycistinorum
×		×	×		Hemiaulus polymorphus var. morsianus
×		×	×	×	Hemiaulus stilwelli
		×			Hemiaulus sp. A
×					Hemiaulus spp.
		×			Hercotheca sp. A
	×		×	×	Hercotheca sp. sensu Kanaya
		×			Hyalodiscus radiatus var. radiatus
		×			Hyalodiscus rossi
		×			Isthmia sp. fragments
		×			Leudugeria janischii
×			×		Liradscus ovalis
		×			Odontella sp. A
	×	×	×	×	Paralia clavigera
	×	×	×	×	Paralia sulcata
		×			Paralia sulcata var. crenulata
		×	×		Poetzkia? sp. Hajos
×			×	×	Proboscia interposita
	×				Pseudopodosira sp.
	×	×	×	×	Pseudopyxilla dubia

[this volume b]. As the focus of this paper is on the Eocene diatom assemblages, the Oligocene and Miocene assemblages are not treated in detail. Reference to, and illustration of, the diatoms listed in Table 2 can be found in diatom reports for drillcores MSSTS-1 [Harwood, 1986], CIROS-1 [Harwood, 1989], CRP-1 [Harwood et al., 1998], CRP-2 [Scherer et al., in press], and RISP Site J-9 cores [Harwood et al., 1989b].

Eocene diatoms are reported separately in Table 3 and illustrated in Plates 1 to 10. The plates are organized to highlight the diatoms in each sample: Sample E-350 in Plate 1; Sample E-345 in Plates 2 and 3; Sample MTD-95 in Plate 4; and Sample D-1 in Plates 5 to 9. Pyritized diatoms in Plate 10 are shown from a thin section of Sample MB-181; diatom assemblages in this erratic could not be extracted through acid preparations. Samples E-350, E-345 and D-1 and MTD-95 yielded the most abundant and well-preserved Eocene diatom assemblages. These four erratics yielded distinctly different diatom floras, that are interpreted to reflect difference in stratigraphic positions in the middle and upper Eocene section in the source area, or a range of environmental settings.. A rich siliceous microfossil assemblage is noted in sample MB-181, but the floras have altered to pyrite (Plate 10), which limits comparison to the other assemblages.

Floral composition is presented as presence/ absence data for the six Eocene erratics (Table 3). The majority of the diatoms in the Eocene samples are planktic, including genera that are common in other Paleogene sections: *Hemiaulus*, *Stephanopyxis*, *Trinacria* and species of the genus *Sheshukovia*, many of which are still incorrectly referred to as *Triceratium* and *Trinacria* in this report; significant taxonomic revision of this group is needed, but is beyond the scope of this paper. Benthic diatoms of the genera *Arachnoidiscus*, *Aulacodiscus*, *Auliscus*, *Biddulphia*, *Briggeria*, *Cocconeis*, and *Hyalodiscus* are also present, but in low number. The presence of benthic diatoms in an assemblage dominantly neritic planktic diatoms most likely reflects proximity to a coastline. Most of the benthic elements are likely transported to this site, perhaps down slope, as the dominance of planktic forms suggests relatively deep water, greater than 75 meters. The Eocene waters in the western Ross Sea were rich in nutrients to support a diverse and abundant flora of dinoflagellates, diatoms, silicoflagellates and ebridians. The fossil bird recovered in erratic E 303 is commonly associated with a fertile marine environment teeming with life [Jones, this volume].

The diatom assemblages from the Eocene erratics also contain a diversity of resting spores of the genera *Xanthiopyxis*, *Goniothecium*, *Dicladia* and *Pterotheca*. Sample E-345 contains *Pterotheca* as the dominant resting spore, in contrast to the assemblage of *Xanthiopyxis* in Samples D-1, MTD 95 and E-350.

The following discussion of each erratic draws attention to the diatoms that can be used to distinguish the assemblages between the different erratics. These taxa have been used in other studies as biostratigraphic markers. Support for the middle to late Eocene age suggested by the dinoflagellate cysts [Levy and Harwood, this volume a] and ebridians and silicoflagellates [Bohaty and Harwood, this volume] is provided by the occurrence of diatoms *Pyxilla reticulata*, *Hemiaulus dissimilis*, *Pterotheca aculeifera*, *Pterotheca danica*, *Stephanopyxis oamaruensis*, *Pyrgupyxis eocena*, *Sceptroneis lingulatus*, *Sphynctolethus pacificus*, and others. Genus *Trigonium*, which first appears in the Eocene, is a common element of several of the erratics.

SYNOPSIS OF DIATOM FLORAS

Erratic: E-350
Lithology: Mudstone (Mmb)
Discussion of diatom flora: This erratic contains at least 42 diatom taxa (Table 3). The occurrence of diatoms *Hemiaulus caracteristicus*, *Sceptroneis lingulatus* and *Sphynctolethus* cf. *pacificus*, which are not present in the other erratics, can be used to characterize this sample and distinguish it from the other erratics.

Age: Late Eocene, possibly equivalent to diatom assemblages from the lower intervals below 366 mbsf of the CIROS-1 drillcore [Harwood, 1989] and ODP Hole 739 [Mahood et al., 1993] based on the presence of *Hemiaulus caracteristicus* and *Sceptroneis lingulatus*.

Erratic: E-345
Lithology: Sandstone (Sm)
Discussion of diatom flora: This erratic contains at least 55 diatom taxa (Table 3). The occurrence of diatoms *Craspedodiscus molleri*, *Hemiaulus stilwelli* and *Chaetoceros didymus*, which are not present in the other erratics, can be used to characterize this sample and distinguish it from the other erratics. If the assumption that *Hemiaulus stilwelli* is ancestral to *Hemiaulus caracteristicus* is correct, then this sample is likely older than Erratic E-345.

Age: Middle to late Eocene?, possibly equivalent to nannofossil Zones CP11-CP13 based on the range of *Craspedodiscus molleri* at DSDP Site 605 [Gombos, 1983].

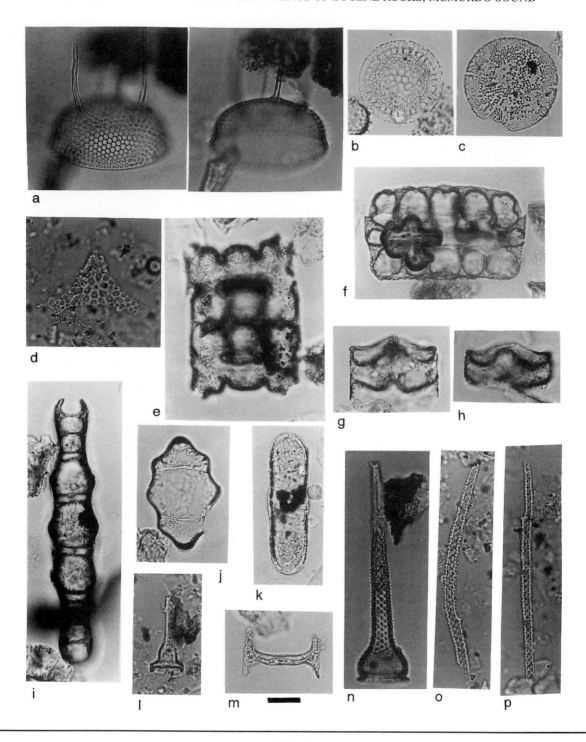

Plate 1

Diatoms from Sample E-350. Scale bar = 10mm. Fig. a. *Stephanopyxis* sp. A, specimen at different focus; Fig. b. *Stephanopyxis superba*; Fig. c. Genus et species indet. A; Fig. d. *Eurossia irregularis* var. *incurvatus*; Fig. e. *Biddulphia*? sp. A; Fig. f. *Biddulphia*? sp. B; Figs. g, h. *Dicladia* sp. 1 of Kanaya; Fig. i. *Biddulphia elegantula*; Fig. j. *Biddulphia*? sp. C; Fig. k. *Xanthiopyxis diaphana*; Fig. l. *Pterotheca aculeifera*; Fig. m. *Hemiaulus caracteristicus*; Fig. n. *Pyxilla reticulata*; Figs. o, p. *Pyxilla* sp. A.

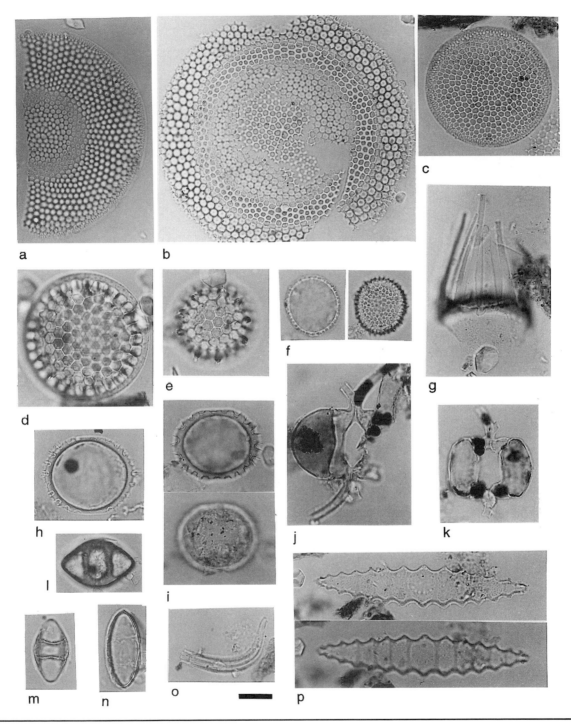

Plate 2

Diatoms from Sample E-345. Scale bar = 10mm. Figs. a, b. *Craspedodiscus molleri*; Fig. c. *Coscinodiscus radiatus*; Fig. d. *Stephanopyxis grunowii*; Fig. e. *Stephanopyxis turris*; Fig. f. *Stephanopyxis*? sp. B, specimen at different focus; Fig. g. *Psuedopyxilla stylifera* comb. nov.; Figs. h, i. *Hercotheca* sp. sensu Kanaya, specimen at different focus; Figs. j, k. *Chaetoceros didymus*; Figs. l-n. *Anaulus* sp. A; Fig. o. *Proboscia interposita*; Fig. p. *Psuedorutilaria levyi* sp. nov. Holotype; CAS #219084; photomicrographs at high and low focus.

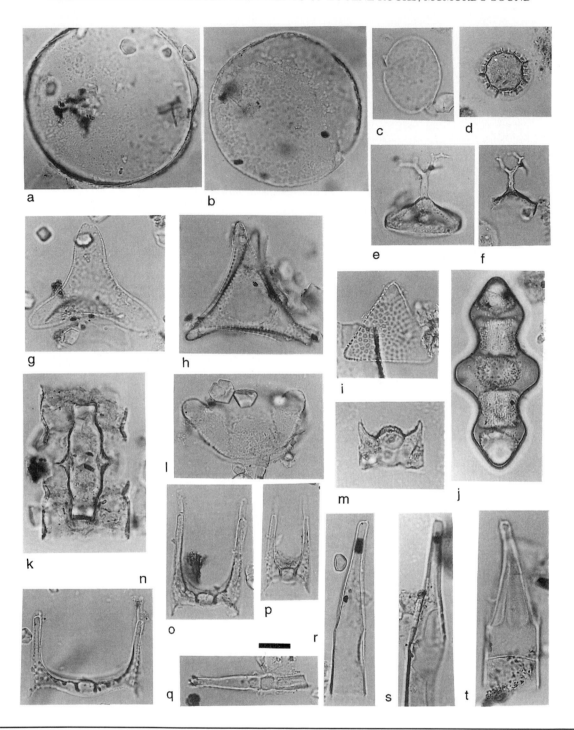

Plate 3

Diatoms from Sample E-345. Scale bar = 10mm. Figs. a, b. Genus et species indet. B; Fig. c. *Dicladia* sp. A; Fig. d. *Poretzkia*? sp. Hajós; Figs, e, f. *Diclada* sp. B; Fig. g. *Triceratium columbi*; Fig. h. *Trinacria lingulata*; Fig. i. *Triceratium americanum*; Fig. j. *Briggeria siberica*; Fig. k. *Triceatium castelliferum*; Fig. l. *Biddulphia*? sp. D; Fig. m. *Hemiaulus hostilis* var. *polaris*; Figs. n-q. *Hemiaulus stilwelli* sp. nov. Holotype, fig. n; Paratype, fig. o; both CAS #219084; Figs. r, s. *Pterotheca minor*; Fig. t. *Pterotheca carinifera*.

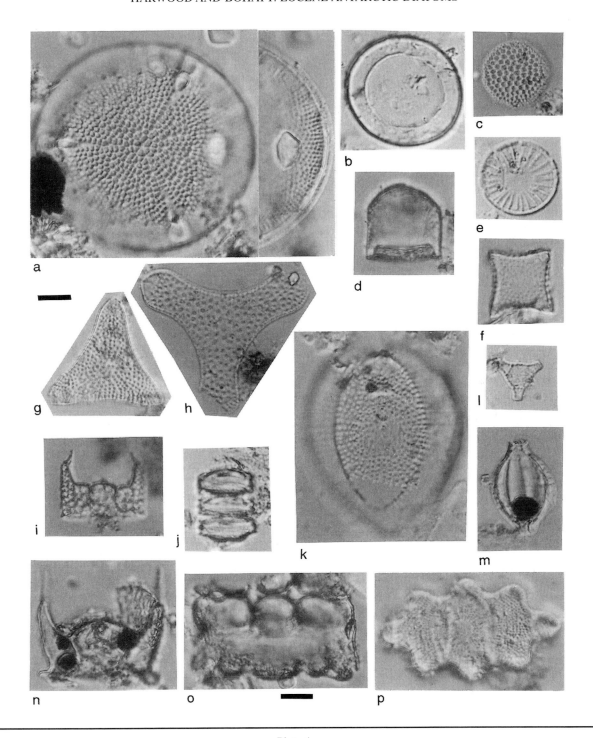

Plate 4

Diatoms from Sample MTD-95. Scale bar = 10mm. Fig. a. *Aulacodiscus rattrayii*, specimen at different focus; Fig. b. *Hercotheca* sp. sensu Kanaya; Fig. c. *Coscinodiscus radiatus*; Fig. d. *Pseudopyxilla dubia*; Fig. e. *Paralia* sp.; Fig. f. *Trinacria cornuta*; Fig. g. *Sheshukovia* sp. A; Fig. h. *Triceratium* cf. *T. russlandicum*; Fig. i. *Hemiaulus polymorphus* var. *morsiana;* Fig. j. *Trochospira spinosa*; Fig. k. *Sphynctolethus* sp. A; Fig. l. *Triceratium inconspicuum* var. *trilobata*; Fig. m. Gn. et sp. indet. (e) Schrader and Fenner; Fig. n. *Hemiaulus hostilis* var. *polaris*; Fig. o. *Biddulphia?* sp. A; Fig. p. *Biddulphia?* sp. E.

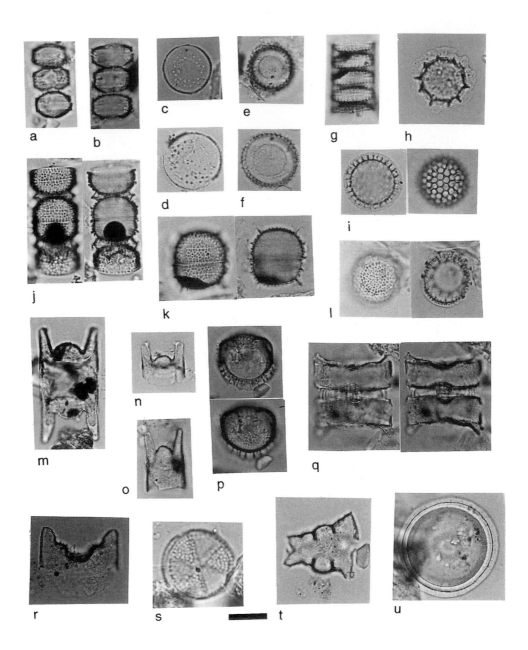

Plate 5

Diatoms from Sample D-1. Scale bar = 10mm. Figs. a-d. *Trochosira* sp. A; Figs. e-g. *Paralia sulcata*; Fig. h. *Poretzia*? sp. Hajos; Fig. i. *Stephanopyxis* sp. C, specimen at different focus; Figs. j-l. *Stephanopyxis*? sp. D, specimens at different focus; Figs. m-o. *Hemiaulus* sp. A; Fig. p. Genus et species indet. C, specimen at different focus; Fig. q. *Spinivinculum* sp. A, specimen at different focus; Fig. r. *Biddulphia tenera*; Fig. s. *Actinoptychus senarius;* Fig. t. *Hemiaulus dissimilis*; Fig. u. *Pseudopodosira* sp.

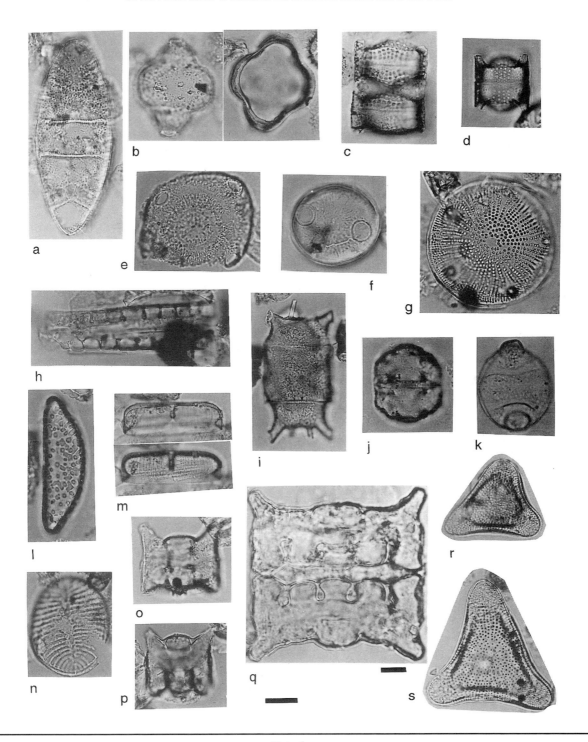

Plate 6

Diatoms from Sample D-1. Scale bar = 10mm. Fig. a. *Triceratium lineatum* Greville var.; Fig. b. *Sphynctolethus* sp. B, photomicrographs at different focus; Figs. c, d. *Trinacria* sp. A; Fig. e. *Glyphodiscus* sp. A; Fig. f. *Auliscus* sp. A; Fig. g. *Aulacodiscus* cf. *huttonii*; Fig. h. *Helminthopsidella ortha*; Fig. i. *Odontella* sp. A; Fig. j. *Goniothecium odontella*; Fig. k. *Biddulphia*? sp. D; Fig. l. *Leudugeria janischii*; Fig. m. Genus et species indet. D, specimen at different focus; Fig. n. *Cocconeis costata*; Figs. o-q. *Biddulphia* sp. F; Figs. r, s. *Sheshukovia* sp. B.

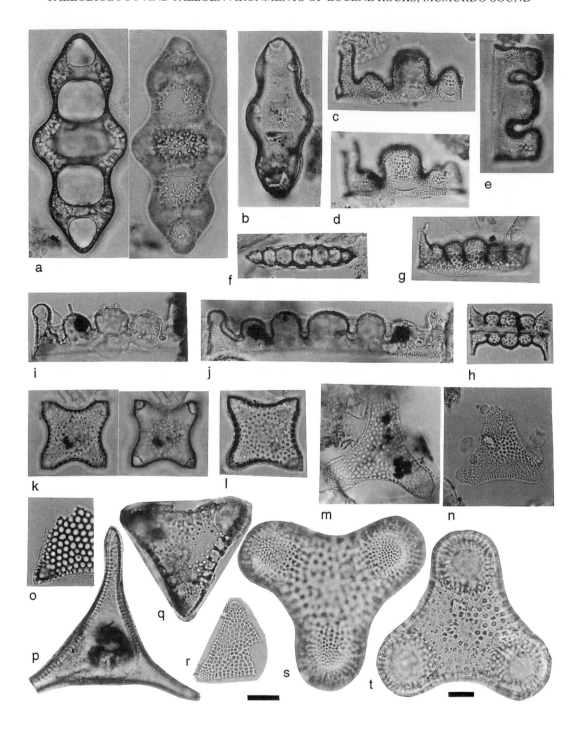

Plate 7

Diatoms from Sample D-1. Scale bar = 10mm. Fig. a. *Briggeria siberica*; Fig. b. *Briggeria* sp. , specimen at different focus; Figs. c, d. *Biddulphia* sp. G; Fig. e. *Bidulphia* sp. H; Figs. f-h. *Hemiaulus polymorphus* var. *morsianus*; Figs. i, j. *Biddulphia rigida*; Figs. k, l. *Trinacria cornuta*, specimen at different focus; Fig. m. *Triceratium* sp. cf. *T. russlandicum*; Fig. n. *Sheshukovia* sp. A; Fig. o. *Triceratium unguiculatum*; Fig. p. *Trinacria fragilis*; Fig. q. *Trinacria acutangulum*; Fig. r. *Triceratium americanum*; Figs. s, t. *Triceratim castellatum* var. *fractum*, same specimen at different focus & rotated 45°.

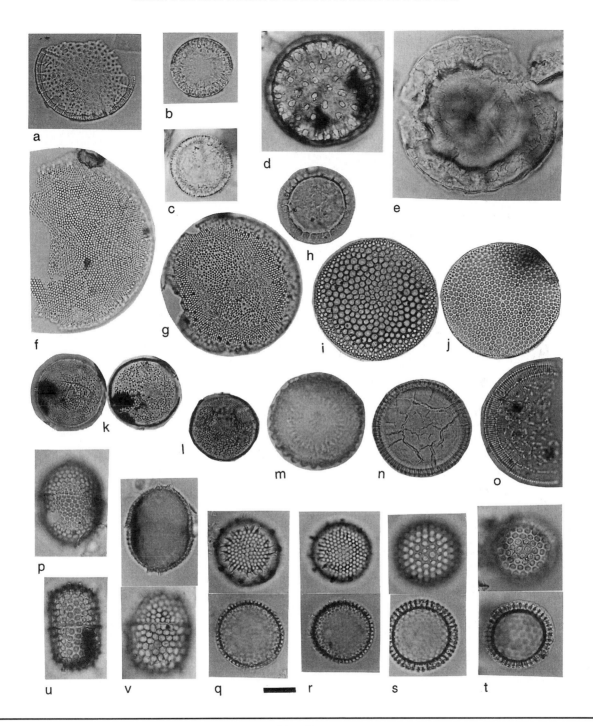

Plate 8

Diatoms from Sample D-1. Scale bar = 10mm. Fig. a. *Stictodiscus hardmanianus*; Figs. b, c. *Distephanosira architecturalis?*; Fig. d. *Stictodiscus californicus* var. *nitida*; Fig. e. *Hercotheca* sp. A.; Figs. f, g. *Stellarima* sp.; Fig. h. Genus et species indet. E; Figs. i, j. *Coscinodiscus radiatus*; Figs. k, l. Genus et species indet. A, specimen at different focus; Fig. m. *Paralia sulcata* var. *crenulata*; Figs. n. *Paralia sulcata*; Fig. o. *Hyalodiscus rossii*; Fig. p. *Stephanopyxis turris* var. A, specimen at different focus; Figs. q, r. *Stephanopyxis* sp. E, specimens at two different focal planes; Figs. s, t. *Stephanopyxis* sp. F; specimens at two different focal planes; Figs. u, v. *Stephanopyxis turris*.

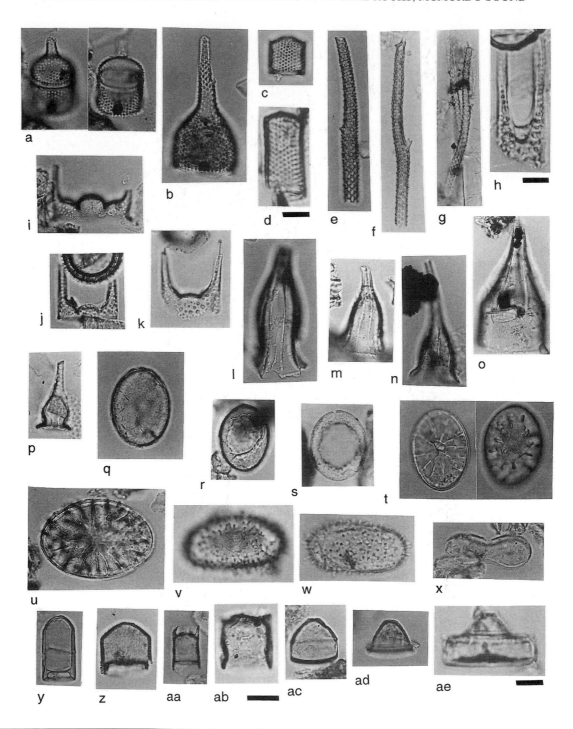

Plate 9

Diatoms from Sample D-1. Scale bar = 10mm. Fig. a. *Pyrgupyxis eocena*; Fig. b. *Pyxilla reticulata* (*johnsonianus* form); Figs. c-f. *Pyxilla reticulata*; Fig. g. *Pyxilla* sp. A; Fig. h. *Hemiaulus altus*; Figs. i, j. *Hemiaulus danicus*; Fig. k. *Hemiaulus* sp.; Figs. l-n. *Pterotheca* sp. A; Fig. o. *Pterotheca carinifera*; Fig. p. *Pterotheca aculeifera*; Figs. q-s. *Dicladia* sp. B; Figs. t, u. *Liradiscus ovalis*; (t) specimen at two focal planes; Figs. v, w. *Xanthiopyxis oblonga*; Fig. x. *Xanthiopyxis panduraeformis*; Figs. y-ab. *Rhizosolenia dubia*; Figs. ac-ae. *Dicladia* sp. C.

Plate 10

Thin-section photomicrograph of Sample MB-181. Pyritized diatoms, silicoflagellates, radiolarians and ebridians are abundant in mudstone clasts. Diatoms of the genera *Hemiaulus*, *Stephanopyxis*, and *Paralia* are visible here, but poor preservation prevents species identification. The flora is similar to that of the Middle to Upper Eocene.

Erratic: D-1
Lithology: Mudstone (Mwb)
Discussion of diatom flora: This erratic contains at least 76 diatom taxa (Table 3). The occurrence of diatoms *Actinocyclus octonarius* var. *tenellus*, *Distephanosira architecturalis*, *Hemiaulus danicus*, *Pyrgupyxis eocenica*, and *Trinacria cornuta*, which are not present in the other erratics, can be used to characterize this sample and distinguish it from the other erratics. The absence of diatoms *Hemiaulus caractericus* and *H. stilwellii*, suggest a stratigraphic level younger than the above two samples.

Age: Middle to early late Eocene based on the presence of *Trinacria cornuta* and many elements common to the Upper Eocene Oamaru Diatomite of New Zealand [Desikachary and Sreelatha, 1989; Edwards, 1991].

Erratic: MTD-95
Lithology: Mudstone (Mmb)

Discussion of diatom flora: This erratic contains at least 16 diatom taxa (Table 3). The occurrence of diatom *Triceratium inconspicuum* var. *trilobata* and *Trochosira spinosa*, which are not present in the other erratics, can be used to characterize this sample and distinguish it from the other erratics.

Age: Middle Eocene based on the presence of *Triceratium inconspicuum* var. *trilobata*, which goes extinct near the end of the middle Eocene [Fenner, 1985].

Erratic: MTD-181
Lithology: Sandstone (Ssg)/ conglomerate (Csgc)
Discussion of diatom flora: This erratic contains 4 identified diatom taxa (Table 3). The diatoms occur in abundance within sedimentary clasts in a conglomerate and are replaced by pyrite. Many taxa are identified only to genus level.

Age: Middle to late Eocene, equivalent to diatom

assemblages from the other erratics, but preservation limits detailed comparison.

Erratic: E-346
Lithology: Diamictite (Dm)
Discussion of diatom flora: This erratic contains at least 11 diatom taxa (Table 2). Although no zonal taxa are present, the association of *Liradiscus* sp. and Fragilariopsis (*Nitzschia*) sp. A of Harwood et al. [1989b] is distinctive.
Age: Early to middle Miocene based on the above mentioned taxa.

Erratic: E-347
Lithology: Diamictite (Dm)
Discussion of diatom flora: This erratic contains at least 17 diatom taxa (Table 2). The assemblage is similar to floras from the MSSTS-1 drillcore and upper part of the CIROS-1 drillcore [Harwood, 1986, 1989], in containing a mixture of planktic and benthic taxa and an abundance of *Stephanopyxis* and *Paralia*.
Age: Late Oligocene to early Miocene.

Erratic: E-351
Lithology: Diamictite (Dm)
Discussion of diatom flora: This erratic contains at least 17 diatom taxa (Table 2). The presence of *Nitzschia maleinterpretaria* and *Eucampia antarctica* characterize this assemblage.
Age: Middle Miocene based on the overlapping range of the above species.

Erratic: MTD-46
Lithology: Mudstone (Mmb)
Discussion of diatom flora: This erratic contains at least 28 diatom taxa (Table 2). The presence of *Nitzschia grossepunctata*, *Nitzschia maleinterpretaria*, and *Denticulopsis maccollumii* characterize this assemblage.
Age: Middle Miocene (14.1 to 15.1 Ma), based on Southern Ocean ranges of the above taxa [Harwood and Maruyama, 1992].

Erratic: MB-244C
Lithology: Mudstone (Mm-d)
Discussion of diatom flora: This erratic contains diatom taxa from the upper Miocene, yet the assemblage composition resembles Pleistocene sea-ice-associated floras, with *Eucampia antarctica*, *Corethron* sp., *Porosira* sp. *Stellarima microtrias*, among others. The assemblage contains *Actinocyclus karstenii*, common *Trinacria* sp. and *Denticulopsis simonsenii*, *D. lauta* (?),

Fragilariopsis sp. A, and '*Tigeria*' sp. (the latter 2 taxa are known from the CRP-1 drillcore and RISP cores [Harwood et al., 1989; 1998]. The assemblage lacks upper Miocene diatoms *Thalassiosira torokina* and *T. oliverana* var. *sparsa*, placing it stratigraphically lower than the *T. torokina* Zone of Harwood and Maruyama [1992] and below the basal sediments of the DVDP-11 drillcore [Winter and Harwood, 1997]. The age of this assemblage is likely ~9 to 11Ma. Diatom assemblages of this age are poorly-known and several new taxa are present.
Age: Early-late Miocene

COMMENTS

The presence of this middle-upper Miocene assemblage in the suite of erratics from McMurdo Sound may help provide information on the minimum age for glacial erosion event that transported the erratics into McMurdo Sound [see Wilson, this volume].

The occurrence of numerous extant sea-ice associated taxa in an upper Miocene assemblage of Erratic MB-244c is significant. Recognition of the first development and presence of sea-ice through the Oligocene to Pliocene is an important paleoenvironmental issue yet to be resolved. Sea-ice forms in waters that are near −1°C or lower. The occurrence of this assemblage in Erratic MB-224c indicates that this community was present in Antarctic waters by at least the late Miocene. However, it was not a permanent feature of the Antarctic from the late Miocene to the present day, as members of this assemblage are significantly reduced to nearly absent numerous times during the Pliocene to early Pleistocene [Winter and Harwood, 1997; Bohaty et al., 1998; Harwood et al., in press]. Additionally, the occurrence of *Fragilariopsis* (*Nitzschia*) sp. A of Harwood et al., [1989 and 1998] in this assemblage indicates this taxon may also have been associated with sea-ice environments in lower stratigraphic levels. This association of *F.* sp. A within an assemblage common in sea-ice environments supports the assumption [Harwood et al., 1989; Scherer et al., in press] that *Fragilariopsis* sp. A was ancestral to the modern sea-ice diatom *Fragilariopsis curta*, common today on the Antarctic shelf.

Stott et al. [1983] reported the presence of Pliocene diatoms in some samples of the erratics. After more thorough and careful sampling and examination it is now believed that this initial report was due to contamination of the coarse sandstone by contact with sea-water and marine sediments during some phase of glacial transport and exposure on the sea-floor. The occurrence of upper

Miocene diatoms in Sample MB-244C, however, indicates that some young rocks are present within the suite of erratics.

Scherer [1991] reports several unknown marine diatoms of possible Paleogene age from beneath the West Antarctic Ice Sheet (WAIS) at Ice Stream B. These unidentified taxa are similar to those in assemblages in the present report, but probably do not represent identical taxa. Future recovery of diatom assemblages in stratigraphic succession from the Antarctic shelf may enable recognition and dating of Paleogene strata beneath the WAIS.

CONCLUSIONS

The distinct diatom floras present within each erratic suggests that the Eocene strata represented by the suite of six erratics (Table 3) represents a considerable amount of time within the middle to late Eocene. Additional information obtained from drilling stratigraphic sections on the Antarctic continental shelf is needed before the full resolving power of diatom biostratigraphy can be applied to the McMurdo Erratics. This descriptive paper will serve to guide future biostratigraphic and taxonomic studies as the Eocene section in Antarctica is recovered through future drilling. Many of the diatoms documented herein are treated informally, though most are illustrated. The McMurdo Erratics provide a unique window on Antarctic paleoenvironment and paleoclimate of the Eocene.

Younger, diatom-bearing marine erratics are of diamictite, whereas the Eocene floras do not occur in diamictite facies. This change in lithology corresponding to age may reflect the shift from a warmer regime with limited glacial influence (Eocene) to an environment where ice at sea-level was a prominent feature (Oligocene and Miocene) [Levy and Harwood, this volume a]. Although the McMurdo Erratics are not within their stratigraphic context, the association of diverse microfossil groups such as diatoms, dinoflagellate cysts, ebridians and silico-flagellates enable age assignment and provide information about stratigraphic sequences hidden beneath the shroud of Antarctic ice.

TAXONOMIC LIST

The following list is of diatoms encountered in the examination of the Eocene erratics (Table 3). Diatoms encountered in the Oligocene and Miocene erratics (Table 2) are not treated below, but reference, synonymy and illustration of these diatoms can be found [Harwood, 1986; 1989; Harwood et al., 1989b; 1998; and Scherer et al., in press].

Actinocyclus octonarius **var.** *tenellus* (Brébisson) Hendey; Hustedt, 1930, p. 530-533, fig. 302.

Actinoptychus senarius Ehrenberg. (Pl. 5, Fig. s).

Anaulus **sp. A**. (Plate 2, Figs. l-n).

Arachnoidiscus **spp.**

Aulacodiscus **cf.** *huttonii* Grove & Sturt; Edwards, 1991, Pl. 6, fig. 68; Desikachary and Sreelatha, 1989, p. 67, pl. 25, fig. 1, 2. (Plate 6, Fig. g).

Aulacodiscus rattrayii Grove & Sturt; Desikachary and Sreelatha, 1989, p. 73, pl. 29, figs. 3-5; Edwards, 1991, pl. 7, figs. 83-86. (Plate 4, Fig. a).

Auliscus **sp. A**. (Plate 6, Fig. f).

Biddulphia elegantula Greville. (Plate 1, Fig. i).

Biddulphia rigida Schmidt; Desikachary and Sreelatha, 1989, p. 103, pl. 43, figs. 10-11. Compare with *Biddulphia fistulosa* Pantocsek and *Biddulphia tuomeyi* (Bailey) Roper. (Plate 7, Figs. i, j).

Biddulphia tenera Grove & Sturt; Desikachary and Sreelatha, 1989, p. 104, pl. 42, fig. 10. (Plate 5, Fig. r)

Biddulphia? **sp. A**. (Plate 1, Fig. e; Plate 4, Fig. o).

Biddulphia? **sp. B**. (Plate 1, Fig. f).

Biddulphia? **sp. C**. (Plate 1, Fig. j).

Biddulphia? **sp. D**. (Plate 3, Fig. l; Plate 6, Fig. k).

Biddulphia? **sp. E**. (Plate 4, Fig. p).

Biddulphia **sp. F**. (Plate 6, Figs. o-q).

Biddulphia **sp. G**. (Plate 7, Figs. c, g).

Biddulphia **sp. H**. (Plate 7, Fig. e).

Briggeria siberica (Grunow) Ross & Sims; Homann, 1991, p. 74, pl. 8, figs. 1-11. (Plate 3, Fig. j; Plate 7, Fig. a).

Briggeria **sp.** (Plate 7, Fig. b).

Chaetoceros didymus Ehrenberg; Desikachary and Sreelatha, 1989, p. 112, pl. 45, figs. 8, 9. *Chaetoceros* sp. Hajós, 1976, p. 828, pl. 9, fig. 5. (Plate 2, Figs. j, k)

Chaetoceros **spp.**

Cocconeis costata Gregory; Harwood, 1989, p. 78. (Plate 6, Fig. n).

Cocconeis **spp.**

Coscinodiscus radiatus Ehrenberg sensu Grunow; Desikachary and Sreelatha, 1989, p. 132-133, pl. 55, fig. 8; pl. 58, fig. 7; Homann, 1991, p. 45, pl. 16, figs. 1-3. *Coscinodiscus marginatus* Gombos, 1983, pl. 4, fig. 14. (Plate 2, Fig. c; Plate 4, Fig. c; Plate 8, Figs. i, j).

Craspedodiscus molleri Schmidt, em. Homann, 1991, p. 47, pl. 17, figs. 1-5; Gombos, 1983, p. 569, pl. 3, figs. 2-4. (Plate 2, Figs. a, b).

Dicladia **sp. A**. (Plate 3, Fig. c; Plate 9, Figs. q-s).

Dicladia **sp. B**. *Pterotheca* sp. 3 Homann, 1991, pl. 53, figs. 28, 28. (Pl. 3 Figs. e, f).

Dicladia sp. C. (Plate 9, Figs. ac-ae).

Dicladia sp. 1 of Kanaya, 1957, p. 119, pl. 8, figs. 18, 19; Dzinoridze, et al., 1978, pl. 9, fig. 16. *Xanthiopyxis* cf. *acrolopha* Forti, illustrated in Hajós, 1976, pl. 17, fig. 10. Gen. et sp. indet. #6 of Schrader and Fenner, 1976, pl. 45, figs. 5, 11, 14; Fenner, 1978, pl. 36, figs. 14-16. (Plate 1, Figs. g, h).

Distephanosira architecturalis? (Brun) Gleser. *Melosira architecturalis* Brun; Gombos and Ciesielski, 1983, p. 602. Specimens are too poorly preserved to identify with confidence. (Plate 8, Figs. b, c.).

Drepanotheca bivittata (Grunow & Pantocsek in Pantocsek) Schrader; Desikachary and Sreelatha, 1989, p. 147, pl. 62, figs. 9-12, 14, 15.

Endictya sp.

Eurossia irregularis var. *incurvatus* Sims in Mahood et al., 1993, p. 256, figs. 37-42, 66-67. *Triceratium macroporum* Hajos sensu Gombos and Ciesielski, 1983, p. 605, pl. 17, fig. 6. *Triceratium polymorphum* Harwood and Maruyama, 1992, pl. 1, fig. 3. (Plate 1, Fig. d).

Glyphodiscus sp. A. (Plate 6, Fig. e).

Goniothecium odontella Ehrenberg; Harwood, 1989, p. 79, pl. 4, fig. 24. (Plate 6, Fig. j).

Helminthopsidella ortha (Schrader) Silva; Desikachary and Sreelatha, 1989, p. 159, pl. 69, fig. 8, 9. (Plate 6, Fig. h).

Hemiaulus altus Hajós in Hajós and Stradner, 1975, p. 931, pl. 5, figs. 17-19. (Plate 9, Fig. h).

Hemiaulus caracteristicus Hajos; Mahood et al., 1993, p. 252-254, figs. 21-22, 25-30, 64. (Plate 1, Fig. m).

Hemiaulus danicus Grunow; Homann, 1991, p. 81-82, pl. 20, figs. 1-10. (Plate 9, Figs. i, j).

Hemiaulus dissimilis Grove & Sturt; Harwood, 1989, p. 79, pl., 4, figs, 3-5, 9; pl. 5, fig. 35. (Plate 5, Fig. t).

Hemiaulus hostilis var. *polaris* Grunow; Krotov and Schibkova, 1959, pl. 4, figs. 4, 5. (Plate 3, Fig. m; Plate 4, Fig. n).

Hemiaulus polycistinorum Ehrenberg; Fenner, 1978, p. 521, pl. 21, figs. 13, 14; pl. 22, figs., 4, 5, 7-10; pl., 23, figs.1-4.

Hemiaulus polymorphus var. *morsiana* Grunow; Homann, 1991, p. 92, pl. 24, figs. 10-14, 19. (Plate 4, fig. i; Plate 7, Figs. f-h).

Hemiaulus stilwelli species nov. Harwood & Bohaty

Description: Valves bipolar, length 10 to 40 mm, with tall, parallel elevations that bear long spines; narrow valve face, hyaline, with transverse costae, at least two, prominent in the central region and weakly developed toward the elevations; areolae rare to absent on the valve face, best developed on base of the elevations, and often enlarged through dissolution.

Discussion: This diatom resembles *Hemiaulus caracteristicus* Hajós (1976) and *Hemiaulus peripterus* Fenner (see Fourtanier, 1991) by (1) the possession of elongate, parallel elevations, (2) a narrow silicified 'bar' that links the two elevations, (3) a vertical costa on the distal side of the elevations that runs down toward the valve margin. These features produce a "H" structure upon dissolution of the weakly-silicified valve wall. The elevations and the central bar are usually preserved, whereas the porous valve face and mantle are not preserved. Mahood et al. (1993) illustrate well-preserved specimens of *Hemiaulus caracteristicus* with a weakly silicified valve of poroid areolae. The specimens illustrated here (Plate 3, figures n, o) show an irregular, dissolved lower margin and siliceous costae on the distal side of the elevations, though not as well developed as in these other taxa. *Hemiaulus. stilwelli* differs from these other taxa by the presence of undulations on the siliceous bar and the presence of distinct transapical costae at the valve center. Small specimens bear gross resemblance to *Hemiaulus polymorphus* var. *frigida* Grunow, yet the entire valve margin is usually visible in this taxon, whereas in *H. stilwelli* the valve ends at the contact between the 'bar' an the thin silicified valve wall. This taxon is named for Jeff Stilwell, co-editor of this volume.

Holotype: Plate 3, Figure n; deposited at the California Academy of Sciences (#219084).

Paratype: Plate 3, Figure o; deposited at the California Academy of Sciences (#219084).

Type locality: Sample E-345, from an Eocene glacial Erratic in McMurdo Sound, Antarctica (Plate 3, Figs n-q).

Hemiaulus sp. A. (Plate 5, Figs. m-o).

Hemiaulus sp. B. (Plate 9, Fig. k).

Hemiaulus spp.

Hercotheca sp. A. Compare with Eocene specimen illustrated in DSDP Site 281, Core 14 CC, Site Report, p. 288, pl. 2, fig. 3. Initial Report of DSDP Leg 29. (Plate 8, Fig. e).

Hercotheca sp. sensu Kanaya, 1957, p.118, pl. 8., figs. 15-17; Dzinoridze et al., 1978, pl. 9, fig. 18. *Melosira truncata* Grove in Schmidt; in Desikachary and Sreelatha, 1989, pl. 79, fig. 3. (Plate 2, Fig. h, i; Plate 4, Fig. b).

Hyalodiscus radiatus var. *radiatus*

Hyalodiscus rossii Desikachary & Sreelatha, 1989, p. 167, pl. 81, figs. 3-5. (Plate 8, Fig. o).

Isthmia spp.

Leudugeria janischii (Grunow in Van Heurck) Van Heurck; Desikachary and Sreelatha, 1989, p. 171, pl. 75, figs. 11-13; Edwards, 1991, pl. 12, fig. 151. (Plate 6. Fig. l).

Liradiscus ovalis Greville; Hajós, 1976, p. 826, pl. 17, figs. 1, 2. (Plate 9, Figs. t, u).

Odontella sp. A. (Plate 6, Fig. i).

Paralia sulcata (Ehrenberg) Cleve; Homann, 1991, p. 51-52. (Plate 5, Figs. e-g; Plate 8, Fig. n).

Paralia sulcata var. *crenulata* Grunow; Homann, p. 53, pl. 31, fig. 8-10. (Plate 8, Fig. m).

Poretzkia? sp. of Hajós, 1976, p. 826, pl. 17, fig. 3. (Pl. 3, Fig. d; Plate 5, Fig. h).

Proboscia interposita (Hajós) Jordan & Priddle, 1991, p. 57. *Rhizosolenia interposita* Hajós, 1976, p. 827, pl. 21, fig. 8. (Plate 2, Fig. o).

Pseudopodosira sp. (Plate 5, Fig. u).

Pseudopyxilla stylifera (Brun) Harwood & Bohaty comb. nov. *Skeletonema stylifera* Brun, 1891, p. 44, pl. 21, fig. 7; Sims, 1994, p. 405, figs. 41-43, 54; Fenner, 1978, p. 531. *Ceratulina praebergoni* Hajós, 1976, p. 828, pl. 14, fig. 13; pl. 15, figs. 5-7; text figure 5. Unknown Form 2 of Homann, 1991, pl. 56, fig. 10, 11. The SEM illustrations in Sims (1994) indicate that this diatom resting spore is unrelated to *Skeletonema* and *Skeletonemopsis*. It is transferred here to the genus *Pseudopyxilla*. (Plate 2, Fig. g).

Pseudorutilaria levyi species nov. Harwood & Bohaty

Description: Valve bipolar, straight, gradually tapering to the apices; margin serrate with concave portion of the margin corresponding to the position of a transverse costa; short linking spine at each apex, for connection with adjacent valve in colony (not observed); central area with curved row of short linking spines or processes; valve surface covered by fine areolae.

Discussion: This diatom is related to the upper Eocene diatom *Pseudorutilaria monile* Grove & Sturt common in the Oamaru deposits. *Pseudorutilaria monile* has a central 'chamber' that is distinctly larger than all of those toward the apices, which are of more or less even size. In *P. levyi*, these "chambers" show a gradual decrease in size toward the apices. This taxon is named for Richard Levy who contributed much effort to the study of the McMurdo Erratics.

Holotype: Plate 2, Figure p; deposited at the California Academy of Sciences (#219084).

Type locality: Sample E-345, from an Eocene glacial Erratic in McMurdo Sound, Antarctica. (Plate 2, Fig. p).

Pterotheca aculeifera (Grunow) Grunow, em. Homann, 1991, p. 135, pl. 35, figs. 15-18. (Plate 1, Fig. l; Plate 9 fig. p).

Pterotheca carinifera Grunow; Harwood, 1988, p. 86, fig. 18.6. (Plate 3, Fig. t; Plate 9, fig. o).

Pterotheca danica (Grunow) Forti; Harwood, 1988, p. 86.

Pterothea minor Harwood, 1998, p. 86, fig. 12.12, 12.13. (Plate 3, Figs. r, s).

Pterotheca sp. A. (Plate 9, Figs. l-n).

Pyrgupyxis eocena Hendey; Hajós, 1976, p. 829, pl. 24, figs. 3-5, 8, 9; Gombos and Ciesielski, 1983, p. 603, pl. 12, figs. 6, 7. (Plate 9, Fig. a).

Pyxilla reticulata Grove & Sturt; Harwood, 1989, p. 80, pl. 3, figs. 7-10. (Plate, 1, Fig. n; Plate 9, Fig. c).

Pyxilla sp. A. This species is more weakly-silicified than other species of *Pyxilla*. It bears long and thin elevations, which are of uniform diameter up to the 'barb", where the elevation curves gently. Only fragments of this diatom were encountered. The figured specimens resemble *Pyrgupyxis* aff. *gracilis* (Tempere and Forti) Hendey, illustrated in Schrader and Fenner, 1976, pl. 43, fig. 23. (Plate 1, Figs. o, p).

Rhizosolenia dubia (Grunow) Homann, 1991, p. 69, pl. 35, figs. 1-8, 11-13. *Pseudopyxilla dubia* Grunow in Van Heurck; Harwood, 1998, p. 85, figs. 17.23, 17.24. (Plate 9, Figs. y-ab)

Sceptroneis lingulatus Fenner; Harwood, 1989, p. 80, pl. 6, fig. 11.

Sheshukovia sp. A. (Plate 4, fig. g; Plate 7, Fig. n).

Sheshukovia sp. B. Compare with Genus and species uncertain #3 of Gombos and Ciesielski, 1983, pl. 25, figs. 8, 9. (Plate 6, Figs. r, s).

Sphynctolethus cf. *pacificus* (Hajós) Sims, 1986, p. 250-252, figs. 29-34; Harwood, 1989, p. 80.

Sphynctolethus sp. A. (Plate 4, Fig. k).

Sphynctolethus sp. B. (Plate 6, Fig. b).

Spinivinculum sp. A. (Plate 5, Fig. q).

Stellarima sp. (Plate 8. Figs. f, g).

Stephanopyxis grunowii Grove & Sturt; Harwood, 1989, p. 81, pl. 2, figs, 5, 6. (Plate 2, Fig. d).

Stephanopyxis megapora Grunow; Hajós, 1976, p. 825, pl. 3, figs. 1, 2.

Stephanopyxis cf. *oamaruensis* Hajós, 1976, p. 825, pl. 19, figs. 5-8; Harwood, 1989, p. 81, pl. 2, figs. 27-29.

Stephanopyxis subantarctica Hajós, 1976, p. 825, pl. 5, figs. 6-8.

Stephanopyxis superba (Greville) Grunow; Harwood, 1989, p. 81, pl. 2, figs 14-20. (Plate 1, Fig. b).

Stephanopyxis turris (Greville & Arnott) Ralfs. (Plate 2, Fig. e; Plate 8, Figs. u, v).

Stephanopyxis turris var. A. Specimens with hollow processes at valve center. (Plate 8, Fig. p).

Stephanopyxis sp. A. Compare with *Stephanopyxis* sp. 2 of Homann, 1991, pl. 39, figs. 1-5. (Plate 1, Fig. a).

Stephanopyxis? sp. B. This small diatom is of uncertain placement. It bears some resemblance to specimens illustrated in Harwood, 1989, pl. 1, figs. 17, 18, as *Thalassiosira*? sp. A from the upper Eocene of CIROS-1. (Plate 2, Fig. f).

***Stephanopyxis* sp. C**. (Plate 5, Fig. i).

***Stephanopyxis*? sp. D**. (Plate 5, Fig. j-l).

***Stephanopyxis* sp. E**. (Plate 8, Figs. q, r).

***Stephanopyxis* sp. F**. (Plate 8, Figs. s, t).

***Stictodiscus californicus* var.** *nitidus* Grove & Sturt; Desikachary and Sreelatha, 1989, p. 234, pl. 10, figs. 1-5, pl. 108, fig. 1; Edwards, 1991, pl. 9, fig. 112. (Plate 8, fig. d).

Stictodiscus hardmanianus Greville; Harwood, 1989, p. 81, pl. 1, fig. 6.

Triceratium americanum Ralfs in Pritchard; Desikachary and Sreelatha, 1989, pl. 15, figs. 5, 6, 8.

Discussion: This diatom should likely be transferred to *Shesukovia* upon SEM examination. (Plate 3, Fig. i; Plate 7, Fig. r).

***Triceratium castellatum* var.** *fractum* (Walker & Chase) Grunow in Schmidt; Desikachary and Sreelatha, 1989, p. 250, pl. 118, figs. 1-5; pl. 122, figs. 1, 2, 5; Edwards, 1991, pl. 15, fig. 195-196. *Triceratium castellatum* West var., in Hajós, 1976, p. 828, pl. 12, figs. 4, 5. (Plate 7, Figs. s, t).

Triceratium castelliferum Grunow in Schmidt; Desikachary and Sreelatha, 1989, p. 250-251, pl. 115, figs. 9, 10; pl. 117, figs. 1-4, 6, 7. (Plate 3, Fig. k)

***Triceratium columbi*?** Witt; Desikachary and Sreelatha, 1989, p. 251, pl. 117, fig. 8. (Plate 3, Fig. g).

***Triceratium inconspicuum* var.** *trilobata* Fenner, 1978, p. 534, pl. 30, figs. 23-26. (Plate 4, Fig. l).

***Triceratium lineatum* Greville var.** in Grove and Sturt, 1886, pl. 2, fig. 2. (Plate 6, fig. a).

Triceratium* sp. cf. *T. russlandicum Forti, in Gombos, 1983, p. 571, pl. 1, fig. 12; pl. 2, fig. 9. (Plate, 4, Fig. h; Plate 7, Fig. m).

Triceratium unguiculatum Greville; Desikachary and Sreelatha, 1989, p. 269; Gombos, 1977, p. 598-599, pl. 33, figs. 1, 3, pl. 34, figs. 1-6; Gombos and Ciesielski, 1983, p. 605, pl. 14, figs. 9-12; pl. 16, figs. 1-4; Sims and Ross, 1990. (Plate 7, Fig. o)

Trigonium arcticum (Brightwell) Cleve.

Trinacria acutangulum (Strelnikova) Harwood, 1988, p. 89, figs. 21.8-21.10, 21.12. (Plate 7, Figs. q).

Trinacria cornuta (Greville) Sims and Ross, 1988, p. 279-282, pl.1; pl. 12, figs. 78, 79; *Trinacria excavata* forma *tetragona* Schmidt; Fenner, 1985, p. 741, figs. 8.29, 8.30. (Plate 4, Fig. f; Plate 7, Figs. k, l).

Trinacria excavata Heiberg.

Trinacria fragilis Grunow in Schmidt; Desikachary and

Sreelatha, 1989, p. 283, pl. 137, fig. 4. (Plate 7, Fig. p).

Trinacria lingulata (Greville) Grove & Sturt; Desikachary and Sreelatha, 1989, p. 283, pl. 137, figs. 3, 6. (Plate 3, Fig. h).

***Trinacria* sp. A**. (Plate 6, Figs. c, d).

***Trochosira spinosa*?** Kitton; Homann, 1991, p. 67, pl. 1, figs. 6-13. (Plate 4 Fig. j).

***Trochosira* sp. A**. (Plate 5, Figs. a-d).

Xanthiopyxis acrolopha Forti; Harwood, 1989, p. 82, pl. 3, fig. 34.

Xanthiopyxis diaphana Forti, Fenner, 1978, pl. 35, figs. 4, 5. (Plate 1, Fig. k).

Xanthiopyxis oblonga Ehrenberg; Hajós, 1976, p. 826, pl. 17, fig. 11; Homann, 1991, p. 143, pl. 57, figs. 5-7, 9-12. (Plate 9, Figs. v, w).

Xanthiopyxis panduraeformis Pantocsek; Hajós, 1976, p. 826, pl. 11, fig. 5, pl. 17, fig. 11. (Plate 9, Fig. x).

Genus et species indet. A. This diatom may belong within, or may be related to genus *Actinocyclus*. (Plate 1, Fig. c; Plate 8, Figs. k, l).

Genus et species indet. B. (Pl. 3, Figs. a, b).

Genus et species indet. C. (Plate 5, Fig. p).

Genus et species indet. D. (Plate 6, Fig. m).

Genus et species indet. E. Possibly a resting spore valve of *Stephanopyxis*. (Plate 8, Fig. h).

CHRYSOPHCEAN CYST

Archaeomonad gen. et sp. indet. (e) of Schrader and Fenner, 1976, pl. 25, fig. 39; Fenner, 1978, pl. 33, fig. 12. (Plate 4, Fig. m).

Acknowledgements. This paper benefited from reviews by R. Levy and J. Barron. We acknowledge the field assistance of J. Francis, R. Graham, X. Jiang, J. Kaser, M. Pole, A. Srivastava, J. Stilwell, D. Watkins, G. Wilson and D. Winter and photographic assistance of Stacie Czyszon. D. Winter was of great aid in final production of this manuscript. This research was supported by NSF grants OPP-9317901 and OPP-9158075 to D. Harwood and through generous donations by the Alumni of the Department of Geology, University of Nebraska-Lincoln.

REFERENCES

Askin, R.A.
(this vol.) Spores and Pollen from the McMurdo Sound Erratics, Antarctica, in *Paleobiology and Paleoenvironments of Eocene Rocks, McMurdo Sound, East Antarctica*, edited by J.D. Stilwell and R.M. Feldmann, Antarctic Research Series, American Geophysical Union.

Barrett, P.J. (ed.)
1989   Antarctic Cenozoic history from the CIROS-1 drillhole, McMurdo Sound. *DSIR Bulletin 245*: 254pp.

Barrett, P.J. and F.J. Davey, (eds.)
1992   Antarctic stratigraphic drilling, Cape Roberts Project Workshop Report. *The Royal Society of New Zealand Miscellaneous Series 23*, Wellington, New Zealand, 38 pp.

Barron, J.A., B. Larsen, and J.G. Baldauf
1991   Evidence for late Eocene to early Oligocene Antarctic glaciation and observations on late Neogene glacial history of Antarctica: results from Leg 119. In J.A. Barron, B. Larsen, *Proceedings of the Ocean Drilling Program, Scientific Results,* College Station, Texas, (Ocean Drilling Program), *119*: 869-891.

Bohaty, S.M., Scherer, R.P., and Harwood, D.M.
1998   Lower Pleistocene diatom biostratigraphy of the CRP-1 drillcore. *Scientific Results of the Cape Roberts Drilling Project, Terra Antarctica, 5* (3): 431-454.

Bohaty, S.M. and D.M. Harwood
(this vol.) Ebridian and silicoflagellate biostratigraphy from Eocene McMurdo Erratics and the Southern Ocean, in *Paleobiology and Paleoenvironments of Eocene Rocks, McMurdo Sound, East Antarctica,* edited by J.D. Stilwell and R.M. Feldmann, Antarctic Research Series, American Geophysical Union.

Brun, J.
1891   Diatomées espèces nouvelles marines, fossiles ou pélagique. *Mémoires de la Societé de Physique et d'Histoire Naturelle de Genève, 31* (2/1): 48pp.

Buckeridge, J., St. J.S.
(this vol.) A new species of *Austrobalanus* (Cirripedia, Thoracica) from Eocene erratics, Mount Discovery, McMurdo Sound, East Antarctica, in *Paleobiology and Paleoenvironments of Eocene Rocks, McMurdo Sound, East Antarctica,* edited by J.D. Stilwell and R.M. Feldmann, Antarctic Research Series, American Geophysical Union.

Cranwell, L.M.
1964   Hystrichospheres as an aid to Antarctic dating with special reference to the recovery of *Cordosphaeridium* in erratics at McMurdo Sound. *Grana Palynologica, 5*(3): 397-405.

Cranwell, L.M., H.J. Harrington and I.G. Speden
1960   Lower Tertiary microfossils from McMurdo Sound, Antarctica. *Nature, 186*(4726): 700-702.

Desikachary, T.V. and P.M. Sreelatha
1989   Oamaru Diatoms. *Bibliotheca Diatomologica,* Band 19, Cramer, Berlin, 475 pp.

Dzinoridze, R.N., A.P. Jousé, G.S. Koroleva-Golikova, G.E. Kozlova, G.S. Nagaeva, M.G. Petrushevskaya and N.I. Strelnikova
1978   Diatom and radiolarian Cenozoic Stratigraphy, Norwegian Basin; DSDP Leg 38. In M.Talwani, G. Udintsev, et al., Initial Reports of the Deep Sea Drilling Project, U.S. Government Printing Office, Washington, D.C., Supplement to Volumes *38-41*: 289-427.

Edwards, A.R. (compiler)
1991   The Oamaru Diatomite. *New Zealand Geological Survey Paleontological Bulletin, 64,* Lower Hutt, 260 pp.

Feldmann, R.M. and W.J. Zinsmeister
1984   First occurrence of fossil decapod crustaceans (Callianassidae) from the McMurdo Sound region, Antarctica. *Journal of Paleontology, 58*(4): 1041-1045.

Fenner, J.
1978   Cenozoic diatom biostratigraphy of the equatorial and southern Atlantic Ocean. In Supko, P. R., K. Perch-Nielsen, et al., Initial Reports of the Deep Sea Drilling Project, U.S. Government Printing Office, Washington, D.C., Supplement to Volumes *38-41*: 491-623.

Fenner, J.
1985   Late Cretaceous to Oligocene planktic diatoms. In H. M. Bolli, J. B. Saunders and K. Perch-Nielsen, (eds.), *Plankton Stratigraphy,* Cambridge (Cambridge Univ. Press), 713-762.

Fourtanier, E.
1991   Paleocene and Eocene diatom biostratigraphy and taxonomy of eastern Indian Ocean Site 752. In J. Weissel, E. Taylor, J. Alt, et al., *Proceedings of the Ocean Drilling Program, Scientific Results,* College Station, Texas, (Ocean Drilling Program), *121*: 171-187.

Francis, J.E.
(this vol.) Fossil wood from Eocene high latitude forests, McMurdo Sound, Antarctica, in *Paleobiology and Paleoenvironments of Eocene Rocks, McMurdo Sound, East Antarctica,* edited by J.D. Stilwell and R.M. Feldmann, Antarctic Research Series, American Geophysical Union.

Gombos, A. M., Jr.
1977   Paleogene and Neogene diatoms from the Falkland Plateau and Malvinas outer basin, Leg 36, Deep Sea Drilling Project. In P.F. Barker, I.W.D. Dalziel, et al., Initial Reports of the Deep Sea Drilling Project, U.S. Government Printing Office, Washington, D.C., *36*: 575-687.

Gombos, A. M., Jr.
1983   Middle Eocene diatoms from the south Atlantic. In W.J. Ludwig, V.A. Krasheninnikov, et al., Initial Reports of the Deep Sea Drilling Project, U.S. Government Printing Office, Washington,

D.C., *71*(part 1): 565-581.

Gombos, A. M., Jr. and P.F. Ciesielski
1983   Late Eocene to early Miocene diatoms from the southwest Atlantic. In W.J. Ludwig, V.A. Krasheninkov, et al., Initial Reports of the Deep Sea Drilling Project, U.S. Government Printing Office, Washington, D.C., *71*(part 1): 583-634.

Grove, E. and G. Sturt
1887   On a fossil marine diatomaceous deposit from Oamaru, Otago, New Zealand. *J. Quekett Miscrosc. Club,* Ser. II, *3*: 7-12; 63-78.

Hajós, M.
1976   Upper Eocene and lower Oligocene Diatomaceae, Archaeomonadaceae, and Silicoflagellatae in southwestern Pacific sediments, DSDP Leg 29. In C.D. Hollister, Craddock, C. et al. (eds.), Initial Reports of the Deep Sea Drilling Project, U.S. Government Printing Office, Washington, D.C., *35*: 817-883.

Hannah M.J., M.B. Cita, R. Coccioni and S. Monechi
1997   The Eocene/Oligocene boundary at 70° South, McMurdo Sound, Antarctica. *Terra Antartica,* *4*(2): 79-87.

Hara, U.
(this vol.) Bryozoan remains from Eocene glacial erratics of McMurdo Sound, East Antarctica, in *Paleobiology and Paleoenvironments of Eocene Rocks, McMurdo Sound, East Antarctica,* edited by J.D. Stilwell and R.M. Feldmann, Antarctic Research Series, American Geophysical Union.

Harrington, H.J.
1969   Fossiliferous rocks in moraines at Minna Bluff, McMurdo Sound. *Antarctic Journal of the United States,* *4*(4): 134-135.

Harwood, D.M.
1986   Diatoms. In: Antarctic Cenozoic history from the MSSTS-1 drillhole, McMurdo Sound. P.J. Barrett, editor. *Bulletin in the Misc. Series of the New Zealand D.S.I.R.,* No. *237,* p. 69-108.

Harwood, D.M.
1988   Upper Cretaceous and lower Paleocene diatom and silicoflagellate biostratigraphy from Seymour Island, eastern Antarctic Peninsula. In M.O. Woodburne, and R.M. Feldmann, (eds.), *The Geology and Paleontology of Seymour Island, Geological Society of America Memoir, 169*: 55-130.

Harwood, D.M.
1989   Siliceous microfossils. In P.J. Barrett (ed.), Antarctic Cenozoic History from the CIROS-1 Drillhole, McMurdo Sound, New Zealand, *DSIR Bulletin, 245*: 67-97.

Harwood, D.M., Barrett, P.J., Edwards, A.R., Rieck, H.J. and Webb, P.-N.
1989a   Biostratigraphy and chronology. In Barrett, P.J.

(ed.), Antarctic Cenozoic History from the CIROS-1 Drillhole, McMurdo Sound, New Zealand, *DSIR Bulletin, 245*: 231-239.

Harwood, D.M., Scherer, R.P. and Webb, P.-N.
1989b   Multiple Miocene marine productivity events in West Antarctica as recorded in upper Miocene sediments beneath the Ross Ice Shelf (Site J-9). *Marine Micropaleontology, 15*: 91-115.

Harwood, D.M. and Maruyama, T.
1992   Middle Eocene to Pleistocene diatom biostratigraphy of Southern Ocean sediments from the Kerguelen Plateau. In S.W. Wise, R. Schlich, et al., *Proceedings of the Ocean Drilling Program, Scientific Results,* College Station, Texas (Ocean Drilling Program), *120* (part 2): 683-733.

Harwood, D.M., S.M. Bohaty and R.P. Scherer
1998   Lower Miocene diatom biostratigraphy of the CRP-1 drillcore. Scientific Results of the Cape Roberts Drilling Project, *Terra Antartica, 5*(3): 499-514.

Harwood, D.M., A. McMinn and P.G. Quilty
(in press) Diatom biostratigraphy and age of the Pliocene Sørsdal Formation, Vestfold Hills, East Antarctica, *Antarctic Science.*

Hertlein, L.G.
1969   Fossiliferous boulder of early Tertiary age from Ross Island, Antarctica. *Antarctic Journal of the United States, 4*(5): 199-201.

Homann, M.
1991   Die Diatomeen der Fur-Formation (Alttertiar, Limfjord/Danemark). *Geologisches Jahrbuch Reihe A*, Heft 123.

Hotchkiss, F.M.C. and B.H. Fell
1972   Zoogeographical implications of a Paleogene echinoid from East Antarctica. *Journal of Geophysical Research, 78*: 3448-3468.

Hustedt, F.
1930-1933 Die Kieselalgen Deutschlands, Osterreichs und der Schweiz unter Beruchsichtigung der F. Ubrigen Lander Europas sowie angrenzenden Meeres-gebiete, Teil 2. *Rabenhorst's Krypotogamen-Flora, Vol. 7:* Leipzig (Akademische Verlagsgesli-schaft).

Jones, C.M.
(this vol.) First fossil bird from East Antarctica, in *Paleobiology and Paleoenvironments of Eocene Rocks, McMurdo Sound, East Antarctica,* edited by J.D. Stilwell and R.M. Feldmann, Antarctic Research Series, American Geophysical Union.

Jordan, R.W. and J. Priddle
1991   Fossil members of the diatom genus *Proboscia. Diatom Research, 6*(1): 55-61.

Kanaya, T.
1957   Eocene diatom assemblages from the Kellogg and Sydney shales, Mt. Diablo Area, California.

Sci. Rep. Tohoku Univ., Ser. 2 (Geology), *28*: 27-124.

Krotov, A. I. and K.G. Shibkova
1959   Species novae diatomacearum e paleogeno montium uralensium. Botan. materialy otdela sporovych rastenii, Botanicheskii Institut Akad. NAUK, SSSR, *12*: 112-129.

Lee, D. and J.D. Stilwell
(this vol.)   Rhynchonellide Brachiopods from late Eocene erratics in the McMurdo Sound region, Antarctica, in *Paleobiology and Paleoenvironments of Eocene Rocks, McMurdo Sound, East Antarctica,* edited by J.D. Stilwell and R.M. Feldmann, Antarctic Research Series, American Geophysical Union.

Levy, R.H.
1998   Middle to late Eocene coastal paleoenvironments of Southern Victoria Land, East Antarctica: a palynological and sedimentological study of glacial erratics. Ph.D. dissertation, University of Nebraska – Lincoln, Lincoln, Nebraska.

Levy, R.H. and D.M. Harwood
(this vol. a)   Sedimentary lithofacies and inferred depositional environments of the McMurdo Sound Erratics, in *Paleobiology and Paleoenvironments of Eocene Rocks, McMurdo Sound, East Antarctica,* edited by J.D. Stilwell and R.M. Feldmann, Antarctic Research Series, American Geophysical Union.

Levy, R.H. and D.M. Harwood
(this vol. b)   Marine palynomorph biostratigraphy and age(s) of the McMurdo Sound Erratics, in *Paleobiology and Paleoenvironments of Eocene Rocks, McMurdo Sound, East Antarctica,* edited by J.D. Stilwell and R.M. Feldmann, Antarctic Research Series, American Geophysical Union.

Long, D.J. and J.D. Stilwell
(this vol.)   Fish remains from the Eocene of Mount Discovery, East Antarctica, in *Paleobiology and Paleoenvironments of Eocene Rocks, McMurdo Sound, East Antarctica,* edited by J.D. Stilwell and R.M. Feldmann, Antarctic Research Series, American Geophysical Union.

Mahood, A.D., J.A. Barron and P.A. Sims
1993   A study of some well preserved Oligocene diatoms from Antarctica. Nova Hedwigia, Beiheft, *106*: 243-267.

McIntyre, D.J. and G.J. Wilson
1966   Preliminary palynology of some Antarctic Tertiary erratics. *New Zealand Journal of Botany, 4*(3): 315-321.

Pole, M., R. Hill, and D.M. Harwood
(this vol.)   Eocene plant macrofossils from erratics, McMurdo Sound, in *Paleobiology and*

Paleoenvironments of Eocene Rocks, McMurdo Sound, East Antarctica, edited by J.D. Stilwell and R.M. Feldmann, Antarctic Research Series, American Geophysical Union.

Ross, R. and P.A. Sims
1985   Some genera of the Biddulphiaceae (diatoms) with interlocking linking spines. *Bulletin of the British Museum of Natural History (Bot.), 13* (3): 277-381.

Rowe, G.H.
1974   A petrographic and paleontologic study of Lower Tertiary erratics from Quaternary moraines, Black Island, Antarctica. B.Sc.(Honors) thesis, Victoria University of Wellington, New Zealand, 91 pp.

Scherer, R.P.
1991   Quaternary and Tertiary microfossils from beneath Ice Stream B: evidence for a dynamic West Antarctic Ice Sheet history. *Palaeogeography, Palaeoclimatology, Palaeoecology (Global and Planetary Change Section), 90*: 395-412.

Scherer, R.P., S.M. Bohaty and D.M. Harwood
(in press)   Oligocene and lower Miocene siliceous microfossil biostratigraphy of Cape Roberts Project Core CRP2/2A, Victoria Land Basin, Ross Sea, Antarctica.

Schrader, H.-J. and Fenner, J.
1976   Norwegian Sea Cenozoic diatom biostratigraphy and taxonomy, Part 1: Norwegian Sea Cenozoic diatom biostratigraphy. In Talwani, M., G. Udintsev, et al., Initial Reports of Deep Sea Drilling Project, U.S. Government Printing Office, Washington, D.C., Supplement to Volumes *38-41*: 921-1098.

Schweitzer, C. and R.M. Feldmann
(this vol.)   *Callichirus? symmetricus* (Decapoda: Thalassinoidea) and associated burrows, Eocene, Antarctica, in *Paleobiology and Paleoenvironments of Eocene Rocks, McMurdo Sound, East Antarctica,* edited by J.D. Stilwell and R.M. Feldmann, Antarctic Research Series, American Geophysical Union.

Sims, P.A.
1986   *Sphynctolethus* Hanna, *Ailurella*, gen. nov. and evolutionary trends within the Hemiauloideae. *Diatom Research, 1*: 241-269.

Sims, P.A.
1994   *Skeletonemopsis,* a new genus based on the fossil species of the genus *Skeletonema* Grev. *Diatom Research, 9*(2): 387-410.

Sims, P.A. and R. Ross
1990   *Triceratium pulvinar* and *T. unguiculatum,* two confused species. *Diatom Research, 5*(1): 155-169.

Stilwell, J.D.
(this vol.) Eocene Mollusca (Bivalvia, Gastropoda and Scaphopoda) from McMurdo Sound: Systematics and paleoecologic significance, in *Paleobiology and Paleoenvironments of Eocene Rocks, McMurdo Sound, East Antarctica,* edited by J.D. Stilwell and R.M. Feldmann, Antarctic Research Series, American Geophysical Union.

Stilwell, J.D., R.H. Levy, R.M. Feldmann and D.M. Harwood
1997 On the rare occurrence of Eocene Callianassid decapods (Arthropoda) preserved in their burrows, Mount Discovery, East Antarctica. *Journal of Paleontology, 71*(2): 284-287.

Stott, L.D., B.C. McKelvey, D.M. Harwood and P.-N. Webb
1983 A revision of the ages of Cenozoic erratics at Mount Discovery and Minna Bluff, McMurdo Sound. *Antarctic Journal of United States*, 1983 Review: 36-38.

Vella, P.
1969 Surficial geological sequence, Black Island and Brown Peninsula, McMurdo Sound, Antarctica. *New Zealand Journal of Geology and Geophysics, 12*(4): 761-770.

Webb, P.-N. and G.S. Wilson (eds.)
1995 Cape Roberts Project: Antarctic Stratigraphic Drilling. Proceedings of a meeting to consider the project science plan and potential contributions by the U.S. science community, 6-7 March, 1994. BPRC Report No. 10, Byrd Polar Research Center, The Ohio State University, Columbus, Ohio, 117 pp.

Willis, P.M.A. and J.D. Stilwell
(this vol.) A possible piscivorous crocodile from Eocene deposits of McMurdo Sound, East Antarctica, in *Paleobiology and Paleoenvironments of Eocene Rocks, McMurdo Sound, East Antarctica,* edited by J.D. Stilwell and R.M. Feldmann, Antarctic Research Series, American Geophysical Union.

Wilson, G.J.
1967 Some new species of Lower Tertiary dinoflagellates from McMurdo Sound, Antarctica. *New Zealand Journal of Botany, 5*(1): 57-83.

Wilson, G.S.
(this vol.) Glacial geology and origin of fossiliferous-erratic-bearing moraines, southern McMurdo Sound, Antarctica – An alternative ice sheet hypothesis, in *Paleobiology and Paleoenvironments of Eocene Rocks, McMurdo Sound, East Antarctica,* edited by J.D. Stilwell and R.M. Feldmann, Antarctic Research Series, American Geophysical Union.

Wilson, G.S., A.P. Roberts, K.L. Verosub, F. Florindo and L. Sagnotti
1998 Magnetobiostratigraphic chronology of the Eocene-Oligocene transition in the CIROS-1 core, Victoria Land margin, Antarctica: Implications for Antarctic glacial history. *GSA Bulletin, 110*(1): 35-47.

Winter, D.M. and D.M. Harwood
1997 Integrated diatom biostratigraphy of late Neogene drillcores in Southern Victoria Land and correlation to Southern Ocean Records. In: *The Antarctic Region: Geological Evolution and Processes,* C.A. Ricci (ed.), Siena, p. 985-992.

Woodburne, M.O. and R.M. Feldmann (eds.)
1988 Geology and Paleontology of Seymour Island, Antarctic Peninsula. *The Geological Society of America Memoir 169,* Boulder, Colorado, 566 pp.

David M. Harwood and Steven M. Bohaty, Dept. of Geosciences, University of Nebraska-Lincoln, Lincoln NE 68588-0340 USA

PALEOBIOLOGY AND PALEOENVIRONMENTS OF EOCENE ROCKS, MCMURDO SOUND, EAST ANTARCTICA
ANTARCTIC RESEARCH SERIES VOLUME 76, PAGES 99–159

EBRIDIAN AND SILICOFLAGELLATE BIOSTRATIGRAPHY
FROM EOCENE McMURDO ERRATICS AND THE SOUTHERN OCEAN

Steven M. Bohaty and David M. Harwood

Department of Geosciences, University of Nebraska - Lincoln, Lincoln, Nebraska 68588-0340

Glacial erratics collected from coastal moraines in southern McMurdo Sound, East Antarctica, contain Paleogene siliceous microfossil assemblages, including diatoms, ebridians, silicoflagellates, endoskeletal dinoflagellates, and chrysophyte cysts. Ebridians are particularly abundant and diverse in these erratics and indicate an age of late middle to late Eocene. This age assignment is based on reference to ebridian biostratigraphic ranges from Ocean Drilling Project (ODP) Hole 748B (Kerguelen Plateau) and the CIROS-1 drillcore (McMurdo Sound). Correlation to nannofossil stratigraphy in these cores suggests an absolute age range of 43.7 to 33.7 Ma for the erratics, and the absence of key ebridian and silicoflagellate taxa in CIROS-1 further constrains the age to > 34.9 Ma. Two ebridian biostratigraphic zones are proposed for the middle Eocene to lower Oligocene section of Hole 748B: the upper middle to upper Eocene *Micromarsupium anceps* Partial Range Zone and the lower Oligocene *Hermesinum geminum* Partial Range Zone. A new ebridian species, *Pseudammodochium lingii*, is described from the CIROS-1 drillcore.

INTRODUCTION

The southern McMurdo Sound region of Antarctica is informally recognized as the area southwest of Ross Island, including Minna Bluff, Mt. Discovery, Brown Peninsula, White Island and Black Island (Figure 1). Today, this area is permanently covered by an extension of the Ross Ice Shelf known as the McMurdo Ice Shelf, which consists of relatively stagnant shelf ice [Wilson, this volume, and references therein]. A number of ice-cored, lateral coastal moraines are present in southern McMurdo Sound along the margins of the McMurdo Ice Shelf. These moraines contain a wide assortment of glacial erratics, derived from both basement and sedimentary units. Several discrete morainal bands along the flanks of Mt. Discovery and Minna Bluff contain abundant fossiliferous erratics (Figure 1).

Fossiliferous erratics of southern McMurdo Sound are collectively referred to as the 'McMurdo Erratics.' A wide range of sedimentary lithologies are represented by these erratics, including mudstones, sandstones, conglom-erates, diamictites, and volcaniclastics [Levy and Harwood, this volume b]. Middle Eocene through Quaternary ages have been interpreted for different suites of these erratics [e.g. Cranwell et al., 1960; Speden, 1962; Harrington, 1969]. Sediments of similar age and lithology are not known to crop out in the Transantarctic Mountains, and thus the McMurdo Erratics provide an opportunity to supplement current knowledge of the Cenozoic sedimentary, paleontological, and paleoenvironmental record of the Ross Sea region and the Antarctic continental shelf.

The source of the McMurdo Erratics is most likely deep sub-glacial basins located south of Minna Bluff [Stilwell et al., 1997; Levy, 1998; Levy and Harwood, this volume b; Wilson, this volume]. A grounded, polythermal ice sheet is interpreted to have eroded and transported the erratics into southern McMurdo Sound in the late Pliocene or early Quaternary [Wilson, this volume]. Subsequent to initial erosion and transport, the erratics were redistributed and emplaced in their present-day position by the advance and grounding of one or more Quaternary ice sheets [Wilson, this volume].

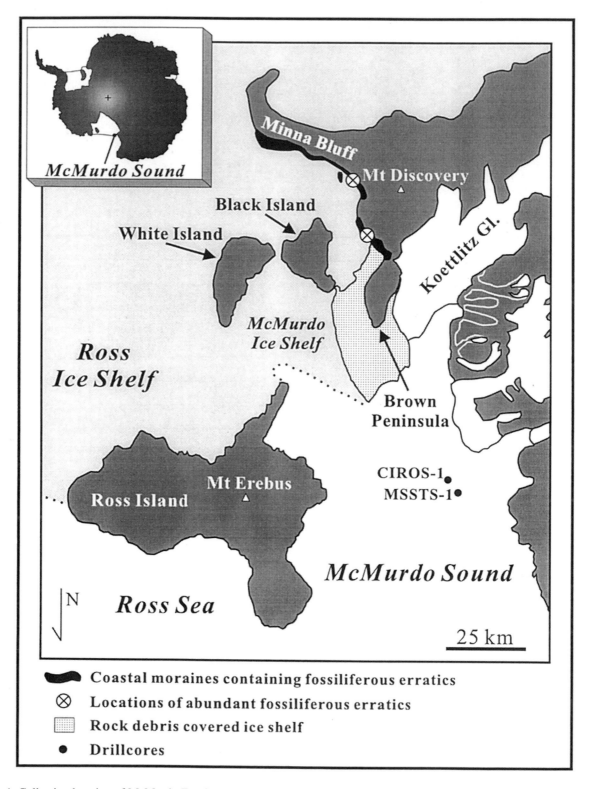

Fig. 1. Collection location of McMurdo Erratics.

A number of expeditions have focused on the collection of McMurdo Erratics over the past forty years. Initial studies assigned an Eocene age to a group of sandstone erratics based on marine palynomorph assemblages [e.g. Cranwell, 1969; McIntyre and Wilson, 1966]. Macrofossil assemblages were also recognized in these erratics in early collections, including gastropods and decapod crustaceans [Hertlein, 1969; Feldmann and Zinsmeister, 1984].

During the austral summer field seasons of 1992-93, 1993-94, and 1995-96, several hundred McMurdo Erratics were collected in southern McMurdo Sound along Mount Discovery and Minna Bluff, and on Black Island (Figure 1). These collections have provided a wealth of fossil material, including marine invertebrates [Stilwell, this volume; Buckeridge, this volume; Lee and Stilwell, this volume], marine vertebrates [Long and Stilwell, this volume] and terrestrial vertebrates [Jones, this volume], terrestrial macroflora [Pole et al., this volume; Francis, this volume], and terrestrial microflora (pollen) [Askin, this volume]. Marine microfossils recovered from the erratics include dinoflagellates [Levy and Harwood, this volume a], diatoms [Harwood and Bohaty, this volume], foraminifera, ebridians, silicoflagellates, chrysophyte cysts, endoskeletal dino-flagellates, and radiolarian fragments. In the present paper, ebridian, silicoflagellate, and chrysophyte cyst assemblages recovered from the McMurdo Erratics are documented and interpreted in a biostratigraphic context.

EBRIDIANS

Siliceous microfossil assemblages in the McMurdo Erratics contain abundant ebridians, a group of marine plankton related to the heterotrophic dinoflagellates. Ebridians are placed close to the dinoflagellates in the Division Pyrrhophyta, Class Ebriophyceae, and Order Ebriales [Loeblich and Loeblich, 1969; Loeblich, 1970]. Ebridian skeletons are typically 10 to 150 (m in length and consist of a framework of solid silica elements. Deflandre [1934] proposed a system of terminology for describing ebridian skeletal elements. This system, however, is derived from sponge spicule taxonomy and is often difficult to apply between morphologically dissimilar genera. The rudimentary and initial structure, common to most ebridian taxa, is the triode (a flat, tri-radial structure) or the triaene (a four-spoked, tetra-axial structure). These structures are interior elements of the ebridian skeleton and form a substructure from which other elements radiate. Although the skeletal elements of ebridians are generally solid, Fenner [1991] notes that the triode elements of some

Paleocene *Ammodochium* species are hollow.

Skeletal designs of fossil ebridians vary considerably and may represent more than one fossil plankton group with solid-silica elements. Many fossil ebridian taxa are known to possess a podamphora or loricate stage. In this stage, the skeleton is heavily silicified and enclosed in solid silica. A bulbous silica projection may also be present. Living specimens in a loricate stage have not been observed; consequently, the function of the lorica is unknown, but may represent a form of encystment. Many fossil species are also commonly found in a double or paired skeleton arrangement.

Only three species of ebridians are validly described and reported from the modern oceans, representing two genera: *Ebria tripartita*, *Hermesinum adriaticum*, and *Hermesinum platense* [Tappan, 1980]. Living ebridians are opportunistic and heterotrophic; diatoms are reported to be a primary source of nourishment [Tappan, 1980]. The ebridian cell does not contain plastids, but symbiotic zooxanthellae algae have been noted in some living specimens of *Ebria tripartita* [Tappan, 1980]. The living cell has a dinokaryotic nucleus (similar to the dinoflagellates) and possesses two unequal flagella for locomotion. Living specimens of *Ebria* and *Hermesinum* propel themselves in a helical fashion, thus the derivation of the generic designation *Ebria* and the group name "ebridians" from the Latin *ebrius* or "drunken" [Tappan, 1980]. Today, ebridians are found in a wide range of environments of varying temperature and salinity, but are not a common plankton group. Most reports of ebridian blooms are from upwelling areas of neritic shelf environments. Additionally, most reports are from cold and temperate latitudes, but ebridians have also been observed in tropical waters [Tappan, 1980].

Ebridians most likely appeared in the Cretaceous, but were not common until the mid-to-late Paleocene. A morphotype of possible ebridian affinity is noted in Lower Cretaceous strata from the Weddell Sea (ODP Site 693) [Harwood, unpubl. data]. Moshkovitz et al. [1983] also report one unknown, small ebridian taxon from Upper Cretaceous strata in Israel, which is the oldest confirmed occurrence of a fossil ebridian. Aside from these two Cretaceous reports, the first abundant appearance of the ebridians is otherwise noted in the Paleocene [Loeblich et al., 1968].

In pre-Quaternary times, ebridians were relatively diverse and widespread throughout the world's oceans. Rich assemblages are commonly preserved in Paleocene through Miocene biosiliceous sediment in neritic, upwelling areas of continental shelves. Globally, peak ebridian diversity is documented in the Eocene to

TABLE 2. McMurdo Erratics prepared for siliceous microfossil examination. Lithologic descriptions from Levy [1998] and Levy and Harwood [this volume b].

Erratic	Lithology	Erratic	Lithology	Erratic	Lithology
E100	Sandstone	E240	Sandy mudstone	E381	Sandstone
E115	Sandy mudstone	E242D	Diamictite	MB97	?
E145	Sandstone	E244	Mudstone	MB109(1)	Sandstone clast
E155	Sandstone	E303(1)	Sandstone	MB181	Sandstone/ Congl.
E163	Sandstone	E303(2)	Sandy mudstone	MB210	Sandstone
E168	Sandstone	E313	Sandstone	MB212K	Mudstone
E169	Sandstone	E317	Sandstone	MB235A	Diamictite
E171	Sandstone	E323	Metased. (?)	MB244C	Mudstone
E181	Sandstone	E331	Sandstone	MB245	Sandy mudstone
E184	Sandstone	E344(1)	Sandy mudstone	MB290G	Mudstone
E185	Sandstone	E345	Sandstone	MB299	Diamictite
E189	Sandstone	E346	Diamictite	MtD1(a)	Sandstone
E191	Sandstone	E347	Diamictite	MtD42	Conglomerate
E192	Sandstone	E350	Sandy mudstone	MtD46	Sandy mudstone
E194	Sandstone	E351	Diamictite	MtD95	Sandy mudstone
E200	Sandstone	E355	Metased. (?)	MtD153(1)	Sandstone
E202	Sandstone	E356	Sandstone	MtD211A	Sandy mudstone
E203	Sandstone	E357	Sandstone	D1	Sandy mudstone
E207	Sandstone	E360	Mudstone	D2	Sandstone
E208	Sandstone	E363	Mudstone	D3	Sandstone
E214	Sandy mudstone	E364	Sandy mudstone	D4	Sandstone
E216	Sandy mudstone	E365(1)	Sandy mudstone	D5	Sandstone
E219	Sandy mudstone	E365(2)	Sandstone	D6	Sandstone
E243	Diamictite				

diatom biostratigraphy [Harwood and Bohaty, this volume] and the absence of dinoflagellates [Levy and Harwood, this volume a].

Hydrochloric acid residues from Erratics D1, MtD95, E345, E350, and E364 were further processed to concentrate siliceous microfossils obtained in initial preparations. These samples were sieved through a 10 μm polyester mesh sieve and washed with a Calgon solution to remove excess clay-sized material. Most ebridians encountered in this study were in the 10 to 30 μm size range, and many would have been lost through a 20 or 25 μm sieve. After sieving, samples were washed with deionized water and centrifuged for 5 minutes at 1500 rpm (repeated three times). Samples were then placed in 50 ml vials and settled for 1 minute, and strewn slide mounts were made on 22x40 mm cover slips from the suspended material.

Light microscope work on Erratics D1, MtD95, E345, E350, and E364 was performed at 750x. Higher magnifications were necessary in order to identify smaller ebridian taxa. The entire 22x40 mm slide was examined for each erratic sample. Siliceous microfossil preservation was rated at Poor (P), Moderate (M), or Good (G). Overall ebridian abundance was determined according to the following scheme [modified from Harwood and Maruyama, 1992]:

B = Barren; no ebridians present

X = Present; 1-10 specimens encountered in 30 traverses

R = Rare; one specimen encountered in 5 to 40 fields of view

F = Frequent; one specimen encountered in 1 to 5 fields of view

C = Common; one specimen in every field of view

A = Abundant; 2-5 specimens in every field of view

V = Very Abundant; more than 5 specimens in every field of view

TABLE 3. Collection location, lithology, and general age assignment of erratics containing siliceous microfossils. Ages of post-Eocene erratics are based on diatom biostratigraphy [Harwood and Bohaty, this volume]. Lithofacies designations from Levy [1998] and Levy and Harwood [this volume b].

Erratic	Collection Location	Lithology	Age
MB244C	Minna Bluff	Mudstone (Mm-d)	Miocene / Plio-Pleistocene
MtD46	Mt. Discovery	Sandy mudstone (Mmb)	middle Miocene
MB235A	Minna Bluff	Diamictite (Dm)	middle(?) Miocene
E351	Minna Bluff	Diamictite (Dm)	middle(?) Miocene
E346	Minna Bluff	Diamictite (Dm)	Miocene
E347	Minna Bluff	Diamictite (Dm)	Oligocene-Miocene
D1	Mt. Discovery	Sandy mudstone (Mwb)	late middle to late Eocene
MtD95	Mt. Discovery	Sandy mudstone (Mwb)	late middle to late Eocene
E345	Mt. Discovery	Sandstone (Sm)	late middle to late Eocene
E350	Minna Bluff	Sandy mudstone (Mmb)	late middle to late Eocene
E364	Mt. Discovery	Sandy mudstone (Mwb)	late middle to late Eocene
MB181	Minna Bluff	Sandstone (Ssg)/ Conglomerate (Csgc)	late middle to late Eocene

Scanning Electron Microscope (SEM) mounts were prepared for Erratics D1, E345, E350, and E364 using the same residues (sieved samples) that were processed for light microscope study. Best results for SEM examination were achieved by allowing several drops of suspended sample to air dry on carbon tape (mounted on a SEM stub). A thin layer of gold-palladium alloy (200 Å) was then applied by sputter coating with a Denton Desk-II Sputter Coater. Examination was performed on a Cambridge Stereoscan 90 scanning electron microscope operated at 15 kV. SEM photomicrographs were taken on Polaroid 55 film. SEM work concentrated on Erratics D1 and E345, which contained abundant and well-preserved ebridians.

ODP Hole 748B

Middle Eocene to upper Oligocene samples from ODP Hole 748B were prepared from the sample set utilized in the diatom study of Harwood and Maruyama [1992]. Samples were taken at an average interval of 1.5 meters between cores 20H and 8H. All samples were first dissolved in a 40% hydrochloric acid solution to remove carbonate material and then washed in deionized water by centrifuging for 5 minutes at 1500 rpm (three repetitions each). Strewn slides of every other sample (3.0 m core spacing) were made on 22x40 mm cover slips directly from the HCl residues. The odd samples (at 3

meter intervals) were sieved through a 20 μm nylon mesh sieve in order to concentrate the larger ebridians. Smaller ebridian taxa such as *Ebrinula paradoxa*, *Ammodochium rectangulare*, and *Pseudammodochium sphericum* were preferentially lost in these preparations, but larger microfossils were significantly more concentrated and less obscured by fine material.

Ebridian and endoskeletal dinoflagellate species abundance from Hole 748B was semi-quantitatively recorded using the method described above for the McMurdo Erratics. Relative diatom and silicoflagellate abundance was also noted using the same criteria. Only complete or nearly complete diatom valves were taken into account for abundance estimates. Light microscope examination was performed at 500x, and identification of smaller ebridian taxa were confirmed at a higher magnification (750x). Thirty traverses were made on each slide (85 fields of view per traverse), which represents approximately 75% of the 22x40 mm cover slip. The entire slide was examined in samples that contained very few ebridians.

CIROS-1

Samples for siliceous microfossil examination from the CIROS-1 drillcore (Figure 1) were chosen from the Harwood [1989] sample set. Representative samples containing well-preserved and abundant siliceous microfossils [Harwood, 1989] were selected from upper Eocene to upper Oligocene sediments of the core [Harwood et al., 1989b; Wilson et al., 1998]. The entire slide from >25μm (20x40mm cover slip) and <25μm (22x22 mm cover slip) size fractions was examined at 500x magnification. Relative ebridian, silicoflagellate, and chrysophyte cyst abundance was determined using the same method described above for the McMurdo Erratics.

RESULTS

McMurdo Erratics

Semi-quantitative ebridian, silicoflagellate, and chrysophyte cyst abundance counts from Erratics D1, E345, E350, E364, and MtD95 are shown in Tables 4a and 4b. Ebridian occurrences noted in thin section from Erratic MB 181 are also listed in Tables 4a and 4b. Common varieties or morphological variations of ebridian species were counted separately. Both whole ebridians and fragments were counted, while only complete silicoflagellates were counted. Due to infrequent occurrence, silicoflagellate counts shown in Tables 4a and 4b include specimens observed in several slides in addition to the slide used for the ebridian count.

Erratics D1, MtD95, and E345 contain the most abundant and well-preserved siliceous assemblages, whereas assemblages in Erratics E350 and E364 are less abundant and more poorly preserved. Erratic E364 contains well-preserved, pyritized ebridians, silicoflagellates, and diatoms in moderate abundance. Siliceous microfossils are abundant in Erratic MB181 but were difficult to identify in thin section.

Similar ebridian assemblages are present in Erratics D1, MtD95, E345, E350, and E364, although some variation is noted. All five erratics contain *Ammodochium rectangulare*, *A. rectangulare* (double skeleton, weakly silicified), *Craniopsis octo*, *Ebriopsis crenulata* (non-loricate), *E. crenulata* (loricate), *Micromarsupium anceps*, *Parebriopsis fallax*, *Pseudammodochium dictyoides*, and *P. sphericum*. Silicoflagellates are less abundant than ebridians in these erratics, but silicoflagellate assemblages are moderately diverse (Table 4b). Chrysophyte cysts are abundant (notably in Erratic D1), but no effort was made to fully classify all morphotypes as done by Perch-Nielsen [1975a]. Three morphologically distinct and possibly age diagnostic species, however, of the chrysophyte-cyst genus *Archaeosphaeridium* [Perch-Nielsen, 1975a; Gombos, 1977a] were tabulated (Table 4b).

Ebridian and silicoflagellate occurrence in post-Eocene erratics (E346, E347, E351, MtD46, MB235A,

and MB244C) is presented in Table 5. These younger erratics contain ebridian assemblages of low species richness, represented only by *Pseudammodochium sphericum* and *Pseudam-modochium lingii* n. sp. Erratic MtD46 contains the most abundant and diverse post-Eocene silicoflagellate assemblage. The age of this erratic is determined to be middle Miocene based on diatom biostratigraphy [Harwood and Bohaty, this volume]. Silicoflagellate assemblages in this erratic are similar to those recovered from middle Miocene sediments beneath the Ross Ice Shelf at RISP Site J-9 [Ling and White, 1979; White, 1980; Harwood et al., 1989a].

ODP Hole 748B

The stratigraphic distribution and relative abundance of ebridians and endoskeletal dinoflagellates from Hole 748B are presented in Tables 6 and 7. Core depths for each sample have not been corrected for expansion after recovery. Non-sieved samples are indicated with an asterisk (*), samples sieved at 20 μm are indicated with a pound sign (#), and non-sieved samples prepared on the ship (ODP Leg 120) are indicated with a solid diamond (w). Most ebridians recovered from the McMurdo Erratics are also present in Hole 748B [indicated with a plus sign (+)]. The ebridians *Adonnadonna primadonna*, *Ebriopsis antiqua antiqua*, *Hermesinella cornuta*, *Hermesinum geminum*, *Micromarsupium curticannum*, *Triskelion gorgon* are present in Hole 748B but were not observed in the erratics.

Many ebridian taxa show restricted stratigraphic ranges in ODP Hole 748B (Figure 3). Ebridian richness peaks in the upper Eocene (Tables 6 and 7) and several last occurrence datums are recorded in the uppermost Eocene and lower Oligocene (Figures 3 and 4). The extinctions of several genera are documented through the lower Oligocene interval, including *Craniopsis*, *Ebrinula*, *Micromarsupium*, *Parebriopsis*, and *Triskelion*. Compilation of global geologic ranges of ebridian genera shows a similar trend of Eocene- Oligocene extinction [Tappan, 1980, Fig. 5.14; Ernissee and McCartney, 1993, Fig. 8.10]. These extinctions, however, have not been previously well documented in continuous drillcore sections.

The middle Eocene to upper Oligocene sediments of Hole 748B are characterized by several pulses in biosiliceous sedimentation. These pulses include radiolarians, sponge spicules, diatoms, silicoflagellates, ebridians, and endoskeletal dinoflagellates. Although several of these groups are unrelated and occupied different levels of the water column, they were most likely responding to silica, nutrient, and temperature variations in the middle Eocene to late Oligocene Southern Ocean. Two major

TABLE 4a. Ebridian abundance counts for Eocene McMurdo Erratics. Species observed in thin section from Erratic MB181 are recorded with an "X."

Ebridians	D1	MtD95	E345	E350	E364	MB181
Abundance	A	A	C	F	C	N/A
Preservation	G	M	M	P	M	N/A
Ammodochium ampulla Deflandre 1934	6	3	3	2	6	
Ammodochium ampulla (double skeleton)	1	0	3	0	0	
Ammodochium ampulla (double skeleton, loricate)	52	17	6	4	0	
Ammodochium danicum Deflandre 1951	0	0	2	0	0	
Ammodochium novum Perch-Nielsen 1978	0	0	2	0	0	
Ammodochium rectangulare (Schulz) Deflandre 1933	22	47	39	8	31	X
Ammodochium rectangulare (loricate)	0	0	0	0	1	
Ammodochium rectangulare (hyper-silicified)	2	0	1	0	0	
Ammodochium rectangulare (double skeleton, weakly silicified)	23	50	11	2	3	
A. rectangulare (double skeleton, anterior and medial areas silic.)	30	35	5	0	0	
Ammodochium speciosum Deflandre 1934	4	1	1	0	2	
Ammodochium sp. 1	1	0	5	1	1	
Craniopsis octo Hovasse ex Frenguelli 1940	23	6	18	2	5	
Craniopsis octo (double skeleton)	0	2	0	0	0	
Ebrinula paradoxa (Hovasse) Deflandre 1950	36	28	89	0	24	
Ebrinula paradoxa (double skeleton)	1	5	2	0	0	
Ebriopsis crenulata (Hovasse) emend. (non-loricate)	105	38	26	6	41	X
Ebriopsis crenulata (hypersilicified, non-loricate)	9	6	2	0	0	
Ebriopsis crenulata (loricate)	118	21	16	4	3	X
Ebriopsis sp. 1 (three-tier structure)	0	0	0	0	2	
Falsebria imitata Deflandre 1950	0	1	5	1	6	
Falsebria sp. 1	2	0	0	0	0	
Hovassebria sinistra Deflandre 1951	0	0	2	0	2	
Micromarsupium anceps Deflandre 1934 (weakly silicified)	2	1	1	2	3	
Micromarsupium anceps (heavily silicified)	30	8	4	3	1	X
Micromarsupium anceps (fragment)	70	10	7	7	2	
Parebriopsis fallax Hovasse 1932	10	3	3	2	6	
Parebriopsis fallax (hyper-silicified crest)	4	1	2	1	1	
Polyebriopsis sp. 1	2	1	1	0	1	
Pseudammodochium dictyoides Hovasse 1932 (single skeleton)	201	96	35	21	9	X
Pseudammodochium dictyoides (single skeleton, fragment)	80	10	7	0	2	
Pseud. dictyoides (double skeleton, openings not connected)	45	7	5	2	2	
Pseud. dictyoides (double skeleton, openings connected)	225	59	25	21	13	X
Pseudammodochium dictyoides (double skeleton, fragment)	125	4	11	0	2	
Pseudammodochium dictyoides (triode)	20	9	8	1	5	
Pseud. sphericum Hovasse 1932 (single skeleton)	12	7	27	6	17	X
Pseudammodochium sphericum (double skeleton)	23	0	14	2	0	
Triskelion gorgon Gombos 1982	0	0	0	1	0	
Unidentified ebridian fragments	82	66	11	24	77	
Unidentifiable / unknown ebridians	29	34	13	2	5	
Total identified complete ebridians	989	435	356	86	176	

TABLE 4*b*. Silicoflagellate, chrysophyte cyst, and endoskeletal dinoflagellate abundance counts for Eocene McMurdo Erratics. Taxa observed in thin section from Erratic MB181 are recorded with an "X."

Silicoflagellates	D1	MtD95	E345	E350	E364	MB181
Bachmannocena? diodon diodon? (Ehrenberg) Bukry 1987	-	-	-	-	2	-
Cannopilus hemisphaericus (Ehrenberg) Haeckel 1887	-	-	-	-	1	-
Corbisema apiculata (Lemmermann) Hanna 1931	2	3	-	-	-	-
Corbisema flexuosa (Stradner) Perch-Nielsen 1975	-	-	-	2	-	-
Corbisema hastata globulata Bukry 1976	1	1	2	-	5	-
Corbisema hastata hastata (Lemmermann) Frenguelli 1940	1	6	-	1	3	-
Corbisema regina Bukry 1984	3	2	-	-	2	-
Corbisema spinosa Deflandre 1950	1	10	11	-	4	X
Corbisema triacantha (Ehrenberg) Hanna 1931 (apical bar)	7	11	7	2	-	-
Corbisema triacantha (apical plate)	4	3	1	2	2	-
Corbisema triacantha cf. *lepidospinosa* Ciesielski 1991	-	-	1	-	-	-
Dictyocha fibula fibula Ehrenberg ex Locker and Martini 1986	-	-	3	-	-	-
Dictyocha frenguellii Deflandre 1950	-	3	-	-	-	X
Dictyocha hexacantha Schulz 1928	2	-	5	-	2	-
Dictyocha pentagona (Schulz) Bukry and Foster 1973	2	-	-	-	3	-
Dictyocha cf. *anguinea* Shaw and Ciesielski 1983	-	-	3	-	2	-
Distephanus speculum (Ehrenberg) Haeckel 1887	-	-	-	-	1	-
Distephanus speculum speculum f. *pseudofibula* Schulz 1928	1	-	-	-	2	-
Distephanus sp. 1 (highly arched apical structure)	-	-	2	-	-	-
Naviculopsis constricta (Schulz) Bukry 1984	5	1	4	2	1	-
Naviculopsis cf. *constricta*		4	-	-	-	-
Naviculopsis eobiapiculata Bukry 1978	1	1	-	-	2	-
Naviculopsis foliacea Deflandre 1950	-	-	-	-	2	-
Naviculopsis foliacea tumida Bukry 1978					1	
Septamesocena apiculata apiculata (Schulz) Bachmann 1970	-	-	2	-	1	-
Chrysophyte Cysts						
Archaeosphaeridium australensis Perch-Nielsen 1975	7	-	6	11	3	-
Arch. australensis Perch-Nielsen 1975 (no spines)	16	85	-	-	-	-
Archaeosphaeridium dumitricae Perch-Nielsen 1975	-	-	3	-	-	-
Archaeosphaeridium tasmaniae Perch-Nielsen 1975	4	9	4	2	-	-
Endoskeletal Dinoflagellates						
Carduifolia gracilis Hovasse 1932	8	5	4	5	-	-

pulses of increased ebridian abundance and species richness are noted (Figure 4): a middle Eocene pulse in Core 19H (169.08 to 158.58 mbsf) and an upper Eocene pulse in Cores 15H and 14H (133.08 to 114.58 mbsf). These distributions are interpreted to reflect variable productivity and sedimentation, but preservational influences or biases may also play a role in the stratigraphic distribution and occurrence of taxa.

CIROS-1

Ebridian, silicoflagellate, and chrysophyte cyst data from CIROS-1 are presented in Table 8. This data set represents selected samples from initial siliceous microfossil preparations [Harwood, 1989]. Ebridians are abundant and relatively well-preserved above ~500 mbsf in CIROS-1. Above ~371 mbsf, however, species richness

TABLE 5. Ebridian and silicoflagellate occurrence in post-Eocene McMurdo Erratics.

Ebridians / Silicoflagellates	Erratic					
	MtD46	E346	E347	E351	MB235A	MB244C
Pseudammodochium lingii n.sp.	X		X	X	X	
Pseud. sphericum Hovasse (single)	X	X	X	X	X	
Pseud. sphericum Hovasse (double)	X					
Distephanus quinquangellus Bukry and Foster	X					
Distephanus speculum (Ehrenberg) Haeckel	X			X	X	X
Distephanus speculum (binoculoid)	X					
Distephanus speculum (seven-sided)						X
Dist. spec. spec. f. *pseudopentagonus* Schulz	X					
Sept. apiculata glabra (Schulz) Desikachary and Prema	X					
Sept. pappii (Bachmann) Desikachary and Prema	X					

abruptly decreases [Harwood, 1989]. This decrease occurs across a significant disconformity at ~366 mbsf, where a ~4 m.y. hiatus is interpreted in the Oligocene [Harwood et al., 1989b; Wilson et al., 1998]. Highest ebridian and silicoflagellate abundance and species richness occurs in the upper Eocene to lower Oligocene section of this core (~500 to 371 mbsf).

PROPOSED EBRIDIAN ZONATION FROM HOLE 748B

An ebridian zonation for the middle Eocene to lower Oligocene is proposed based on data from ODP Hole 748B (Figure 3; Tables 6 and 7). Until this zonation can be tested at other Southern Ocean sites with abundant Paleogene ebridian assemblages, effects of preservational biases on biostratigraphic ranges of these taxa cannot be evaluated. Ebridian study of other sites may also allow the construction of a higher resolution zonation, when more datums are applied in addition to those utilized in the present study. Comparison of results from Hole 748B and reported Southern Ocean occurrences of taxa suggest the following datums may be biostratigraphically useful: the first occurrence (FO) of *Hovassebria brevispinosa/ Falsebria ambigua*; the last occurrences (LO) of *Ammodochium speciosum, Ebrinula paradoxa, Hermesinum geminum, Parebriopsis fallax,* and *Pseudammodochium dictyoides;* and the full ranges of *Ammodochium ampulla* (double, loricate), *Craniopsis octo, Ebriopsis crenulata* (loricate), and large, heavily-silicified varieties of *Micromarsupium anceps* (Figure 3). Several of these datums were chosen to divide the mid-

dle Eocene to lower Oligocene of Hole 748B into two ebridian zones (Figure 3; Tables 6 and 7). Selection of specific datums was also based on abundance and age distributions from other Southern Ocean sites [Perch-Nielsen, 1975a; McCartney and Wise, 1990, Locker and Martini, 1986a]. This zonal scheme should be considered preliminary until additional ebridian data can be gathered at other sites. Corresponding ages and zonal assignments in Tables 6 and 7 are derived from nannofossil, planktic foraminifer, diatom, radiolarian, and silicoflagellate biostratigraphy [Wei et al., 1992; Berggren, 1992; Harwood and Maruyama, 1992; McCartney and Harwood, 1992; Takemura, 1992].

MICROMARSUPIUM ANCEPS Partial Range Zone

Definition: Interval from the FO of *Craniopsis octo* at the base up to the LO *Micromarsupium anceps* (large, heavily-silicified forms) at the top.

Age: Late middle Eocene to late Eocene (~42.0 to 33.7 Ma).

Correlative nannofossil zones: The FO of *Craniopsis octo* occurs near the base of the *Cribrocentrum reticulatum* Zone (defined by the FO of *C. reticulatum* at 42.0 Ma), and the LO of *Micromarsupium anceps* occurs at the top of the *Reticulofenestra oamaruensis* Zone (defined by the LO of *R. oamaruensis* at 33.7 Ma) [Wei et al., 1992; Berggren et al., 1995].

Common ebridian taxa: Ammodochium ampulla (double, loricate), all varieties of *Ammodochium rectangulare, Ammodochium speciosum, Craniopsis octo,*

TABLE 6. Ebridian and endoskeletal dinoflagellate occurrence and abundance for the middle Eocene interval of ODP Hole 748B. Nannofossil stratigraphy from Wei et al. [1992]; planktonic formaminifer stratigraphy from Berggren [1992]; diatom stratigraphy from Harwood and Maruyama [1992]; radiolarian stratigraphy from Takemura [1992]; and silicoflagellate stratigraphy from McCartney and Harwood [1992]. See text for descriptions of symbols and definitions of proposed ebridian zones.

Sample Interval ODP 748B	20H-7,47-49*	20H-6,47-49#	20H-5,47-49*	20H-2,47-49*	20H-1,47-49*	20H-1,47-49#	19H-6,47-49*	19H-5,47-49*	19H-5,47-49#	19H-4,47-49*	19H-3,116-118◆	19H-3,47-49*	19H-3,47-49#	19H-2,47-49*	19H-1,47-49*	19H-1,47-49#	18H-7,47-49#	18H-6,47-49*	18H-5,47-49#	18H-4,47-49*	18H-3,47-49#	18H-2,47-49*	18H-1,47-49#	17H-7,47-49#	17H-6,47-49*	17H-5,47-49#	17H-4,47-49*
Age	middle Eocene																										
Nannofossil Biozones	*R. umbilica*						*C. reticulatum*																			*D.sai.*	
Planktic Foram Biozones	*A. collactea*																				*G. index*						
Diatom Biozones	Unzoned																										
Silicoflagellate Biozones	*Dictyocha grandis*																										
Radiolarian Biozones	Unzoned																										
Proposed Ebridian Zonation	Unzoned											*Micromarsupium anceps*															
Depth (mbsf)	180.58	179.08	177.58	173.08	171.58	171.58	169.08	167.58	167.58	166.08	165.27	164.58	164.58	163.08	161.58	161.58	161.58	160.08	158.58	157.08	155.58	154.08	152.58	151.08	150.58	149.08	147.58
Ebridian Abundance	B	B	B	X	B	B	R	R	R	F	C	C	A	R	F	F	C	F	R	X	X	R	R	X	X	X	X
Preservation	-	-	-	P	-	-	M	M	M	G	G	G	G	M	M	M	P	P	M	P	P	P	P	M	P	P	P
Species (Varietal) Richness	0	0	0	1	0	0	9	4	3	10	8	12	10	7	8	6	4	3	4	1	1	3	1	1	7	2	1
Adonnadonna primadonna																											
Ammodochium ampulla +																											
A. ampulla (single, loricate)+																											
A. ampulla (double, weakly silic.)+										X		X	X														
A. ampulla (double, loricate)+										X		X	R	X		X						X			X		
A. danicum +										X																	
A. novum +																											
A. rectangulare +							R	R		R	R	F	X		X			X				X			X		X
A. rectangulare (dbl., weakly silic.)+												X	X	X													
A. rectangulare (double, ant. silic.)+				X							R	X	X														
A. speciosum +							X	X	X	X	X	X			X	X									X		
Craniopsis octo +												X	X		X										X		
Craniopsis octo (double)																											
Ebrinula paradoxa +																											
Ebriopsis antiqua antiqua														X													
E. crenulata (non-loricate)+							R	R	X	R	F	R	R	R	R	F	X	R	R	X		X			X	X	
E. crenulata (loricate)+							X	X	R	R	R	R	R	R	R	R	X		R						X		
Haplohermesinum cornuta																											
Hermesinum geminum																											
Hovassebria brevispinosa +																											
Micromarsupium anceps +							X			R	F	F	A	R	X	X	C	F	R		X				X	X	
M. curticannum							X																				
Parebriopsis fallax +																											
P. fallax (hyper-silicified)+																											
Podamphora tenuis																											
Pseudammodochium dictyoides (single)+							X			X		R															
P. dictyoides (double, not conn.)+							X																				
P. lingii +																											
P. sphericum (single)+							X			X	X	R	X	X	X	R	X		X								
P. sphericum (double)+															X												
Triskelion gorgon											X												R	X			
Gen. et sp. indet. 1																											
Gen. et sp. indet. 2																											
Actiniscus elongatus																											
A. pentasterias																											
Carduifolia gracilis +									X	X	X	R			X												
Foliactiniscus mirabilis																											
F. pannosus																											

Sponge Spicule Nannofossil Ooze

TABLE 7. Ebridian and endoskeletal dinoflegellate occurrence and abundance for the upper Eocene to upper Oligocene interval of ODP Hole 748B. Nannofossil stratigraphy from Wei et al. [1992]; planktonic formaminifer stratigraphy from Berggren [1992]; diatom stratigraphy from Harwood and Maruyama [1992]; radiolarian stratigraphy from Takemura [1992]; and silicoflagellate stratigraphy from McCartney and Harwood [1992]. See text for descriptions of symbols and definitions of proposed ebridian zones.

late Eocene		early Oligocene		late Olig.	Age
C. oamaruensis / I.rec. / R. oamar.		B. spin. / D. dav. / C. altus		late Olig.	Nannofossil Biozones
G. suteri		S. angiporoides / C. cubensis		G. lab. / G. euap.	Planktic Foram Biozones
Unzoned		R. olig. / C. jous. / R. vigilans		A. gem. / L. orn.	Diatom Biozones
Bach. paulschulzii		N. trispinosa		C. archang.	Silicoflagellate Biozones
Eucyrtidium spinosum		Axoprunum(?) irregularis		L. con.	Radiolarian Biozones
Micromarsupium anceps		Hermesinum geminum		Unzoned	Proposed Ebridian Zonation

Sample Interval ODP 748B

Sample intervals: 17H-3,47-49# ; 17H-2,47-49# ; 16H-7,47-49# ; 16H-6,47-49# ; 16H-5,47-49# ; 16H-4,47-49# ; 16H-3,47-49# ; 16H-2,47-49# ; 16H-1,47-49# ; 15H-6,47-49# ; 15H-5,47-49# ; 15H-4,47-49# ; 15H-3,47-49# ; 15H-2,47-49* ; 15H-1,47-49* ; 14H-6,47-49* ; 14H-5,47-49* ; 14H-4,47-49* ; 14H-3,47-49* ; 14H-2,47-49* ; 14H-1,47-49* ; 13H-6,47-49* ; 13H-4,47-49* ; 13H-2,47-49* ; 13H-1,47-49* ; 12H-6,47-49* ; 12H-4,47-49* ; 12H-2,47-49* ; 11H-4,47-49* ; 11H-2,47-49* ; 11H-1,47-49* ; 10H-6,47-49* ; 10H-4,47-49* ; 10H-2,47-49#

Depth (mbsf): 146.08 ; 144.58 ; 142.58 ; 142.08 ; 141.08 ; 139.58 ; 138.08 ; 136.58 ; 135.08 ; 133.08 ; 131.58 ; 130.08 ; 128.58 ; 128.58 ; 127.08 ; 125.58 ; 123.58 ; 122.08 ; 120.58 ; 119.08 ; 117.58 ; 116.08 ; 114.58 ; 112.58 ; 109.58 ; 106.58 ; 104.58 ; 103.08 ; 100.08 ; 97.08 ; 97.08 ; 93.58 ; 90.58 ; 87.58 ; 87.58 ; 84.08 ; 81.08 ; 78.08 ; 78.08

Ebridian Abundance / Preservation / Species (Varietal) Richness

Taxa recorded:

- Adonnadonna primadonna
- Ammodochium ampulla +
- A. ampulla (single, loricate)+
- A. ampulla (double, weakly silic.)+
- A. ampulla (double, loricate)+
- A. danicum +
- A. novum +
- A. rectangulare +
- A. rectangulare (dbl., weakly silic.)+
- A. rectangulare (double, ant. silic.)+
- A. speciosum +
- Craniopsis octo +
- Craniopsis octo (double)
- Ebrinula paradoxa +
- Ebriopsis antiqua antiqua
- E. crenulata (non-loricate)+
- E. crenulata (loricate)+
- Haplohermesinum cornuta
- Hermesinum geminum
- Hovassebria brevispinosa +
- Micromarsupium anceps +
- M. curticannum
- Parebriopsis fallax +
- P. fallax (hyper-silicified)+
- Podamphora tenuis
- Pseud. dictyoides (single)+
- P. dictyoides (double, not conn.)+
- P. lingii +
- P. sphericum (single)+
- P. sphericum (double)+
- Triskelion gorgon
- Gen. et sp. indet. 1
- Gen. et sp. indet. 2
- Actiniscus elongatus
- A. pentasterias
- Carduifolia gracilis +
- Foliactiniscus mirabilis
- F. pannosus

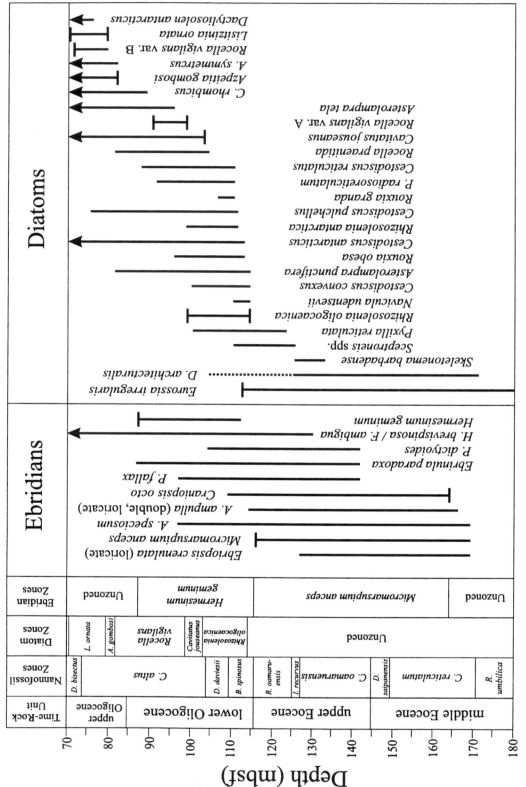

Fig. 3. Proposed ebridian zonation from ODP Hole 748B and ranges of selected ebridian and diatom taxa plotted against core depth. Nannofossil zonation from Wei et al. [1992] and diatom data and zonation from Harwood and Maruyama [1992].

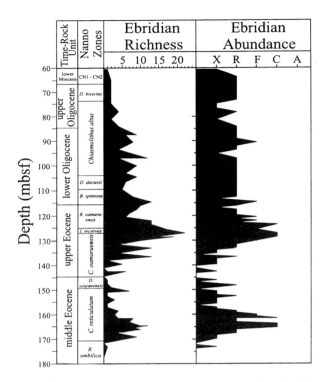

Fig. 4. Ebridian varietal richness and relative abundance for the middle Eocene to upper Oligocene interval of ODP Hole 748B. X = Present, R = Rare, F = Frequent, C = Common, and A = Abundant.

Ebrinula paradoxa, Ebriopsis crenulata (non-loricate), *E. crenulata* (loricate), *Hovassebria brevispinosa, Micromarsupium anceps, Parebriopsis fallax, Pseudammodochium dictyoides* (single), *Pseudammodochium sphericum,* and *Triskelion gorgon.*

Comments: The LO of *Craniopsis octo* may be used as a secondary datum to approximate the top of this zone.

HERMESINUM GEMINUM Partial Range Zone

Definition: Interval from the LO of *Micromarsupium anceps* at the base up to the LO of *Hermesinum geminum* at the top.

Age: Early Oligocene (~33.7 to 28.5 Ma).

Correlative nannofossil zones: The LO of *Micromarsupium anceps* occurs at the base of the *Blackites spinosus* Zone (defined by the LO of *R. oamaruensis* at 33.7 Ma), and the LO of *Hermesinum geminum* occurs within the *Chiasmolithus altus* Zone (defined from LO of *Reticulofenestra umbilica* to the LO of *Chiasmolithus altus,* 31.3 to 26.1 Ma) [Wei et al., 1992; Berggren et al., 1995].

Correlative diatom zones: The ebridian *Hermesinum geminum* Zone spans the *Rhizosolenia oligocaenica, Cavitatus jouseanus,* and *Rocella vigilans* Zones of Hole 748B [Harwood and Maruyama, 1992].

Common ebridian taxa: All varieties of *Ammodochium rectangulare, Ebrinula paradoxa, Ebriopsis crenulata* (non-loricate), *Hermesinum geminum, Micromarsupium curticannum,* and *Pseudammodochium sphericum* (single).

Comments: The LO of *Ebrinula paradoxa* may be used as a secondary datum to approximate the top of this zone.

EBRIDIAN AGE ASSIGNMENT FOR EOCENE McMURDO ERRATICS

The ranges of several ebridian taxa are identified as biostratigraphically restricted in ODP Hole 748B (Figure 3). Taxa with ranges restricted to the middle Eocene to lowermost Oligocene include *Ammodochium ampulla* (double, loricate), *Craniopsis octo, Ebriopsis crenulata* (loricate), and *Micromarsupium anceps* (large, heavily-silicified forms). Both *Craniopsis octo* and *Micromarsupium anceps* are present in Erratics D1, MtD95, E345, E350, and E364, which suggest a general assignment to the middle to upper Eocene *Micromarsupium anceps* Zone (Figure 3; Tables 6 and 7). As noted above, the *Micromarsupium anceps* Zone in Hole 748B correlates to the Southern Ocean nannofossil *Cribrocentrum reticulatum* through *Reticulofenestra oamaruensis* Zones [Wei et al., 1992], which ranges in age from 42.0 to 33.7 Ma [Berggren et al., 1995].

Comparison with other Southern Ocean drillcores shows an apparent discrepancy in the FO *Craniopsis octo.* Data from DSDP Hole 512 [Bohaty, unpubl. data] suggest the FO of *Craniopsis octo* occurs in the Southern Ocean nannofossil *Reticulofenestra umbilica* Zone (see notes in systematic paleontology section), which is one zone lower than observed in ODP Hole 748B. Therefore, taking into account the known occurrence of *Micromarsupium anceps* and *Craniopsis octo* in the Southern Ocean, a conservative estimate for the maximum age range of Erratics D1, MtD95, E345, E350, and E364 is ~43.7 to 33.7 Ma.

In addition to the above ebridians, several silicoflagellates provide biostratigraphic age constraint for the erratics; these taxa include *Dictyocha hexacantha, Naviculopsis foliacea,* and *Naviculopsis constricta* (see notes in systematic paleontology section regarding the Southern Ocean ranges of these taxa). *Naviculopsis constricta* is present in Erratics D1, MtD95, E345, E350, and

TABLE 8. Ebridian, silicoflagellate, and chrysophyte cyst occurrence from selected CIROS-1 samples. Age assignments are from Wilson et al. [1998]. See methods section in text for abundance designations.

Age	Sample Depth (mbsf) CIROS-1	Ebridian Preservation	Ebridian Abundance	Ebridian Species (Varietal) Richness	Ammodochium rectangulare	A. rectangulare (dbl., weakly silic.)	A. rectangulare (double, ant. silic.)	Ebrinula paradoxa	Ebriopsis crenulata (non-loricate)	Ebriopsis crenulata (loricate)	Hovassebria brevispinosa/ Falsebria ambigua	M. curticannum	Parebriopsis fallax	P. fallax (hyper-silicified)	Pseud. dictyoides (single)	P. lingii	P. sphericum (single)	P. sphericum (double)	Corbisema triacantha (apical bars)	Corbisema triacantha (apical plate)	Dictyocha deflandrei	Dictyocha frenguellii	Dictyocha pentagona	Dictyocha spp.	Distephanus speculum	Septamesocena apiculata apiculata	Archaeosphaeridium australensis	A. australensis (spineless)	A. tasmaniae	A. tasmaniae (spineless)
late Oligocene	173.26	M	R	2													R	X								X				
	224.91	P	X	6	X	X	X			X							X	X								X				
	296.68	P	R	3	X												R	X												
	309.38	M	R	4	X					X							R	X				X								
	335.17	M	X	4	X	X											X	X												
	342.16	M	R	4	X	X											R	X									Unconformity			
early Olig.	374.77	M	R	9	X	X		X	X	X	X	X	X			X			X	X			X				X			R
	387.68	M	F	8	R	X		X	X	X		X				R	X				R		X		X	X		X	R
	404.41	M	R	6	R	X		X		X						X	X	X			R	X	X			X		X	X
late Eocene	428.00	G	R	6	X	X			X	X	X					R					X					X			X
	484.95	G	R	5	X	X				R	X					X		X			X					F	X		X
	494.52	G	R	7	X	X		X	X	X	X		X								X					R	X		
	498.04	M	R	7	X	X			X	X	X	X	X					X			X	X	X			X		X	X
	500.14	M	R	2	X					X			X								X		X					X	X
	524.63	P	X	1	X																					X	X		X
	644.66	P	X	5					X			X	X			X	X		X		X								
	661.13	P	X	2	X							X																	

E364. Ciesielski [1991] reports the Last Abundant Appearance Datum (LAAD) of *Naviculopsis constricta* approximately at the C13/C15 boundary at ODP Site 703, Meteor Rise. This reversal is dated at 34.7 Ma [Berggren et al., 1995] and suggests an age assignment of ∆34.7 Ma for Erratics D1, MtD95, E345, E350, and E364.

As cautioned with ebridian distributions in Hole 748B, the occurrence of ebridian and silicoflagellate taxa in the McMurdo Erratics may be influenced by preservational factors as well as chronostratigraphic differences between the erratics. For more specific age interpretations, the erratics will be considered separately below, as each may be of different age within the middle to late Eocene [Harwood and Bohaty, this volume]. Similar,

broad age assignments of late middle to late Eocene (43.7 to 34.7 Ma), however, are interpreted for Erratics D1, MtD95, E345, E350, and E364, based on the common presence of several key ebridian and silicoflagellate taxa.

Erratics D1 and MtD95 (sandy mudstones) are characterized by high ebridian abundance and good siliceous microfossil preservation. These erratics contain four ebridian taxa with last occurrence datums in the upper Eocene or lowermost Oligocene of Hole 748B (Figure 3; Table 4a): *Ammodochium ampulla* (double, loricate), *Craniopsis octo*, *Ebriopsis crenulata* (loricate), and *Micromarsupium anceps*. As noted above, overlapping ranges of these taxa (Figure 3; Tables 6 and 7) suggest Erratics D1 and MtD95 were derived from middle to upper Eocene strata, spanning the Southern Ocean nannofossil *Reticulofenestra umbilica* to *Reticulofenestra oamaruensis* Zones (43.7 to 33.7) [Wei et al., 1992; Berggren et al., 1995]. In support of this age assignment, the presence of the silicoflagellate *Dictyocha hexacantha* suggests an age range of 43.7 to 34.3 Ma (see notes in systematic paleontology), and the presence of *Naviculopsis constricta* suggests an age older than 34.7 Ma.

Ebridians in Erratic E345 (sandstone) are common and moderately preserved. *Ammodochium ampulla* (double, loricate), *Craniopsis octo*, *Micromarsupium anceps*, *Dictyocha hexacantha*, and *Naviculopsis constricta* are present in this erratic (Tables 4a and 4b) suggesting a late middle to late Eocene age (Figure 3; Tables 6 and 7). Based on the correlations to nannofossil and paleomagnetic stratigraphy noted above, the presence of these taxa suggest an age range of 43.7 to 34.7 Ma for this erratic [Berggren et al., 1995]. Erratic E345 also contains the chrysophyte cyst *Archaeosphaeridium dumitricae*. This taxon was not recorded in Hole 748B, but Perch-Nielsen [1975a] and Gombos [1977a] report a restricted upper Eocene occurrence for this species.

Siliceous microfossils in Erratic E350 (sandy mudstone) are poorly preserved and present in low abundance. Although rare in occurrence, the following key taxa are present: *Ammodochium ampulla* (double, loricate), *Craniopsis octo*, *Micromarsupium anceps*, *Dictyocha hexacantha*, and *Naviculopsis constricta* (Tables 4a and 4b). The presence of these taxa suggest an age range of 43.7 to 34.7 Ma [Berggren et al., 1995].

Erratic E364 (sandy mudstone) is characterized by poor to moderate siliceous microfossil preservation (well-preserved specimens are pyritized). Again, the age-diagnostic ebridians *Craniopsis octo* and *Micromarsupium anceps* (Table 4a) suggest a late middle to late Eocene age (43.7 to 33.7 Ma) for this erratic. The silicoflagellates *Dictyocha hexacantha*, *Naviculopsis constricta*, and *Naviculopsis foliacea* are also present, which have a reported occurrence in middle to upper Eocene sediments of Southern Ocean drillcores [Bukry, 1975b; Perch-Nielsen, 1975b; Hajós, 1976; Shaw and Ciesielski, 1983; McCartney and Wise, 1987; Bukry, 1987; Ciesielski, 1991; McCartney and Harwood, 1992]. The presence of *Naviculopsis constricta* further suggests this erratic is older than 34.7 Ma [Ciesielski, 1991; Berggren et al., 1995].

A few pyritized remains of ebridians were observed in thin section of Erratic MB181 (interbedded sandstone and conglomerate). *Micromarsupium anceps* and *Ebriopsis crenulata* (loricate) are present suggesting a

middle to late Eocene age based on occurrence in ODP Hole 748B (Figure 3; Tables 4a, 6, and 7). The first occurrences of these taxa, however, are not well known, and an early Eocene age cannot be ruled out from ebridian data alone.

SILICEOUS MICROFOSSIL OCCURRENCE IN ROSS SEA DRILLCORES

The lowermost, well-preserved siliceous microfossils in the CIROS-1 drillcore occur in an interval from 500 to 485 mbsf [Harwood, 1989]. Abundant and diverse siliceous microfossil assemblages, indicating open-marine conditions, are present in this interval [Harwood, 1989], the base of which is dated at ~34.9 Ma [Wilson et al., 1998]. The ebridian and silicoflagellate assemblage in this interval is characterized by *Ammodochium rectangulare*, *Hovassebria brevispinosa*, *Ebriopsis crenulata*, *Parebriopsis fallax*, *Dictyocha deflandrei*, and *Dictyocha frenguellii* (Table 8). Several key taxa identified in the McMurdo Erratics are not present in this interval of CIROS-1 (or in any overlying intervals). These taxa include *Ammodochium ampulla* (double, loricate), *Craniopsis octo*, *Micromarsupium anceps*, *Pseudammodochium dictyoides* (double), *Archaeosphaeridium dumitricae*, *Corbisema spinosa*, *Dictyocha hexacantha*, and *Naviculopsis constricta* (Table 8). The absence of these taxa in CIROS-1 is assumed to represent chronostratigraphic differences between the 500 to 485 mbsf interval in CIROS-1 and the McMurdo Erratics, rather than environmental differences, as ebridians and silicoflagellates were likely widespread throughout open-marine to neritic environments of the Eocene Antarctic shelf. The presence of *Micromarsupium anceps* and *Craniopsis octo* in Erratics D1, MtD95, E345, E350, and E364 and their absence in CIROS-1, therefore, is interpreted to indicate an age Δ34.9 Ma for these erratics.

In summary, a general age assignment of 43.7 to 34.9 Ma is interpreted for Erratics D1, MtD95, E345, E350, and E364. These ages are based on ebridian and silicoflagellate biostratigraphic distributions in Eocene sediments of the Southern Ocean and occurrences in the CIROS-1 drillcore. The maximum oldest age for the erratics is derived from the first known occurrence of *Craniopsis octo* in the Southern Ocean, which occurs in the nannofossil *Reticulofenestra umbilica* Zone at DSDP Site 512 [Bohaty, unpubl. data]. The base of this nannofossil zone is defined by the first occurrence of *Reticulofenestra umbilica* and is dated at 43.7 Ma [Berggren et al., 1995]. The maximum youngest age for the erratics is derived from the lowermost, open-marine sediments of the CIROS-1 drillcore (at ~500 mbsf) [Harwood, 1989]. This level of the drillcore is dated at ~34.9 Ma [Wilson et al., 1998], and the absence of several key siliceous microfossil taxa in this interval suggests an age greater than 34.9 Ma. This youngest age interpretation is further supported by the presence of *Naviculopsis constricta* in the erratics, which has a LAAD in the Southern Ocean at 34.7 Ma [Ciesielski, 1991; Berggren et al., 1995].

DISCUSSION

Ages derived from ebridian and silicoflagellate biostratigraphy for Erratics D1, MtD95, E345, E350, and E364 are in agreement with age assignments determined from other microfossil groups (Table 9). From dinoflagellate biostratigraphy, Erratic D1 is assigned a late middle to late Eocene age, and Erratics E345, E350, E364, and MB181 are assigned middle to late Eocene ages [Levy, 1998; Levy and Harwood, this volume a]. Erratic MtD95 was not prepared for dinoflagellate examination. Similarly, middle to late Eocene ages are assigned to these erratics, based on diatom biostratigraphy [Harwood and Bohaty, this volume]. Distinct diatom assemblages are present in several of the Eocene erratics [Harwood and Bohaty, this volume], suggesting an age separation between these erratics that is not resolvable by ebridian or silicoflagellate biostratigraphy. Restricted ages within the middle to late Eocene are interpreted for Erratics D1, E350, and MtD95, based on Southern Ocean ranges of key diatom taxa [Harwood and Bohaty, this volume]. Erratic D1 is assigned a middle to early late Eocene age; Erratic E350 is assigned a late Eocene age; and Erratic MtD95 is assigned a middle Eocene age [Harwood and Bohaty, this volume].

The source strata from which Erratics D1, MtD95, E345, E350, E364, and MB181 were derived are interpreted to have been deposited during the middle to late Eocene, prior to major ice buildup in East Antarctica [Levy, 1998]. Integrated facies interpretations and biostratigraphic age constraints for a large suite of McMurdo Erratics provide data bearing on this issue [Levy, 1998; Levy and Harwood, this volume b]. Erratics assigned middle to late Eocene ages do not contain definitive sedimentological evidence of glacial activity, such as the presence of outsized clasts. Many younger erratics (those assigned Oligocene through Miocene ages), however, are of diamictite and mudstone facies, interpreted to have been deposited in a glaciomarine environment [Levy, 1998].

BASIS FOR FUTURE WORK

A number of ebridian, silicoflagellate, and chrysophyte cyst taxa show restricted occurrences and age distributions in Ross Sea drillcores (Figure 5). Refined correlation of current Ross Sea drillcores and the recovery of Eocene sections in the future will provide a framework from which to evaluate ranges of specific taxa. Future drilling will also enable construction of a Ross Sea siliceous microfossil zonation, using many of these taxa. Several trends are noted at the present time, which may eventually form the basis of a zonation:

1. Eight taxa are restricted to the McMurdo Erratics (presently assigned late middle to late Eocene ages) and have not been observed in any Ross Sea drillcores (Figure 5). These taxa include *Ammodochium ampulla* (double, loricate), *Craniopsis octo*, *Micromarsupium anceps*, *Pseudammodochium dictyoides* (double), *Corbisema spinosa*, *Dictyocha hexacantha*, and *Naviculopsis constricta*.

2. Six taxa are present in the McMurdo Erratics and also in the lower section of CIROS-1 (702 to 366 mbsf) and CRP-2/2A (624 to 444 mbsf), indicating a combined age range for these taxa of middle Eocene to early Oligocene (Figure 5) [Wilson et al., 1998; Cape Roberts Science Team, 1999]. These taxa include *Pseudammodochium dictyoides* (single), *Ebrinula paradoxa*, *Ebriopsis crenulata* (loricate), *Parebriopsis fallax*, *Archaeosphaeridium australensis*, and *Archaeosphaeridium tasmaniae*.

3. Four taxa are not present in the McMurdo Erratics, but are reported from the CIROS-1, CRP-1, CRP-2/2A, and DSDP Hole 272 drillcores (Figure 5). These taxa include *Hovassebria brevispinosa/ Falsebria ambigua*, *Pseudammodochium lingii*, *Dictyocha deflandrei*, and *Septamesocena pappii*. Age assignments for these cores range from the early Oligocene to middle Miocene [Savage and Ciesielski, 1983; Harwood et al., 1989b; Wilson et al., 1998; Cape Roberts Science Team, 1998; Cape Roberts Science Team, 1999].

CONCLUSION

Siliceous microfossil assemblages recovered from McMurdo Erratics represent Eocene through Pleistocene ages based on diatom, dinoflagellate, ebridian, and silicoflagellate biostratigraphic age assignments [Harwood and Bohaty, this volume; Levy and Harwood, this volume a; this paper]. Several erratics in the older suite, Erratics D1, MtD95, E345, E350, E364, and MB181, contain diverse ebridian, silicoflagellate, and chryso-

TABLE 9. Microfossil age summary for Eocene erratics considered in the present study. Dinoflagellate age assignments are from Levy and Harwood [this volume a], and diatom age assignments are from Harwood and Bohaty [this volume].

Erratic	Ebridians & Silicoflagellates	Dinoflagellates	Diatoms
D1	43.7 to 34.9 Ma	late middle to late Eocene	middle to early late Eocene
MtD95	43.7 to 34.9 Ma	-	middle Eocene
E345	43.7 to 34.9 Ma	middle to late Eocene	middle to late Eocene
E350	43.7 to 34.9 Ma	middle to late Eocene	late Eocene
E364	43.7 to 34.9 Ma	middle to late Eocene	middle Eocene
MB181	Δ 33.7 Ma	middle to late Eocene	middle to late Eocene

phyte cyst assemblages and are assigned a middle to late Eocene age from ebridian and silicoflagellate biostratigraphy. Ebridian assemblages in these erratics correlate to the proposed *Micromarsupium anceps* Zone from ODP Hole 748B, defined by the FO of *Craniopsis octo* and the LO of *Micromarsupium anceps*. This ebridian zone corresponds to the Southern Ocean nannofossil *Reticulofenestra umbilica* through *Reticulofenestra oamaruensis* Zones [Wei et al. 1992] and suggests a maximum age of 43.7 Ma based on the FO of *R. umbilica* and a minimum age of 33.7 Ma based of the LO of *R. oamaruensis* [Berggren et al., 1995].

The interpreted age for Erratics D1, MtD95, E345, E350, and E364 is further constrained by comparison to siliceous microfossil occurrences and distributions in the CIROS-1 drillcore. The absence of *Micromarsupium anceps* and *Craniopsis octo* in well-preserved assemblages at ~500 mbsf in CIROS-1 suggests these erratics are derived from older sequences. Sediments in CIROS-1 at ~500 mbsf are assigned an age of ~35.0 Ma [Wilson et al., 1998], thereby constraining the interpreted age range for these erratics to 43.7 to 35.0 Ma.

With continued work, ebridians may prove to be of greater utility in Paleogene biostratigraphy and contribute to age control provided by other microfossil groups. Further documentation of the stratigraphic ranges of ebridian and silicoflagellate taxa will enable the construction of an Eocene through Miocene zonal scheme for the Ross Embayment. Siliceous microfossil biostratigraphy may also prove valuable in correlating McMurdo Erratics to future Paleogene cores recovered on the Antarctic shelf and help place the paleontological information documented in the McMurdo Erratics in a firmer stratigraphic context.

SYSTEMATIC PALEONTOLOGY

Ebridians

Ebridian taxonomy applied in the present study closely follows Perch-Nielsen [1975a] and, where original references are not available, relies on descriptions and illustrations collected by Loeblich et al. [1968]. Tappan [1980] gives detailed synopses of ebridian families and associated genera, but the relationships of ebridian taxa above the generic level are not well understood; most ebridian taxa are known only as fossils. Extensive ebridian taxonomic and/or biostratigraphic work in the Southern Ocean has not been undertaken since Perch-Nielsen [1975a]. Other reports of Southern Ocean ebridian occurrences are documented in Busen and Wise [1977], White [1980], Gombos [1982], Ling [1984], Harwood [1986a,b,c], Harwood [1989], and McCartney and Wise [1990]. Many ebridian taxa recognized in the McMurdo Erratics and ODP Hole 748B were first described from the upper Eocene Oamaru Diatomite Member of the Waiareka Volcanics Formation in New Zealand [see Edwards, 1991]. Original descriptions and plate references are given with secondary references that contain descriptions, plates, and figures used in ebridian identification in the present study.

Varieties of many ebridian species are separated in the abundance tables and in the following section on systematic paleontology. These morphologies are designated in parentheses following the species names. "Single" refers to single skeleton morphologies, "double" refers to double skeletons, and "loricate" refers to those specimens with a silicified chamber or lorica. Some varietal forms, such as the double skeleton, lori-

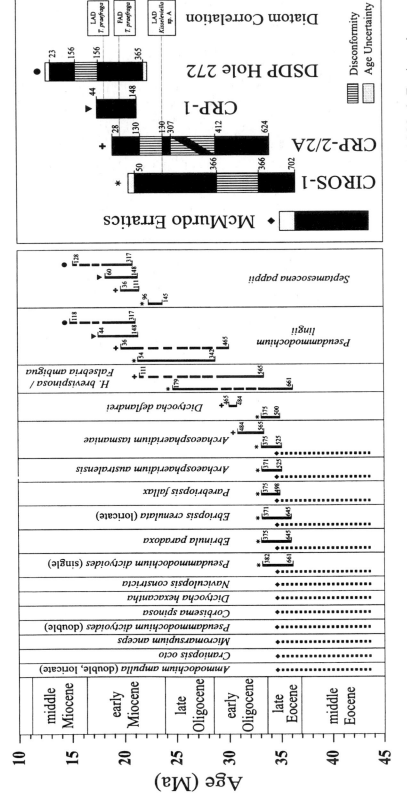

Fig. 5. Summary of observed and reported ranges of selected ebridian, silicoflagellate, and chrysophyte cyst taxa from the McMurdo Erratics and Ross Sea drillcores. Individual taxon ranges are plotted for each core, where V = CIROS-1, 9 = CRP-2/2A, t = CRP-1, 1 = DSDP Hole 272, and u = occurrence in McMurdo Erratics D1, E345, E350, E364, and MtD95. Core depths (in mbsf) are noted for each datum. Shaded regions represent uncertainty in age correlation, and ruled areas represent age gaps due to disconformities. Ages in the upper section of CIROS-1 (366 to 50 mbsf) are currently under revision. Age data and taxon ranges compiled from data in this paper and from White [1980], Savage and Ciesielski [1983], Harwood [1989], Harwood et al., [1989b], Wilson et al. [1998], Harwood et al. [1998], Cape Roberts Science Team [1998], Harwood et al. [1998], Cape Roberts Science Team [1999].

cate morphology of *Ammodochium ampulla*, appear to be more stratigraphically restricted than the entire range of the nominative taxon.

Genus *Adonnadonna* Gombos, 1982

Adonnadonna primadonna Gombos, 1982
Plate 9, fig. 7

Adonnadonna primadonna Gombos, 1982, p. 446, pl. 1, figs. 8 and 9; McCartney and Wise, 1990, p. 751, pl. 7, fig. 4; Lurvey et al., 1998, p. 194, pl. 4, fig. 1.

Remarks. Although similar in structure to *Triskelion gorgon*, this taxon is characterized by a central circular ring from which apical elements radiate. Specimens of *Adonnadonna primadonna* observed in the present study also possess narrower basal rims than *T. gorgon*. *Adonnadonna primadonna* is morphologically distinct from *T. gorgon* and merits continued designation as a separate species. These taxa, however, should probably be included under the same generic designation, based on morphological similarities [K. McCartney, pers. comm., 1997].

Occurrence. *Adonnadonna primadonna* is recorded in one sample in the upper Eocene of ODP Hole 748B, where it is associated with *T. gorgon*. It also occurs with *T. gorgon* in the middle Eocene on the Falkland Plateau, southwest Atlantic Ocean [Gombos, 1982], and in the upper Eocene of ODP Hole 689B, Maud Rise, Weddell Sea (in the nannofossil *Reticulofenestra oamaruensis* Zone) [McCartney and Wise, 1990; Wei and Wise, 1990].

Genus *Ammodochium* Hovasse, 1932a

Ammodochium ampulla Deflandre, 1934
Plate 1, fig. 1; Plate 3, figs. 1-3; Plate 4, figs. 4 and 11; Plate 10, figs. 13 and 16

Ammodochium ampulla Deflandre, 1934, p. 77, fig. 2; Perch-Nielsen, 1975a, p. 880, pl. 4, figs. 17, 18, and 29, pl. 5, figs. 23-26.

Remarks. Although similar to *Ammodochium rectangulare*, this taxon is identified by its small pro-clade and opisthoclade windows. Both single and weakly-silicified double skeletons of *Ammodochium rectangulare* are identified in the present study. A third varietal group, which possess a heavily-silicified, double skeleton with a lorica is also noted. Due to silicification, the diagnostic opisthoclade and proclade windows could not

be identified on all *A. ampulla* (double, loricate) specimens. Double skeleton loricate varieties of *A. ampulla* are typically much larger than single skeleton specimens and may represent a separate taxon.

Occurrence. Single skeleton varieties of *A. ampulla* are recorded in middle Eocene to lower Oligocene sediments of ODP Hole 748B, but are known to range into the Neogene [McCartney and Wise, 1990]. *Ammodochium ampulla* (double, loricate) may have a more restricted range, occurring only in the middle to upper Eocene of Hole 748B in the nannofossil *Cribrocentrum reticulatum* through *Blackites spinosus* Zones, 42. 0 to 31.8 Ma [Wei et al., 1992; Berggren et al., 1995]. *Ammodochium ampulla* (double, loricate) also occurs in the middle Eocene of DSDP Hole 512 [Bohaty, unpubl. data] in the nannofossil *Reticulofenestra umbilica* Zone (43.7 to 40.2 Ma) [Wise, 1983; Berggren et al., 1995]. Perch-Nielsen [1975a] did not separate the different forms of *A. ampulla*, but reported a similar range of all forms from the upper Eocene to lower Oligocene of DSDP sites 277, 283, and 281, southwestern Pacific Ocean.

Ammodochium danicum Deflandre, 1951
Plate 3, figs. 9 and 10

Ammodochium danicum Deflandre, 1951, p. 53, figs. 13 and 14; Locker, 1996, p. 114, pl. 5, fig. 1.

Remarks. Unlike *Ammodochium ampulla*, this taxon possesses only anterior proclade windows.

Occurrence. *Ammodochium danicum* was rare in the present study, occurring only in Erratic E345 and in one sample in the middle Eocene of ODP Hole 748B.

Ammodochium novum Perch-Nielsen, 1978

Ammodochium novum Perch-Nielsen, 1978, p. 152, pl. 8, figs. 13 and 14; Locker, 1996, p. 114, pl. 5, fig. 13.

Remarks. *Ammodochium novum* is characterized by anterior-posterior asymmetry and wide proclade and opisthoclade elements. Three pores are also present on each proclade and opisthoclade element.

Occurrence. This taxon has been previously documented only in Eocene sediments of the Norwegian-Greenland Sea [Perch-Nielsen, 1978; Locker, 1996].

Ammodochium rectangulare (Schulz) Deflandre, 1933
Plate 1, figs. 2 and 3; Plate 3, figs. 4-8; Plate 4, figs. 7 and 8; Plate 5, fig. 9; Plate 9, fig. 11; Plate 10, fig. 14

Ebria antiqua var. *rectangularis* Schulz, 1928, p. 274, figs. 72a-d.

Ammodochium rectangulare (Schulz) Deflandre, 1933, pp. 517-518, figs. 5-7; Locker and Martini, 1986a, p. 943, pl. 1, fig. 7; Harwood, 1989, p. 82, pl. 6, fig. 21; Locker, 1996, p. 114, pl. 5, fig. 2.

Remarks. Locker and Martini [1986a] restrict the use of *Ammodochium rectangulare* to robust morphologies with oval-shaped openings between the triode and surrounding elements. This morphology is reported to occur only in Paleogene sediments [Locker and Martini, 1986a]. *Ammodochium serotinum* is applied to similar Neogene *Ammodochium* morphologies that are more delicate and possess semi-circular openings between the triode and surrounding elements [Locker and Martini, 1986a].

Several varietal forms of *A. rectangulare* are noted from the McMurdo Erratics and ODP Hole 748B. Most single skeletons show smooth surface ornamentation. Some morphologies, however, are hyper-silicified with crenulate to spiny surface ornamentation. One loricate, single skeleton was also observed in Erratic E364. Double skeletons of *A. rectangulare* range from weakly-silicified to heavily-silicified and vary significantly in size. Some heavily-silicified double skeletons posses anterior silicification that may represent the initial silica added to form a loricate stage. Double skeletons of *A. rectangulare* with silicification on both the anterior and posterior ends were noted in Sample 14H-5, 47-49 (>20 µm), in Hole 748B. Loricate, double-skeleton *A. rectangulare* specimens were not observed in the present study [see Ling, 1985a, pl. 2, figs. 11 and 12], but may have been misidentified as *A. ampulla* (double, loricate).

Ammodochium speciosum Deflandre, 1934
Plate 3, figs. 13 and 14; Plate 11, figs. 1-5

Ammodochium speciosum Deflandre, 1934, pp. 92-94, figs. 37 and 38; Perch-Nielsen, 1975a, p. 880, pl. 5, figs. 1 and 2.

Remarks. Both *Ammodochium speciosum* and *Craniopsis octo* are similar in structure, possessing mesoclade windows opposed at 180°. *Ammodochium speciosum* can be distinguished from other *Ammodochium* spp. by its wide apical elements, with large anterior and posterior windows, and its box-like morphology. The central area of the apical element is either a solid-silica bar (see pl. 11, fig. 1) or perforated by two elongate pores (see pl. 11, fig. 4).

Occurrence. *Ammodochium speciosum* occurs in middle Eocene to lower Oligocene sediments of ODP Hole 748B. Perch-Nielsen [1975a] also reports this taxon in upper Eocene sediments of DSDP Site 277, southwest Pacific Ocean.

Ammodochium sp. 1
Plate 3, figs. 15 and 16

Remarks. This unknown morphology is noted in both the McMurdo Erratics and ODP Hole 748B, although it is extremely rare. It is characterized by curved proclades and opisthoclades with windows present on the anterior and posterior ends, respectively, of these elements. Few well-preserved specimens were observed and the characteristics of this morphology are not fully apparent. Several unknown varieties of *Ammodochium* species may be represented in this group.

Genus *Craniopsis* Hovasse *ex* Frenguelli, 1940

Craniopsis octo Hovasse *ex* Frenguelli, 1940
Plate 1, fig. 10; Plate 4, fig. 6; Plate 11, figs. 8 and 9

Craniopsis octo Hovasse *ex* Frenguelli, 1940, p. 95, figs. 31a,b; Perch-Nielsen, 1975a, p. 880, pl. 4, figs. 1-10; White, 1980, p. 156, pl. 8, fig. 3; McCartney and Wise, 1990, p. 751, pl. 7, fig. 5.

Remarks. Both single and double skeleton varieties of *Craniopsis octo* are identified in ODP Hole 748B. Various degrees of silicification are also noted in this taxon, as illustrated by Perch-Nielsen [1975a]. Some

---

Plate 1.

SEM photomicrographs, scale bars equal 10 µm. Figure 1. Ammodochium ampulla Deflandre; (1) Loricate double skeleton, Erratic D1. Figures 2-3. Ammodochium rectangulare (Schulz) Deflandre; (2) Double skeleton with anterior and medial silicification, Erratic D1; (3) Single skeleton, Erratic D1. Figures 4-9. Ebriopsis crenulata Hovasse emend.; (4, 5) Erratic D1; (6) Loricate, Erratic D1; (7) Loricate, Erratic E345; (8, 9) Loricate, Erratic D1. Figure 10. Craniopsis octo Hovasse ex Frenguelli; (10) Erratic D1. Figure 11. Corbisema spinosa Deflandre; (11) Erratic E345. Figure 12. Distephanus speculum speculum f. pseudofibula Schulz; (12) Erratic D1.

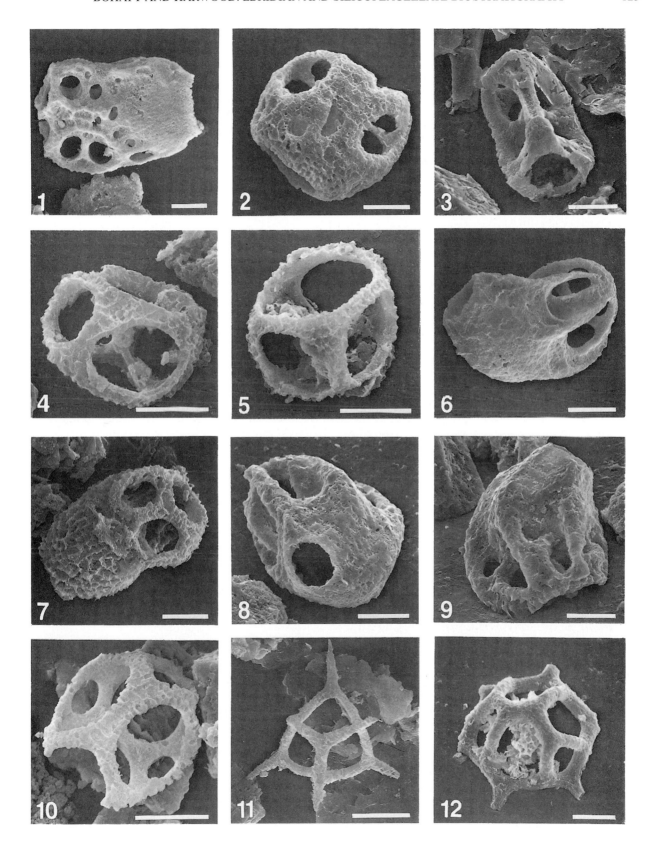

specimens of *C. octo* possess a single, posterior pore on the apical element. The anterior synclade element often breaks off on many specimens of this taxon, giving a two-pronged appearance. *Craniopsis octo* can be distinguished from *Ammodochium speciosum* by its anterior-posterior asymmetry (see notes under *A. speciosum*).

Occurrence. *Craniopsis octo* was originally described and illustrated from the upper Eocene Oamaru Diatomite in New Zealand [Frenguelli, 1940]. In Southern Ocean sediments, it is distributed through the middle Eocene to lowermost Oligocene. In ODP Hole 748B, it occurs in the middle Eocene to upper Eocene and in one sample in the lowermost Oligocene. Perch-Nielsen [1975a] reports this species in the upper Eocene from DSDP sites 277 and 283, southwestern Pacific Ocean. It is also present in upper Eocene to lower Oligocene sediments of ODP Hole 689B, Maud Rise, Weddell Sea (in the nannofossil *Isthmolithus recurvus* through *Blackites spinosus* Zones) [McCartney and Wise, 1990; Wei and Wise, 1990].

The first occurrence datum of *Craniopsis octo* is used in the present study to define the base of the *Micromarsupium anceps* Zone. In ODP Hole 748B, the first occurrence of this taxon occurs within the lower part of the nannofossil *Cribrocentrum reticulatum* Zone (42.0 to 40.4 Ma) [Wei et al., 1992; Berggren et al., 1995]. *Craniopsis octo* is also present in the middle Eocene of DSDP Site 512, Maurice Ewing Bank [Bohaty, unpubl. data], in the nannofossil *Reticulofenestra umbilica* Zone (43.7 to 42.0 Ma) [Wise, 1983; Berggren et al., 1995]. This discrepancy in the first occurrence of *Craniopsis octo* results from either diachroneity between the two drillcores or from preservational factors, where its true first occurrence is not represented in Hole 748B.

Genus *Ebrinula* Deflandre, 1950a

Ebrinula paradoxa (Hovasse) Deflandre, 1950a
Plate 2, fig. 1; Plate 3, figs. 11 and 12; Plate 4, fig. 5; Plate 11, figs. 13 and 14

Ammodochium prismaticum var. *paradoxum* Hovasse, 1932c, p. 462, fig. 11.
Ebrinula paradoxa (Hovasse) Deflandre, 1950a, p. 1780, figs. 1-4.

Remarks. Both single and double-skeleton varieties of *Ebrinula paradoxa* are noted in samples from the McMurdo Erratics. Under low magnification, the rectangular outline of *E. paradoxa* resembles *Ammodochium rectangulare*, and, depending on the orientation of the

specimen, it may be difficult to distinguish between these two species. *Ebrinula paradoxa*, however, is characterized by a fourth, free element that is not attached to the triode, which is distinctly apparent in apical or antapical view. The free element of *E. paradoxa* is commonly broken off, and, consequently, broken specimens of *E. paradoxa* are difficult to distinguish from *A. rectangulare*.

Occurrence. This taxon is consistently found in the upper Eocene to lower Oligocene of Hole 748B. It is also present in the middle Eocene of DSDP Site 512 [Bohaty, unpubl. data] in the Southern Ocean nannofossil *Reticulofenestra umbilica* Zone [Wise, 1983; Berggren et al., 1995]. *Ebrinula paradoxa* has not been previously reported in DSDP or ODP literature, but may have been misidentified as *A. rectangulare*.

Genus *Ebriopsis* Hovasse, 1932a

Ebriopsis antiqua antiqua (Schulz) Ling, 1977
Plate 5, figs. 4 and 7

Ebria antiqua Schulz, 1928 (in part), pp. 273, fig. 69a-f; Ling, 1971 (in part), p. 693, pl. 1, figs. 21-23; Ling, 1972 (in part), p. 197, pl. 32, figs. 8-10; Perch-Nielsen, 1975a, p. 880, pl. 4, fig. 15; Ling, 1980, p. 380, pl. 2, fig. 16.
Ebriopsis antiqua antiqua (Schulz) Ling, 1977, p. 215, pl. 3, figs. 17 and 18.
non *Ebriopsis antiqua antiqua* (Schulz) Ling; McCartney and Wise, 1987, p. 807, pl. 5, figs. 10, 12, and 13; McCartney and Wise, 1990, p. 751, pl. 7, fig. 6.

Remarks. Although similar to *Ebriopsis crenulata*, specimens identified in the present study as *E. antiqua antiqua* have flattened apices where the elements join together at the margins (see remarks under *E. crenulata*). A range of morphologies between the designated end members of *E. antiqua antiqua*, *E. crenulata*, and *Parebriopsis fallax*, however, occur in Core 15H of ODP Hole 748B.

Occurrence. *Ebriopsis antiqua antiqua* was observed only in the upper Eocene of ODP Hole 748B. This taxon, however, also ranges into the Neogene [Ling, 1992].

Ebriopsis crenulata (Hovasse) emend.
Plate 1, figs. 4-9; Plate 5, figs. 10 and 11; Plate 10, figs. 9, 10, and 15

Ebriopsis crenulata Hovasse, 1932b, p. 281, fig. 4; Ling, 1972, pp. 197-198, pl. 32, figs. 13-18; Perch-Nielsen, 1975a, p. 880, pl. 4, figs. 11-14, pl. 5, figs. 18-21; Ling, 1980, p. 380, pl. 1, figs. 17-19; Martini, 1981, pl. 1, fig. 2; Locker and Martini, 1986a, p. 943, pl. 1, figs. 10 and 11; McCartney and Wise, 1987, p. 807, pl. 5, fig. 11; Harwood, 1989, p. 82, pl. 6, figs. 26 and 27.

Ebriopsis antiqua Schulz; Ling, 1972 (in part), p. 197, pl. 32, figs. 11 and 12.

Ebriopsis mesnili Deflandre; Dzinoridze et al., 1978, pl. 11, figs. 15-16.

Ebriopsis antiqua antiqua (Schulz) Ling; McCartney and Wise, 1987, p. 807, pl. 5, figs. 10, 12, and 13; McCartney and Wise, 1990, p. 751, pl. 7, fig. 6.

Remarks. In the present study, *Ebriopsis crenulata* is emended to include rounded, robust morphologies with both smooth and crenulate surface ornamentation. Most workers have previously only designated morphologies with crenulate surface texture as *E. crenulata* [Ling, 1972; McCartney and Wise, 1990]. Crenulate surface texture on many specimens of this taxon may be due to dissolution, so it is necessary to separate *Ebriopsis* taxa based on structural concepts. In ODP Hole 748B, *E. crenulata* is distinguished from *E. antiqua antiqua* by its more rounded nature without flattened apices and, commonly, its more silicified elements.

Both non-loricate and loricate stages of *E. crenulata* occur in the McMurdo Erratics and Hole 748B. Specimens identified as hyper-silicified varieties of *E. crenulata* may represent broken loricate stages or the beginning stage of lorica development. In non-loricate specimens, both *E. crenulata* and *E. antiqua antiqua* are difficult to recognize in side view due to their flattened, sub-spherical nature [see Tappan, 1980, fig. 5.16].

Occurrence. Non-loricate forms of *E. crenulata* are distributed through the entire middle Eocene to upper Oligocene study interval of ODP Hole 748B. Loricate stages of *E. crenulata* show a more restricted distribution, occurring only in middle to upper Eocene sediments of Hole 748B and in late Eocene to early Oligocene sediments of CIROS-1. Loricate and non-loricate forms of *Ebriopsis crenulata* are also present in upper Eocene to lower Oligocene sediments (645.29 to 371.06 mbsf) of the CIROS-1 drillcore (see Figure 5) [Harwood, 1989; Harwood et al., 1989b; Wilson et al., 1998; this paper]

Ebriopsis sp. 1
Plate 6, figs. 8-10

Remarks.—Several specimens of unknown *Ebriopsis* morphologies (similar to *Ebriopsis crenulata*) were identified in Erratic E364. These morphologies have additional elements arranged in a tri-radial manner that result in a three-tiered, spherical structure.

Genus *Falsebria* Deflandre, 1951

Falsebria ambigua Deflandre, 1951

Falsebria ambigua Deflandre, 1951, pl. 37, figs. 19-21.

Remarks. In the present study, *Falsebria ambigua* is grouped with *Hovassebria brevispinosa*. Some overlap exists in the taxonomic definition of these two taxa. Both possess three radiating elements (tri-radial symmetry) joined at 120° in the same plane. Two of the elements in *Hovassebria brevispinosa* are commonly closed by a loop, and the loop is absent in *Falsebria ambigua*.

Occurrence. The stratigraphic distribution of the *Hovassebria brevispinosa/ Falsebria ambigua* group in the Southern Ocean is upper Eocene through lower Miocene. *Falsebria ambigua*, however, is originally described from Paleocene sediments of the Fuur Formation in Denmark [Deflandre, 1951], and *Hovassebria brevispinosa* is originally described from the upper Eocene Oamaru Diatomite in New Zealand [Hovasse, 1932b]. This age separation may suggest the forms observed in the present study represent varietal forms of *Hovassebria brevispinosa*. This is consistent with original illustrations of *Hovassebria brevispinosa* [Hovasse, 1932b], showing forms with and without the closed loop structure.

Falsebria imitata Deflandre, 1950b

Falsebria imitata Deflandre, 1950b, p. 159, fig. 8.

Remarks. This taxon may represent incompletely formed stages or broken parts of other species such as *Ebriopsis crenulata* or *Parebriopsis fallax*.

Falsebria sp. 1

Remarks. Heavily-silicified specimens similar to *Falsebria* morphologies reported by Perch-Nielsen [1978, pl. 8, figs. 9-11] occur in Erratic D1.

Genus *Haplohermesinum* Hovasse, 1943

Haplohermesinum cornuta (Dumitrica and Perch-Nielsen) Locker, 1996.

Ebriopsis cornuta Dumitrica and Perch-Nielsen in Perch-Nielsen, 1975a, p. 880, text-fig. 2, pl. 7, figs. 8 and 9; Perch-Nielsen, 1978, p. 153, pl. 9, figs. 1-5, pl. 10, fig. 13; Ling, 1985b, p. 85, pl. 11, fig. 24, pl. 13, fig. 4; Dell'Agnese and Clark, 1994, text-fig. 5, fig. #8.

Haplohermesinum cornuta (Dumitrica and Perch-Nielsen) Locker, 1996, p. 114, pl. 5, fig. 19.

Remarks. To avoid confusion with the homonymic *Ebriopsis cornuta* (Ling) Locker and Martini 1986a, this taxon was reassigned to the genus *Haplohermesinum* by Locker [1996]. *Haplohermesinum cornuta* (Dumitrica and Perch-Nielsen) Locker is similar in primary structure to *Haplohermesinum transversa* Deflandre 1934, but possesses an additional anterior synclade (with or without a spine) and a posterior spine.

In ODP Hole 748B, both delicately-silicified and robust, heavily-silicified forms of *H. cornuta* are recorded.

Occurrence. Only a few specimens of this taxon were observed in upper Eocene samples of Hole 748B. Perch-Nielsen [1975a] also records *H. cornuta* in the upper Eocene of DSDP Site 277, southwest Pacific Ocean, and Perch-Nielsen [1978] records a similar occurrence in upper Eocene sediments of DSDP sites 338 and 340, Vøering Plateau, Norwegian Sea. *Haplohermesinum cornuta* is also reported from Eocene sediments recovered in the Arctic Ocean [Magavern et al., 1996] and in Eocene to Oligocene sediments of the Norwegian-Greenland Sea [Locker, 1996]. *Hermesinella transversa*, a morphologically similar taxon, occurs in upper Eocene sediments of DSDP Site 281 [Perch-Nielsen, 1975a].

Genus *Hermesinum* Zacharias, 1906

***Hermesinum geminum* Dumitrica and Perch-Nielsen in Perch-Nielsen, 1975a**
Plate 9, fig. 3

Hermesinum geminum Dumitrica and Perch-Nielsen in Perch-Nielsen, 1975a, pp. 880-881, text-figs. 3-5, pl. 4, fig. 16, pl. 8, figs. 1-16; Perch-Nielsen, 1978, p. 153, pl. 10, fig. 12; White, 1980, pp. 160-161, pl. 8, figs. 10 and 11.

Remarks. All specimens of *Hermesinum geminum* recorded in the present study are double skeletons - an observation which is consistent with reports from other Southern Ocean drillcores [Perch-Nielsen, 1975a, 1978].

Occurrence.—In the present study, *Hermesinum geminum* is recorded only in the lower Oligocene of ODP Hole 748B. From other Southern Ocean sites, this taxon is reported in upper Eocene to lower Oligocene sediments in the southwestern Pacific Ocean [Perch-Nielsen, 1975a]. From northern high latitude cores in the Norwegian-Greenland Sea, *Hermesinum geminum* occurs in middle Oligocene [Perch-Nielsen, 1978] and upper Oligocene [Locker, 1996] sections.

Small morphologies of *H. geminum*, different from those observed in the present study, occur in one sample in the upper Oligocene of DSDP Site 278 [Perch-Nielsen, 1975a]. The stratigraphic distribution of the small and large varieties of this taxon is unknown at the present time.

The last occurrence datum of large *Hermesinum geminum* morphologies (>100 μm in length) in the lower Oligocene may be a useful biostratigraphic marker in the Southern Ocean. This datum is used in the present study to define the top of the ebridian *Hermesinum geminum* Zone and occurs in the middle of the nannofossil *Chiasmolithus altus* Zone, 31.3 to 26.1 Ma [Wei et al., 1992; Berggren et al., 1995].

Genus *Hovassebria* Deflandre, 1936

***Hovassebria brevispinosa* (Hovasse) Deflandre, 1936**
Plate 5, fig. 5

Cornua brevispinosa Hovasse, 1932b, p. 281, fig. 5.
Hovassebria brevispinosa (Hovasse) Deflandre, 1936, p. 73, fig 131 (I-III); Harwood, 1989, p. 82, pl. 6, fig. 25.

Remarks. Morphologies designated as *Hovassebria brevispinosa* possess three radiating elements in the same plane (tri-radial symmetry). Commonly, two of the ele-

---

Plate 2.

SEM Photomicrographs, scale bars equal 10 μm. Figure 1. *Ebrinula paradoxa* (Hovasse) Deflandre; (1) Erratic E345. Figures 2-3. *Parebriopsis fallax* Hovasse; (2) Erratic D1; (3) Hyper-silicified crest, Erratic D1. Figures 4-5, 10. *Pseudammodochium dictyoides* Hovasse; (4) Single skeleton, Erratic D1; (5) Broken double skeleton (note internal triodes), Erratic D1; (10) Double skeletons, Erratic D1. Figures 6-9. *Micromarsupium anceps* Deflandre; (6) Erratic D1; (7) Erratic D1; (8) Erratic D1; (9) Erratic E345.

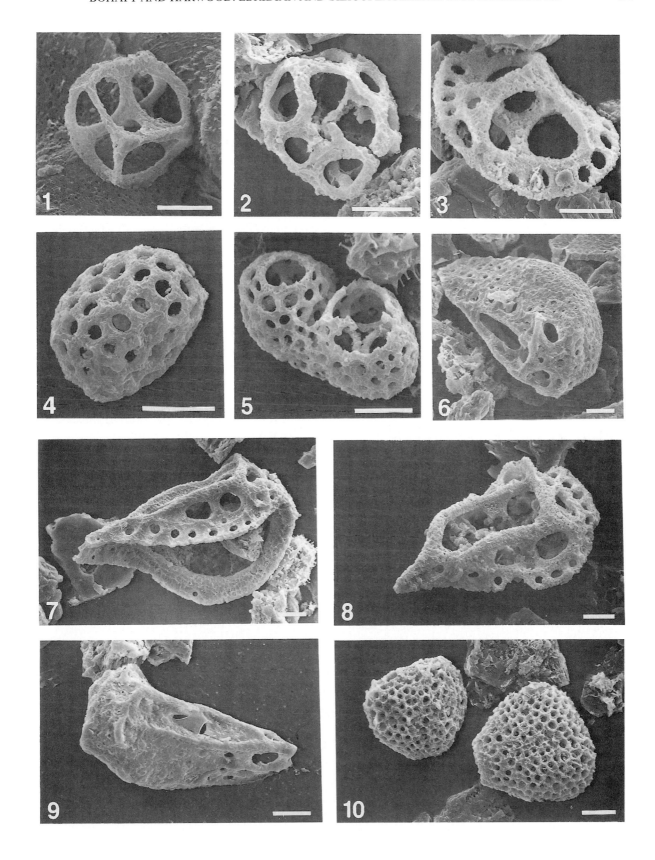

ments (or actines) are connected by a closed loop, which may be filled with silica. Broken pieces of *Parebriopsis fallax* or *Ebriopsis crenulata* may resemble *H. brevispinosa* but are not as heavily silicified. In the present study, *Hovassebria brevispinosa* and *Falsebria ambigua* are grouped together (see notes under *Falsebria ambigua*). *Falsebria ambigua* is similar in basic morphology to *H. brevispinosa* but does not possess a loop structure.

Occurrence. The Southern Ocean stratigraphic distribution of the *Hovassebria brevispinosa/ Falsebria ambigua* group is upper Eocene through lower Miocene. These forms occur in the upper Eocene to lower Oligocene of ODP Hole 748B and are also recorded in upper Oligocene to lower Miocene sediments of the MSSTS-1 drillcore [Harwood, 1986c], in upper Eocene to upper Oligocene sediments of the CIROS-1 drillcore [Harwood, 1989], and Oligocene to lower Miocene sediments of the CRP-2/2A drillcore [Cape Roberts Science Team, 1999].

Hovassebria sinistra Deflandre, 1951

Hovassebria sinistra Deflandre, 1951, pp. 34 and 75, figs. 95-98.

Remarks. This morphology may represent broken or incompletely formed ebridian fragments derived from various taxa.

Genus *Micromarsupium* Deflandre, 1934

Micromarsupium anceps Deflandre, 1934
Plate 2, figs. 6-9; Plate 6, figs. 3, 4, 6, and 7;
Plate 7, figs. 1-12

Micromarsupium anceps Deflandre, 1934, pp. 86-88, figs. 20-32; Perch-Nielsen, 1975a (in part), p. 881, pl. 6, figs. 3, 4, and 6-11, pl. 7, figs. 1-5, 11, and 13; White, 1980, p. 162, pl. 8, fig. 12; Locker and Martini, 1986a, p. 994, pl. 1, figs. 8 and 9.

Remarks. Although *Micromarsupium anceps* is one of the more distinctive ebridian taxa, a wide range of morphological variation is displayed by this taxon. Both heavily and weakly ("framework") silicified morphologies have been documented [see Perch-Nielsen, 1975a]. Heavily-silicified specimens are more easily preserved, and, commonly, weakly-silicified specimens occur only as fragments. This taxon is typically large (50 to 130 μm in length) and elongate, forming a tear-drop shape. Due to heavy silicification, many loricate *M. anceps* specimens lack a visible internal structure and are identified on gross morphology. Also, many loricate specimens exhibit a perforated, linear plate that runs parallel to the long-axis of the skeleton. Detailed SEM work is needed on *Micromarsupium anceps*, particularly on specimens from well-preserved, deep-sea samples. Future SEM observations may warrant the division of this taxon into more than one variety or species.

Occurrence. The first and last occurrence datums of *Micromarsupium anceps* may be useful stratigraphic markers in the Southern Ocean. Large, heavily-silicified varieties of *M. anceps* occur in the middle and upper Eocene of ODP Hole 748B and do not range into the Oligocene. Perch-Nielsen [1975a], however, reports *M. anceps* from upper Eocene to lower Oligocene sediments of DSDP sites 283, 281 and 280. The reported presence of this species in the lower Oligocene of Site 280 may represent forms identified in the present study as *Micromarsupium curticannum*. Locker and Martini [1986a] report *M. anceps* in the middle Eocene (nannofossil Zones NP15 to NP16) of DSDP Hole 588C, southwest Pacific Ocean. The presence of *Reticulofenestra umbilica* and *Chiasmolithus solitus* in the middle Eocene of Hole 588C [Martini, 1986] places this section in the *Reticulofenestra umbilica* through *Cribrocentrum reticulatum* Zones of Wise [1983] and Wei et al. [1992]. The maximum age range for these zones is defined by the first occurrence of *Reticulofenestra umbilica* at 43.7 Ma and the last occurrence of *Chiasmolithus solitus* at 40.4 Ma [Berggren et al., 1995]. *Micromarsupium anceps* is

---

Plate 3.

Scale bars equal 20 μm. Figures 1-3. *Ammodochium ampulla* Deflandre; (1, 2) Single skeleton, low/ high focus, Sample 748B-14H-1, 47-49; (3) Double skeleton, Sample 748B-14H-1, 47-49. Figures 4-8. *Ammodochium rectangulare* (Schulz) Deflandre; (4) Double skeleton, Sample 748B-15H-3, 47-49; (5, 6) Single skeleton, low/ high focus, Sample 748B-13H-1, 47-49; (7, 8) Hypersilicified, low/ high focus, Erratic D1. Figures 9-10. *Ammodochium danicum* Deflandre; (9, 10) Low/ high focus, Sample 748B-14H-1, 47-49. Figures 11-12. *Ebrinula paradoxa* (Hovasse) Deflandre; (11, 12) Pyritized, high/ low focus, Erratic E364. Figures 13-14. *Ammodochium speciosum* Deflandre; (13, 14) Pyritized with broken apical element, low/ high focus, Erratic E364. Figures 15-16. *Ammodochium* sp. 1; (15, 16) Low/ high focus, Erratic D1.

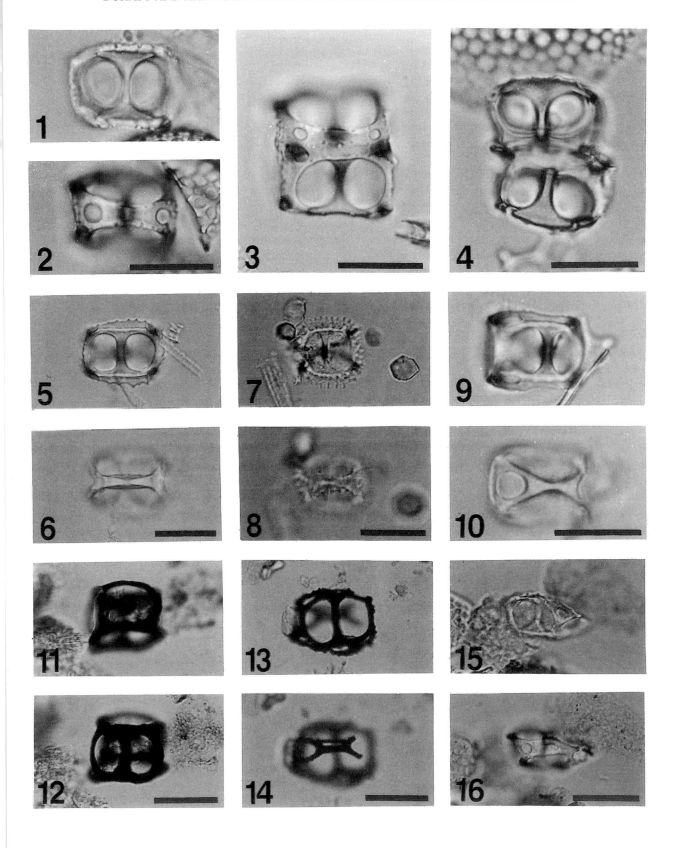

Genus *Podamphora* Gemeinhardt, 1931

Podamphora tenuis (Hovasse) Deflandre, 1951
Plate 11, fig. 7

Parebria tenuis Hovasse, 1932c, p. 459, fig. 3.
Podamphora tenuis (Hovasse) Deflandre, 1951, pp. 72-
73, figs. 2 and 137-140.

Occurrence. Two specimens of this taxon were observed in the lower Oligocene of ODP Hole 748B in Sample 748B-13H-1, 47-49.

Genus *Polyebriopsis* Hovasse, 1932c

Polyebriopsis sp. 1

Remarks. Large, morphologically indistinct ebridian specimens, possibly belonging to the genus *Polyebriopsis*, were observed in Erratics D1, E345, and E364.

Genus *Pseudammodochium* Hovasse, 1932c

Pseudammodochium dictyoides Hovasse, 1932c
Plate 2, figs. 4, 5, and 10; Plate 10, figs. 1-6

Pseudammodochium dictyoides Hovasse, 1932c, p. 463,
figs. 12-15; Dzinoridze et al., 1978, pl. 11, fig. 14;
Ling, 1984, p. 159, text-fig. 2, figs. #4 and #12;
Ling, 1985b, p. 85, pl. 11, figs. 25-27; Harwood,
1989 (in part), p. 82, pl. 6, fig. 22; Dell'Agnese and
Clark, 1994, text-fig. 5, fig. #9; Locker, 1996, p.
114, pl. 5, fig. 7.

Remarks. This taxon is characterized by a perforated, oval-shaped skeleton with a distinctive internal triode. Triodes broken out of *Pseudammodochium dictyoides*, often with fragments of the outer wall still attached, have a characteristic triangular appearance (see pl. 10, fig. 6). Both single and double skeleton specimens are common in the McMurdo Erratics and ODP Hole 748B. Many double skeleton forms are connected by a perforate silica collar. It is unclear whether double skeletons without this collar are broken specimens or originally did not possess a collar.

Occurrence. This combined Southern Ocean range of all varietal forms of *Pseudammodochium dictyoides* is middle Eocene to early Oligocene. In ODP Hole 748B, *P. dictyoides* ranges from the middle Eocene to lower Oligocene. It is present in middle Eocene sediments of DSDP Site 512 [Bohaty, unpubl. data] in the nannofossil *Reticulofenestra umbilica* Zone (43.7 to 42.0 Ma) [Wise, 1983; Berggren et al., 1995]. Perch-Nielsen [1977] also reports *P. dictyoides* from middle Eocene sediments of DSDP Site 356 in the southwestern Atlantic Ocean.

In the Ross Sea, single-skeleton morphologies of *Pseudammodochium dictyoides* are longer ranging than double skeleton morphologies (Figure 5). Double-skeleton morphologies are present only in the Eocene McMurdo Erratics and have not been observed in drillcores. Single-skeleton morphologies are present in the erratics and in the upper Eocene to lower Oligocene interval of the CIROS-1 drillcore.

From high northern latitudes, Perch-Nielsen [1978, pl. 6, figs. 18-20] reports double *Pseudammodochium* sp. morphologies in upper Eocene sediments of DSDP Site 339, Norwegian Sea. Double and single *P. dictyoides* morphologies were also observed in Eocene sediments from the Arctic Ocean [Ling, 1985b; Magavern et al., 1996].

Pseudammodochium lingii n. sp.
Plate 5, fig. 1; Plate 8, figs. 1-10

Pseudammodochium sp. cf. *P. dictyoides* Hovasse; Ling,
1984, text-fig. 2, figs. #5, #6, and #13; Harwood,
1986c, p. 87, pl. 2, figs. 16 and 17; Harwood et al.,
1989a, pl. 4, fig. 13.

Description. Skeletons consist of a pair of hollow, siliceous spheres connected by a lorica, which represents

---

Plate 5.

Scale bar for Figs. 1-8 equals 20 μm; scale bar for Figs. 9-11 equals 15 μm. Figure 1. *Pseudammodochium lingii* n. sp.; (1) Sample 748B-12H-4, 47-49. Figures 2-3. *Actiniscus elongatus* Dumitrica; (2) Sample 748B-12H-6, 47-49; (3) Sample 748B-12H-6, 47-49. Figures 4, 7. *Ebriopsis antiqua antiqua* (Schulz) Ling; (4) Sample 748B-14H-6, 47-49; (7) Sample 748B-15H-4, 47-49. Figure 5. *Hovassebria brevispinosa* (Hovasse) Deflandre; (5) Sample 748B-14H-6, 47-49. Figure 6. *Pseudammodochium sphericum* Hovasse; (6) Single skeleton, Sample 748B-15H-4, 47-49. Figure 8. *Micromarsupium curticannum* Deflandre; (8) Sample 748B-13H-2, 47-49. Figure 9. *Ammodochium rectangulare* (Schulz) Deflandre; (9) Apical view of broken single skeleton, Sample 748B-15H-3, 47-49. Figures 10-11. *Ebriopsis crenulata* Hovasse emend.; (10) High focus, Sample 748B-19H-1, 47-49; (11) Low focus.

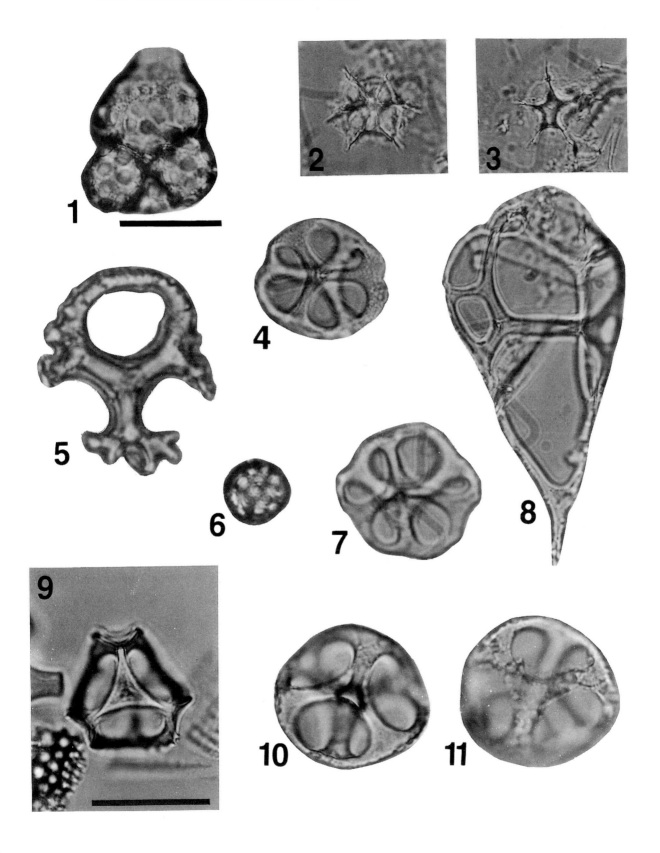

a third chamber. The spheres are arranged in either a slightly oblique or parallel fashion relative to one another. The wall structure of the spheres and lorica ranges from porous to coarsely silicified. Heavily-silicified specimens are ornamented with numerous echinate projections. The lorica is bulbous to elongate and constricts anteriorly to an opening, which may bear a short neck. A narrow lip around opening of the lorica is present on some specimens.

Dimensions. Holotype: 34 μm (width) x 36 μm (height). Average dimensions of twenty measured specimens: 31 μm (width) x 38 μm (height) x 19 μm (thickness)

Holotype. Plate 8, fig. 1.

Type Level and Locality. CIROS-1 drillhole, 100.46 mbsf, upper Oligocene [Harwood, 1989; Harwood et al., 1989b].

Type specimen. Holotype deposited in the California Academy of Sciences microfossil collection, accession #68033.

Remarks. Although closely related to *P. sphericum*, the original description of *P. sphericum* includes only non-loricate single and double skeletons [Hovasse, 1932c]. Double-skeleton morphologies of *P. sphericum* (middle Eocene to middle Miocene) may have given rise to *P. lingii* during the early Oligocene.

Derivation of Name. Named in honor of Dr. Hsin Y. Ling whose contributions to ebridian and silicoflagellate study have greatly improved our taxonomic and biostratigraphic understanding of these groups.

Occurrence. Pseudammodochium lingii is a common component of Oligocene through Miocene siliceous microfossil assemblages of the Ross Sea (Figure 5). It is present in sediments recovered from beneath the Ross Ice Shelf at RISP Site J-9 [Ling, 1984; Harwood et al., 1989a], in DSDP Hole 272 [White, 1980], in the MSSTS-1 drillcore [Harwood, 1986c], in the CIROS-1 drillcore [Harwood, 1989], in the CRP-1 drillcore [Cape Roberts Science Team, 1998; Harwood et al., 1998], in the CRP-2/2A drillcore [Cape Roberts Science Team, 1999], and in several post-Eocene McMurdo Erratics [this paper].

Known Geologic Range. Pseudammodochium lingii occurs in lower Oligocene to middle Miocene sediments of the Southern Ocean. In the Ross Embayment (Figure 5), the first occurrence of *P. lingii* is noted in the lower Oligocene of the CRP-2/2A drillcore [Cape Roberts Science Team, 1998], and a last occurrence is recorded in the middle Miocene of DSDP Hole 272 [White, 1980].

Pseudammodochium sphericum Hovasse, 1932c
Plate 5, fig. 6; Plate 10, figs. 7 and 8

Pseudammodochium sphericum Hovasse, 1932c, p. 463, fig. 16; Perch-Nielsen, 1975a, p. 881, pl. 1, figs. 17 and 18.

Pseudammodochium cf. *sphaericum* Hovasse; Harwood, 1986c, p. 87, pl. 2, figs. 18 and 19.

Pseudammodochium sphaericum Hovasse; Harwood, 1989, p. 82, pl. 6, fig. 24.

Pseudammodochium dictyoides Hovasse; Harwood, 1989 (in part), p. 82, pl. 6, fig. 23.

Remarks. In relation to *Pseudammodochium dictyoides*, *P. sphericum* is smaller, more spherical and does not possess an internal triode. The internal space of *P. sphericum* is hollow. Both single and double skeletons of this taxon were observed in the McMurdo Erratics and ODP Hole 748B. A wide variation in the number and size of pores was noted on specimens observed in Erratics D1, E345, and E364. Broken radiolarian fragments often show a resemblance to this taxon, but *P. sphericum* can be distinguished by its smooth or weakly ornamented surface texture.

Occurrence. Pseudammodochium sphericum ranges through the entire interval of study (middle Eocene to upper Oligocene) in ODP Hole 748B. Perch-Nielsen [1975a] reports a similar occurrence from DSDP Leg 29 Sites in the Southern Ocean. This taxon is also recorded in upper Oligocene to lower Miocene sediments of the MSSTS-1 drillcore [Harwood, 1986c] and in lower Oligocene to lower Miocene sediments of the CIROS-1 and CRP-2/2A drillcores [Harwood, 1989; Cape Roberts Science Team, 1999].

Plate 6.

Scale bars equal 20 μm. Figures 1-2, 5. *Micromarsupium curticannum* Deflandre; (1, 2) High/ low focus, Sample 748B-15H-3, 47-49; (5) Sample 748B-15H-3, 47-49. Figures 3-4, 6-7. *Micromarsupium anceps* Deflandre; (3) Pyritized and weakly silicified, Erratic E364; (4) Pyritized and loricate, Erratic E364; (6) Weakly silicified, Erratic D1; (7) Loricate, Erratic D1. Figures 8-10. *Ebriopsis* sp. 1; (8, 9, 10) Pyritized, high/ middle/ low focus, Erratic E364.

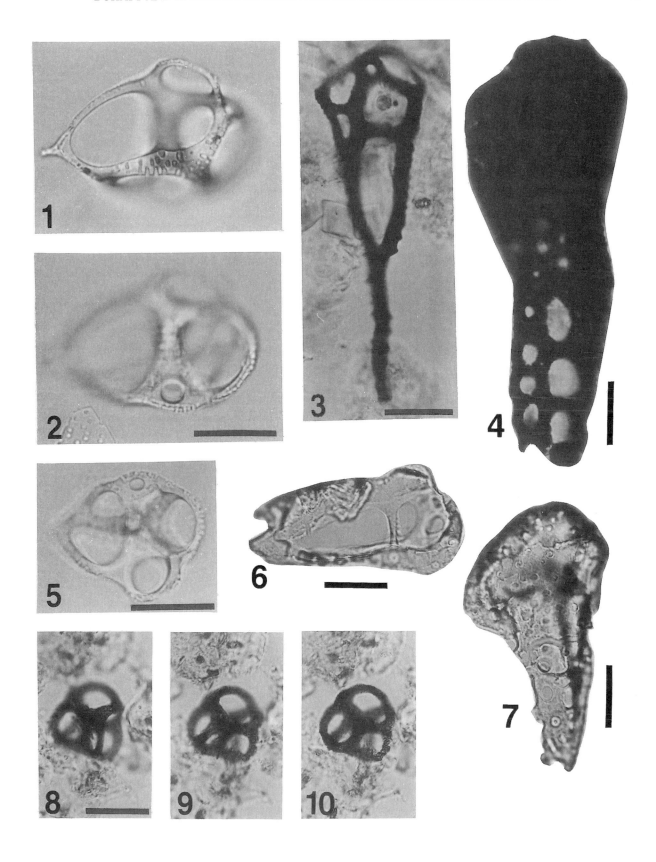

Genus *Triskelion* Gombos, 1982

***Triskelion gorgon* Gombos, 1982**
Plate 9, figs. 1, 2, and 10

Triskelion gorgon Gombos, 1982, pp. 446-448, pl. 1, figs. 1-7; McCartney and Wise, 1990, p. 751, pl. 7, figs 1-3; Lurvey et al., 1998, p. 194, pl. 4, figs. 2-4.

Remarks. *Triskelion gorgon* morphologies in the middle to upper Eocene of ODP Hole 748B are rounded to tear-drop in shape and heavily silicified. Consequently, some specimens of this taxon are difficult to distinguish from heavily-silicified morphologies of *Micromarsupium anceps*. Although these taxa are approximately the same size, *T. gorgon* possesses an open lattice of elements in the central area and a wide, pitted outer rim [as illustrated by Gombos, 1982]. The central areas of *T. gorgon* commonly break out, leaving only thick rims. The *Triskelion gorgon* morphology may only represent one half of the ebridian skeleton, analogous to the two halves of a walnut shell, that were easily separated upon death of the organism. No two specimens, however, were observed in this position opposed to one another.

Occurrence. *Triskelion gorgon* occurs in the middle Eocene to lower Oligocene of ODP Hole 748B. Gombos [1982] first reported *T. gorgon* from the middle Eocene of the southwest Atlantic Ocean, and it is also reported from lower Oligocene sediments in Hole 689B, Maud Rise, Weddell Sea (in the nannofossil *Blackites spinosus* Zone) [McCartney and Wise, 1990; Wei and Wise, 1990]. The occurrence of *T. gorgon* is also noted in middle Eocene sediments of DSDP Site 512, Falkland Plateau [Bohaty, unpubl. data].

Gen. et sp. indet. 1
Plate 9, figs. 8 and 9

? "Unknown Ebridian" Lurvey et al., 1998, pl. 4, fig. 5.

Remarks. An unknown ebridian taxon with a medium-sized (~80 μm long-axis diameter), ellipsoid skeleton was observed in Sample 14H-5, 47-49 (>20 μm), in the upper Eocene of ODP Hole 748B. These forms are similar in structure to *Triskelion gorgon*, but are smaller and

have an apical structure that consists of perforate silica rather than a lattice-work of elements (see pl. 9, fig. 8). The basal plate is a thick rim of silica with a constricted, elongate opening (see pl. 9, fig. 9).

Gen. et sp. indet. 2
Plate 9, figs. 4 and 5

Remarks. Several double-skeleton ebridian morphologies of unknown affinity were observed in Sample 12H-2, 47-49 (>20 μm) in the lower Oligocene of ODP Hole 748B. Each single skeleton possesses a central triode with ornamented proclade and opisthoclades, possibly placing this morphology in the genus *Ammodochium*. The double-skeleton arrangement is surrounded by a flange ornamented with crenulations and several small pores (see pl. 9, fig. 5).

Silicoflagellates

In the following silicoflagellate systematics, a complete bibliography with synonymies is not given for each taxon. References for original descriptions are given with recent references that contain descriptions and plates used in silicoflagellate identification in the present study. Current silicoflagellate taxonomic concepts at the genus level are described and illustrated in Desikachary and Prema [1996].

Genus *Bachmannocena* Locker, 1974, emend. Bukry, 1987

***Bachmannocena? diodon diodon?* (Ehrenberg) Bukry, 1987**
Plate 11, fig. 12

Mesocena diodon Ehrenberg, 1844, p. 71 and p. 84; Ehrenberg, 1854, pl. 33, fig. 18.
Bachmannocena diodon diodon (Ehrenberg) Bukry, 1987, p. 403; McCartney et al., 1995, p. 143, pl. 4, fig. 8.

Remarks. *Bachmannocena diodon diodon* is commonly found in Miocene to Pliocene biosiliceous sediments [McCartney et al., 1995]. Paleogene morphologies observed in Erratic 364 are characterized by two promi-

---

Plate 7.

Scale bar equals 20 μm. Figures 1-12. *Micromarsupium anceps* Deflandre; (1-12) Sample 748B-19H-3, 47-49.

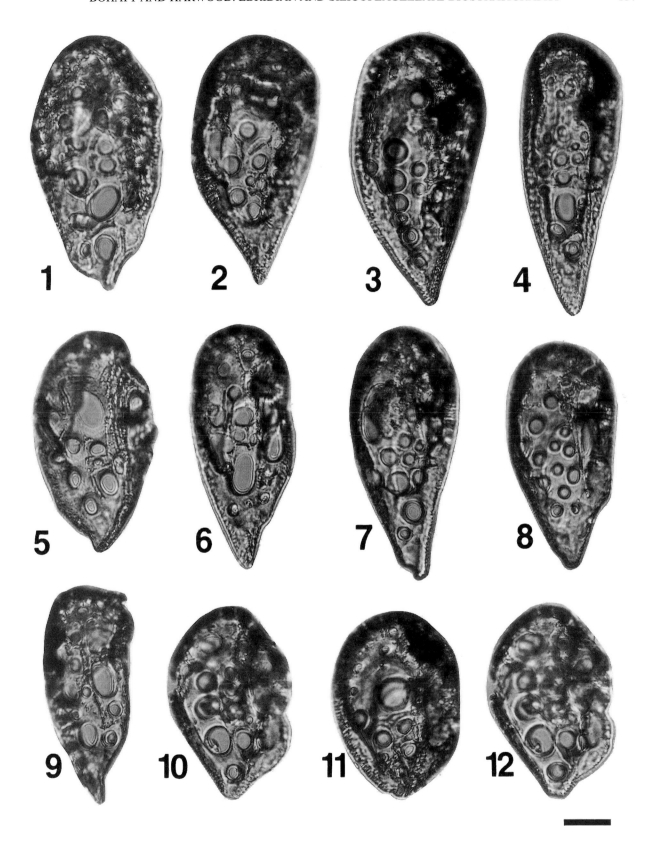

nent basal ring pikes (accessory spines), 90° from the basal ring spines, and have a smooth surface ornamentation. These forms may represent a separate taxon, unrelated to the Neogene *Bachmannocena diodon diodon*.

Genus *Cannopilus* Haeckel, 1887

Cannopilus hemisphaericus (Ehrenberg) Haeckel, 1887
Plate 12, fig. 13

Dictyocha hemisphaerica Ehrenberg, 1844, p. 258 and p. 266; Lemmermann, 1901, pl. 11, fig. 21.
Cannopilus hemisphaericus (Ehrenberg) Haeckel, 1887, p. 1569; Ling, 1972, pp. 147-148, pl. 23, figs. 1-5; Ciesielski, 1975, p. 654, pl. 2, figs. 2 and 3; Perch-Nielsen, 1975b, p. 685, pl. 1, figs, 10-12; Desikachary and Prema, 1996, pp. 208-209, pl. 65, fig. 7.

Genus *Corbisema* Hanna, 1928

Corbisema apiculata (Lemmermann) Hanna, 1931
Plate 12, fig. 3; Plate 13, fig. 11

Dictyocha triacantha var. *apiculata* Lemmermann, 1901, p. 259, pl. 10, figs. 19 and 20.
Corbisema apiculata (Lemmermann) Hanna, 1931, p. 198, pl. D, fig. 2; Ling, 1972, pp. 151-152, pl. 23, figs. 13-17; Ciesielski, 1975, p. 654, pl. 2, figs. 4-11; Perch-Nielsen, 1975b, p. 685, pl. 2, figs, 15, 16, and 19, pl. 3, figs. 19, 20, and 24, pl. 15, figs. 1 and 2; Busen and Wise, 1977, p. 711, pl. 1, figs. 1 and 2; Shaw and Ciesielski, 1983, p. 706, pl. 1, figs. 1-3; Ling, 1985b, p. 81, pl. 10, figs. 1 and 2.

Corbisema flexuosa (Stradner) Perch-Nielsen, 1975b
Plate 11, fig. 10

Corbisema triacantha var. *flexuosa* Stradner, 1961, p. 89, pl. 1, figs. 1-8, fig. 1c; Ling, 1972, p. 157-158, pl. 24, figs. 14-17; Ciesielski, 1975, p. 655, pl. 3, fig. 8.
Corbisema flexuosa (Stradner) Bukry, 1975b, p. 853, pl. 1, figs. 4 and 5.
Corbisema flexuosa (Stradner) Perch-Nielsen, 1975b, p. 685, pl. 3, fig. 10; Shaw and Ciesielski, 1983, p. 709, pl. 1, figs. 7 and 9.

Remarks. Shaw and Ciesielski [1983] include both apical-plate and apical-bar morphologies within this taxon. Specimens observed in the present study possess apical bars.

Corbisema hastata globulata Bukry, 1976a
Plate 12, figs. 1 and 2

Corbisema hastata globulata Bukry, 1976a, p. 892, pl. 4, figs. 1-8; Bukry, 1977a, p. 831, pl. 1, fig. 2; Ciesielski, 1991, p. 76, pl. 6, figs. 13 and 14.

Remarks. The basal ring of *Corbisema hastata globulata* is isosceles to equilateral in shape, with rounded apices. The basal-ring elements are indented where the apical bars join the basal ring, and the basal-ring spines are short. The basal-ring diameter of this taxon is typically ~35-45 µm, which is approximately twice the basal-ring diameter of *Corbisema triacantha* and *Corbisema regina*.

Corbisema hastata hastata (Lemmermann) Frenguelli, 1940
Plate 10, fig. 20

Dictyocha triacantha var. *hastata* Lemmermann, 1901, p. 259, pl. 10, fig. 16.
Corbisema hastata hastata (Lemmermann) Frenguelli, 1940 (in part), p. 62, figs. 12b and 12c; Bukry, 1975b, pp. 853-854, pl. 1, fig. 9; Ciesielski, 1975, p. 655, pl. 2, figs. 12-15; Perch-Nielsen, 1975b, p. 685, pl. 3, figs. 2-4, 8, and 21, pl. 15, fig. 3; Busen and Wise, 1977, p. 711, pl. 2, figs. 1-4, and 10, pl. 10, figs. 1 and 4; Shaw and Ciesielski, 1983, p. 709, pl. 2, figs. 2-4; Desikachary and Prema, 1996, pp. 134-135, pl. 36, figs. 1, 5, and 9, pl. 38, figs. 1 and 2.

Corbisema regina Bukry in Barron et al., 1984
Plate 11, fig. 11; Plate 12, fig. 4

Corbisema regina Bukry in Barron et al., 1984, p. 150, pl. 2, figs. 5-13; Bukry, 1987, p. 406, pl. 5, fig. 4; Desikachary and Prema, 1996, p. 142, pl. 32, figs. 4 and 7.

Plate 8.

Scale bar equals 20 µm. Figures 1-10. *Pseudammodochium lingii* n. sp.; (1) Holotype, CIROS-1, 100.46 m; (2-8) CIROS-1, 100.46 m; (9) Side view, CIROS-1, 100.46 m; (10) Anterior view of lorica opening, CIROS-1, 100.46 m.

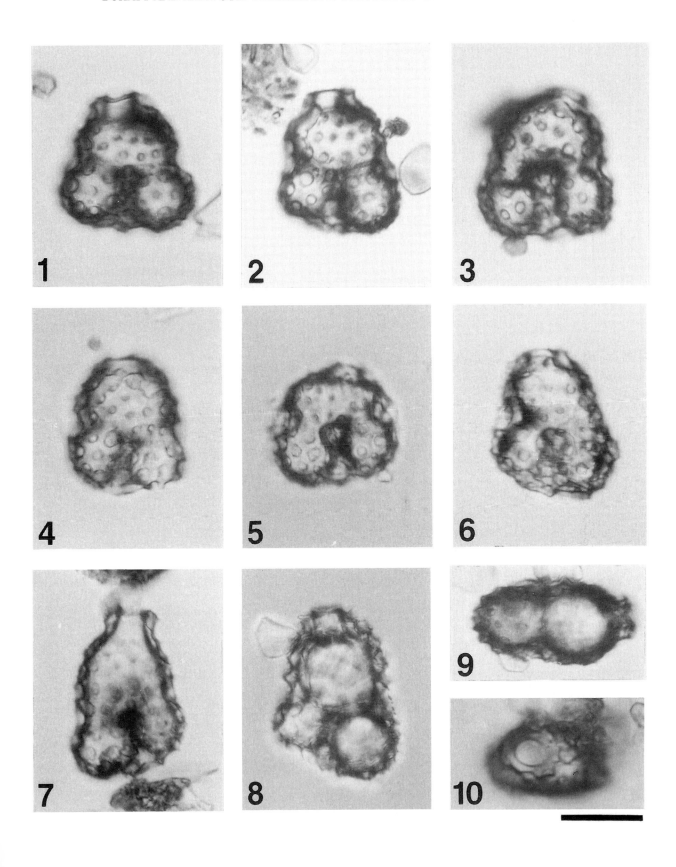

Remarks. *Corbisema regina* is similar in basal-ring size to *C. triacantha* but possesses an apical plate and has indented sides at the apical-bar/ basal-ring junctions.

Occurrence. *Corbisema regina* was originally described from the middle Eocene Kellogg Shale, California, and also occurs in middle Eocene sediments of DSDP Site 356, South Atlantic Ocean [Barron et al., 1984]. It is reported from middle to upper Eocene sediments of DSDP sites 612 and 613 off the coast of New Jersey [Bukry, 1987] and in upper Eocene sediments of DSDP Hole 406 on the Rockall Plateau, North Atlantic Ocean [Bukry, 1985]. Desikachary and Prema [1996] report this taxon in the middle Eocene of DSDP Site 212 in the Indian Ocean.

Corbisema spinosa **Deflandre, 1950c**
Plate 1, fig. 11; Plate 12, figs. 5 and 6

Corbisema spinosa Deflandre, 1950c, p. 193, figs. 178-182; Perch-Nielsen, 1975b, p. 686, pl. 3, fig. 23; Ciesielski, 1975, p. 655, pl. 3, fig. 9; Shaw and Ciesielski, 1983, p. 712, pl. 5, figs. 7 and 8; McCartney and Wise, 1990, p. 748, pl. 2, fig. 2.

Remarks. This species is distinguished from *Dictyocha hexacantha* by spines that extend from the apical bars that are not in the plane of the basal ring. Also, *Corbisema spinosa* is commonly smaller in basal-ring diameter than *D. hexacantha*.

Occurrence. *Corbisema spinosa* is distributed through Eocene and Oligocene sediments of the Southern Ocean [Perch-Nielsen, 1975b; Ciesielski, 1975, 1991; Shaw and Ciesielski, 1983]. Its first occurrence is used a zonal indicator in the lower Eocene in low-latitudes [Bukry, 1981].

Corbisema triacantha **(Ehrenberg) Hanna, 1931**
Plate 11, figs. 15 and 16

Dictyocha triacantha Ehrenberg, 1844, p. 80; Lemmer-

mann, 1901, p. 258, pl. 10, fig. 10.
Corbisema triacantha (Ehrenberg) Bukry and Foster, McCartney et al., 1995, p. 145, pl. 4, figs. 4 and 5.
Corbisema triacantha (Ehrenberg) Hanna, 1931, p. 198, pl. D, fig. 1; Ling, 1972, pp. 156-157, pl. 24, figs. 8-13; Bukry and Foster, 1973, p. 826, pl. 2, fig. 3; Ciesielski, 1975, p. 655, pl. 3, figs. 3-6; Perch-Nielsen, 1975b, p. 686, pl. 3, figs. 11, 15, and 16; Busen and Wise, 1977, p. 712-713, pl. 3, figs. 3-8; Shaw and Ciesielski, 1983, p. 709, pl. 2, figs. 5-7.

Remarks. Two morphologies of *Corbisema triacantha* are distinguished in the McMurdo Erratics. Both varieties possess equilateral, straight basal-ring sides, but are grouped separately based on the presence of an apical plate (pl. 10, fig. 15) or apical bar (pl. 10, fig. 16). The basal ring of this taxon typically measures ~15 to 25 μm.

Corbisema triacantha **cf.** *lepidospinosa* **Ciesielski, 1991**

cf. *Corbisema triacantha lepidospinosa* Ciesielski, 1991, pp. 77-78, pl. 4, figs. 9-14.

Remarks. One specimen resembling *Corbisema triacantha lepidospinosa* Ciesielski 1991 was observed in Erratic E345. This specimen possesses long basal-ring spines and a small basal-ring diameter, but may represent an aberrant variation of *Corbisema triacantha*.

Occurrence. Ciesielski [1991] reports *Corbisema triacantha lepidospinosa* only from upper Paleocene sediments of ODP Hole 700B, South Atlantic Ocean.

Genus *Dictyocha* **Ehrenberg, 1837, emend. Frenguelli, 1940**

Dictyocha **cf.** *anguinea* **Shaw and Ciesielski, 1983**
Plate 12, fig. 9; Plate 13, figs. 1-4

Dictyocha anguinea Shaw and Ciesielski, 1983, p. 710, pl. 7, figs. 1-5, pl. 8, figs. 1 and 3.

Plate 9.

Scale bars equal 20 μm. Figures 1-2, 10. *Triskelion gorgon* Gombos; (1, 2) High/ low focus, Sample 748B-18H-1, 47-49; (10) Sample 748B-15H-4, 47-49. Figure 3. *Hermesinum geminum* Dumitrica and Perch-Nielsen; (3) Sample 748B-13H-1, 47-49. Figures 4-5. Gen. et sp. indet.; (2; 4, 5) High/ low focus, Sample 748B-12H-2, 47-49. Figure 6. *Parebriopsis fallax* Hovasse; (6) Hyper-silicified, Sample 748B-15H-1, 47-49. Figure 7. *Adonnadonna primadonna* Gombos; (7) Sample 748B-15H-1, 47-49. Figures 8-9. Gen. et sp. indet. 1; (8, 9) High/ low focus, Sample 748B-14H-5, 47-49. Figure 11. *Ammodochium rectangulare* (Schulz) Deflandre; (11) Anterior, posterior, and medial silicification, Sample 748B-14H-5, 47-49.

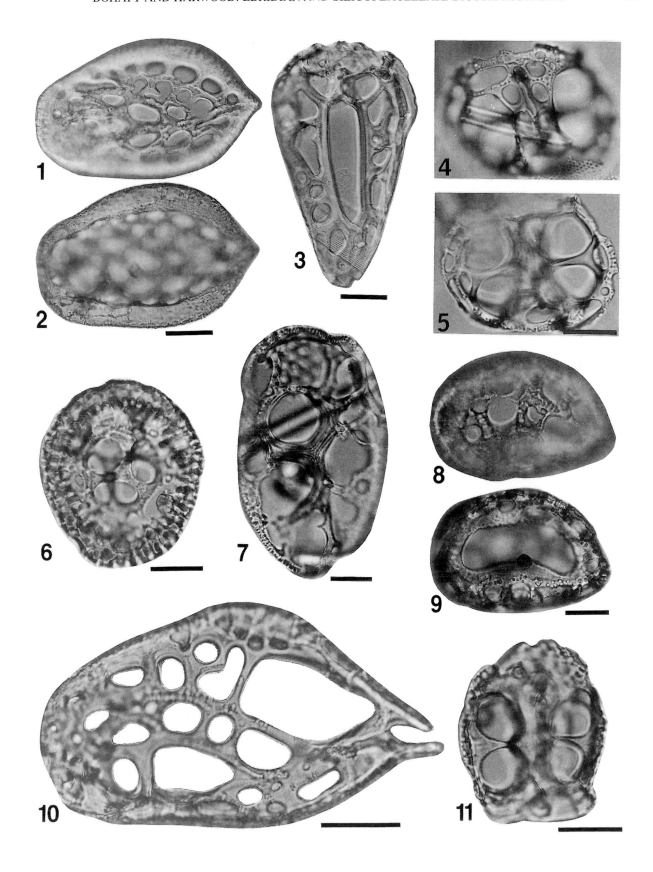

Remarks. *Dictyocha anguinea* is characterized by four to six sides and accessory spines that rise in an apical direction from the apical-bar/ basal-ring junctions. Similar morphologies were observed in Erratics E345 and E364. These specimens, however, are smaller in diameter (~20-40 µm) than those described by Shaw and Ciesielski [1983], which have a basal ring diameter of ~70-90 µm.

Dictyocha deflandrei **Frenguelli**
ex **Glezer, 1966 (1970)**
Plate 13, fig. 9

Dictyocha deflandrei Frenguelli, 1940 (in part), p. 65, figs. 14a and 14d; Ciesielski, 1975, pl. 4, figs. 7-9.
Dictyocha deflandrei Frenguelli *ex* Glezer, 1966 (1970), p. 262, pl. 12, figs. 13 and 16, pl. 32, fig. 4; Bukry, 1975b, p. 854, pl. 2, figs. 9-13; Hajós, 1976, p. 830, pl. 25, fig. 18; Busen and Wise, 1977, p. 713, pl. 3, fig. 10; Shaw and Ciesielski, 1983, p. 711, pl. 6, fig. 8; Harwood, 1989, p. 82, pl. 6, figs. 17 and 18; McCartney and Wise, 1990, p. 748, pl. 2, figs. 5a and 5b; Desikachary and Prema, 1996, p. 69, pl. 82, figs. 4, 5, and 7.

Remarks. Subspecies of *Dictyocha deflandrei* were not separated in the present study, as done by Glezer [1966] and Ciesielski [1991].

Dictyocha fibula fibula **Ehrenberg *ex* Locker and**
Martini, 1986b
Plate 12, fig. 10; Plate 13, figs. 10, 12, and 13

Dictyocha fibula Ehrenberg, 1839, p. 129; Locker, 1974, p. 636, pl. 1, fig. 6 (lectotype).
Dictyocha aspera (Lemmermann) Bukry and Foster, 1973 (in part), p. 826, pl. 2, fig. 4.
Dictyocha fibula fibula Ehrenberg *ex* Locker and Martini,

1986b, p. 904, pl. 5, figs. 1 and 2, pl. 11, figs. 8 and 9; McCartney et al., 1995, p. 147, pl. 2, fig. 1, pl. 5, fig. 5.

Remarks. See McCartney et al. [1995] for discussion concerning *Dictyocha fibula fibula* and designation of a lectotype by Locker [1974].

Dictyocha frenguellii **Deflandre, 1950c**
Plate 13, fig. 5

Dictyocha frenguellii Deflandre, 1950c, p. 194, figs. 188-193; Ciesielski, 1975, pp. 658-659, pl. 6, figs. 3-9; Bukry, 1975a, pl. 1, figs. 11 and 12; Perch-Nielsen, 1975b, p. 686, pl. 4, figs. 14 and 17, pl. 5, fig. 1; Desikachary and Prema, 1996, pp. 76-77, pl. 82, figs. 1-3.
Dictyocha fischeri Bukry, 1976a, p. 894; Harwood, 1989, p. 82, pl. 6, figs. 9 and 10.

Remarks. *Dictyocha frenguellii*, as applied in the present study, includes forms designated as *Dictyocha fischeri* by Bukry [1976a]. See McCartney and Wise [1990] for additional notes regarding *D. frenguellii*.

Dictyocha hexacantha **Schulz, 1928**
Plate 10, fig. 19

Dictyocha hexacantha Schulz, 1928, p. 255, fig. 43; Bukry, 1975b, p. 855, pl. 4, figs. 1 and 2; Ciesielski, 1975, p. 659, pl. 6, figs. 10 and 11; Shaw and Ciesielski, 1983, p. 711, pl. 4, figs. 8 and 9; Barron et al., 1984, p. 154, pl. 4, figs. 1-3.
Corbisema hexacantha Deflandre; Perch-Nielsen, 1975b, p. 685, pl. 3, figs. 13 and 14.
Remarks. *Dictyocha hexacantha* is distinguished from *Corbisema spinosa* by spines that originate from

---

Plate 10.

Scale bar equals 20 µm. Figures 1-6. *Pseudammodochium dictyoides* Hovasse; (1) Double skeleton with chambers connected, Erratic D1; (2) Double skeleton with chambers not connected, Erratic D1; (3) Single skeleton, Erratic D1; (4) Broken double skeleton, apical view, Erratic D1; (5) Single skeleton, apical view, Erratic D1; (6) Triode, Erratic D1. Figures 7-8. *Pseudammodochium sphericum* Hovasse; (7) Double skeleton, Erratic D1; (8) Single skeleton, Erratic D1. Figures 9-10, 15. *Ebriopsis crenulata* Hovasse emend.; (9) Loricate, Erratic D1; (10) Loricate, Erratic D1; (15) Side view, Erratic D1. Figure 11. *Archaeosphaeridium australensis* Perch-Nielsen; (11) CIROS-1, 484.95 m. Figures 12, 18. *Archaeosphaeridium tasmaniae* Perch-Nielsen; (12) Erratic E345; (18) Erratic E345. Figures 13, 16. *Ammodochium ampulla* Deflandre; (13) Double skeleton, loricate, Erratic D1; (16) Double skeleton, loricate, Erratic D1. Figure 14. *Ammodochium rectangulare* (Schulz) Deflandre; (14) Double skeleton with anterior and medial silicification, Erratic D1. Figure 17. *Naviculopsis constricta* (Schulz) Bukry; (17) Erratic E350. Figure 19. *Dictyocha hexacantha* Schulz; (19) Erratic E364. Figure 20. *Corbisema hastata hastata* (Lemmermann) Frenguelli; (20) Erratic E364.

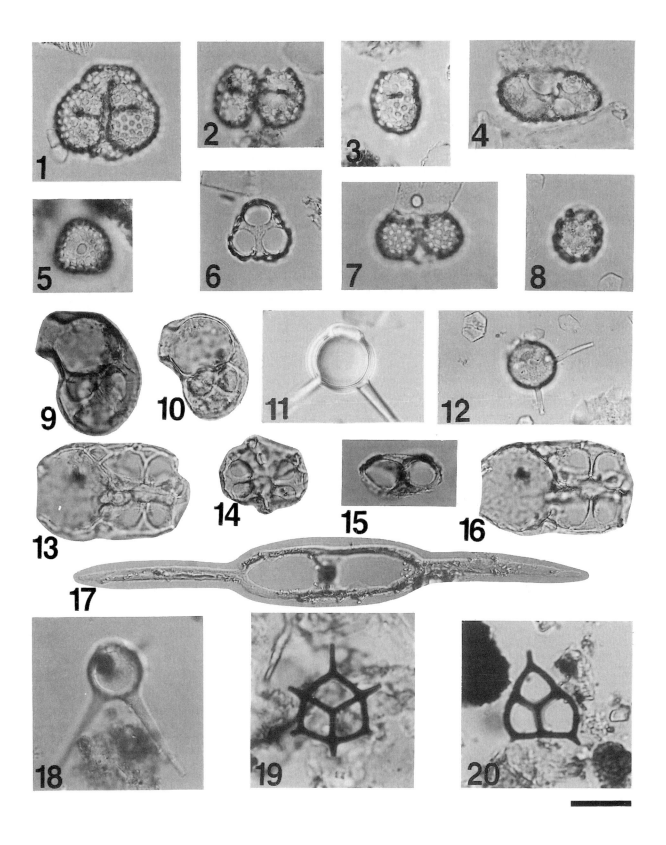

the apical-bar/ basal-ring junction that are in the plane of the basal ring. Also, *Dictyocha hexacantha* is commonly larger than *C. spinosa*.

Occurrence. The first and last occurrence datums of *Dictyocha hexacantha* are used as low-latitude zonal markers in the upper Eocene [Bukry, 1977a, 1981]. These datums have not been applied in the southern high latitudes, where *D. hexacantha* is rare in Paleogene sediments. *Dictyocha hexacantha*, however, shows a consistent occurrence (in low abundance) in upper Eocene sediments in the Southern Ocean [Perch-Nielsen, 1975b; Bukry, 1975b; Hajós, 1976; Busen and Wise, 1977; Shaw and Ciesielski, 1983; Ciesielski, 1991]. Ciesielski [1991] notes the rare occurrence of *D. hexacantha* within the nannofossil *Chiasmolithus oamaruensis* to *Isthmolithus recurvus* Zones (37.0 to 35.4 Ma) in ODP Hole 703A. Infrequent and rare lower Oligocene occurrences have also been reported in the Southern Ocean [Ciesielski, 1975; Busen and Wise, 1977] but may represent reworking or poor chronostratigraphic control in these cores. In the North Atlantic Ocean at DSDP Site 612, Bukry [1987] determined *D. hexacantha* to be a late middle Eocene to late Eocene marker, ranging through nannofossil Zones CP14a to CP15b (43.7 to 34.3 Ma).

Dictyocha pentagona (Schulz) Bukry and Foster, 1973

Dictyocha fibula var. *pentagona* Schulz, 1928, p. 255, figs. 41a-b.
Dictyocha pentagona (Schulz) Bukry and Foster, 1973, p. 827, pl. 3, fig. 10; Ciesielski, 1975, p. 659, pl. 7, figs. 6-7; Perch-Nielsen, 1975b, p. 687, pl. 5, fig. 11, pl. 15, fig. 4; Shaw and Ciesielski, 1983, p. 711, pl. 4, figs. 12-13; Barron et al., 1984, p. 154, pl. 4, fig. 4; Dumoulin, 1984, p. 45, pl. 1, fig. 7.

Remarks. *Dictyocha pentagona* morphologies observed in the present study are similar to those designated as "subspecies B" by Dumoulin [1984]. These

specimens have a small basal-ring diameter (~20 μm) and straight basal-ring sides. Larger morphologies [e.g. Bukry and Foster, 1973, pl. 3, fig. 10] are most likely derived from a lineage separate from that observed in the present study. Bukry [1976a] suggests *D. pentagona* should be considered a polyphyletic form group, as the *Dictyocha pentagona* morphotype appears to have arisen several times from different lineages during the Cenozoic.

Genus *Distephanus* Stöhr, 1880

Distephanus quinquangellus Bukry and Foster, 1973

Distephanus quinquangellus Bukry and Foster, 1973, p. 828, pl. 5, fig. 4; Perch-Nielsen, 1975b, p. 688, pl. 6, figs. 12 and 13, pl. 7, figs. 11, 14, and 15; Martini and Müller, 1976, p. 872, pl. 3, fig. 1, pl. 9, fig. 4; McCartney et al., 1995, p. 149, pl. 8, fig. 3, pl. 10, fig. 3.

Distephanus speculum (Ehrenberg) Haeckel, 1887

Dictyocha speculum Ehrenberg 1839, p. 129, pl. 4, fig. 4.
Distephanus speculum (Ehrenberg) Haeckel, 1887, p. 1565; Ling, 1972, p. 166, pl. 26, figs. 23 and 24, pl. 27, figs. 1 and 2; Bukry and Foster, 1973, p. 828, pl. 5, fig. 8; Ciesielski, 1975, p. 660, pl. 9, figs. 11 and 12, pl. 10, figs. 1-3; Perch-Nielsen, 1975b, p. 688, pl. 6, figs. 12 and 13, pl. 7, figs. 16-18 and 23.

Distephanus speculum speculum f. *pseudofibula* Schulz, 1928
Plate 1, fig. 12; Plate 12, figs. 7 and 8

Distephanus speculum speculum f. *pseudofibula* Schulz, 1928, p. 262, fig. 51a,b; McCartney and Wise, 1990, p. 749, pl. 5, figs. 1-4, pl. 6, figs. 2 and 3; McCartney and Harwood, 1992, p. 825, pl. 3, fig. 6.

Plate 11.

Scale bars equal 20 μm. Figures 1-5. *Ammodochium speciosum* Deflandre; (1, 2, 3) Low/ middle/ high focus, Sample 748B-14H-1, 47-49; (4, 5) High/ low focus, Sample 748B-15H-1, 47-49. Figure 6. *Parebriopsis fallax* Hovasse; (6) Hyper-silicified, Sample 748B-15H-3, 47-49. Figure 7. *Podamphora tenuis* (Hovasse) Deflandre; (7) Sample 748B-13H-1, 47-49. Figures 8-9. *Craniopsis octo* Hovasse *ex* Frenguelli; (8, 9) High/ low focus, Erratic D1. Figure 10. *Corbisema flexuosa* (Stradner) Perch-Nielsen; (10) Erratic E350. Figure 11. *Corbisema regina* Bukry; (11) Erratic D1. Figure 12. *Bachmannocena? diodon diodon?* (Ehrenberg) Bukry; (12) Erratic E364. Figures 13-14. *Ebrinula paradoxa* (Hovasse) Deflandre; (13) Apical view, high/ middle focus, Erratic E345. Figures 15-16. *Corbisema triacantha* (Ehrenberg) Hanna; (15) Apical-plate morphology, Sample 748B-14H-1, 47-49; (16) Apical-bar morphology, Erratic D1.

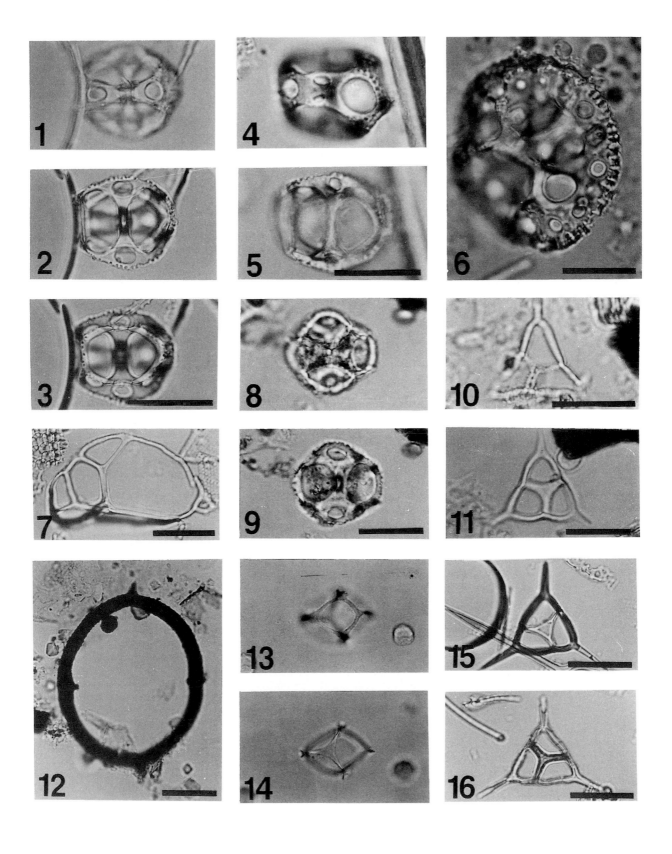

Remarks. In the present study, *Distephanus speculum speculum* f. *pseudofibula* is used as a form taxon or morphotype grouping. Specimens observed in Erratics D1 and E364 are smaller and have shorter spines than upper Miocene and lower Pliocene forms. The Paleogene morphologies most likely represent six-sided variations of *Dictyocha pentagona* (or a similar taxon) and have no phylogenetic relation to upper Neogene forms recorded and illustrated by McCartney and Wise [1990].

Distephanus speculum speculum f. *pseudopentagonus* McCartney and Wise, 1990

Distephanus speculum speculum f. *pseudopentagonus* McCartney and Wise, 1990, p. 750, pl. 5, fig. 6.

Distephanus sp. 1
Plate 12, fig. 12

Remarks. Two specimens with highly arched apical structures and six-sided basal rings were observed in Erratic E345.

Genus *Naviculopsis* Frenguelli, 1940

Naviculopsis constricta (Schulz) Bukry emend. in Barron et al., 1984
Plate 10, fig. 17

Dictyocha navicula var. *constricta* Schulz, 1928, p. 245, fig. 21.
Naviculopsis constricta (Schulz) Frenguelli; Ling, 1972, pp. 183-184, pl. 30, figs. 5-8; Shaw and Ciesielski, 1983, p. 715, pl. 15, figs. 4-8; McCartney and Wise, 1990, p. 750, pl. 1, fig. 5; McCartney and Harwood, 1992, p. 825, pl. 1, fig. 5.
Naviculopsis constricta (Schulz) Stradner; Perch-Nielsen, 1975b, p. 689, pl. 12, figs. 16, 17, and 23.
Naviculopsis constricta (Schulz) Bukry emend. in Barron et al., 1984, pp. 151-152, pl. 5, fig. 6.

Occurrence. *Naviculopsis constricta* is reported from upper Paleocene to lower Oligocene sediments in the Southern Ocean [Shaw and Ciesielski, 1983; McCartney and Wise, 1990; Ciesielski, 1991; McCartney and Harwood, 1992]. In ODP Hole 703A, Cieisielski [1991] records the Last Abundant Appearance Datum (LAAD) of *Naviculopsis constricta* at 40 cm above the C13/C15 boundary in the nannofossil *Chiasmolithus oamaruensis* Zone (35.4 to 33.7 Ma) [Madile and Monechi, 1991; Berggren et al., 1995]. The C13/C15 paleomagnetic reversal is dated at 34.7 Ma [Berggren et al., 1995]. The last abundant appearance of *N. constricta* may be related to cooling of Southern Ocean surface waters at the Eocene-Oligocene boundary, although *N. constricta* is generally considered a "cool-water" indicator in the mid-to-low latitudes [Bukry, 1987]. Significant cooling across the Eocene-Oligocene boundary, however, may have restricted the biogeographic distribution of *N. constricta*.

Naviculopsis cf. *constricta* (Schulz)
Plate 13, fig. 7

cf. *Naviculopsis constricta* (Schulz) Bukry emend. in Barron et al., 1984, pp. 151-152, pl. 5, fig. 6.

Remarks. Several specimens of a *Naviculopsis* sp., similar to *N. constricta*, were observed in Erratic MtD95 (see pl. 13, fig. 7). These forms possess short basal-ring spines and a relatively elongate basal ring.

Naviculopsis eobiapiculata Bukry, 1978a

Naviculopsis eobiapiculata Bukry, 1978a, p. 787, pl. 4, figs. 9-16; McCartney and Wise, 1987, p. 807, pl. 5, figs. 5-8; Ciesielski, 1991, p. 82, pl. 9, figs. 17-18, pl. 10, figs. 4-5; McCartney and Harwood, 1992, p. 825, pl. 1, figs. 2 and 4.

Naviculopsis foliacea Deflandre, 1950d

Plate 12.

Scale bar equal 20 μm. Figures 1-2. *Corbisema hastata globulata* Bukry; (1) Erratic D1; (2) Pyritized, Erratic E364. Figure 3. *Corbisema apiculata* (Lemmermann) Hanna; (3) Erratic D1. Figure 4. *Corbisema regina* Bukry; (4) Erratic D1. Figures 5-6. *Corbisema spinosa* Deflandre; (5, 6) High/ low focus, Erratic E345. Figures 7-8. *Distephanus speculum speculum* f. *pseudofibula* Schulz; (7, 8) Pyritized, high/ low focus, Erratic E364. Figure 9. *Dictyocha* cf. *aguinea* Shaw and Ciesielski; (9) Pyritized, Erratic E364. Figure 10. *Dictyocha fibula fibula* Ehrenberg *ex* Locker and Martini; (10) Erratic E345. Figure 11. *Septamesocena apiculata apiculata* (Schulz) Bachmann; (11) Erratic E345. Figure 12. *Distephanus* sp. 1; (12) Side view, Erratic E345. Figure 13. *Cannopilus hemisphaericus* (Ehrenberg) Haeckel; (13) Pyritized, Erratic E364.

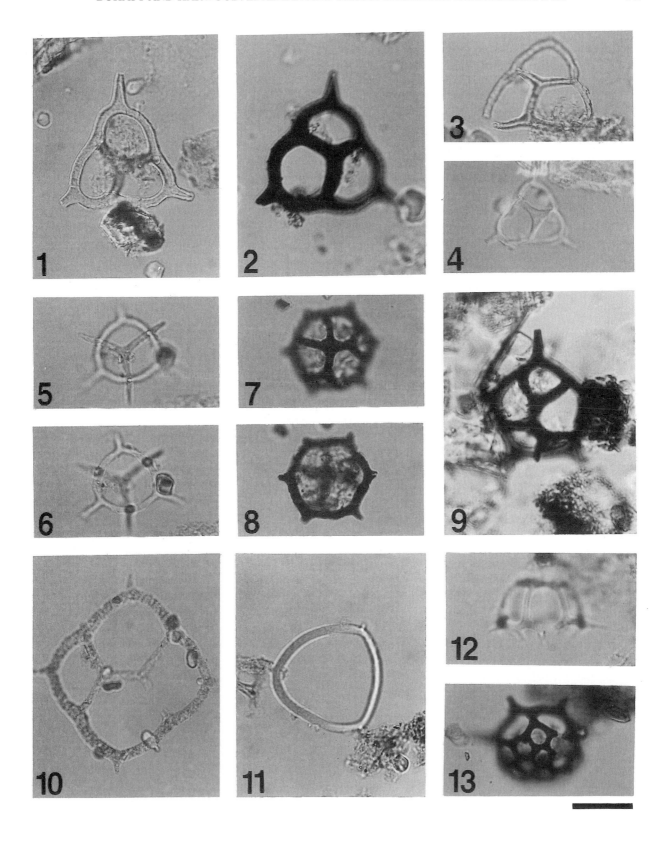

Naviculopsis foliacea Deflandre, 1950d, p. 204, figs. 235-240; Ling, 1972, pp. 184-185, pl. 30, figs. 9-11; Perch-Nielsen, 1975b, p. 689, pl. 12, fig. 15; Bukry, 1976b, p. 849, pl. 2, fig. 11; Shaw and Ciesielski, 1983, p. 715, pl. 16, figs. 1-7, 10, and 12; Barron et al., 1984, p. 154, pl. 5, fig. 7.

Remarks. *Naviculopsis foliacea* is considered a warm-temperate paleotemperature indicator relative to *Naviculopsis constricta* [Bukry, 1987].

Occurrence. In reports from the Southern Ocean, the range of *N. foliacea* is restricted to the middle to upper Eocene. *Naviculopsis foliacea* is reported from the upper Eocene of DSDP sites 277, 281, and 283 [Perch-Nielsen, 1975b], the upper Eocene of DSDP Hole 328 [Busen and Wise, 1977], the middle Eocene of DSDP holes 512 and 512A [Shaw and Ciesielski, 1983], the middle to upper Eocene of ODP holes 702B and 703A [Ciesielski, 1991], and the middle Eocene of ODP Hole 748B [McCartney and Harwood, 1992]. Outside of the Southern Ocean, *N. foliacea* may range into the Oligocene; Desikachary and Prema [1996] report this taxon in the lower Oligocene at DSDP Site 236 in the Indian Ocean. The first occurrence of *N. foliacea* is used as a low-latitude zonal indicator in the lower Eocene [Bukry, 1977a, 1981].

Naviculopsis foliacea tumida Bukry, 1978b

Naviculopsis foliacea tumida Bukry, 1978b, p. 820, pl. 8, figs. 1-8, pl. 17, figs. 11-12.

Genus *Septamesocena* Bachmann, 1970

Septamesocena apiculata apiculata (Schulz) Bachmann, 1970
Plate 12, fig. 11

Mesocena oamaruensis var. *apiculata* Schulz, 1928, p. 240, fig. 11.

Septamesocena apiculata apiculata (Schulz) Bachmann, 1970, p. 13; Ling, 1972, pp. 193-194, pl. 29, figs. 11-15.

Mesocena apiculata (Schulz) Hanna; Ciesielski, 1975, p. 661, pl. 11, figs. 1-5; Busen and Wise, 1977, p. 715, pl. 7, figs. 1 and 5; Shaw and Ciesielski, 1983, p. 714, pl. 12, figs. 1-7.

Bachmannocena apiculata apiculata (Schulz) Bukry, 1987, p. 403, pl. 1, fig. 1; McCartney and Wise, 1990, p. 747, pl. 2, figs. 6-10; Ciesielski, 1991, pp. 66-67, pl. 8, fig. 15.

Septamesocena apiculata glabra (Schulz) Desikachary and Prema, 1996

Mesocena polymorpha var. *triangula* f. *glabra* Schulz, 1928, p. 237, figs. 3b and 3c.

Mesocena apiculata glabra (Schulz) Bukry, 1977b, p. 698, pl. 2, figs. 14 and 15.

Bachmannocena apiculata glabra (Schulz) Bukry, 1987, p. 404.

Septamesocena apiculata glabra (Schulz) Desikachary and Prema, 1996, pp. 179-180.

Remarks. *Septamesocena apiculata glabra* is characterized by two concave basal-ring sides and one convex basal-ring side.

Septamesocena pappii (Bachmann) Desikachary and Prema, 1996

Mesocena pappii Bachmann, 1962, p. 380, pl. 1, figs. 1-9; Ling, 1973, p. 753, pl. 3, figs. 5 and 6; Ciesielski, 1975, p. 661, pl. 12, fig. 8; Bukry, 1975a, pl. 2, fig. 7, pl. 3, fig. 1; Perch-Nielsen, 1975b, p. 688, pl. 10, figs. 4 and 9.

Bachmannocena pappii (Bachmann) Bukry, 1987, p. 404.

Septamesocena pappii (Bachmann) Desikachary and Prema, 1996, pp. 181-182, pl. 45, fig. 5.

Plate 13.

Scale bar equals 20 μm. Figures 1-4. *Dictyocha* cf. *aguinea* Shaw and Ciesielski; (1, 2) Low/ high focus, Erratic E345; (3, 4) Low/ high focus, Erratic E364. Figure 5. *Dictyocha frenguellii* Deflandre; (5) Erratic MtD95. Figures 6, 8. *Dictyocha* sp.; (6) Erratic E345; (8) Erratic E345. Figure 7. *Naviculopsis* cf. *constricta* (Schulz) Bukry; (7) Erratic MtD95. Figure 9. *Dictyocha deflandrei* Frenguelli *ex* Glezer; (9) CIROS-1, 500.14 m. Figures 10, 12-13. *Dictyocha fibula fibula* Ehrenberg *ex* Locker and Martini; (10) Aberrant morphology, Erratic E345; (12) Erratic D1; (13) Broken, aberrant morphology, Erratic D1. Figure 11. *Corbisema apiculata* (Lemmermann) Hanna; (11) Erratic MtD95.

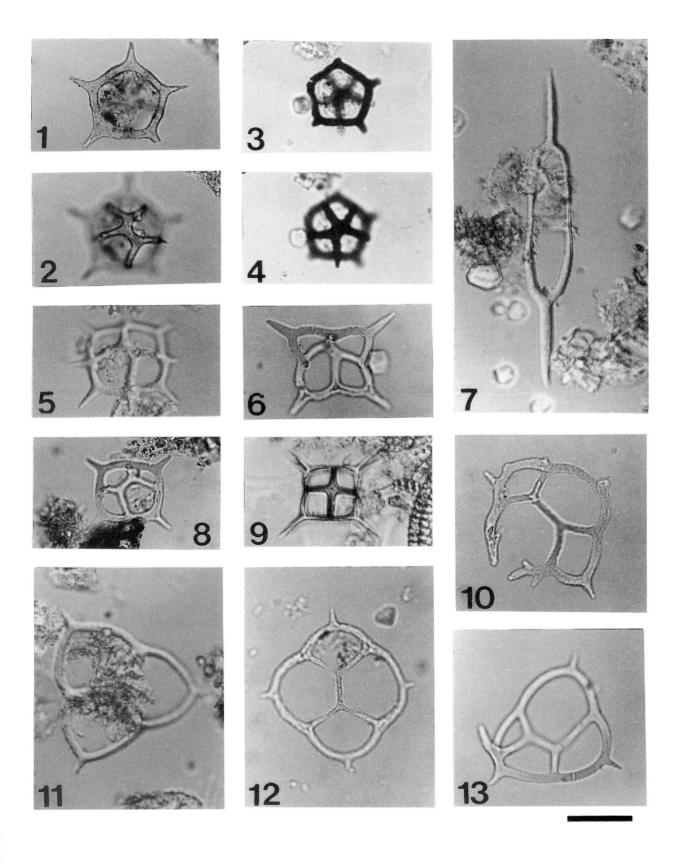

Chrysophyte Cysts

Genus *Archaeosphaeridium* Deflandre, 1932

Archaeosphaeridium australensis
Perch-Nielsen, 1975a
Plate 10, fig. 11

Archaeosphaeridium australensis Perch-Nielsen, 1975a,
    p. 878, pl. 2, figs. 1-10; Gombos, 1977a, p. 690,
    pl. 1, figs. 1-8; Harwood, 1989, p. 83, pl. 6, figs. 28
    and 29.

Remarks. This taxon is characterized by a wide pore
with a diameter over half that of the cyst. The pore has a
short neck, and the surface of the cyst is smooth. One to
three long, downward pointing spines are also characteristic of this species. Spineless morphologies with a similar cyst structure to *A. australensis* were observed in
Erratic D1 and were grouped separately.

Occurrence. Perch-Nielsen [1975a] reports *A. australensis* in upper Eocene to Oligocene sediments of
DSDP Site 328, southwest Pacific Ocean. Gombos
[1977a] reports this taxon in upper Eocene to upper
Oligocene sediments of DSDP Site 328, south Atlantic
Ocean, and Harwood [1989] records *A. australensis* in the
upper Eocene to lower Oligocene of the CIROS-1 drillcore (695.58 to 371.06 m depth). Spineless varieties are
noted in middle Eocene sediments of DSDP Site 512
[Bohaty, unpubl. data] on the Falkland Plateau. In the present study, *Archaeosphaeridium australensis* was not
observed in middle Eocene to upper Oligocene sediments
of ODP Hole 748B, but it is present in several McMurdo
Erratics and in the CIROS-1 drillcore (Figure 5).

Archaeosphaeridium dumitricae **Perch-Nielsen, 1975a**

Archaeosphaeridium dumitricae Perch-Nielsen, 1975a,
    p. 878, pl. 2, figs. 11-17; Gombos, 1977a, p. 690,
    pl. 1, fig. 9.

Remarks. The species is characterized by a small
pore and several short to medium-sized spines with inter-
connected bases, which are oriented in all directions.

Occurrence. *Archaeosphaeridium dumitricae* is
reported in upper Eocene sediments at DSDP sites 281
and 283, southwest Pacific Ocean [Perch-Nielsen,
1975a] and at DSDP Site 328, south Atlantic Ocean
[Gombos, 1977a]. This taxon may be restricted to the
upper Eocene in the Southern Ocean [Gombos, 1977a].
In the present study, *Archaeosphaeridium dumitricae*

was not observed in middle Eocene to upper Oligocene
sediments of ODP Hole 748B.

Archaeosphaeridium tasmaniae **Perch-Nielsen, 1975a**
Plate 10, figs. 12 and 18

Archaeosphaeridium tasmaniae Perch-Nielsen, 1975a,
    p. 878, pl. 2, figs. 18-23, pl. 3, figs. 1-10, pl. 12, figs.
    1-3; Gombos, 1977a, p. 690, pl. 1, fig. 10, pl. 2, figs.
    1-5.

Remarks. This species is characterized by a smooth
surface ornamentation and a pore diameter less than half
that of the cyst. Two to six long spines are present, which
are oriented in all directions.

Occurrence. *Archaeosphaeridium tasmaniae* is reported from Oligocene sediments of DSDP Site 280,
southwest Pacific Ocean [Perch-Nielsen, 1975a] and in
upper Eocene to lower Oligocene sediments of DSDP
Site 328, south Atlantic Ocean [Gombos, 1977a]. This
taxon was not observed in middle Eocene to upper
Oligocene sediments of ODP Hole 748B in the present
study. It is present, however, in several McMurdo
Erratics [this paper] and in the CIROS-1 and CRP-2/2A
drillcores [Cape Roberts Science Team, 1999] (Figure 5).

Endoskeletal Dinoflagellates

Genus *Actiniscus* Ehrenberg, 1854

Actiniscus elongatus **Dumitrica, 1968**
Plate 5, figs. 2 and 3

Actiniscus elongatus Dumitrica, 1968, p. 240, pl. 4, figs.
    22 and 26; Dumitrica, 1973, p. 822, pl. 3, figs. 6-12,
    pl. 5, figs. 10 and 11; Perch-Nielsen, 1975a, p. 882,
    pl. 10, figs. 11-13; Perch-Nielsen, 1978, p. 154, pl.
    5, figs. 13 and 14; Locker and Martini, 1986a, p.
    945, pl. 3, figs. 1 and 2, pl. 4, figs. 1 and 2.

Remarks. Both five and six-arm varieties of
Actiniscus elongatus were noted in the present study.
Some five-arm morphologies are similar in structure and
symmetry to *Foliactiniscus mirabilis*. These specimens
were separated based on the presence of apical ridges/
crests (*F. mirabilis*) or an apical plate (*A. elongatus*).
Dumitrica [1973] limits the genus *Foliactiniscus* to morphologies with median crests that join in the absence of a
central protuberance or plate. Based on the presence of a
central plate, *Actiniscus elongatus* should remain in the
genus *Actiniscus*, even though it is bilaterally symmetrical.

Actiniscus pentasterias (Ehrenberg) Ehrenberg, 1854
Plate 4, fig. 2

Dictyocha pentasterias Ehrenberg, 1840, p. 111 and p. 149.

Actiniscus pentasterias (Ehrenberg) Ehrenberg, 1854, pl. 18, fig. 61, pl. 19, fig. 45, pl. 20, fig. 48, pl. 33, fig. 1, pl. 35A, fig. 1, pl. 36, fig. 36; Frenguelli, 1940, p. 109, fig. 38A; Dumitrica, 1973, p. 822, pl. 2, figs. 2, 3, 6-11, and 14, pl. 3, figs. 13 and 14, pl. 5, figs. 6-8; Perch-Nielsen, 1975a, p. 882, pl. 10, figs. 2-10 and 16; Perch-Nielsen, 1978, p. 154, pl. 5, figs. 1-7 and 9-11, pl. 6, figs. 9 and 13-16; Locker and Martini, 1986a, p. 945, pl. 3, figs. 1-14, pl. 4, figs. 8 and 9; Locker, 1995, p. 115, pl. 6, fig. 15.

Genus *Carduifolia* Hovasse, 1932a

Carduifolia gracilis Hovasse, 1932a
Plate 4, fig. 1

Carduifolia gracilis Hovasse, 1932a, p. 127, figs. 10a-c; Frenguelli, 1940, p. 86, fig. 25b; Dumitrica, 1968, p. 236, pl. 3, fig. 15; Dumitrica, 1973, p. 824, pl. 4, figs. 21 and 26; Perch-Nielsen, 1975a, p. 883, pl. 10, figs. 19-25; Perch-Nielsen, 1978, p. 154, pl. 5, fig. 20; Locker, 1995, p. 115, pl. 6, fig. 19.

Remarks. On well-preserved specimens of *Carduifolia gracilis*, two spines or barbs are commonly present on the distal end of each arm.

Genus *Foliactiniscus* Dumitrica, 1973

Foliactiniscus mirabilis Dumitrica, 1973
Plate 4, fig. 3

Foliactiniscus mirabilis Dumitrica, 1973, p. 823, pl. 1, figs, 12, 13, and 20, pl. 2, figs. 4, 12, and 13; Locker and Martini, 1986a, p. 946, pl. 3, figs. 5 and 6.

Foliactiniscus pannosus Dumitrica, 1973

Foliactiniscus pannosus Dumitrica, 1973, p. 823, pl. 1, figs. 18, 19, and 21-23, pl. 2, figs. 1 and 5.

Acknowledgments. This paper benefitted from reviews by Dr. Hsin Y. Ling and Dr. Kevin McCartney. Dr. Kitt Lee is thanked for SEM technical assistance and helpful advice in sample preparation. The authors would also like to acknowledge Richard Levy, Richard Graham, Ruth Ford, Anne Rogers, Jean Self-Trail, and Efthimia Papastavros for assistance in preparation of this manuscript. The International Ocean Drilling Program provided core samples upon request. This work was supported by NSF Grants OPP-9317901 and OPP-9158075 and generous donations by the Geology Alumni of the University of Nebraska.

REFERENCES

Ahlbach, W.J., and K. McCartney
1992   Siliceous sponge spicules from Site 748, in *Proceedings of the Ocean Drilling Program, Scientific Results*, edited by S.W. Wise, Jr., R. Schlich et al., College Station, Texas (Ocean Drilling Program), 120 (part 2), 833-837.

Askin, R.A.
This vol.   Terrestrial palynomorphs, in *Paleobiology and Paleoenvironments of Eocene Rocks, McMurdo Sound, East Antarctica*, edited by J.D. Stilwell and R.M. Feldmann, Antarctic Research Series, American Geophysical Union.

Aubry, M.P.
1992   Paleogene calcareous nannofossils from the Kerguelen Plateau, Leg 120, in *Proceedings of the Ocean Drilling Program, Scientific Results*, edited by S.W. Wise, Jr., R. Schlich et al., College Station, Texas (Ocean Drilling Program), 120 (part 2), 471-491.

Bachmann, A.
1962   Eine neue *Mesocena*-Art (Silicoflagellidae) aus dem kalifornischen Eözan, *Verh. Geol. Bundes Anstalt*, 378-383.

Bachmann, A.
1970   Flagellata (Silicoflagellata), in *Catalogus Fossilium Austriae*, edited by H. Zapfe, Oesterreichischen Akademie der Wissenschaften, Heft 1b, 1-28.

Barrett, P.J.
1989   Introduction, in *Antarctic Cenozoic History from the CIROS-I Drillhole, McMurdo Sound*, edited by P.J. Barrett, DSIR Bulletin, 245, 5-6.

Barron J.A., and A.D. Mahood
1993   Exceptionally well-preserved early Oligocene diatoms from glacial sediments of Prydz Bay, East Antarctica, *Micropaleontology*, 39(1), 29-45.

Barron, J.A., D. Bukry, and R.Z. Poore
1984   Correlation of the middle Eocene Kellogg Shale of northern California, *Micropaleon-tology*, 30(2), 138-170.

Barron, J.A., B. Larsen, and J.G. Baldauf
1991   Evidence for late Eocene to early Oligocene Antarctic glaciation and observations on Late Neogene glacial history of Antarctica: results from Leg 119, in *Proceedings of the Ocean Drilling Program, Scientific Results*, edited by

J.A. Barron, B. Larsen et al., College Station, Texas (Ocean Drilling Program), 119, 869-891.

Berggren, W.A.
1992   Paleogene planktonic foraminifer magneto-biostratigraphy of the southern Kerguelen Plateau (Sites 747-749), in *Proceedings of the Ocean Drilling Program, Scientific Results*, edited by S.W. Wise, Jr., R. Schlich et al., College Station, Texas (Ocean Drilling Program), 120 (part 2), 551-568.

Berggren, W.A., D.V. Kent, C.C. Swisher, III, and M.-P. Aubry
1995   A revised Cenozoic geochronology and chronostratigraphy, in *Geochronology, Time Scales and Global Stratigraphic Correlation*, edited by W.A. Berggren, D.V. Kent, M.-P. Aubry, and J.A. Hardenbol, Special Publication - SEPM, Society for Sedimentary Geology, 54, 129-212.

Buckeridge, J., St. J.S.
This vol.   A new species of *Austrobalanus* (Cirripedia, Thoracica) from Eocene erratics, Mount Discovery, McMurdo Sound, East Antarctica, in *Paleobiology and Paleoenvironments of Eocene Rocks, McMurdo Sound, East Antarctica*, edited by J.D. Stilwell and R.M. Feldmann, Antarctic Research Series, American Geophysical Union.

Bukry, D.
1975a   Coccolith and silicoflagellate stratigraphy near Antarctica, Deep Sea Drilling Project, Leg 28, in *Initial Reports of the Deep Sea Drilling Project*, edited by D.E. Hayes, L.A. Frakes et al., U.S. Government Printing Office, Washington, D.C., 28, 709-723.

Bukry, D.
1975b   Silicoflagellate and coccolith stratigraphy, Deep Sea Drilling Project, Leg 29, in *Initial Reports of the Deep Sea Drilling Project*, edited by J.P. Kennett, R.E. Houtz et al., U.S.Government Printing Office, Washington, D. C., 29, 845-872.

Bukry, D.
1976a   Cenozoic silicoflagellate and coccolith stratigraphy South Atlantic Ocean, Deep Sea Drilling Project Leg 36, in *Initial Reports of the Deep Sea Drilling Project*, edited by C.D. Hollister, C. Craddock et al., U.S. Government Printing Office, Washington, D.C., 35, 885-917.

Bukry, D.
1976b   Silicoflagellate and coccolith stratigraphy, Norwegian-Greenland Sea, Deep Sea Drilling Project Leg 38, in *Initial Reports of the Deep Sea Drilling Project*, edited by M. Talwani, G. Udintsev et al., U.S. Government Printing Office, Washington, D.C., 38, 843-855.

Bukry, D.
1977a   Coccolith and silicoflagellate stratigraphy, South Atlantic Ocean, Deep Sea Drilling Project Leg 39, in *Initial Reports of the Deep Sea Drilling Project*, edited by P.R. Supko, K. Perch-Nielsen et al., U.S. Government Printing Office, Washington, D.C., 39, 825-839.

Bukry, D.
1977b   Cenozoic coccolith and silicoflagellate stratigraphy, offshore northwest Africa, Deep Sea Drilling Project Leg 41, in *Initial Reports of the Deep Sea Drilling Project*, edited by Y. Lancelot, E. Seibold et al., U.S. Government Printing Office, Washington, D.C., 41, 689-707.

Bukry, D.
1978a   Cenozoic silicoflagellate and coccolith stratigraphy, northwestern Atlantic Ocean, Deep Sea Drilling Project Leg 43, in *Initial Reports of the Deep Sea Drilling Project*, edited by W.E. Benson, R.E. Sheridan et al., U.S. Government Printing Office, Washington, D.C., 44, 775-805.

Bukry, D.
1978b   Cenozoic coccolith, silicoflagellate, and diatom stratigraphy, Deep Sea Drilling Project Leg 44, in *Initial Reports of the Deep Sea Drilling Project*, edited by W.E. Benson, R.E. Sheridan et al., U.S. Government Printing Office, Washington, D.C., 44, 807-863.

Bukry, D.
1981   Synthesis of silicoflagellate stratigraphy for Maestrichtian to Quaternary marine sediment, in *The Deep Sea Drilling Project: A Decade of Progress*, edited by J.E. Warme, R.G. Douglas, and E.L. Winterer, Special Publication - SEPM, Society for Sedimentary Geology, 32, 433-444.

Bukry, D.
1985   Cenozoic silicoflagellates from Rockall Plateau, Deep Sea Drilling Project Leg 81, in *Initial Reports of the Deep Sea Drilling Project*, edited by D.G. Roberts, D. Schnitker et al., U.S. Government Printing Office, Washington, D.C., 81, 547-563.

Bukry, D.
1987   Eocene siliceous and calcareous phyto-plankton, Deep Sea Drilling Project Leg 95, in *Initial Reports of the Deep Sea Drilling Project*, edited by C.W. Poag, A.B. Watts et al., U.S. Government Printing Office, Washington, D.C., 95, 395-415.

Bukry, D., and J.H. Foster
1973   Silicoflagellate and diatom stratigraphy, Leg 16, Deep Sea Drilling Project, in *Initial Reports of the Deep Sea Drilling Project*, edited by T.H. van Andel, G.R. Heath et al., U.S. Government Printing Office, Washington, D.C., 16, 815-871.

Busen, K.E., and S.W. Wise, Jr.
1977   Silicoflagellate stratigraphy, in *Initial Reports of*

the Deep Sea Drilling Project, edited by P.F. Barker et al., U.S. Government Printing Office, Washington, D.C., 36, 697-743.

Cape Roberts Science Team
1998   Initial Report on CRP-1, Cape Roberts Project, Antarctica, *Terra Antartica*, 5(1), 1-187.

Cape Roberts Science Team
1999   Studies from the Cape Roberts Project, Ross Sea, Antarctica, Initial Report of CRP-2/2A, *Terra Antartica*, 6(1/2), 1-173.

Ciesielski, P.F.
1975   Biostratigraphy and paleoecology of Neogene and Oligocene silicoflagellates from cores recovered during Antarctic Leg 28, Deep Sea Drilling Project, in *Initial Reports of the Deep Sea Drilling Project*, edited by D.E. Hayes, L.A. Frakes et al., U.S. Government Printing Office, Washington, D.C., 28, 625-691.

Ciesielski, P.F.
1991   Biostratigraphy of diverse silicoflagellate assemblages from the early Paleocene to early Miocene of Holes 698A, 700B, 702B, and 703A: subantarctic South Atlantic, in *Proceedings of the Ocean Drilling Program, Scientific Results*, edited by P.F. Ciesielski, Y. Kristoffersen et al., College Station, Texas (Ocean Drilling Program), 114, 49-96.

Ciesielski, P.F., Y. Kristoffersen et al.
1988   *Proceedings of the Ocean Drilling Program, Initial Reports*, College Station, Texas (Ocean Drilling Program), 114.

Cranwell, L.M.
1969   Antarctic and circum-Antarctic palynological contributions, *Antarctic Journal of the United States*, 4(4), 197-198.

Cranwell, L.M., H.J. Harrington, and I.G. Speden
1960   Lower Tertiary microfossils from McMurdo Sound, Antarctica, *Nature*, 186(4726), 700-702.

Deflandre, G.
1932   Archaeomonadaceae, une famille nouvelle de Protistes fossiles à loge siliceux, *C.R. Acad. Sci., Paris*, 194, 1859-1861.

Deflandre, G.
1933   Enkystment et stade loriqué chez les ébriacées, *Bulletin de la Société Zoologique de France*, 57, 514-523.

Deflandre, G.
1934   Nomenclature du squelette des ébriacées et description de quelques formes nouvelles, *Annales de Protistologie*, 4, 75-96.

Deflandre, G.
1936   Les flagellés fossiles, *Aperçu Biologique et Paléontologique: Role Géologique*, Paris, Hermann and Cie., Actualités Scient. et Ind., 98 pp.

Deflandre, G.
1950a   Analyse du squelette d'Ebria et relations de ce genre avec les Ammodochiidae, *Acad. Sci. Paris*, 230, 1780-1782.

Deflandre, G.
1950b   Sur une tendance évolutive des ébriédiens, *Acad. Sci. Paris*, 231, 158-160.

Deflandre, G.
1950c   Contribution a l'étude des silicoflagellidés actuels et fossiles, *Microscopie*, 2, 72-108.

Deflandre, G.
1950d   Contribution a l'étude des silicoflagellidés actuels et fossiles, *Microscopie*, 2, 191-210.

Deflandre, G.
1951   Recherches sur les Ébriédiens: Paléobiologie, Évolution, et Systématique, *Bulletin Biolo-gique de la France et de la Belgique*, 85, 1-84.

Deflandre, G.
1952   Classe des ébridiens, in *Traité de Paleontologie*, edited by J. Piveteau, Paris, 1, 125-128.

Dell'Agnese, D.J., and D.L. Clark
1994   Siliceous microfossils from the warm Late Cretaceous and early Cenozoic Arctic Ocean, *Journal of Paleontology*, 68(1), 31-47.

Desikachary, T.V., and P. Prema
1996   Silicoflagellates (Dictyochophyceae), *Bibliotheca Phycologica*, Band 100, 298 pp.

Dumitrica, P.
1968   Consideratii micropaleontologice aspura orizontului argilos cu radiolari din Tortonianul regiunii carpatice, *Stud. Cerc. Geol. Geofiz. Geogr.*, Ser. Geol. Bucuresti, 13(1), 227-241.

Dumitrica, P.
1973   Cenozoic endoskeletal dinoflagellates in southwestern Pacific sediments cored during Leg 21 of the DSDP, in *Initial Reports of the Deep Sea Drilling Project*, edited by R.E. Burns, J.E. Andrews et al., U.S. Government Printing Office, Washington, D.C., 21, 819-835.

Dumoulin, J.A.
1984   Silicoflagellate biostratigraphy and paleoecology of the upper Kreyenhagen Formation (Eocene-Oligocene), California, in *Kreyenhagen Formation and Related Rocks*, edited by J.R. Blueford, Pacific Section of the Society of Economic Paleontologists and Mineralogists, 29-49.

Dzinoridze, R.N., A.P. Jousé, G.S. Koroleva-Golikova, G.E. Kozlova, G.S. Nagaeva, M.G. Petrushevskaya, and N.I. Strelnikova
1978   Diatom and radiolarian Cenozoic stratigraphy, Norwegian Basin, DSDP Leg 38, in *Initial Reports of the Deep Sea Drilling Project*, edited by M. Talwani, G. Udintsev et al., U.S. Government Printing Office, Washington, D.C., Supplement to Volumes 38-41, 289-427.

Edwards, A.R. (compiler)
1991   *The Oamaru Diatomite*, New Zealand Geological Survey Paleontological Bulletin, 64, 260 pp.

Ehrenberg, C.G.
1837   Eine briefliche Nachricht des Hrn. Agassiz in Neuchatel über den ebenfalls aus mikroskopischen Kiesel-Organismen gebil-deten Polirschiefer von Oran in Afrika, *Ber. Verh. K. Preuss Akad. Wiss.*, 59-61.

Ehrenberg, C.G.
1839   Über die Bildung der Kreidefelsen und des Kriedemergels durch unsichtbare Organismen, *K. Acad. Wiss. Ber., Abh.* 1838, 59-148.

Ehrenberg, C.G.
1840   Über noch jetzt zahlreich lebende Thierarten der Kreidebildung und den Organismus der Polythalamien, *K. Acad. Wiss. Ber., Abh.* 1839, 81-174.

Ehrenberg, C.G.
1844   Mittheilung über zwei neue Lager von Gebirgsmassen aus Infusorien als Meeres-Absatz in Nord-Amerika und eine Vergleichung derselben mit den organischen Kreide-Gebilden in Europa und Afrika, *Ber. Verh. K. Preuss Akad. Wiss.*, 57-97.

Ehrenberg, C.G.
1854   Mikrogeologie, das Erden und Felsen schaffende Wirken des unsichtbar kleinen selbständigen Lebens auf der Erde, Liepzig, Leopold Voss, 1-374.

Elliot, D.H., and T.A. Trautman
1982   Lower Tertiary strata on Seymour Island, Antarctic Peninsula, in *Antarctic Geoscience*, edited by C. Craddock, University of Wisconsin Press, Madison, Wisconsin, 287-297.

Ernissee, J.J., and K. McCartney
1993   Ebridians, in *Fossil Prokaryotes and Protists*, edited by J. Lipps, Blackwell Scientific Publications, Boston, Massachusetts, 131-140.

Ernissee, J.J., and K. McCartney
1995   Ebridians and endoskeletal dinoflagellates, *Siliceous Microfossils*, edited by C.D. Blome et al., Paleontological Society Short Courses in Paleontology, 8, 177-185.

Feldmann, R.M., and W.J. Zinsmeister
1984   First occurrence of fossil decapod crustaceans (Callianassidae) from the McMurdo Sound region, Antarctica, *Journal of Paleontology*, 58(4), 1041-1045.

Fenner, J.
1991   Rare and unknown noncalcareous microfossils recovered from Leg 114 sites, in *Proceedings of the Ocean Drilling Program, Scientific Results*, edited by P.F. Ciesielski, Y. Kristoffersen et al., College Station, Texas (Ocean Drilling Program), 114, 303-310.

Francis, J.E.
This vol.   Fossil wood from Eocene high latitude forests, McMurdo Sound, Antarctica, in *Paleobiology and Paleoenvironments of Eocene Rocks, McMurdo Sound, East Antarctica*, edited by J.D. Stilwell and R.M. Feldmann, Antarctic Research Series, American Geophysical Union.

Frenguelli, J.
1940   Consideraciones sobre los silicoflagélados fósiles, *Revista del Museo de la Plata (Ser. 2)*, 2(7), 37-112.

Gemeinhardt, K.
1931   Organismenformen auf der Grenze zwischen Radiolarien und Flagellaten, *Ber. Deutsch. Bot. Ges.*, 49, 103-110.

Glezer, Z.I.
1970   Silicoflagellatophyceae, *Cryptogamic Plants of the U.S.S.R.*, edited by M.M. Gollerbakh, Academy of Sciences of the U.S.S.R., Komarov Institute of Botany, 1966. (Translated from Russian, Israel Program for Scientific Translations, Keter Press, Jerusalem, 7, 1-363).

Gombos, A.M., Jr.
1977a   Archaeomonads as Eocene and Oligocene guide fossils in marine sediments, in *Initial Reports of the Deep Sea Drilling Project*, edited by P.F. Barker et al., U.S. Government Printing Office, Washington, D.C., 36, 689-695.

Gombos, A.M., Jr.
1977b   Paleogene and Neogene diatoms from the Falkland Plateau and Malvinas Outer Basin, Leg 36, Deep Sea Drilling Project, edited by P.F. Barker et al., U.S. Government Printing Office, Washington, D.C., 36, 575-687.

Gombos, A.M., Jr.
1982   Three new and unusual genera of ebridians from the southwest Atlantic Ocean, *Journal of Paleontology*, 56(2), 444-448.

Gombos, A.M., Jr.
1983   Middle Eocene diatoms from the South Atlantic, in *Initial Reports of the Deep Sea Drilling Project*, edited by W.J. Ludwig, V.A. Krasheninnikov et al., U.S. Government Printing Office, Washington, D.C., 71, 565-581.

Gombos, A.M., Jr., and P.F. Ciesielski
1983   Late Eocene to early Miocene diatoms from the southwest Atlantic, in *Initial Reports of the Deep Sea Drilling Project*, edited by W.J. Ludwig, V.A. Krasheninnikov et al., U.S. Government Printing Office, Washington, D.C., 71, 583-634.

Haeckel, E.H.P.A.
1887   Report on the radiolaria collected by H.M.S.
           Challenger during the years 1873-1876, *Rept.
           Sci. Results Voyage of the H.M.S. Challenger
           During the Years 1873-1876*, 18, 1-1803.
Hajós, M.
1976   Upper Eocene and lower Oligocene
           Diatomaceae, Archaeomonadaceae, and
           Silicoflagellatae, in southwestern Pacific sedi-
           ments, DSDP Leg 29, in *Initial Reports of the
           Deep Sea Drilling Project*, edited by C.D.
           Hollister, C. Craddock et al., U.S. Government
           Printing Office, Washington, D.C., 35, 817-883.
Hambrey, M.J., and P.J. Barrett
1993   Cenozoic sedimentary and climatic record, Ross
           Sea region, Antarctica, in *The Antarctic
           Paleoenvironment: A Perspective on Global
           Change, Part Two*, edited by J.P. Kennett and
           D.A. Warnke, Antarctic Research Series,
           American Geophysical Union, Washington,
           D.C., 60, 91-124.
Hanna, G.D.
1928   Silicoflagellata from the Cretaceous of
           California, *Journal of Paleontology*, 1, 259-263.
Hanna, G.D.
1931   Diatoms and silicoflagellates of the Kreyenhagen
           Shale, *Mining in California*, 197-201.
Harrington, H.J.
1969   Fossiliferous rocks in moraines at Minna Bluff,
           McMurdo Sound, *Antarctic Journal of the
           United States*, 4(4), 134-135.
Harwood, D.M.
1986a   Recycled siliceous microfossils from the Sirius
           Formation, *Antarctic Journal of the United
           States*, 21(5), 101-103.
Harwood, D.M.
1986b   Seymour siliceous microfossil biostratigraphy
           (abstr.), *Abstracts with Programs - Geological
           Society of America*, 20th Annual Meeting,
           18(4), 292.
Harwood, D.M.
1986c   Diatoms, in *Antarctic Cenozoic History from the
           MSSTS-1 Drillhole, McMurdo Sound*, edited by
           P.J. Barrett, DSIR Bull. N.Z., 237, 69-107.
Harwood, D.M.
1989   Siliceous microfossils, in *Antarctic Cenozoic
           History from the CIROS-I Drillhole, McMurdo
           Sound*, edited by P.J. Barrett, DSIR Bulletin,
           245, 67-97.
Harwood, D.M., and S.M. Bohaty
This vol. Marine diatom assemblages from Eocene errat-
           ics, McMurdo Sound, Antarctica, in
           *Paleobiology and Paleoenvironments of Eocene
           Rocks, McMurdo Sound, East Antarctica*, edited
           by J.D. Stilwell and R.M. Feldmann, Antarctic

Research Series, American Geophysical Union.
Harwood, D.M., and T. Maruyama
1992   Middle Eocene to Pleistocene diatom biostratig-
           raphy of Southern Ocean sediments from the
           Kerguelen Plateau, Leg 120, in *Proceedings of
           the Ocean Drilling Program, Scientific Results*,
           edited by S.W. Wise, Jr., R. Schlich et al.,
           College Station, Texas (Ocean Drilling
           Program), 120 (part 2), 683-733.
Harwood, D.M., R.P. Scherer, and P.-N. Webb
1989a   Multiple Miocene productivity events in West
           Antarctica as recorded in upper Miocene sedi-
           ments beneath the Ross Ice Shelf (Site J-9),
           Marine Micropaleontology, 15, 91-115.
Harwood, D.M., P.J. Barrett, A.R. Edwards, H.J. Rieck, and P.-
           N. Webb
1989b   Biostratigraphy and chronology, in *Antarctic
           Cenozoic History from the CIROS-I Drillhole,
           McMurdo Sound*, edited by P.J. Barrett, DSIR
           Bulletin, 245, 231-239.
Harwood, D.M., S.M. Bohaty, and R.P. Scherer
1998   Lower Miocene diatom biostratigraphy of the
           CRP-1 drillcore, McMurdo Sound, Antarctica, in
           *Studies from the Cape Roberts Project, Ross Sea,
           Antarctica, Scientific Report of CRP-1*, edited by
           M.J. Hambrey and S.W. Wise, *Terra Antartica*,
           5(3), 499-514.
Hertlein, L.G.
1969   Fossiliferous boulder of Early Tertiary age from
           Ross Island, Antarctica, *Antarctic Journal of the
           United States*, 4(4), 199-201.
Hovasse, R.
1932a   Note préliminaire sur les ébriacées, *Bulletin de
           la Société Zoologique de France*, 57, 118-131.
Hovasse, R.
1932b   Sconde note sur les ébriacées, *Bulletin de la
           Société Zoologique de France*, 57, 276-283.
Hovasse, R.
1932c   Troisième note sur les ébriacées, *Bulletin de la
           Société Zoologique de France*, 57, 457-476.
Hovasse, R.
1943   Nouvelles recherches sur les flagellés à squelette
           siliceux: ébriidés et silicoflagellés fossiles de la
           diatomite de Saint-Laurent-La-Vernède (Gard),
           *Bulletin Biologique de la France et de la
           Belgique*, 77, 285-294.
Jones, C.M.
This vol. First fossil bird from East Antarctica, in
           *Paleobiology and Paleoenvironments of Eocene
           Rocks, McMurdo Sound, East Antarctica*, edited
           by J.D. Stilwell and R.M. Feldmann, Antarctic
           Research Series, American Geophysical Union.
Kastner, M., J.B. Keene, and J.M. Gieskes
1977   Diagenesis of siliceous oozes—I, Chemical
           controls on the rate of Opal-A to Opal-CT trans-

formation - an experimental study, *Geochimica et Cosmochimica Acta*, 41, 1041-1059.

Lemmermann, E.
1901   Silicoflagellatae, *Deutsche Bot. Ges.*, 19, 247-271.

Lee, D.E., and J.D. Stilwell
This vol.   Rhynchonellide brachiopods from late Eocene erratics in the McMurdo Sound region, Antarctica, in *Paleobiology and Paleoenvironments of Eocene Rocks, McMurdo Sound, East Antarctica*, edited by J.D. Stilwell and R.M. Feldmann, Antarctic Research Series, American Geophysical Union.

Levy, R.H.
1998   Middle to late Eocene coastal paleoenvironments of Southern Victoria Land, East Antarctica: a palynological and sedimentological study of glacial erratics, Ph.D. Dissertation, Universtiy of Nebraska - Lincoln, Lincoln, Nebraska.

Levy, R.H., and D.M. Harwood
This vol a.   Marine palynomorph biostratigraphy and age(s) of the McMurdo Sound erratics, in *Paleobiology and Paleoenvironments of Eocene Rocks, McMurdo Sound, East Antarctica*, edited by J.D. Stilwell and R.M. Feldmann, Antarctic Research Series, American Geophysical Union.

Levy, R.H., and D.M. Harwood
This vol b.   Sedimentary lithofacies and inferred depositional environments of the McMurdo Sound Erratics, in *Paleobiology and Paleoenvironments of Eocene Rocks, McMurdo Sound, East Antarctica*, edited by J.D. Stilwell and R.M. Feldmann, Antarctic Research Series, American Geophysical Union.

Ling, H.Y.
1971   Silicoflagellates and ebridians from the Shinzan diatomaceous mudstone member of the Onnagawa Formation, (Miocene), northeast Japan, in *Proceedings of the 2nd Planktonic Conference*, edited by A. Farinacci, 2, 689-704.

Ling, H.Y.
1972   Upper Cretaceous and Cenozoic silicoflagellates and ebridians, *Bulletins of American Paleontology*, 62(273), 135-229.

Ling, H.Y.
1973   Silicoflagellates and ebridians from Leg 19, in *Initial Reports of the Deep Sea Drilling Project*, edited by J.S. Creager, D.W. Scholl et al., U.S. Government Printing Office, Washington, D.C., 19, 751-775.

Ling, H.Y.
1975   Silicoflagellates and ebridians from Leg 31, in *Initial Reports of the Deep Sea Drilling Project*, edited by D.E. Karig, J.C. Ingle, Jr. et al., U.S.

Government Printing Office, Washington, D.C., 31, 763-777.

Ling, H.Y.
1977   Late Cenozoic silicoflagellates and ebridians from the eastern North Pacific region, in *Proceedings of the First International Congress on Pacific Neogene Stratigraphy, Tokyo, 1976*, 205-233.

Ling, H.Y.
1980   Silicoflagellates and ebridians from Leg 55, in *Initial Reports of the Deep Sea Drilling Project*, edited by E.D. Jackson, I. Koisumi et al., U.S. Government Printing Office, Washington, D.C., 55, 375-385.

Ling, H.Y.
1984   Occurrence of *Pseudammodochium* from the subbottom sediments of the Ross Ice Shelf, *Antarctic Journal of the United States*, 18(5), 159-161.

Ling, H.Y.
1985a   Paleogene silicoflagellates and ebridians from the Goban Spur, Northeastern Atlantic, in *Initial Reports of the Deep Sea Drilling Project*, edited by P.C. de Graciansky, C.W. Poag et al., U.S. Government Printing Office, Washington, D.C., 80, 663-668.

Ling, H.Y.
1985b   Early Paleogene silicoflagellates and ebridians from the Arctic Ocean, *Transactions and Proceedings of the Palaeontological Society of Japan*, New Series, 138, 79-93.

Ling, H.Y.
1992   Late Neogene silicoflagellates and ebridians from Leg 128, Sea of Japan, in *Proceedings of the Ocean Drilling Program, Scientific Results*, edited by K.A. Pisciotto, J.C. Ingle, Jr., M.T. von Breymann, J.A. Barron et al., College Station, Texas (Ocean Drilling Program), 127/128, 237-248.

Ling, H.Y., and R.J. White
1979   Silicoflagellate *Mesocena pappii* identified in R.I.S.P. Site J-9 core sediments, *Antarctic Journal of the United States*, 14, 126-127.

Locker, S.
1974   Revision der Silicoflagellaten aus der Mikrogeologischen Sammlung von C.G. Ehrenberg, *Ecologae Geol. Helv.*, 67, 631-646.

Locker, S.
1995   Silicoflagellates, ebridians, and actiniscidians from Pliocene and Quaternary sediments off southern Chile, ODP Leg 141, in *Proceedings of the Ocean Drilling Program, Scientific Results*, edited by Lewis, S.D., J.H. Behrmann, R.J. Musgrave, and S.C. Cande, College Station, Texas (Ocean Drilling Program), 141, 223-233.

Locker, S.
1996   Cenozoic siliceous flagellates from the Fram
           Strait and the East Greenland Margin: biostrati-
           graphic and paleoceanographic results, in
           *Proceedings of the Ocean Drilling Program,
           Scientific Results*, edited by J. Thiede, A.M.
           Myhre, J.V. Firth, G.L. Johnson, and W.F.
           Ruddiman, College Station, Texas (Ocean
           Drilling Program), 151, 101-124.

Locker, S., and E. Martini
1986a   Ebridians and actiniscidians from the southwest
           Pacific, in *Initial Reports of the Deep Sea
           Drilling Project*, edited by J.P. Kennett, C.C. von
           der Borch et al., U.S. Government Printing
           Office, Washington, D.C., 90, 939-951.

Locker, S., and E. Martini
1986b   Silicoflagellates and some sponge spicules the
           southwest Pacific, DSDP 90, in *Initial Reports of
           the Deep Sea Drilling Project*, edited by J.P.
           Kennett, C.C. von der Borch et al., U.S.
           Government Printing Office, Washington, D.C.,
           90, 887-924.

Locker, S., and E. Martini
1989   Cenozoic silicoflagellates, ebridians, and actinis-
           cidians from the Vøering Plateau (ODP Leg 104),
           in *Proceedings of the Ocean Drilling Program,
           Scientific Results*, edited by O. Eldholm, J.
           Thiede, E. Taylor et al., College Station, Texas
           (Ocean Drilling Program), 104, 543-585.

Loeblich, A.R., III
1970   The Amphiesma or dinoflagellate cell covering,
           *Proceedings of the North American
           Paleontological Convention, Chicago*, Part G,
           867-929.

Loeblich, A.R., Jr., and A.R. Loeblich, III
1969   Index to the genera, subgenera, and sections of
           the Pyrrhophyta, III, *Journal of Paleon-tology*,
           43(1), 193-198.

Loeblich, A.R., III, L.A. Loeblich, H. Tappan, and A.R.
           Loeblich, Jr.
1968   *Annotated Index of Fossil and Recent
           Silicoflagellates and Ebridians, with
           Descriptions and Illustrations of Validly
           Proposed Taxa*, Geological Society of America
           Memoir 106, Boulder, Colorado, 319 pp.

Long, D.J., and J.D. Stilwell
This vol. Fish remains from the Eocene of Mount
           Discovery, East Antarctica, in *Paleobiology and
           Paleoenvironments of Eocene Fossil-iferous
           Erratics, McMurdo Sound, East Antarctica*, edit-
           ed by J.D. Stilwell and R.M. Feldmann, Antarctic
           Research Series, American Geophysical Union.

Lurvey, L.K., K. McCartney, and W. Wei
1998   Siliceous sponge spicules, silicoflagellates, and
           ebridians from Hole 918D, continental rise of the

           Greenland Margin, in *Proceedings of the Ocean
           Drilling Program, Scientific Results*, edited by
           A.D. Saunders, H.C. Larsen, and S.W. Wise, Jr.,
           College Station, Texas (Ocean Drilling Program),
           152, 191-199.

Mackensen, A., and W.A. Berggren
1992   Paleogene benthic foraminifers from the southern
           Indian Ocean (Kerguelen Plateau): biostratigra-
           phy and paleoecology, in *Proceedings of the
           Ocean Drilling Program, Scientific Results*, edit-
           ed by S.W. Wise, Jr., R. Schlich et al., College
           Station, Texas (Ocean Drilling Program), 120
           (part 2), 603-630.

Madile, M., and S. Monechi
1991   Late Eocene to early Oligocene calcareous nan-
           nofossil assemblages from sites 699 and 703,
           subantarctic South Atlantic Ocean, in
           *Proceedings of the Ocean Drilling Program,
           Scientific Results*, edited by P.F. Ciesielski, Y.
           Kristoffersen et al., College Station, Texas
           (Ocean Drilling Program), 114, 179-192.

Magavern, S., D.L. Clark, and S.L. Clark
1996   87/86Sr, phytoplankton, and the nature of the
           Late Cretaceous and Early Cenozoic Arctic
           Ocean, *Marine Geology*, 133, 183-192.

Martini, E.
1981   Silicoflagellaten in Paläogen von Nord-deutsch-
           land, *Senckenbergiana Lethaea*, 62(2/6), 277-
           284.

Martini, E.
1986   Paleogene calcareous nannoplankton from the
           southwest Pacific Ocean, Deep Sea Drilling
           Project, Leg 90, in *Initial Reports of the Deep
           Sea Drilling Project*, edited by J.P. Kennett, C.C.
           von der Borch et al., U.S. Government Printing
           Office, Washington, D.C., 90, 747-761.

Martini, E., and C. Müller
1976   Eocene to Pleistocene silicoflagellates from the
           Norwegian-Greenland Sea (DSDP leg 38), in
           Initial Reports of the Deep Sea Drilling Project,
           edited by M. Talwani, G. Udintsev et al., U.S.
           Government Printing Office, Washington, D.C.,
           38, 857-895.

McCartney, K., and D.M. Harwood
1992   Silicoflagellates from Leg 120 on the Kerguelen
           Plateau, southeast Indian Ocean, in *Proceedings
           of the Ocean Drilling Program, Scientific
           Results*, edited by S.W. Wise, Jr., R. Schlich et
           al., College Station, Texas (Ocean Drilling
           Program), 120 (part 2), 811-831.

McCartney, K., and S.W. Wise, Jr.
1987   Silicoflagellates and ebridians from the New
           Jersey transect, Deep Sea Drilling project Leg
           93, Sites 604 and 605, in *Initial Reports of the
           Deep Sea Drilling Project*, edited by J.E. van

Hinte, S.W. Wise, Jr. et al., U.S. Government Printing Office, Washington, D.C., 93, 801-814.

McCartney, K., and S.W. Wise, Jr.
1990   Cenozoic silicoflagellates and ebridians from ODP Leg 113: biostratigraphy and notes on morphologic variability, in *Proceedings of the Ocean Drilling Program, Scientific Results*, edited by P.F. Barker, J.P. Kennett et al., College Station, Texas (Ocean Drilling Program), 113, 729-760.

McCartney, K., S. Churchill, and L. Woestendiek
1995   Silicoflagellates and ebridians from Leg 138, eastern equatorial Pacific, in *Proceedings of the Ocean Drilling Program, Scientific Results*, edited by N.G. Pisias, L.A. Mayer, T.R. Janecek, A. Palmer-Julson, and T.H. van Andel, College Station, Texas (Ocean Drilling Program), 138, 129-162.

McIntyre, D.J., and G.J. Wilson
1966   Some new species of Lower Tertiary dinoflagellates from McMurdo Sound, Antarctica, *New Zealand Journal of Botany*, 5(2), 223-240.

Moshkovitz, S., A. Ehrlich, and D. Soudry
1983   Siliceous microfossils of the Upper Cretaceous Mishash Formation, central Negev, Israel, *Cretaceous Research*, 4, 173-194.

Perch-Nielsen, K.
1975a   Late Cretaceous to Pleistocene archaeo-monads, ebridians, endoskeletal dinoflagel-lates, and other siliceous microfossils from the subantarctic southwest Pacific, DSDP, Leg 29, in *Initial Reports of the Deep Sea Drilling Project*, edited by J.P. Kennett, R.E. Houtz et al., U.S. Government Printing Office, Washington, D.C., 29, 873-907.

Perch-Nielsen, K.
1975b   Late Cretaceous to Pleistocene silicoflagellates from the southern southwest Pacific, DSDP, Leg 29, in *Initial Reports of the Deep Sea Drilling Project*, edited by J.P. Kennett, R.E. Houtz et al., U.S. Government Printing Office, Washington, D.C., 29, 677-721.

Perch-Nielsen, K.
1977   Tertiary silicoflagellates and other siliceous microfossils from the western South Atlantic, Deep Sea Drilling Project, Leg 39, in *Initial Reports of the Deep Sea Drilling Project*, edited by P.R. Supko, K. Perch-Nielsen et al., U.S. Government Printing Office, Washington, D.C., 39, 863-867.

Perch-Nielsen, K.
1978   Eocene to Pliocene archaeomonads, ebridians, and endoskeletal dinoflagellates from the Norwegian Sea, in *Initial Reports of the Deep Sea Drilling Project*, edited by M. Talwani et al., U.S. Government Printing Office, Washington, D.C.,

Supplement to Volumes 38-41, 147-175.

Pole, M., B. Hill, and D.M. Harwood
This vol.   Eocene plant macrofossils from erratics, McMurdo Sound, in *Paleobiology and Paleoenvironments of Eocene Rocks, McMurdo Sound, East Antarctica*, edited by J.D. Stilwell and R.M. Feldmann, Antarctic Research Series, American Geophysical Union.

Savage, M.L. and P.F. Ciesielski
1983   A revised history of glacial sedimentation in the Ross Sea region, in *Antarctic Earth Science*, edited by R.L. Oliver, P.R. James, and J.B. Jago, Australian Academy of Science, Canberra, 555-559.

Scherer, R.P.
1991   Quaternary and Tertiary microfossils from beneath Ice Stream B: evidence for a dynamic West Antarctic Ice Sheet history, *Palaeogeography, Palaeoclimatology, Palaeo-ecology (Global and Planetary Change Section)*, 90, 395-412.

Schlich, R., S.W. Wise, Jr. et al.
1989   *Proceedings of the Ocean Drilling Program, Initial Reports*, College Station, Texas (Ocean Drilling Program), 120.

Schrader, H.-J.
1978   Quaternary through Neogene history of the Black Sea, deduced from the paleoecology of diatoms, silicoflagellates, ebridians, and chrysomonads, in *Initial Reports of the Deep Sea Drilling Project*, edited by D.A. Ross, N.P. Neprochnov et al., U.S. Government Printing Office, Washington, D.C., 42 (part 2), 789-901.

Schulz, P.
1928   Beiträge zur Kenntnis fossiler und rezenter Silicoflagellaten, *Bot. Arch.*, 21, 225-292.

Shaw, C.A., and P.F. Ciesielski
1983   Silicoflagellate biostratigraphy of middle Eocene to Holocene subantarctic sediments recovered by Deep Sea Drilling Project Leg 71, in *Initial Reports of the Deep Sea Drilling Project*, edited by W.J. Ludwig, V.A. Krasheninnikov et al., U.S. Government Printing Office, Washington, D.C., 71, 687-737.

Speden, I.G.
1962   Fossiliferous Quaternary marine deposits in the McMurdo Sound region, *New Zealand Journal of Geology and Geophysics*, 5(5), 746-777.

Stilwell, J.D.
This vol.   Eocene mollusca (Bivalvia, Gastropoda, and Scaphopoda) from McMurdo Sound: systematics and paleoecologic significance, in *Paleobiology and Paleoenvironments of Eocene Rocks, McMurdo Sound, East Antarctica*, edited by J.D. Stilwell and R.M. Feldmann, Antarctic Research Series, American Geophysical Union.

Stilwell, J.D., R.H. Levy, R.M. Feldmann, and D.M. Harwood
1997   On the rare occurrence of Eocene Callianassid decapods (Arthropoda) preserved in their burrows, Mt. Discovery, East Antarctica, *Journal of Paleontology*, 284-287.

Stöhr, E.
1880   Die Radiolarienfauna der Tripoli von Grotte Provinz Girgenti in Sicilien, *Palaeontographica*, 26, 69-124.

Stradner, H.
1961   Über fossile Silicoflagelliden und die Möglichkeit ihrer Verwendung in der Erdölstratigraphie, *Erdöl und Kohle*, 14, 87-92.

Takemura, A.
1992   Radiolarian Paleogene biostratigraphy in the Southern Indian Ocean, Leg 120, in *Proceedings of the Ocean Drilling Program, Scientific Results*, edited by S.W. Wise, Jr., R. Schlich et al., College Station, Texas (Ocean Drilling Program), 120 (part 2), 735-756.

Tappan, H.
1980   Chapter 5: Ebridians, *The Paleobiology of Plant Protists*, W.H. Freeman and Company, San Francisco, 463-489.

Wei, W.
1992   Updated nannofossil stratigraphy of the CIROS-1 core from McMurdo Sound (Ross Sea), in *Proceedings of the Ocean Drilling Program, Scientific Results*, edited by S.W. Wise, Jr., R. Schlich et al., College Station, Texas (Ocean Drilling Program), 120 (part 2), 1105-1117.

Wei, W., and S.W. Wise, Jr.
1990   Middle Eocene to Pleistocene calcareous nannofossils recovered by Ocean Drilling Program Leg 113 in the Weddell Sea, in *Proceedings of the Ocean Drilling Program, Scientific Results*, edited by P.F. Barker, J.P. Kennett et al., College Station, Texas (Ocean Drilling Program), 113, 639-666.

Wei, W., G. Villa, and S.W. Wise, Jr.
1992   Paleoceanographic implications of Eocene-Oligocene calcareous nannofossils from sites 711 and 748 in the Indian Ocean, in *Proceedings of the Ocean Drilling Program, Scientific Results*, edited by S.W. Wise, Jr., R. Schlich et al., College Station, Texas (Ocean Drilling Program), 120 (part 2), 979-999.

White, R.J.
1980   Southern Ocean silicoflagellate and ebridian biostratigraphy, the opening of the Drake Passage, and the Miocene of the Ross Sea, Antarctica, Northern Illinois University, DeKalb, Illinois, Master's Thesis, 205 pp.

Wilson, G.S.
This vol.   Glacial geology processes and history of fossiliferous erratic-bearing moraines, southern McMurdo Sound, Antarctica, in *Paleobiology and Paleoenvironments of Eocene Rocks, McMurdo Sound, East Antarctica*, edited by J.D. Stilwell and R.M. Feldmann, Antarctic Research Series, American Geophysical Union.

Wilson, G.S., A.P. Roberts, K.L. Verosub, F. Florindo, and L. Sagnotti
1998   Magnetobiostratigraphic chronology of the Eocene-Oligocene transition in the CIROS-1 core, Victoria Land margin, Antarctica: implications for Antarctic glacial history, *GSA Bulletin*, 110(1), 35-47.

Wise, S.W., Jr.,
1983   Mesozoic and Cenozoic calcareous nannofossils recovered by Deep Sea Drilling Project Leg 71 in the Falkland Plateau region, southwest Atlantic Ocean, in *Initial Reports of the Deep Sea Drilling Project*, edited by W.J. Ludwig, V.A. Krasheninnikov et al., U.S. Government Printing Office, Washington, D.C., 71, 481-550.

Zacharias, O.
1906   Eine neue Dictyochide aus den Mittelmeer, Hermesinum adriaticum n.g., n.sp., *Arch. Hydrobiol. u. Planktonk.*, Stuttgart, 1, 394-398.

Steven M. Bohaty, Department of Geosciences, University of Nebraska - Lincoln, Lincoln, Nebraska 68588-0340

David M. Harwood, Department of Geosciences, University of Nebraska - Lincoln, Lincoln, Nebraska 68588-0340

SPORES AND POLLEN FROM THE MCMURDO SOUND ERRATICS, ANTARCTICA

Rosemary A. Askin

Byrd Polar Research Center, The Ohio State University, Columbus, Ohio 43210

Terrestrial palynomorphs are associated with dinoflagellate cysts in fossiliferous glacial erratics from the McMurdo Sound area, Ross Ice Shelf, Antarctica. The terrestrial assemblage includes over 49 spore and pollen taxa derived from land plants, along with fungal remains. Reworked into the Cenozoic erratics are Permian-Triassic and rare Cretaceous palynomorphs. The spore and pollen assemblages are characterized by diverse and common *Nothofagidites* pollen, diverse and less common podocarpaceous conifer pollen, diverse though uncommon Proteaceae pollen, rare representatives of other angiosperm families, and rare cryptogram spores. Erratics identified as middle to upper Eocene and upper middle to upper Eocene are notably richer in species diversity and numbers of specimens than younger erratics and reflect a *Nothofagus*-podocarpaceous conifer-Proteaceae vegetation with other angiosperms and a few cryptogams growing in temperate climate conditions. Spore and pollen occurrences support the Eocene ages previously derived from dinoflagellate cyst and siliceous microfossil data. Erratics identified as ?early Oligocene and post-Eocene show a major drop in species diversity, consistent with hypothesized deteriorating (colder) climates near the end of the Eocene. In addition, erratics derived from Permian-Triassic strata of the Beacon Supergroup are recognized.

INTRODUCTION

This paper describes the spore and pollen assemblages recovered from a selected set of glacial erratics from McMurdo Sound, East Antarctica. General descriptive and age information for the erratics, plus a map of the area, are provided in the introductory chapter of this volume and are not repeated here. The palynological preparations were carried out for study of the dinoflagellate cysts by R. H. Levy, and detailed sample information and methods are described by Levy and Harwood [this volume]. A subset of microslides of 35 samples (one slide per sample except for samples E214, E345 and D1, where two slides were provided) was selected by R.H. Levy for examination of the spore and pollen flora. As shown in Table 2 of Levy and Harwood [this volume], terrestrial palynomorphs are typically rare or absent in McMurdo erratics examined in this study, although asso-

ciated marine dinoflagellate cysts are often common to abundant, consistent with the marine depositional environment of the entombing sediments of those erratics. Palynomorph content is variable, however: one previously described McMurdo erratic from Black Island does contain abundant (55%) terrestrial palynomorphs dominated by pollen of *Nothofagus* [Wilson, 1967].

Paleogene sedimentary rocks do not crop out in the ice-free margins of the Ross Sea. In Antarctica such outcrops occur only in the northern Antarctic Peninsula area and South Shetland Islands. Prior to recent drilling projects in the Ross Sea-McMurdo Sound area, our knowledge of the Paleogene terrestrial vegetation in the surrounding land areas was restricted to two sources of indirect information: palynomorphs recovered in early studies [e.g. Cranwell et al., 1960; McIntyre and Wilson, 1966] of some McMurdo Sound erratics, and recycled spore and pollen assemblages that lack good age control

161

from Quaternary surficial seafloor sediments [e.g. Wilson, 1968; Truswell, 1983].

Dinoflagellate cysts in the palynomorph assemblages from the McMurdo erratics reported by Cranwell et al. [1960], and subsequently by Cranwell [1964; 1969], McIntyre and Wilson [1966] and Wilson [1967], provided an Eocene age [McIntyre and Wilson, 1966] for the erratics. Levy and Harwood [this volume] give a comprehensive discussion on the dinoflagellate cyst floras and age interpretations, and suggest the erratics are derived from sediments showing a range of Paleogene ages. They recognize four age groupings for the dinoflagellate cyst assemblages: middle to late Eocene, late middle to late Eocene, ?early Oligocene, and post-Eocene (these are also possibly Oligocene). In addition, erratics derived from the Beacon Supergroup are recognized [Levy and Harwood, this volume; and this study].

Previously Reported Paleogene Spore and Pollen Assemblages from Ross Sea Area

Among the sparse terrestrial palynomorphs from early preparations of McMurdo Sound erratics, Cranwell et al. [1960], Cranwell [1969], McIntyre and Wilson [1966], and Wilson [1967] recorded relatively common pollen of *Nothofagus* (southern beech). Also recorded are other angiosperm taxa (particularly pollen of Proteaceae, plus occasional Liliaceae, Gunneraceae, Sterculiaceae, Myrtaceae, Loranthaceae, Pedaliaceae); various species of podocarpaceous conifer pollen; cycad/*Ginkgo* pollen; cryptogam spores (particularly of ferns and lycopods); and fungal spores, hyphae and microthyriaceous fruiting bodies.

The recycled terrestrial palynomorphs catalogued by Truswell [1983] from the Ross Sea included many taxa that elsewhere are long-ranging from the Late Cretaceous through the Paleogene, and some restricted Tertiary forms. Truswell illustrated diverse cryptogam, gymnosperm and angiosperm taxa, including many previously unreported from Antarctica. Important additions to angiosperm families possibly represented in the vegetation adjacent to the Ross Sea area (or from more inland parts of Antarctica) were Sapindaceae (Cupanieae Tribe), Ericaceae/ Epacridaceae, Casuarinaceae, Euphorbiaceae, Restionaceae, and Sparganiaceae/ Typhaceae.

Recent drilling projects in the Ross Sea area have greatly increased our knowledge of Antarctic Cenozoic history. Those that encountered palynomorph-bearing sediments of Paleogene age include the lower parts of the drillholes MSSTS-1 (McMurdo Sound Sediment and Tectonic Studies project, Barrett, 1986) and CIROS-1

(Cenozoic Investigations in the western Ross Sea project, Barrett, 1989), and much of CRP-2/2A (Cape Roberts Project, Cape Roberts Science Team, 1999). Palynomorphs interpreted as Paleogene in age but recycled into Neogene sediments were recovered from other cores, including DSDP Site 270 (Ross Sea), RISP site J9 (Ross Ice Shelf Project), and both Cape Roberts cores CRP-1 and CRP-2/2A. MSSTS-1 penetrated Upper Oligocene sediments in its basal part. Truswell [1986] interpreted the sparse palynomorphs from these Oligocene strata and the overlying sediments as reworked, noting they were associated with reworked Eocene dinocysts, and suggested palynomorphs in coeval sediments in DSDP Site 270 were most likely also reworked. Assemblages described from MSSTS-1 [Truswell, 1986] and Site 270 [Kemp, 1975; Kemp and Barrett, 1975] are typically dominated by pollen of *Nothofagus*, along with various podocarpaceous conifer pollen, relatively common Proteaceae, and a few other angiosperm pollen of Myrtaceae and unknown affinities, with rare cryptogam spores. Truswell [1990] added the occurrence of Casuarinaceae and Liliaceae in the MSSTS-1 samples.

Mildenhall [1989] recovered abundant *Nothofagus* pollen, some in clumps (and thus from contemporaneous nearby vegetation), from what were then believed to be Oligocene sediments of CIROS-1 (this core is now known to include Upper Eocene strata in its lower part, Wilson et al., 1998), with Podocarpaceae pollen, some cryptogam spores, and a variety of other angiosperm taxa including the significant additions to the Antarctic Paleogene of Chenopodiaceae and Onagraceae. The presence of Chenopodiaceae pollen was also used to support the contemporaneous nature (rather than reworked) of part of the assemblage as these pollen do not occur prior to the Oligocene in other southern hemisphere localities. Likelihood of a reworked origin from older Eocene or Paleocene sediments was noted for many of the other palynomorphs.

A reworked Paleogene origin was suggested for many, and likely all as interpreted by Wrenn [1981], of the spores and pollen associated with Eocene dinocysts in Miocene sediments of the J9 core. Brady and Martin [1979] had earlier concluded that the low diversity assemblage, which included *Nothofagus*, Proteaceae and podocarpaceous conifer pollen, and some cryptogam spores, possibly represented low diversity Miocene vegetation. Similarly, Jiang and Harwood [1993] believed their assemblage of spores and pollen from J9 represented contemporaneous Miocene vegetation, because it was recovered from a Miocene diatomite unlikely to include reworked material.

The above drillhole results and discussions illustrate the difficulty in these samples of distinguishing with certainty reworked from likely "in-place" (contemporaneous with deposition) Cenozoic spores and pollen, and the problems associated with sparseness of material and circular reasoning invoking a largely glaciated and unvegetated post-Eocene landscape and thus a reworked origin for any spores and pollen.

The latest drillholes, CRP-1 and CRP-2/2A of the Cape Roberts Project, provide a section from the base of the Oligocene through the early Miocene, albeit with many unconformities [Cape Roberts Science Team, 1998; 1999]. Unambiguous differentiation of reworked from contemporaneous palynomorph specimens is still a problem, however a better understanding of the vegetational composition and diversity trends in the land areas adjacent to the Ross Sea area during the late Paleogene and into the Neogene is beginning to emerge. It appears that a sparse and low diversity tundra vegetation survived there essentially unchanged for much of the Oligocene and Miocene (and probably well into the Pliocene, based on Sirius Group data from the Transantarctic Mountains). Among the *Nothofagus* species in this vegetation is *N. lachlaniae* which also characterizes Sirius Group deposits where it represents a prostrate tundra plant with habit similar to the Arctic dwarf willow (*Salix arctica*) and the subalpine-alpine *N. gunnii* of Tasmania [Hill and Truswell, 1993; Francis and Hill, 1996]. Other species of *Nothofagus* also occur, especially in the older parts of the Oligocene, and a few species of podocarpaceous conifers, other angiosperms, and cryptogams, mainly Marchantiaceae (liverworts) and mosses. Raine [1998] suggested the herb-moss tundra reflected by the sparse CRP-1 Miocene palynomorph assemblages grew in a climate with summer temperatures similar to that of islands in the vicinity of the Antarctic Convergence today. He noted two intervals in the Miocene with slightly more diverse assemblages that may reflect woody vegetation growing in warmer sites or times, though could also result from reworked older material. The late Oligocene vegetation is closely similar, while palynomorph assemblages encountered in the early Oligocene of CRP-2/2A are more diverse and with more common specimens, suggestive of a slightly richer woody vegetation during early Oligocene time [Cape Roberts Science Team, 1999]. Lithologic and other fossil evidence [Cape Roberts Science Team, 1999] also indicate that early Oligocene climatic conditions, while still largely glacial, are somewhat milder than during the late Oligocene-Miocene, consistent with the palynomorph evidence. Assemblages in the basal sediments of CRP-2/2A, however, never reach the diversity and abundance of the floristically rich assemblages characteristic of the southern high latitudes Eocene. Palynomorphs described here from the McMurdo erratics provide an additional glimpse into this Eocene vegetation and some of the taxa that grew in the warmer, more temperate Eocene climates.

RESULTS

Distribution of spore and pollen taxa recognized in the 35 samples of McMurdo Sound erratics is shown in Table 1, with samples arranged in the same order as in Levy and Harwood [this volume]. The reader is referred to the Levy and Harwood paper for detailed sample location and lithologic information. Preservation ranges from very good to poor. Spores and pollen are sparse to rare, and, except for those erratics that probably have a Beacon Supergroup provenance, recovered assemblages have several notable compositional traits in common:

- pollen of *Nothofagidites* spp. are typically fairly diverse (7+ species, some "*fusca*" group species are not differentiated) and common
- podocarpaceous conifer pollen are typically common
- pollen of Proteaceae, while not usually common, are typically diverse (13+ species in total, including *Propylipollis*, *Proteacidites*, and probably *Peninsulapollis*; up to 7+ species in one sample)
- aside from the Nothofagaceae and Proteaceae, representatives of other angiosperm families are rare
- cryptogam spores are rare and of low diversity

(N.B. the term "diverse" used here for these McMurdo Sound assemblages is relative: elsewhere in lower latitudes they would not be considered diverse).

There are differences in composition between groups of erratics that follow the age-groupings of Levy and Harwood [this volume]. The most significant trend is the major drop in spore and pollen species diversity (Table 2) in erratics that have been identified as ?lower Oligocene and post-Eocene. This diversity decrease supports the post-Eocene ages of these two groups in that it is consistent with hypothesized deteriorating (colder) climates near the end of the Eocene, with resulting diversity decreases in terrestrial vegetation. Ice build-up during the late Eocene and Oligocene in the McMurdo Sound area is attested to by the geologic record of glacially-derived sediments encountered in CIROS-1 [summarized by Barrett et al., 1989; Wilson et al., 1998] and CRP-2/2A [Cape Roberts Science Team, 1999].

Compositional trends for the different groups of erratics, correlated to their age groupings, as defined by

TABLE 1. Occurrence data for spores and pollen, fungal remains, and Permian/Triassic spores and pollen.

	MTD 1	MTD 42	MTD 153(1)	MTD 153(2)	MTD 189	MTD 190	MB 80	MB 103(2)	MB 109(1)	MB 109(2)	MB 181(2)	MB 212K	MB 217A	MB 235A	MB 235C	MB 244C	MB 245	E 153	E 184	E 202	E 208	E 214	E 215	E 242D	E 243	E 244	E 313	E 345	E 350	E 364	SV 3	SV 12	SIM 1	SIM 11	D 1
Cryptogam spores																																			
Baculatisporites comaumensis		X			X																														
Coptospora cf.sp.A of Dettmann 1963												X											X												
Cyathidites minor		X			X																									X					
Cyathidites cf.*subtilis*					X																														X
Cyathidites sp.of Askin 1990																								X						X					
Laevigatosporites ovatus		X	X		X	X	X							X				X				X						X	X						X
Osmundacidites wellmanii		X			X																														
Retitriletes austroclavatidites		X								X	X																			X					
Retitriletes cf.*eminulus*																		X										X							
Retitriletes spp.		X									X																								
Rugulatisporites cf.*trophus*		X																								X									
Stereisporites antiquasporites		X			X													X								X		X	X						X
Gymnosperm pollen																																			
Araucariacites australis					X															X															
Cupressaceae/Taxodiaceae		X																												X					X
Dacrycarpites australiensis		X									X										X									X					X
Dacrydiumites praecupressinoides																		X	X																X
Microcachryidites antarcticus		X	X		X																							X	X	X					X
Microalatidites paleogenicus																		X																	
Microalatidites varisaccatus	X	X				X	X			X								X					X					X							
Phyllocladidites mawsonii		X			X									X				X						X				X		X					X
Podocarpidites ellipticus	X	X	X	X	X					X														X		X		X	X	X					X
Podocarpidites marwickii	X	X	X			X			X	X	X							X												X					X
Podocarpidites cf.*exiguus*																							X					X							
Podocarpidites cf.*torquatus*		X												X				X										X		X					X
Podocarpidites spp.	X	X	X			X	X	X										X	X		X	X	X	X	X		X	X	X	X				X	X
Angiosperm pollen																																			
Ericipites scabratus																												X							
Haloragacidites harrisii	X		X																																
Malvacipollis cf.*subtilis*		X																																	
Nothofagidites asperus		X			X																														X
Nothofagidites flemingii	X	X	X		X		X											X	X				X					X	X	X					X
Nothofagidites lachlaniae	X	X	X	X	X		X	X	X	X		X						X	X	X		X	X	X			X	X	X	X			X	X	X
Nothofagidites mataurensis		X	X	X	X					X								X	X				X					X		X			X	X	X
Nothofagidites suggatei		X			X																							X							
Nothofagidites sp. 1	X			X			X							X									X					X	X	X					X
Nothofagidites spp. (*fusca* group)	X	X	X	X	X	X	X	X	X	X		X	X					X	X	X	X	X	X	X	X			X	X	X			X	X	X
Peninsulapollis askiniae	X	X																																	
Peninsulapollis gillii (rew.?)		X																																	X
Peninsulapollis ?truswelliae																												X		X					
Propylipollis ambiguus																		X																	X
Propylipollis crassimarginis		X	X																																
Propylipollis pseudomoides	X	X																										X		X					X
Propylipollis reticuloscabratus					X						X																								X
Propylipollis subscabratus	X	X	X		X			X										X																	
Propylipollis sp. 1		X																																	
Propylipollis spp.	X	X						X										X										X							X
Proteacidites parvus				X				X																	X			X							X
Proteacidites simplex					X																														
Proteacidites sp. 1		X						X																											
Proteacidites spp.	X	X	X		X			X										X																	X
Rhoipites sp.																												X		X					
Tricolpites spp.	X	X			X									X																					
Tricolporites sp.		X																																	
Triporopollenites sp.1		X																																	
Triporopollenites spp.					X													X					X					X							X

TABLE 1 (continued)

	MTD 1	MTD 42	MTD 153(1)	MTD 153(2)	MTD 189	MTD 190	MB 80	MB 103(2)	MB 109(1)	MB 109(2)	MB 181(2)	MB 212K	MB 217A	MB 235A	MB 235C	MB 244C	MB 245	E 153	E 184	E 202	E 208	E 214	E 215	E 242D	E 243	E 244	E 313	E 345	E 350	E 364	SV 3	SV 12	SIM 1	SIM 11	D 1
Fungal remains																																			
Fungal spores	X	X	X	X		X	X		X		X				X	X												X	X						X
Microthyriaceous fruiting bodies	X	X	X					X	X	X				X	X	X					X	X						X	X	X					X
Reworked Permian/Triassic spores and pollen																																			
Apiculatisporis sp.																										X									
Brevitriletes parmatus													X																						
Deltoidospora directa																								X	X										
Lophotriletes sp.													X																						
Osmundacidites wellmanii																										X									
trilete spores (indeterminate)				X									X													X							X		
Alisporites spp.	X	X		X					X				X													X									
bisaccates (frag./indet.)			X	X			X	X			X	X																					X		
Protohaploxypinus spp.				X					X																								X		
Striatopodocarpidites cancellatus				X																															
taeniate bisaccates (frag./indet.)				X			X	X				X												X	X								X		
Cannanoropollis sp.																																	X		
Plicatipollenites sp.																																	X		
Potonieisporites sp.																																	X		
monosaccates (frag./indet.)		X																								X							X		
Cycadopites sp.																										X									

Levy and Harwood [this volume, summarized in their Table 4], or their presumed provenance, are summarized in Tables 1 and 2, and discussed below. Environmental and climatic interpretations will be discussed further in a future multiauthored paper that considers all the sedimentologic and fossil evidence.

Group 1 (middle to upper Eocene) and Group 2 (upper middle to upper Eocene)

Group 1 includes the majority of the samples, namely MTD 1, 42, 153(1), 153(2), 189, 190; MB 80, 103(2), 109(1), 109(2), 181(2), 235A (reworked assemblage), 245; E 153, 184, 208, 214, 215, 345, 350, 364; SIM 11. Group 2 includes two samples, MB 235C and D 1. These two groups are discussed together because they represent a similar age range and have similar compositional characteristics, the only difference being the slightly more precise age for Group 2 based on dinocyst evidence [Levy and Harwood, this volume]. Higher spore and pollen species diversities are seen in these samples, although the number of taxa does not approach diversities observed in Eocene terrestrial assemblages from lower latitudes, or in some Eocene assemblages from Seymour Island, northern Antarctic Peninsula [e.g. Askin, 1997]. This is, in part, a result of the low total recovery from many of these samples (and the single

microslide available), and may be due, in part, to other factors such as lithology, although there seems to be no consistent correlation for Groups 1 and 2 between palynomorph recovery and lithofacies based on data in Tables 1 and 4 of Levy and Harwood [this volume]. Samples such as E214 and MB 235C (and others to a lesser extent) probably reflect depositional environments with a greater marine influence: they have substantially higher dinocyst diversities compared with the few recorded spore and pollen species. At present, there is insufficient material and data on distribution of the terrestrial palynomorphs to justify further speculations on relationship of the erratics to each other within Groups 1 and 2.

Generally, Group 1 and 2 samples have a relatively high number of presumed in-place Cenozoic spore and pollen taxa compared with those of Groups 3 and 4. They include a diversity of Proteaceae and other angiosperm and podocarp species not seen in Groups 3 and 4. This is consistent with derivation from a floristically richer vegetation for at least some of the erratics of Groups 1 and 2, and by inference they reflect a time of temperate moist climates before major glaciation and decimation of the land vegetation. In New Zealand the main vegetational change in response to cooling occurred near the top of the Upper Eocene Kaiatan Stage (~36-35 Ma) [Pocknall, 1989; Hollis et al., 1997]. When compared to the sparse Oligocene assemblages from the glacially-derived

TABLE 2. Diversity data for samples, arranged according to their age groupings. The second column provides Cenozoic spore and pollen ("s&p") diversity, and the third column Reworked Permian/Triassic ("P/Tr") diversity. In the Fungal column "s" denotes fungal spores and "m" microthyriaceous fruiting bodies. Dinocyst diversity figures are from Levy and Harwood [this volume]. * Sample MB 235A is considered a reworked Eocene assemblage by Levy and Harwood [this volume].

Sample	Cenozoic s&p diversity	Reworked P/Tr diversity	Fungal sp+mic	Dinocyst diversity
Group 1	(middle to upper Eocene)			
MTD 1	15	1	s m	24
MTD 42	24	2	s m	26
MTD 153(1)	24	1	s m	22
MTD 153(2)	6	1	s	12
MTD 189	3	5	.	6
MTD 190	24	.	s	11
MB 80	4	2	s	12
MB 103(2)	2	3	.	3
MB 109(1)	11	1	s m	28
MB 109(2)	4	1	m	15
MB 181(2)	8	1	s m	25
MB 235A *	2	4	m	8
MB 245	14	.	s m	14
E 153	11	.	s m	11
E 184	4	.	m	14
E 208	5	.	.	5
E 214	7	.	m	22
E 215	7	.	m	11
E 345	19	.	s m	13
E 350	14	.	s m	16
E 364	18	.	m	18
SIM 11	4	.	.	10
Group 2	(upper middle to upper Eocene)			
MB 235C	6	1	.	20
D 1	27	.	s m	25
Group 3	(?lower Oligocene)			
E 202	1	.	.	2
E 313	2	.	.	4
Group 4	(post-Eocene)			
MB 212K	1	1	.	1
MB 217A	.	.	.	1
MB 244C	.	1	.	.
E 242D	4	5	.	1
E 243	3	4	.	1
Group 5	(Paleozoic/Mesozoic)			
SV 3	.	.	.	.
SIM 1	.	8	.	.
Group 6	(Unknown age)			
E 244	.	1	s	.
SV 12	.	.	.	.

sequences of the CRP-2/2A core, the Group 1 and 2 erratics samples are also consistent with derivation from an older (Eocene) flora growing in more favorable conditions. These palynomorph assemblages thus support the broad age categories assigned by Levy and Harwood [this volume]. Further support for their age assignments is available from occurrences of some terrestrial species. Most of the cryptogam and gymnosperm and some angiosperm taxa are long ranging throughout the Cenozoic, though some species have a more restricted age in Australia and New Zealand, or at least indicate an age no older than Eocene. Examples of the latter are *Nothofagidites lachlaniae*, *N. matauraensis*, *Propylipollis crassi-marginis* and *Proteacidites simplex* [e.g. Dudgeon, 1983; Pocknall and Mildenhall, 1984]. The combined stratigraphic ranges of terrestrial species are consistent with Eocene age for these samples, and likely middle to late Eocene. In New Zealand, the *Nothofagidites matauraensis* Zone is considered late Eocene to late Oligocene [Couper, 1960; Pocknall and Mildenhall, 1984], although *N. matauraensis* first appears in the middle Eocene Bortonian. *Nothofagidites matauraensis* is not usually abundant until the late Kaiatan, except in southernmost New Zealand where it becomes common in probable Bortonian-Kaiatan strata (J. I. Raine, personal communication, 1999).

Fungal remains, including a variety of spores and microthyriaceous fruiting bodies (Tables 1 and 2), and hyphae that were noted in Table 2 of Levy and Harwood [this volume], occur in these samples. They are particularly common in samples MTD 42 and E 364. Spores and fruiting bodies are noticeably absent from the younger samples of Groups 3 and 4, perhaps a reflection of a less diverse vegetation and drier climates, though recovered palynomorphs were very rare overall with low probability of realistic representative recovery.

Group 3 (?lower Oligocene)

Two samples are included in this group, E 202 and E 313. Rare specimens of *Nothofagidites* spp. (*fusca* group, including *N. lachlaniae*) and *Podocarpidites* spp. were observed in these samples.

Group 4 (post-Eocene)

The samples in this group are MB 212K, 217A, 244C; E 242D, 243. Terrestrial palynomorphs are either very rare or absent in these samples, and like the ?lower Oligocene samples above include *Nothofagidites lachlaniae* (in one sample, MB 212K), and *Podocarpidites* spp.

and cryptogam spores in E 242 D and 243.

The record from Groups 3 and 4 erratics suggests a much reduced species richness in the vegetation in post-Eocene Antarctica. Only *Nothofagus*, podocarpaceous conifers and a few cryptogams are recorded, although the sparcity of palynomorphs does not give a representative sampling of the vegetation. *Nothofagidites lachlaniae* was noted above as occurring in the Oligocene and Miocene of the CRP-1 and CRP-2/2A cores, and it is the sole *Nothofagus* species reported from Sirius Group deposits of Pliocene age from the Transantarctic Mountains. Its common occurrence in Groups 1 and 2 erratics suggests that this species was relatively common in the Eocene of the Ross Sea area of Antarctica, growing in temperate conditions as it was also in New Zealand, along with a variety of other *Nothofagus* species. It is one of the few species present in the Groups 3 and 4 erratics, and judging from Sirius Group occurrences it may have been the only *Nothofagus* species to survive the deteriorating climate to the Pliocene, perhaps adapting to periglacial conditions during the Oligocene and Neogene.

Group 5 (Permian/Triassic)

Samples SV 3 and SIM 1 are believed to represent erratics derived from the Beacon Supergroup, the upper part (Victoria Group) of which includes Permian and Triassic sediments. Sample SV 3, which is barren of palynomorphs, was assigned to this group by Levy and Harwood [this volume] because of its lithology. Sample SIM 1 is assigned here to this group because the only recognizable palynomorphs are Permian. The SIM 1 assemblage is probably Early Permian because of the "relatively abundant" monosaccate pollen, characteristic of Lower Permian Victoria Group sediments from the adjacent Transantarctic Mountains [e.g. Kyle and Schopf, 1982]. Alternatively these specimens might be reworked into younger otherwise barren sediments.

As shown in Tables 1 and 2, spores and pollen are reworked from the Beacon Supergroup into many of the Eocene and younger erratics. These reworked specimens reflect erosion of both Permian and Triassic strata, though, judging from the more frequently occurring taeniate bisaccate and monosaccate specimens and *Brevitriletes parmatus* (an Early Permian form in Australia, e.g. Backhouse, 1991) the eroded rocks are mainly Permian. Some of the *Alisporites* spp. and spores (e.g. *Osmundacidites wellmanii*) may be Triassic. In some samples the reworked Beacon palynomorphs show a range of exinal colors or Thermal Alteration Index (TAI), or at least two sets of TAI values, suggestive of erosion of Beacon sediments that have been variously affected by thermal metamorphism by Jurassic dolerite intrusion.

Group 6 (Unknown age)

Samples E 244 and SV 12 are of unknown age, assigned to this group by Levy and Harwood [this volume]. No definitive evidence of age for these erratics could be provided by this study. Table 2 shows that fungal spores are associated with Eocene Groups 1 and 2 spore and pollen assemblages (though their presence in rocks of other ages is possible also). Sample E 244 contains a single fungal spore, while the taeniate bisaccate pollen fragment indicates Beacon Supergroup provenance, though it is uncertain whether this is a primary occurrence or reworked.

Note on Reworked Cretaceous Palynomorphs

Possibly reworked specimens of *Peninsulapollis gillii* (a typically Late Cretaceous [Campanian-Maastrichtian] species, though ranging to early Eocene, e.g. Dettmann & Jarzen, 1988) were noted in samples MTD 153(1) and D 1. Fragments of the Cretaceous dinoflagellate cysts *Odontochitina operculata* in E 364 and an *Isabelidinium ?cretaceum* in D 1 were also observed, supporting the occurrence of extremely rare reworking of Cretaceous sediments into these erratics.

TAXONOMIC NOTES

The following section provides notes on the sparse Cenozoic spore and pollen taxa recovered from the selected subset of microslides of McMurdo erratics. For comparative purposes (with larger more complete populations from other areas), the listing includes, where appropriate, remarks on morphological characters, plus some size data. Where possible, affinity with extant plant families (or higher taxonomic group) is also provided.

Occurrence data is given in Table 1, and palynomorph taxa are listed below in the order given in the Table, which is alphabetical within each of the three major groups: cryptogam spores, gymnosperm pollen and angiosperm pollen. Fungal remains and recycled Permian/Triassic taxa are not included below – these are merely listed in Table 1 and illustrations are provided of a few selected forms. Microslides including the illustrated specimens will be housed in the Department of Geosciences, University of Nebraska-Lincoln.

Cryptogam Spores

Baculatisporites comaumensis (Cookson) Potonié 1956
Plate 1, fig. f

Triletes comaumensis Cookson 1953: 470, pl. 2, figs. 27, 28
Baculatisporites comaumensis (Cookson) Potonié 1956: 23

Size: Equatorial diameter (excluding processes) 31, 39 µm (2 specimens)
Affinity: Osmundaceae/Hymenophyllaceae

Coptospora cf. sp. A of Dettmann 1963
Plate 1, fig. j

Coptospora sp. A in Dettmann 1963: 89,90, pl. 20, figs. 6-8
Coptospora sp. cf. *C.* sp. A of Dettmann 1963, in Truswell 1983: 144, pl.2, figs.4,5,9

Remarks: This form is the same as those described by Truswell (1983) and, like those, is smaller than Dettmann's Australian specimens and has more elongate verrucae which are almost rugulate.
Size: Equatorial diameter 44, 48 µm (2 specimens)
Affinity: Bryophyta

Plate 1.

Photomicrographs, magnification ~ x1440. Erratic sample number and microslide number (if applicable) provided, along with England Finder coordinates. a. *Cyathidtes* cf. *subtilis*, MTD 190, H28/2; b. *Stereisporites antiquasporites*, MTD 153(1), T26/2; c. *Rugulatisporites* cf. *trophus*, E242D/1, J32; d. *Retitriletes* cf. *eminulus*, E 345 >10 µm/1, H15/4; e. *Retitriletes austroclavatidites*, MB 245, P30/1; f. *Baculatisporites comaumensis*, MTD 153(1), W41/4; g. *Dacrydiumites praecupressinoides*, MB 245, R39; h. Cupressaceae/Taxodiaceae, E364 >10 µm/2, D17; i. *Laevigatosporites ovatus*, MTD 189, Q32/4; j. *Coptospora* cf. sp. A of Dettmann 1963, MB 235A, O35; k. *Microalatidites paleogenicus*, E 153 >10 µm/1, T31/1; l. *Microalatidites varisaccatus*, MTD 190, N32/3; m. *Microcachryidites antarcticus*, E 345 >10 µm/1, Q10/1

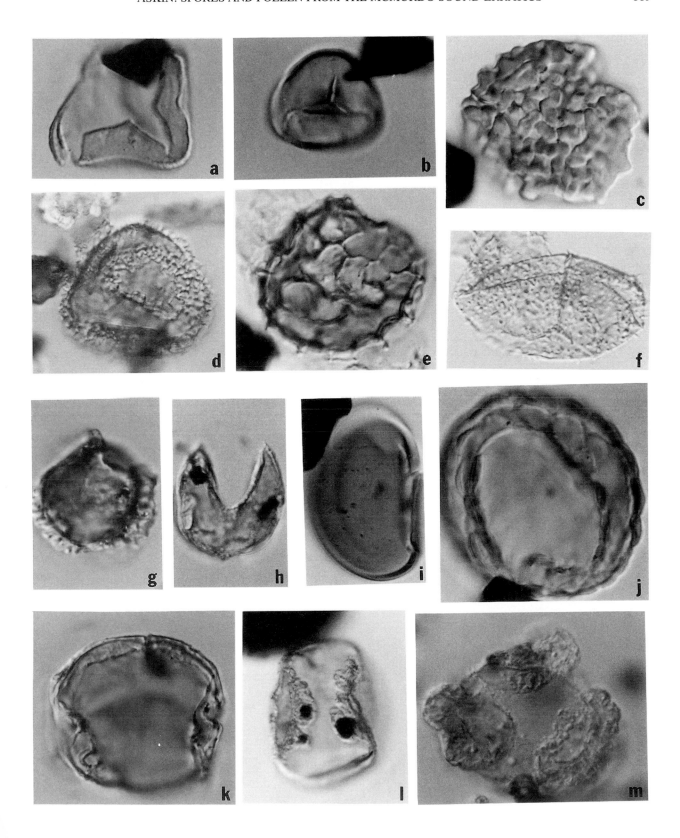

Cyathidites minor Couper 1953
(not illustrated)

Cyathidites minor Couper 1953: 28, pl. 2, fig. 13

Size: Equatorial diameter: 25-34 μm (4 specimens)
Affinity: Cyathaceae, and/or other filiceans

Cyathidites cf. *subtilis* Partridge *in* Stover & Partridge 1973
Plate 1, fig. a

Cyathidites subtilis Partridge in Stover & Partridge 1973: 247, pl. 13, figs. 1,2

Remarks: These three specimens could be included in *Cyathidites minor*, were it not for their finely sculptured exines. They differ from the similar sculptured form, *C. subtilis*, in their finer granulate ornament (<0.5 μm) which looks finely spinulate in places (this is visible along the equator). These specimens are a little corroded, and their exines do not appear reticulate as in *C. subtilis* (Stover and Partridge, 1973).
Size: Equatorial diameter 29-31 μm (3 specimens)
Affinity: Cyathaceae, and/or other filiceans

Cyathidites sp. of Askin 1990
(not illustrated)

Cyathidites sp. in Askin 1990: 146, pl. 1, fig. 8

Size: Equatorial diameter 31, 38 μm (2 specimens)
Affinity: Cyathaceae, and/or other filiceans

Laevigatosporites ovatus Wilson & Webster 1947
Plate 1, fig. i.

Laevigatosporites ovatus Wilson & Webster 1947: 153, pl. 2, fig. 9

Size: Length 27 (34) 46 μm, width 16 (22) 30 μm (21 specimens)
Affinity: ?Schizeaceae/Polypodiaceae/Gleicheniaceae

Osmundacidites wellmanii Couper 1953
(not illustrated)

Osmundacidites wellmannii Couper 1953: 20, pl. 1, fig. 5

Size: Equatorial diameter 41 μm (1 specimen)
Affinity: Osmundaceae/Hymenophyllaceae

Retitriletes austroclavatidites (Cookson) Döring, Krutzsch, Mai & Schulz *in* Krutzsch 1963
Plate 1, fig. e

Lycopodium austroclavatidites Cookson 1953: 469, pl. 2, fig. 35
Retitriletes austroclavidites (Cookson) Döring, Krutzsch, Mai & Schulz in Krutzsch 1963: 16

Size: Equatorial diameter 32-52 μm (7 specimens)
Affinity: Lycopodiaceae

Retitriletes cf. *eminulus* (Dettmann) Srivastava 1975
Plate 1, fig. d

Lycopodiumsporites eminulus Dettmann 1963: 45, pl.7, figs. 8-12
Retitriletes eminulus (Dettmann) Srivastava 1975: 58

Remarks: These specimens, which are proximally smooth, have a finer more irregular reticulum than the Australian type material. The nature of the reticulum is the same as the Seymour Island specimens of Askin (1990) but the latter tend to a more triangular shape.
Size: Equatorial diameter 32-36 μm (3 specimens)
Affinity: Lycopodiaceae

Rugulatisporites cf. *trophus* Partridge 1973 *in* Stover & Partridge 1973
Plate 1, fig. c

Rugulatisporites trophus Partridge in Stover & Partridge 1973: 250, pl. 15, fig. 4

Remarks: These specimens are smaller than the Australian type material described and illustrated in Stover and Partridge (1973), and the distal rugulate sculpture is not as coarse and the exine less thick. The proximal face bears greatly reduced sculpture with a tendency to a radial alignment.
Size: Equatorial diameter 30, 35 μm (2 specimens)
Affinity: Unknown

Stereisporites antiquasporites (Wilson & Webster) Dettmann 1963
Plate 1, fig. b

Sphagnum antiquasporites Wilson & Webster 1947: 273, fig. 2
Stereisporites antiquasporites (Wilson & Webster) Dettmann 1963: 25, pl. 1, figs. 20, 21

Size: Equatorial diameter 21 (27) 36µm (14 specimens)
Affinity: Sphagnaceae

Gymnosperm Pollen

Araucariacites australis Cookson 1947
(not illustrated)

Granulonapites (Araucariacites) australis Cookson 1947: 130, pl. 13, figs. 1-4

Size: 64, 72 µm (2 specimens)
Affinity: Araucariaceae

Cupressaceae/Taxodiaceae
Plate 1, fig. h

Remarks: Three specimens included here are the simple pollen type characteristic of the families Cupressaceae and Taxodiaceae. The exine is finely scabrate to smooth. Mildenhall (1994) included such fossil pollen from the Chatham Islands in the species *Taxodiaceaepollenites hiatus* (Potonié) Kremp.

Dacrycarpites australiensis Cookson & Pike 1953
(not illustrated)

Dacrycarpites australiensis Cookson & Pike 1953: 78, pl. 2, figs. 27-31; pl. 3, figs. 46-51

Size: Overall size 48-73 µm, corpus diameter 29-49 µm (5 specimens)
Affinity: Podocarpaceae, *Dacrycarpus*-type

Dacrydiumites praecupressinoides (Couper) Truswell 1983
Plate 1, fig. g

Dacrydium praecupressinoides Couper 1953: 35, pl. 4, figs, 36, 37
Dacrydiumites praecupressinoides (Couper) Truswell 1983: 147, pl. 2, fig. 16

Size: Total breadth 30, 41 µm (2 specimens)
Affinity: Podocarpaceae, *Dacrydium*-type

Microcachryidites antarcticus Cookson 1947
Plate 1, fig. m

Microcachryidites antarcticus Cookson 1947: 132, pl. 13, figs. 12-15; pl. 14, figs. 16-19

Size: Corpus diameter 27-34 µm (8 specimens)
Affinity: Podocarpaceae, *Microcachrys*-type

Microalatidites paleogenicus (Cookson & Pike) Mildenhall & Pocknall 1989
Plate 1, fig. k

Phyllocladus paleogenicus Cookson & Pike 1954: 63, 64, pl. 2, figs. 1-6
Microalatidites paleogenicus (Cookson & Pike) Mildenhall & Pocknall 1989: 34, pl. 4, figs. 1,2

Remarks: Although the corpus length of this specimen is not greater than its width, as is more typical of this form, it still falls with the original concept of Cookson and Pike (1954, see especially their Pl. 2, fig.6).
Size: Total breadth 35 µm (1 specimen)
Affinity: Podocarpaceae, *Phyllocladus*-type

Microalatidites varisaccatus Mildenhall & Pocknall 1989
Plate 1, fig. l

Microalatidites varisaccatus Mildenhall & Pocknall 1989: 34, 35, pl. 4, figs. 3-6

Remarks: Following Mildenhall and Pocknall (1989), specimens similar to *Phyllocladidites mawsonii* but with a thin zone at the base of the sacci and lacking the tubercles, are included in *Microalatidites varisaccatus*.
Affinity: Podocarpaceae, *Lagarostrobus/Phyllocladus* -type

Phyllocladidites mawsonii Cookson 1947 ex Couper 1953
(not illustrated)

Phyllocladidites mawsonii Cookson 1947: 133, pl. 14, figs. 22-28
Phyllocladidites mawsonii Cookson 1947 *ex* Couper 1953: 38, pl. 9, fig. 135

Affinity: Podocarpaceae, *Lagarostrobus*-type

Podocarpidites ellipticus Cookson 1947
Plate 2, fig. a

Podocarpidites ellipticus Cookson 1947: 131, 132, pl. 13, figs. 5-7

Size: Total breadth 43 (55) 70 µm (14 specimens)
Affinity: Podocarpaceae

Podocarpidites marwickii Couper 1953
Plate 2, fig. b

Podocarpidites marwickii Couper 1953: 36, pl. 4, fig. 39

Remarks: The typically larger size (though overlapping with the size range of *P. ellipticus*), along with coarser sacci reticulum of rather delicate sacci (reticulum is finer though typically difficult to discern in *P. ellipticus*), lack of a marginal crest on the corpus and often thicker cappa, is used to separate this species from *P. ellipticus*.
Size: Total breadth 55 (71) 97 µm (22 specimens)
Affinity: Podocarpaceae

Podocarpidites cf. *exiguus* Harris 1965
Plate 2, fig. c

Podocarpidites exiguus Harris 1965: 85, pl. 26, figs. 11,12

Remarks: These small, thin-walled, delicate bisaccate pollen, with indistinct sacci reticulum and folds radiating from the bases of the sacci, have thinner exine and more crescentic, narrower sacci than typical for *P. exiguus*. A similar specimen with narrow, crescentic, radially folded sacci from the Ross Sea area was assigned to *P. exiguus* by Truswell (1983).
Size: Total breadth 32-43 µm (3 specimens)
Affinity: Podocarpaceae

Podocarpidites cf. *torquatus* Mildenhall & Pocknall 1989
Plate 2, figs. d-h

Podocarpidites torquatus Mildenhall & Pocknall 1989: 36, pl. 5, figs. 4-8; pl. 6, fig. 1

Remarks: These specimens have the characteristic frilled crest (composed of high, tightly-folded rugulae) around the corpus and coarsely rugulate cappa of *P. torquatus* Mildenhall & Pocknall. The delicate sacci have a coarse incomplete reticulum that is radially elongate towards the bases, as in the type material. However, the cappa and sacci exine in these Antarctic specimens is not obviously scabrate, appearing smooth on well-preserved specimens. Some specimens (e.g. Plate 2, figs. d, e) are somewhat corroded thus the presence of a fine, surficial scabrate sculpture is difficult to discern. Coarse radial folding at the sacci bases does not occur. A single specimen (Plate 2, figs. g, h) has three smaller folded sacci, but because of its distinctive rugulate cappa and wide, frilled, marginal crest it is included here as an aberrant member of this population rather than a separate species.
Size: Total breadth 36 (48) 66 µm (7 specimens)
Affinity: Podocarpaceae

Angiosperm Pollen

Ericipites scabratus Harris 1965
Plate 2, fig. j

Ericipites scabratus Harris 1965: 13, pl. 29, figs. 22,23

Size: Tetrad diameter 30 µm (1 specimen)
Affinity: Epacridaceae/Ericaceae

Haloragacidites harrisii (Couper) Harris *in* Mildenhall & Harris 1971
Plate 2, fig. i

Triorites harrisii Couper 1953: 96, pl. 7, fig. 111
Haloragacidites harrisii (Couper) Harris in Mildenhall & Harris 1971: 304, 305, figs. 8-13

Size: 29, 41 µm (2 specimens)
Affinity: Casuarinaceae

Malvacipollis cf. *subtilis* Stover *in* Stover & Partridge 1973
Plate 2, fig. k

Malvacipollis subtilis Stover in Stover & Partridge 1973: 272, pl. 26, figs. 7-9

Remarks: This single specimen has large conical to tapered spines, thickened at the base, some of which are

---

Plate 2

Photomicrographs, magnification ~ x1440. a. *Podocarpidites ellipticus*, E 364 >10 µm/2, H25; b. *Podocarpidites marwickii*, E 364 >10 µm/2, G21/4; c. *Podocarpidites* cf. *exiguus*, E 350 >10 µm/2, W35/4; d-h. *Podocarpidites* cf. *torquatus*, d, e. specimen at different focal levels, D 1/3, N11, f. E 345 >10 µm/1, U33/3, g, h. specimen with 3 sacs at different focal levels, E 364 >10 µm/2, K16/4; i. *Haloragacidites harrisii*, MTD 153(2), O28/2; j. *Ericipites scabratus*, E 350 >10 µm/2, L26/4; k. *Malvacipollis* cf. *subtilis* MTD 153(1), L37

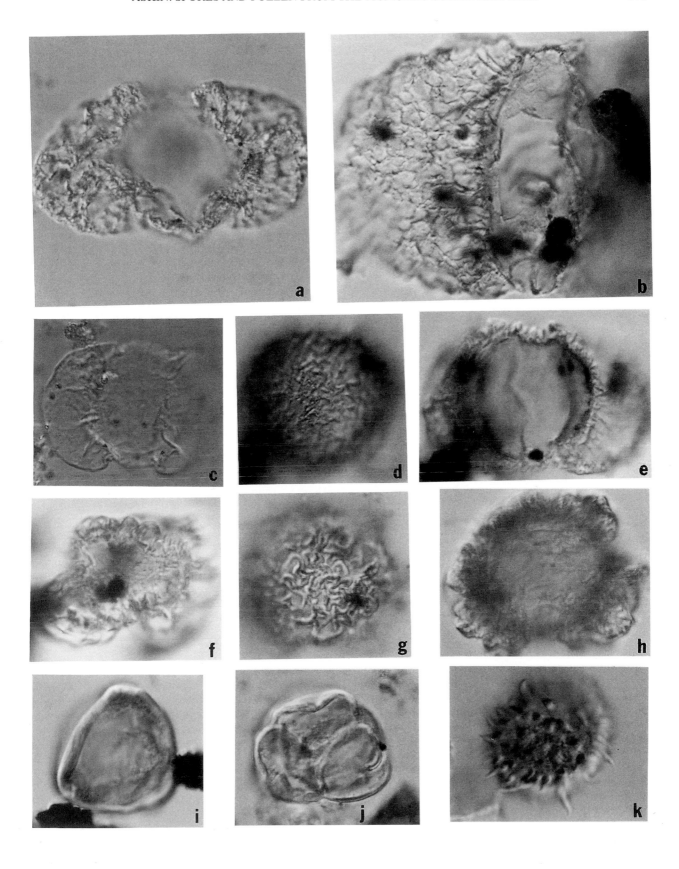

bent, some are stout and some more tapered, with occasional bacula (or broken spines?), about 1.5-2 μm basal diameter and 3-6 μm high. The spines are coarser and longer than in the Australian populations of *M. subtilis*.

Size: Equatorial diameter (excluding ornament) 25 μm (1 specimen)

Affinity: Euphorbiaceae

Nothofagidites asperus (Cookson) Romero 1973
Plate 3, fig. a

Nothofagus asperus Cookson 1959: 25, pl. 4, fig. 1,2
Nothofagidites asperus (Cookson) Romero 1973: 300

Remarks: The few specimens observed in these samples are thin-walled and delicate, with deep simple colpi, and are torn and crumpled.

Size: Equatorial diameter 42, 44, 44 μm (3 specimens)

Affinity: Nothofagaceae, *Nothofagus menziesii*-type

Nothofagidites flemingii (Couper) Potonié 1960
Plate 3, fig. b

Nothofagus flemingii Couper 1953: 47, pl. 6, fig. 72; pl. 9, fig. 139
Nothofagus cincta Cookson 1959: 26, 30, pl. 4, fig. 3
Nothofagidites flemingii (Couper) Potonié 1960: 132, pl. 9, fig. 196

Size: Equatorial diameter 30 (36) 44 μm (10 specimens)

Affinity: Nothofagaceae, *Nothofagus fusca*-type (*fusca* type b of Dettmann et al., 1990)

Nothofagidites lachlaniae (Couper) Pocknall & Mildenhall 1984
Plate 3, fig. c

Nothofagus lachlanae Couper 1953: 50, pl. 6, fig. 79
Nothofagidites lachlaniae (Couper) Pocknall & Mildenhall 1984: 30,31, pl. 11, figs.4,5

Size: Equatorial diameter 21 (26) 30 μm (17 specimens)

Affinity: Nothofagaceae, *Nothofagus fusca*-type (*fusca* type b of Dettmann et al., 1990)

Nothofagidites matauraensis (Couper) Hekel 1972
Plate 3, fig. d

Nothofagus matauraensis Couper 1953: 49, pl. 6, fig. 78; pl. 9, fig.142
Nothofagus astra Couper 1953: 49, pl. 6, fig. 77; pl. 9, fig.141
Nothofagidites matauraensis (Couper) Hekel 1972: 11, pl. 6, figs. 1, 3, 6

Size: Equatorial diameter 24 (34) 42 μm (12 specimens)

Affinity: Nothofagaceae, *Nothofagus brassii*-type (*brassii* type a of Dettmann et al., 1990)

Nothofagidites suggatei (Couper) Hekel 1972
Plate 3, fig. e

Nothofagus suggatei Couper 1953: 48, pl. 6, fig. 74
Nothofagidites suggatei (Couper) Hekel 1972: 10, pl. 6, fig. 12

Remarks: The simple V-shaped colpi are not as deep as in specimens assigned to *N. asperus*, and the overall size is smaller. Sculpture ranges from finely granulate to spinulate. The figured specimen has the coarsest sculpture observed; the other specimens with their thin exines are not well-preserved.

Size: Equatorial diameter 25-36 (4 specimens)

Affinity: Nothofagaceae, *Nothofagidites brassii*-type (*brassii* type b of Dettmann et al., 1990)

Nothofagidites sp. 1
Plate 3, figs. g, h

Remarks: All of the specimens assigned to this taxon have 7 apertures, except two (which have 6). The U-

---

Plate 3

Photomicrographs, magnification ~ x1440. a. *Nothofagidites asperus*, MTD 153(1), Y35/4; b. *Nothofagidites flemingii*, E 350 >10 μm/2, N21/3; c. *Nothofagidites lachlaniae*, E 153 >10 μm/1, F36/3; d. *Nothofagidites matauraensis*, E 153 >10 μm/1, U31; e. *Nothofagidites suggatei*, MTD 42/2, D8; f. *Peninsulapollis askiniae*, MTD 42/2, S20/4; g, h. *Nothofagidites* sp. 1; g. E 350 >10 μm/2, K33/2, h. MTD 189, G28; i. *Peninsulapollis ?truswelliae*, E 345 >10 μm/1, H39; j. *Rhoipites* sp., E 345 >10 μm/1, O15/3; k. *Triporopollenites* sp. 1 of Truswell 1983, MTD 42/2, Q28/2; l. *Peninsulapollis gillii*, MTD 153(1), N38/2

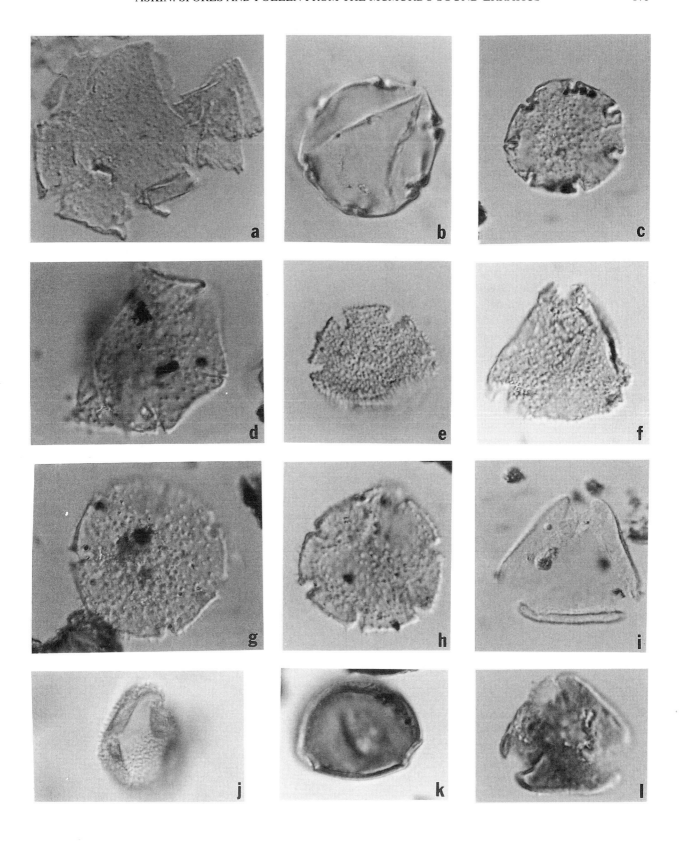

shaped apertures do not have pronounced exinal thickening, and the thin exine bears conspicuous spinules, ~0.5 μm basal diameter, and typically up to 1 μm high (and occasionally up to 1.5 μm high) in the equatorial area.

Size: Equatorial diameter 23 (28) 35 μm (11 specimens)

Affinity: Nothofagaceae, *Nothofagidites fusca*-type (*fusca* type b of Dettmann et al., 1990)

Peninsulapollis askiniae Dettmann & Jarzen 1988
Plate 3, fig. f

Peninsulapollis askiniae Dettmann & Jarzen 1988: 225, figs. 6 I,J, 7A-D

Remarks: Two poor specimens, one folded and torn, and one (figured) not fully expanded, exhibit the characteristic exinal and colpoidate apertural features. Although these specimens, and the *P. ?truswelliae* listed below, could be reworked, they are of similar exinal appearance and color to the rest of the presumed contemporaneous assemblage, and they are also known from the Eocene on Seymour Island (Askin, 1997).

Size: Equatorial diameter 30, 32 μm (2 specimens)

Affinity: Questionably Proteaceae

Peninsulapollis gillii (Cookson) Dettmann & Jarzen 1988
Plate 3, fig. l

Tricolpites gillii Cookson 1957: 49, pl. 10, figs. 12-15
Peninsulapollis gillii (Cookson) Dettmann & Jarzen 1988: 223, 5, figs. 4H-L, 6A-D

Remarks: Two specimens of this typically Late Cretaceous species have corroded or slightly darker exines than the presumed in-place assemblage. It is assumed that they are probably reworked (see comments in previous section on other possibly reworked Cretaceous specimens).

Size: Equatorial diameter 25, 29 μm (2 specimens)

Affinity: Questionably Proteaceae

Peninsulapollis ?truswelliae Dettmann & Jarzen 1988
Plate 3, fig. i

Peninsulapollis truswelliae Dettmann & Jarzen 1988: 225-226, figs. 4M-Q, 6E-H

Remarks: Two specimens, one of which is torn and twisted, have the characteristic colpoidate apertures with ragged margins. The exines are thin (<1 μm thick) , however, with a rather delicate sexine that appears foveolate and finely scabrate with fine coni to spinulae only apparent in places on one specimen. For this reason they are questionably assigned to *P. truswelliae*.

Size: Equatorial diameter 29, ~30 μm (2 specimens)

Affinity: Questionably Proteaceae

Propylipollis ambiguus (Stover) Dettmann & Jarzen 1996
Plate 4, fig. a

Triporopollenites ambiguus Stover in Stover & Partridge 1973: 269, pl. 21, fig. 7
Propylipollis ambiguus (Stover) Dettmann & Jarzen 1996: 144, figs. 25P, Q

Size: Equatorial diameter 35, 40 μm (2 specimens)

Affinity: Proteaceae

Propylipollis crassimarginis Dudgeon 1983
Plate 4, fig. f

Propylipollis crassimarginis Dudgeon 1983: 347, fig. 8

Size: Equatorial diameter 15, 17, 18, 24 μm (4 specimens)

Affinity: Proteaceae

Propylipollis pseudomoides (Stover) Dettmann & Jarzen 1996
Plate 4, fig. b

Proteacidites pseudomoides Stover in Stover & Partridge 1973: 266-267, pl. 25, fig. 3

Plate 4

Photomicrographs, magnification ~ x1440, except l at ~ x520. a. *Propylipollis ambiguus*, E 153 >10 μm/1, R31/2; b. *Propylipollis pseudomoides*, E 364 >10 μm/2, M22/2; c. *Propylipollis reticuloscabratus*, D 1/1, T11/2; d. *Propylipollis* sp. 1, MTD 42/2, Q41/4; e. *Propylipollis subscabratus*, MB 245, C29/4; f. *Propylipollis crassimarginis*, MTD 42/2, D39/1; g. *Proteacidites parvus*, E 345 >10 μm/1, O12; h. *Proteacidites* sp. 1, MTD 153(1), U24/2; i. *Proteacidites simplex*, MTD 190/1, T35; j. Fungal spore, E 345 >10 μm/1, Q42; k. *Striatopodocarpidites cancellatus* (reworked Permian pollen grain), MTD 189, R31; l. Microthyriaceous fruiting body, E 364, >10 μm/2, F31/3

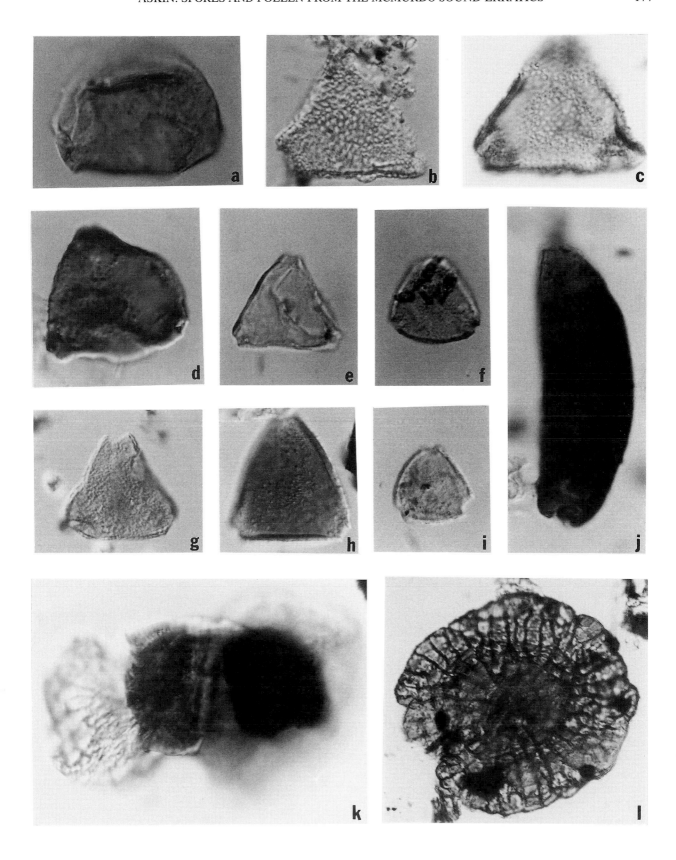

Propylipollis pseudomoides (Stover) Dettmann & Jarzen
   1996: 149, figs. 30A-C

Size: Equatorial diameter 27 (30) 32 μm (6 specimens)
Affinity: Proteaceae

Propylipollis reticuloscabratus (Harris) Martin &
Harris 1974
Plate 4, fig. c

Proteacidites reticuloscabratus Harris 1965: 93, pl. 28,
   figs. 20, 21
Propylipollis reticuloscabratus (Harris) Martin & Harris
   1974: 109

Size: 23, 34, 35 (3 specimens)
Affinity: Proteaceae

Propylipollis subscabratus (Couper) comb. Nov.
Plate 4, fig. e

Proteacidites subscabratus Couper 1960: 52, pl. 6, figs.
   8-10

Remarks: This small simple form is best included
within the genus *Propylipollis* Martin & Harris 1974
because of the nexinal thickening at the pores, although this
thickening is not very pronounced. This species has finely
scabrate to almost smooth exine and straight to slightly
convex sides. The exinal layers are not clearly differentiated, although many of these specimens are somewhat corroded and the sexine tends to be lost in parts.
Size: Equatorial diameter 17 (21) 26 μm (14 specimens)
Affinity: Proteaceae

***Propylipollis* sp. 1**
Plate 4, fig. d

Remarks: This single, slightly distorted specimen is
poorly preserved and corroded, with parts of its exine
missing, but it is noted here because of its distinctive
scattered small (<0.5 μm) grana and coni, and well-
developed nexinal thickening at the pores characteristic
of *Propylipollis*. The sexine is apparently very thin and
delicate (or lost) and the nexine 1.5 μm thick, thickening
to crassimarginate pores with entire margins. Exinal
thickening at the pores is greater than in *P. ambiguus*, the
exine is thicker, and the sculptural elements more abundant and smaller.

Size: Equatorial diameter 30 μm (1 specimen)
Affinity: ?Proteaceae

Proteacidites parvus Cookson 1950
Plate 4, fig. g

Proteacidites parvus Cookson 1950:175, pl.3, fig. 29

Remarks: This species, which has a similar shape
and size (to slightly larger) to those included in
Propylipollis subscabratus and finely scabrate surface
texture, has a clearly differentiated exine (sexine is thicker than the nexine) which thins towards simple pores
with ragged margins. Although the pore structure seems
to be similar to that described by Dettmann and Jarzen
(1996) for their new genus *Lewalanipollis* (pore type 2),
in the absence of detailed TEM and SEM study to confirm this, the species is here retained in *Proteacidites*.
Size: 19 (23) 32 μm (10 specimens)
Affinity: Proteaceae

Proteacidites simplex Dudgeon 1983
Plate 4, fig. i

Proteacidites simplex Dudgeon 1983: 355, fig. 17

Remarks: Three specimens of this small simple
form were observed in one sample.
Size: Equatorial diameter 16, 17, 18 (3 specimens)
Affinity: Proteaceae

***Proteacidites* sp. 1**
Plate 4, fig. h

Remarks: Two specimens were observed with
straight to slightly convex sides, and an exine up to 1.5
μm thick which is not well differentiated but appears to
have a relatively thick nexine which thins towards entire-
margined pores. The thin delicate sexine, which is at
least partly destroyed, bears scattered small coni and
grana (<0.5 μm basal diameter, ~0.5 μm high).
Size: Equatorial diameter 28, 30 μm (2 specimens)
Affinity: ?Proteaceae

***Rhoipites* sp.**
Plate 3, fig. j

Remarks: Two poorly preserved, laterally compressed specimens that are tricolporate with a distinctly
columellate and reticulate (lumina ~0.5 μm across) exine
are included in this category. There is slight nexinal

thickening at the pores and the sexinal layer thickens at the poles.

Size: Polar diameter 26, 30 µm, equatorial diameter 16, 22 µm (2 specimens)

Affinity: Unknown angiosperm

Tricolpites and *Tricolporites* spp.
(not illustrated)

Remarks: This category is used for a few small nondescript tricolpate and tricolporate pollen with no distinguishing features and thin, smooth to finely- scabrate exines.

Size: Equatorial diameter 15-24 µm (6 specimens)

Affinity: Unknown angiosperms

Triporopollenites sp. 1 of Truswell 1983
Plate 3, fig. k

Remarks: These specimens are identical to those described by Truswell (1983) except that the sides are less convex in one specimen. Rare small scattered grana occur on the otherwise smooth exine, although these are not visible at the focus level of the figured specimen.

Size: Equatorial diameter 24, 27, 28 µm (3 specimens)

Affinity: Unknown angiosperm

Triporopollenites spp.
(not illustrated)

Remarks: This category is used for a few small nondescript triporate pollen, most poorly preserved and corroded.

Size: Equatorial diameter 19-25 µm (6 specimens)

Affinity: Unknown affinity, some may include poorly preserved Proteaceae pollen.

Acknowledgements. This study was supported by NSF grant OPP 9527013. I thank Richard Levy and David Harwood for the opportunity to make the study, and for the microslides. This paper has greatly benefited from the thorough reviews and helpful comments of Ian Raine, and the reviewers Alan Graham and Graham L. Williams.

REFERENCES

Askin, R. A.
1990   Cryptogam spores from the upper Campanian and Maastrichtian of Seymour Island, Antarctica, *Micropaleont.*, *36*,141-156.

Askin, R. A.
1997   Eocene - ?earliest Oligocene terrestrial paly-nology of Seymour Island, Antarctica, in *The Antarctic Region: Geological Evolution and Processes*, edited by C. A. Ricci, pp. 993-996, Terra Antarctica, Sienna, Italy, 1997.

Backhouse, J.
1991   Permian palynostratigraphy of the Collie Basin, Western Australia, *Rev. Palaeobot. Palynol.*, *67*, 237-314.

Barrett, P.J.(editor)
1986   Antarctic Cenozoic history from the MSSTS-1 drillhole, McMurdo Sound, *DSIR Bull.*, *237*, 174 pp.

Barrett, P.J., (editor)
1989   Antarctic Cenozoic history from the CIROS-1 drillhole, McMurdo Sound, *DSIR Bull.*, *245*, 254 pp.

Barrett, P. J., M. J. Hambrey, D. M. Harwood, A. R. Pyne, and P.–N. Webb
1989   Synthesis, *in Antarctic Cenozoic history from the CIROS-1 drillhole, McMurdo Sound, Antarctica*, P. J. Barrett, ed., pp. 241-251, *DSIR Bull.*, *245*.

Brady, H., and H. Martin
1979   Ross Sea region in the Middle Miocene: a glimpse into the past, *Science*, *203*, 437-438.

Cape Roberts Science Team
1998   Initial Report on CRP-1, Cape Roberts Project, Antarctica, *Terra Antartica*, *5*, 187 pp.

Cape Roberts Science Team
1999   Studies from the Cape Roberts Project, Ross Sea Antarctica, Initial Report on CRP-2/2A, *Terra Antartica*, *6*, 173 pp.

Cookson, I. C.
1947   Plant microfossils from the lignites of the Kerguelen Archipelago, *B.A.N.Z. Antarctic Research Expedition 1929-1931*, *Report A2*, 127-142.

Cookson, I. C.
1950   Fossil pollen grains of proteaceous type from Tertiary deposits in Australia, *Aust. Jour. Sci.*, *Series B, 3*, 166-177.

Cookson, I. C.
1953   Difference in microspore composition of some samples from a bore at Comaum, South Australia, *Aust. Jour. Botany*, *1*, 462-473.

Cookson, I. C.
1957   On some Australian Tertiary spores and pollen grains that extend the geological and geographical distribution of living genera, *Proc. Royal Soc. Victoria*, *69*, 41-53.

Cookson, I. C.
1959   Fossil pollen grains of *Nothofagus* from Australia, *Proc. Royal Soc. Victoria*, *71*, 25-30.

Cookson, I. C., and K. M. Pike
1953   The Tertiary occurrence and distribution of *Podocarpus* (section *Dacrycarpus*) in Australia and Tasmania, *Aust. Jour. Botany, 1,* 71-82.

Cookson, I. C., and K. M. Pike
1954   The fossil occurrence of *Phyllocladus* and two other podocarpaceous types in Australia, *Aust. Jour. Botany, 2,* 60-68.

Couper, R. A.
1953   Upper Mesozoic and Cainozoic spores and pollen grains from New Zealand, *N.Z. Geol. Surv. Palaeontol. Bull., 22,* 77 pp.

Couper, R. A.
1960   New Zealand Mesozoic and Cainozoic plant microfossils, *N.Z. Geol. Surv. Palaeontol. Bull., 32,* 88 pp.

Cranwell, L. M.
1964   Hystrichospheres as an aid to Antarctic dating with special reference to the recovery of *Cordosphaeridium* in erratics at McMurdo Sound, *Grana palyn., 5,* 397-405.

Cranwell, L. M.
1969   Antarctic and Circum-Antarctic palynological contributions, *Antarc. Jour. U.S., 4,* 197-198.

Cranwell, L. M., H. J. Harrington, and I. G. Speden
1960   Lower Tertiary microfossils from McMurdo Sound, Antarctica, *Nature, 186,* 700-702.

Dettmann, M. E.
1963   Upper Mesozoic microfloras from south-eastern Australia, *Proc. Royal Soc. Victoria, 77,* 11-48.

Dettmann, M. E., and D. M. Jarzen
1988   Angiosperm pollen from uppermost Cretaceous strata of southeastern Australia and the Antarctic Peninsula, *Assoc. Australas. Palaeont. Memoir, 5,* 217-237.

Dettmann, M. E., and D. M. Jarzen
1996   Pollen of proteaceous-type from latest Cretaceous sediments, southeastern Australia, *Alcheringa, 20,* 103-160.

Dettmann, M. E., D. T. Pocknall, E. J. Romero, and M. del C. Zamaloa
1990   Nothofagidites Erdtman ex Potonié; a catalogue of species with notes on the paleogeographic distribution of Nothofagus Bl. (Southern Beech), *N. Z. Geol. Surv. Paleont. Bull., 60:* 79 pp.

Dudgeon, M. J.
1983   Eocene pollen of probably proteaceous affinity from the Yaamba Basin, central Queensland., *Assoc. Australas. Palaeont. Memoir, 1,* 339-362.

Francis, J. E., and R. S. Hill
1996   Fossil plants from the Pliocene Sirius Group, Transantarctic Mountains: Evidence for climate from growth rings and fossil leaves, *Palaios, 11,* 389-396.

Harris, W. K.
1965   Basal Tertiary microfloras from the Princetown area, Victoria, Australia, *Palaeontogr. Abt. B, 115,* 75-106.

Hekel, H.
1972   Pollen and spore assemblages from Queensland Tertiary sediments, *Geol. Surv. Qld., Publ. 355, Palaeont. Pap. 30,* 34 pp.

Hill, R. S., and E. M. Truswell
1993   *Nothofagus* fossils in the Sirius Group, Transantarctic Mountains: leaves and pollen and their climatic implications, in *The Antarctic Paleoenvironment: A Perspective on Global Change,* edited by J. P. Kennett and D. A. Warnke, *AGU Antarc. Research Series, 60,* 67-73.

Hollis, C. J., D.B. Waghorn, C.P. Strong, and E.M. Crouch
1997   Integrated Paleogene biostratigraphy of DSDP site 277 (Leg 29): foraminifera, calcareous nannofossils, Radiolaria, and palynomorphs, *IGNS Sci. Report, 97/07,* 73 pp.

Jiang, X., and D. M. Harwood
1993   A glimpse of early Miocene Antarctic forests: palynomorphs from RISP diatomite, *Antarc. Jour. U.S., 27,* 3-6.

Kemp, E..M.
1975   Palynology of Leg 28 drillsites, Deep Sea Drilling Project, in *Initial Reports of the Deep Sea Drilling Project, Leg 28,* edited by D. E. Hays, L. A. Frakes, et al., pp. 599-623, U.S. Govt. Print. Office, Washington, D.C.

Kemp, E. M., and P. J. Barrett
1975   Antarctic glaciation and Early Tertiary vegetation, *Nature, 258,* 507-508.

Krutzsch, W.
1963   Atlas der mittel- und jungtertiaren dispersen Sporen- und Pollen- sowie der Microplanktonformen des nordlichen Mitteleuropas, Liefg II, Die Sporen des Anthocerotaceae und der Mitteleuropas, Liefg II, Die Sporen des Anthocerotaceae und der Lycopodiaceae. *Deutscher Verlag Wissenschaften, Berlin,* 141 pp.

Kyle, R. A. , and J. M. Schopf
1982   Permian and Triassic palynostratigraphy of the Victoria Group, Transantarctic Mountains, *Antarctic Geoscience,* C. Craddock, ed., pp. 649-659, Univ. Wisconsin Press, Madison.

Levy, R. H., and D. M. Harwood
(this vol.)   Marine palynomorph stratigraphy and age(s) of the McMurdo Sound erratics.

Martin, A. R. H., and W. K. Harris
1974   Reappraisal of some palynomorphs of supposed Proteaceous affinity, *Grana, 14,* 108-113.

McIntyre, D. J., and G. J. Wilson
1966   Preliminary palynology of some Antarctic Tertiary erratics. *N. Z. Jour. Botany, 4,* 315-321.

Mildenhall, D. C.
1989   Terrestrial palynology, Antarctic Cenozoic history from the CIROS-1 drillhole, McMurdo Sound, Antarctica, edited by P. J. Barrett, pp. 119-127, *DSIR Publ., DSIR Bull., 245.*

Mildenhall, D. C.
1994   Palynological reconnaissance of Early Cretaceous to Holocene sediments, Chatham Island, New Zealand, *I.G.N.S. Monograph 7*, 206 pp.

Mildenhall, D. C., and W. F. Harris
1971   Status of *Haloragacidites* (al. *Triorites*) *harrisii* (Couper) Harris comb. nov. and *Haloragacidites trioratus* Couper, 1953, *N. Z. Jour. Botany, 9*, 297-306.

Mildenhall, D. C., and D. T. Pocknall
1989   Miocene – Pleistocene spores and pollen from Central Otago, South Island, New Zealand, *N.Z. Geol. Surv. Paleont. Bull., 59*, 128 pp.

Pocknall, D. T.
1985   Palynology of Waikato Coal Measures (Late Eocene-Late Oligocene) from the Raglan area, North Island, New Zealand, *N. Z. Jour. Geol. Geophys., 28*, 329-349.

Pocknall, D. T.
1989   Late Eocene to Early Miocene vegetation and climate history of New Zealand, *Jour. Royal Soc. N.Z., 19*, 1-18.

Pocknall, D. T., and D. C. Mildenhall
1984   Late Oligocene-Early Miocene spores and pollen from Southland, New Zealand, *N.Z. Geol. Surv. Paleont. Bull., 51*, 66 pp.

Potonié, R.
1956   Synopsis der Gattungen der Sporen dispersae. I. Teil: Sporites, *Beih. Geol. Jahrb, 23*, 103 pp.

Potonié, R.
1960   Synopsis der Gattungen der Sporen dispersae. III. Teil: Nachträge sporites, fortsetzung pollenites, *Beih. Geol. Jahrb, 39*, 189 pp.

Raine, J.I.
1998   Terrestrial palynomorphs from Cape Roberts Project drillhole CRP-1, Ross Sea, Antarctica, *Terra Antarctica, 5*, 539-548.

Romero, E. J.
1973   Polen fósil de *"Nothofagus"* (*"Nothofagidites"*) del Cretácico y Paleoceno de Patagonia, *Rev. Museo La Plata, 7*, 291-303.

Stover, L. E., and A. D. Partridge
1973   Tertiary and Late Cretaceous spores and pollen from the Gippsland Basin, southeastern Australia, *Proc. Royal Soc. Victoria, 85*, 237-286.

Srivastava, S.K.
1975   Microspores from the Fredericksburg Group (Albian) of the southern United States, *Paleobiologie continentale, 6*, 119 pp., 1973.

Truswell, E. M.
1983   Recycled Cretaceous and Tertiary pollen and spores in Antarctic marine sediments: a catalogue, *Palaeontogr. Abt. B, 186*, 121-174.

Truswell, E. M.
1986   Palynology, in *Antarctic Cenozoic history from the MSSTS-1 drillhole, McMurdo Sound*, edited by P. J. Barrett, pp. 131-134, *DSIR Bull. 237.*

Truswell, E. M.
1990   Cretaceous and Tertiary vegetation of Antarctica: a palynological perspective, in *Antarctic Paleobiology, Its Role in the Reconstruction of Gondwana*, edited by T. N. Taylor and E. L. Taylor, pp. 71-88, Springer-Verlag, New York.

Wilson, G. J.
1967   Some new species of Lower Tertiary dinoflagellates from McMurdo Sound, Antarctica, *N. Z. Jour. Botany, 5*, 57-83.

Wilson, G. J.
1968   On the occurrence of fossil microspores, pollen grains, and microplankton in bottom sediments of the Ross Sea, Antarctica, *N.Z. Jour. Marine & Freshwater Res., 2*, 381-389.

Wilson, G. S., A. P. Roberts, K. L. Verosub, F. Florindo, and L. Sagnotti,
1998   Magnetostratigraphic chronology of the Eocene-Oligocene transition in the CIROS-1 core, Victoria Land margin, Antarctica: Implications for Antarctic glacial history, *G. S. A. Bull., 110*, 35-47.

Wilson, L. R., and R. M. Webster
1947   Plant microfossils from a Fort Union coal of Montana. *Amer. Jour. Botany, 33*, 271-278.

Wrenn, J.H.
1981   Preliminary palynology of the RISP site J-9, Ross Sea, *Antarc. Jour. U.S., 16*, 72-74.

Rosemary A. Askin, Byrd Polar Research Center, Ohio State University, Columbus, OH 43210

PALEOBIOLOGY AND PALEOENVIRONMENTS OF EOCENE ROCKS, MCMURDO SOUND, EAST ANTARCTICA
ANTARCTIC RESEARCH SERIES VOLUME 76, PAGES 183–242

TERTIARY MARINE PALYNOMORPHS FROM THE MCMURDO SOUND ERRATICS, ANTARCTICA

Richard H. Levy and David M. Harwood

Department of Geosciences, University of Nebraska-Lincoln, Lincoln, NE 68588, U.S.A.

Sedimentological and paleontological data acquired from fossiliferous glacial erratics present in coastal moraines in McMurdo Sound, East Antarctica, provide an indirect means to obtain information on strata hidden beneath the Antarctic ice sheet. The erratics provide a record of Paleogene paleoenvironmental conditions and fossil biotas that are unknown from East Antarctic outcrops and drillcores at the present time. Relatively rich assemblages of marine palynomorphs present in many erratics aid in the determination of age relationships and paleoenvironmental setting. Forty-three dinoflagellate cyst (dinocyst) taxa were recovered including the following new species: *Glaphyrocysta radiata*, *Phelodinium harringtonii*, *Selenopemphix prionota*, *Turbiosphaera sagena* and *Vozzhennikovia netrona*. Fifteen acritarch species and three prasinophyte taxa were also recovered. Many of the assemblages are most similar to Eocene palynomorph floras documented from Seymour Island, Antarctic Peninsula. This paper compiles available data on Eocene-Oligocene Southern Ocean dinoflagellate cyst biostratigraphy in order to determine the age of the erratics. Three chronostratigraphic groups are recognized based upon the marine palynomorphs: middle to upper Eocene; ?lower Oligocene; and post-Eocene.

INTRODUCTION

The McMurdo Erratics - an Overview

Geoscientists working to reconstruct the Early Cenozoic geologic history of East Antarctica are hindered by the lack of outcrop, as strata of this age lie hidden beneath the East Antarctic ice sheet and ice shelves. Fossiliferous glacial erratics (the McMurdo Erratics) in coastal moraines in McMurdo Sound, East Antarctica (Figure 1), provide a record of these strata. Although the erratics are removed from their stratigraphic context, they do contain sufficient paleontologic and sedimentologic information to allow comment on Eocene paleoclimate, paleobiogeography and paleoecology of the East Antarctic coast.

Lithologies represented in the McMurdo Erratics include a suite of nearshore and continental shelf siliciclastic sediments [Levy and Harwood, this volume]. Medium to coarse-grained sandstones, sandy mudstones, and conglomerates comprise the majority of fossil rich rocks. Diamictites and mudstones of probable glacial origin also occur. The suite of rocks provides a record of both ice-free coastal environments and glacial-marine depositional environments.

Fossils present in the McMurdo Erratics include marine microflora (dinoflagellate cysts, acritarchs, prasinophytes, diatoms, silicoflagellates, ebridians), marine microfauna (foraminifera), marine macrofauna (molluscs, brachiopods, arthropods), marine and terrestrial vertebrates, terrestrial microflora (pollen and spores), and terrestrial macroflora (wood and leaves). Marine palynomorphs (dinoflagellate cysts, acritarchs, and prasinophytes) are the focus of this paper. Data on other fossils are presented in Stilwell and Feldmann [this volume]. Marine palynomorph assemblages indicate that the McMurdo Erratics can be separated into three chronostratigraphic groups, middle to upper Eocene, ?lower Oligocene and post-Eocene.

Fig. 1. Location of sites from which erratics were collected in McMurdo Sound.

The McMurdo Erratics span an important transition from pre-glacial to glacial conditions in Antarctica. Oxygen isotope records from deep sea cores provide a proxy climatic record of both paleotemperature and ice-volume fluctuations. These records indicate that the Earth underwent a transition from warm 'ice-free' conditions in the early Eocene to an early Oligocene 'icehouse' environment in which dynamic ice sheets were present on East Antarctica [see Miller et al., 1991; Miller, 1992]. Glaciomarine sediments recovered from the East Antarctic continental shelf support the inferred climatic conditions determined from the proxy oxygen isotope records, but indicate that ice sheets were probably present at sea-level during the late Eocene [Barrett, 1989; Wilson et al., 1998; Barron et al., 1989; Wise et al., 1991; 1992]. Ice-caps and mountain glaciers may have existed on East Antarctica prior to the late Eocene [Hambrey et al., 1989; Hambrey and Barrett, 1993], yet unequivocal evidence for the presence of ice at sea-level has not been uncovered [Wise et al., 1991]. The middle to upper Eocene McMurdo Erratics provide a record of the biota that existed along probable ice-free coastal environments of East

Antarctica, rendering them a useful starting place to compare and contrast with the glacial record preserved in younger sediments. We can document changes in the biotas of coastal East Antarctica as they responded to the effects of climatic cooling.

The middle to late Eocene was a time of increased rate in northward drift of Australia away from Antarctica [e.g. Veevers et al., 1991 and references therein; Lawver et al., 1992]. By comparing and contrasting the flora and fauna recovered from the McMurdo Erratics with fossils present in contemporaneous deposits from the surrounding southern continents, we can try to find solutions to such questions as: (a) how did the East Antarctic flora and fauna evolve as the Antarctic continent became progressively isolated?; (b) what is the history of paleobiogeographic provinces in high southern latitudes?; and (c) can we trace the development of surface-water currents as ocean basins opened?

In order to utilize data obtained from the McMurdo Erratics to document changes in Paleogene East Antarctica environments and biota, we require information regarding the age of each erratic. The focus of this paper is to establish the age of the McMurdo Erratics based on marine palynomorph assemblages. The biostratigraphic age inferences are made through comparison to marine palynomorph biostratigraphy from Antarctic and Southern Ocean records.

Previous Palynological Reports on the McMurdo Erratics

Members of the 1958-59 New Zealand Geological Survey Antarctic Expedition first recognized the value of the McMurdo Erratics. During this expedition, fragments of gray calcareous mudstones were collected from moraines near Minna Bluff and on White Island in McMurdo Sound (Figure 1). From these erratics, Cranwell et al. [1960] reported the first fossil marine palynomorphs from Antarctica and tentatively suggested a Late Cretaceous to Early Tertiary age for the erratics based on lithological, microfloral and microfaunal data. Cranwell [1964] confirmed the identification of *Enneadocysta partridgei* (as *Cordosphaeridium diktyoplokus*) and constrained the age to the Paleogene. McIntyre and Wilson [1966] reported fifteen species of microplankton from five erratics collected from Black Island (Figure 1). An Eocene age was inferred for these erratics based upon the known ranges of the organic-walled microplankton. Wilson [1967] revised this work, erecting seven new species. Many of these species are included in an Early Tertiary high latitudinal flora (transantarctic flora) that extended from south of Australia and New Zealand to southernmost South America [Haskell and Wilson, 1975].

An early Eocene age was determined by Cranwell [1969] for erratics collected from Minna Bluff and White Island by Harrington [1969]. The age interpretation was based on similarity of the palynomorph assemblages characterized by *Enneadocysta partridgei* (as *C. diktyoplokus*), *Turbiosphaera filosa* (as *C. filosum*), *Arachnodinium antarcticum* (as *Aiora fenestrata)*, and *Deflandrea* spp., to those of the Eocene Leña Dura Formation in southern Chile reported by Cookson and Cranwell [1967].

The generally accepted Eocene age for the dinocyst bearing erratics was questioned by Stott [1982] and Stott et al. [1983] based on the presence and age ranges of Neogene diatoms found in association with the dinocysts. However, the *in situ* occurrence of the younger diatoms is questioned by Harwood and Bohaty [this volume] who suggest that the diatoms reported by Stott [1982] and Stott et al. [1983] are contaminants.

In this study, our goal is to document and illustrate marine palynomorphs recovered from a variety of erratics collected from McMurdo Sound, and to assign ages to the erratics based upon current Southern Ocean marine palynomorph biostratigraphy.

Southern Ocean Eocene to Oligocene Marine Palynomorph Biostratigraphy

At the time of the initial studies on the McMurdo Erratics [e.g. Cranwell et al., 1960; Cranwell, 1964; McIntyre and Wilson, 1966, Wilson, 1967], Southern Ocean marine palynomorph biostratigraphy was not well-established. Age determinations were limited due to the dominance of endemic taxa with poorly known biostratigraphic ranges. Southern Ocean marine palynomorph biostratigraphy has subsequently improved due to recovery of reference sections from key Southern Ocean Deep Sea Drilling (DSDP) and Ocean Drilling Program (ODP) sites and studies on the La Meseta Formation, Seymour Island, Antarctic Peninsula and CIROS-1 in McMurdo Sound [see Haskell and Wilson, 1975; Kemp, 1975; Hall, 1977; Goodman and Ford, 1983; Wrenn and Hart, 1988; Askin, 1988, 1997; Wilson, 1989; Cocozza and Clark, 1992; Crouch and Hollis, 1996]. Nevertheless, difficulties still exist in establishing high-resolution age control for these Antarctic stratigraphic sequences.

Well-developed, high-resolution Paleogene dinocyst zonations from New Zealand and Australia allow both intrabasinal and interbasinal correlation of strata in

Australasia [Partridge, 1976; Wilson, 1984; 1987; 1988]. However, application and correlation of these zonations to Antarctic and sub-Antarctic sequences is limited. Key taxa in the New Zealand and Australian zonations include dinoflagellate genera from the subfamily Wetzelielloidea (Vozzhennikova 1961) Bujak and Davies 1983. These genera are poorly represented in Antarctic Paleogene sequences. Wrenn and Beckmann [1982] recorded *Apectodinium homomorphum* from cored sediments at Ross Ice Shelf Project (RISP) site J-9 obtained from beneath the Ross Ice Shelf, representing the only published occurrence of Wetzelielloidea from Antarctica.

Studies on Eocene-?Oligocene deltaic sediments of the La Meseta Formation, Seymour Island, James Ross Basin, Antarctic Peninsula (Figure 2), provide the only detailed documentation of Paleogene marine palynomorph biostratigraphy for Antarctica [Hall, 1977; Wrenn and Hart, 1988; Askin, 1988; 1997; Cocozza and Clark, 1992]. A comprehensive review of the biostratigraphic ranges for marine palynomorph taxa recovered here is presented in Wrenn and Hart [1988]. A preliminary zonation was proposed for the La Meseta Formation by Askin [1988; 1997]. The lowermost biozone is characterized by abundant specimens of the dinoflagellate cyst *Enigmadinium cylindrifloriferum*, and is assigned a

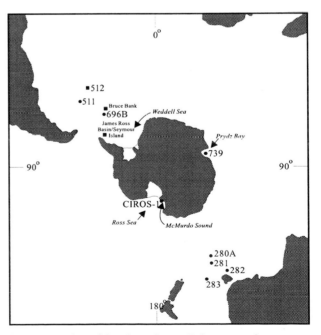

Fig. 2. Location of Southern Ocean Paleogene reference sections. Numbered sites are Deep Sea Drilling Project and Ocean Drilling Program sites. • Indicates sites at which biostratigraphic data are available for both marine palynomorphs and calcareous nannofossils.

late early Eocene age. Overlying zones are defined by terrestrial palynofloral datums and decreasing assemblage diversity [Askin, 1997]. Application of this zonation to regions beyond the James Ross Basin may prove to be problematic as the zones may be facies controlled, thus limiting application in deeper water settings, and the zonation is not calibrated by other fossil groups or magnetostratigraphy.

In order to develop a high-resolution zonation for the Antarctic and sub-Antarctic region, further studies on key DSDP and ODP sites are necessary. At present, documentation of biostratigraphic distributions for many DSDP sites are limited to initial reports [e.g. Haskell and Wilson, 1975; Goodman and Ford, 1983]. Biostratigraphic studies on existing material, as well as sequences recovered in the future, need to be a focus for future work if improved age resolution from dinocyst biostratigraphy is to be achieved. Correlation of key palynomorph datums to calcareous nannofossils, diatoms, silicoflagellates, radiolarians, ebridians and planktonic foraminifera will enhance both chronologic age control and regional correlations.

Despite the limitations of existing Southern Ocean marine palynomorph biostratigraphy, certain key bioevents identified in circum-Antarctic reference sections enable correlation at sub-epoch level. These bioevents include both first and last occurrence datums of palynomorphs and variations in dinocyst species richness.

First and last occurrence datums of key taxa. Several common endemic dinocyst taxa, including *Arachnodinium antarcticum*, *Deflandrea antarctica*, *Spinidinium macmurdoense*, and *Vozzhennikovia antarctica*, first occur in the lower Eocene and range into the lower Oligocene. These species are characteristic members of a transantarctic flora recognized by Haskell and Wilson [1975] and may be widely utilized to infer an age no older than Eocene.

A maximum age of middle Eocene may be inferred for sediments based upon the following bioevents. The last occurrence of *Enigmadinium cylindrifloriferum* [Askin, 1988; 1997; Cocozza and Clark, 1992] and the first appearances of *Enneadocysta* sp. 2 (as *Areosphaeridium* sp. A) and *Pyxidinopsis* sp. A [Cocozza and Clark, 1992] characterize the transition from the lower to middle Eocene in the James Ross Basin, Antarctic Peninsula. *Enneadocysta* sp. 2 (as *Enneadocysta* sp. a) and *Pyxidinopsis* sp. A (as *Pyxidinopsis* sp. a) first appear in the middle Eocene in DSDP Hole 280A [Crouch and Hollis, 1996]. Another useful species is *Vozzhennikovia netrona* n. sp., which first occurs in middle Eocene sediments in the Weddell

Sea [Mohr, 1990], (as *V. apertura*) and DSDP Hole 280A [Crouch and Hollis, 1996], (as *Vozzhennikovia* sp. a).

Several marine palynomorph taxa that have last appearance datums near the Eocene/Oligocene boundary can be used to constrain the age to the Eocene. These include *Hystrichosphaeridium truswelliae*, *Operculodinium bergmannii*, *Impletosphaeridium clavus*, *Spinidinium colemanii* and *Cyclopsiella trematophora* [see Wrenn and Hart, 1988 and references therein]. The stratigraphic position of last occurrence datums of several other Antarctic species is debated in the literature and discussed below.

The global range for *Alterbidinium asymmetricum* was reported as Maastrichtian to Eocene by Wrenn and Hart [1988]. However, in a recent review of Southern Ocean dinocyst biostratigraphy, Raine et al. [1997] reported that *A. asymmetricum* ranged into the lower Oligocene but indicated that the stratigraphic position for this datum needs revision.

Arachnodinium antarcticum was first reported as *Aiora fenestrata* from the McMurdo Erratics by McIntyre and Wilson [1966] and was subsequently recovered from the middle to upper Eocene Rio Turbio Formation [Archangelsky, 1969a]. Kemp [1975] initially suggested that *Arachnodinium antarcticum* (as *Aiora fenestrata*) extended into the lower Oligocene, based on its occurrence in DSDP Hole 274, yet Wrenn and Hart [1988], in their extensive review of Southern Ocean Paleogene dinocyst biostratigraphy, restrict this taxon to the Eocene.

Octodinium askiniae was described by Wrenn and Hart [1988] from Eocene sediments of the La Meseta Formation, Seymour Island. Raine et al. [1997] reported a last occurrence datum (LAD) for this taxon near the Eocene/Oligocene boundary but indicated that the stratigraphic position for this datum needs revision.

Turbiosphaera filosa was first reported (as *Cordosphaeridium* cf. *C. inodes*) by McIntyre and Wilson [1966] from glacial erratics collected in McMurdo Sound. *T. filosa* was initially proposed as *Cordosphaeridium filosum* by Wilson [1967], who noted a similarity to "Forma F" of Evitt [1961], reported from Eocene strata in North America. *T. filosa* was subsequently recovered from middle to upper Eocene strata of the Rio Turbio Formation in Argentina [Archangelsky, 1969a; Archangelsky and Fasola, 1971]. *T. filosa* ranges into the Oligocene in the Northern Hemisphere [Evitt and Pierce, 1975; Jan du Chéne, 1977].

Paleomagnetic age control now available from several Ocean Drilling Program sites and association with Paleogene calcareous nannofossil biostratigraphy enables us to address some of the uncertainty regarding the stratigraphic position of the last appearance datums

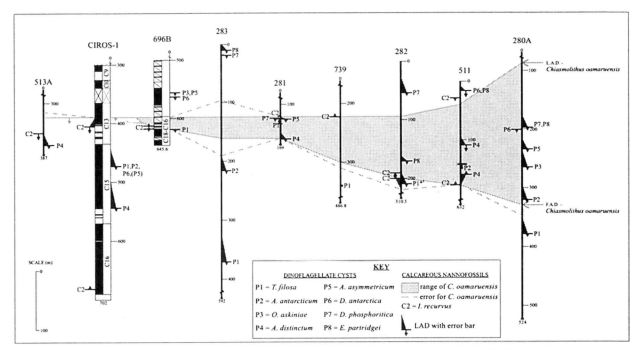

Fig. 3. Biostratigraphic distribution of key dinocyst datums and key calcareous nannofossil datums for Southern Ocean reference sections.

(LADs) of *Alterbidinium asymmetricum*, *Arachnodinium antarcticum*, *Octodinium askiniae*, and *Turbiosphaera filosa*. Of particular significance to this study are the reasonably well-constrained distributions of calcareous nannofossil species *Chiasmolithus oamaruensis* and *Isthmolithus recurvus* [Wei and Wise, 1990; Wei, 1992], which have been used to identify the approximate position of the Eocene-Oligocene boundary. In an attempt to determine the reliability of these dinocyst datums, we compare the calcareous nannofossil and dinocyst biostratigraphic data from Paleogene reference sections in the Southern Ocean and Antarctic shelf (Figure 2). The stratigraphic position of these dinocyst datums relative to the better calibrated calcareous nannofossil biostratigraphy is presented in Figure 3 and reviewed below.

Palynomorph data for DSDP holes 280A, 281, 282 and 283 were obtained from Haskell and Wilson [1975]. Additional data for DSDP holes 280A and 281 were obtained from Crouch and Hollis [1996]. Calcareous nannofossil distributions were reported by Edwards and Perch-Nielsen [1975]. In DSDP Hole 280A, the Last Appearance Datum (LAD) for *Turbiosphaera filosa* (P1) occurs below the First Appearance Datum (FAD) of *Chiasmolithus oamaruensis* (Figure 3), which indicates a probable late middle Eocene age for P1. The LADs of *Arachnodinium antarcticum* (P2), *Octodinium askiniae*

(P3), and *Alterbidinium asymmetricum* (P5) all occur above the FAD of *Chiasmolithus oamaruensis*, but well below the LAD of *C. oamaruensis*, which Wei and Wise [1990] suggested occurs within chron C13r (late Eocene to earliest Oligocene) of the magnetic polarity timescale [Berggren et al., 1995]. This stratigraphic relationship indicates that the LADs for *Arachnodinium antarcticum* (P2), *Octodinium askiniae* (P3) and *Alterbidinium asymmetricum* (P5) in DSDP Hole 280A all occur below the Eocene/Oligocene boundary and within the range of calcareous nannofossil *Chiasmolithus oamaruensis*.

Turbiosphaera filosa (P1) occurs in one sample in DSDP Hole 281 [Crouch and Hollis, 1996], below the first recorded appearance of *Isthmolithus recurvus* (C2) and just below the LAD of *Chiasmolithus oamaruensis* (Figure 3). The LAD of *Alterbidinium asymmetricum* (P5) occurs at the same interval in DSDP Hole 281. The LAD of *A. ?distinctum* (P4) occurs below the FAD of *Chiasmolithus oamaruensis*; either the range of *C. oamaruensis* is short at this site or *Isthmolithus recurvus* (C2) has a younger FAD relative to other sites. Based on the distribution of these nannofossils in DSDP Hole 281, *Turbiosphaera filosa* (P1) and the LADs of *Alterbidinium asymmetricum* (P5) and *A. ?distinctum* (P4) occur within the *Chiasmolithus oamaruensis* zone (lower upper Eocene) of Wei and Wise [1990].

Haskell and Wilson [1975] recorded *Turbiosphaera* cf. *T. filosa* in Hole 282. The LAD for this taxon (P1*) occurs below the FAD of *Chiasmolithus oamaruensis*, indicating a probable earliest late Eocene age for the LAD of *Turbiosphaera* cf. *T. filosa* (P1*). If one accounts for the possible error due to sample spacing (see Figure 3), the LAD of *T*. cf. *T. filosa* (P1*) is still well below the LAD of *Chiasmolithus oamaruensis*, indicating a position below the Eocene/Oligocene boundary.

The LADs of both *Turbiosphaera filosa* (P1) and *Arachnodinium antarcticum* (P2) were recorded in DSDP Hole 283 below the FAD of *Chiasmolithus oamaruensis*, indicating a probable middle Eocene age for these events at this site.

Palynomorph data for DSDP Hole 511 were obtained from Goodman and Ford [1983] and nannofossil distributions were reported by Wise [1983]. Goodman and Ford [1983] recorded *Arachnodinium antarcticum* (P2) in one sample, above the FADs of calcareous nannofossils *Chiasmolithus oamaruensis* and *Isthmolithus recurvus* (C2), but well below the LADs of these nannofossil taxa. Therefore, *Arachnodinium antarcticum* is restricted to the Eocene at this site. The range of *Alterbidinium ?distinctum* (P4) occurs within the ranges of both *Chiasmolithus oamaruensis* and *Isthmolithus recurvus* (C2), supporting an upper Eocene range for *Alterbidinium ?distinctum* as proposed by Crouch and Hollis [1996].

Palynomorph data for ODP Hole 696B were obtained from Mohr [1990], nannofossil data were taken from Wei and Wise [1990]. The LAD for *Turbiosphaera filosa* (P1) occurs below the FADs of both *Chiasmolithus oamaruensis* and *Isthmolithus recurvus* (C2) in this core. Paleomagnetic chrons C18 to C16 are considered to be represented in the interval in which *T. filosa* occurs, which indicates a middle to early late Eocene age for the LAD of *T. filosa*. The LADs of both *Octodinium askiniae* (P3) and *Alterbidinium asymmetricum* (P5) are notably younger in this hole as they occur above the LAD of *Chiasmolithus oamaruensis*. This indicates that *Octodinium askiniae* and *Alterbidinium asymmetricum* range into the lower Oligocene, and are not restricted to the Eocene, as reported in Wrenn and Hart [1988].

Truswell [1991] recorded *Turbiosphaera filosa* (P1) from near the base of ODP Hole 739. This occurrence is below the FADs of calcareous nannofossils *C. oamaruensis* and *I. recurvus*, indicating a probable middle Eocene age, although age control in this core is questionable as fossil recovery is poor [Barron et al., 1989; Wise et al, 1991].

In CIROS-1, Wilson [1989] reported the LADs of *Turbiosphaera filosa* (P1) and *Arachnodinium antarcticum* (P2) at a core depth of 473.25m, within the range of calcareous nannofossil *Isthmolithus recurvus* [Wei, 1992] and 50m below the Eocene/Oligocene boundary, as identified by Wilson and others [1998].

The relative position of the dinocyst biostratigraphic datums considered here are summarized in Figure 4 with respect to the calcareous nannofossil zonation for the Southern Ocean [Wei and Wise, 1990]. Based upon the dinocyst distributions described above, the following inferences are drawn:

(1) *Turbiosphaera filosa*, *Arachnodinium antarcticum* and *Alterbidinium ?distinctum* occur consistently below the LAD of *Chiasmolithus oamaruensis* (Figure 4). As the reported LAD for *C. oamaruensis* is Priabonian (C13r), we infer that the LADs of *Turbiosphaera filosa*, *Arachnodinium antarcticum* and *Alterbidinium ?distinctum* are restricted to the upper Eocene in the Southern Ocean.

(2) The reported FAD for calcareous nannofossil *Isthmolithus recurvus* is upper Eocene (C16n.2n). The only occurrence of *Turbiosphaera filosa* at a stratigraphic level above the FAD of *Isthmolithus recurvus* is in the CIROS-1 drillcore, which is the highest latitude site (Figure 2). It appears that *Turbiosphaera filosa* may have become extinct in the middle Eocene in lower latitudes but persisted into the upper Eocene in southern high latitudes.

(3) The upper Eocene range for *Alterbidinium ?distinctum* determined by Crouch and Hollis [1996] is supported by the reported range from DSDP Hole 511. *A. ?distinctum* may have limited use as a regional biomarker as reports of this taxon from the circum-Antarctic are few [see Goodman and Ford, 1983; Wilson, 1989; Crouch and Hollis, 1996].

(4) The occurrence of *Alterbidinium asymmetricum* and *Octodinium askiniae* above the LAD of *Isthmolithus recurvus* in ODP Hole 696B (Figure 4) indicates that these taxa range at least into the lower Oligocene (Rupelian), although these specimens may be reworked.

Species richness. Temporal decline in species richness within high latitude dinoflagellate cyst assemblages may also be used as an aid in age assignment. Lower middle Eocene sediments (calcareous nannofossil Zone CP13b of Okada and Bukry, [1980]] recovered from the Bruce Bank, Scotia Sea, Antarctica (Figure 2) contain diverse dinocyst assemblages consisting of up to forty-two species per sample [Mao and Mohr, 1995]. Dinocyst assemblages in upper Eocene sediments from the CIROS-1 drillcore (Figure 2) contain no more than fourteen species [see Figure 5, this paper; Wilson, 1989; Hannah, this volume]. Lower Oligocene samples recovered in CIROS-1 and ODP Hole 696B (Figure 2) contain

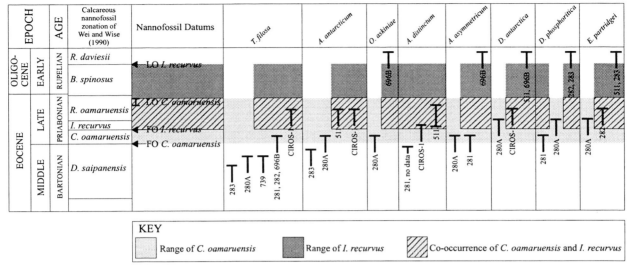

Fig. 4. Summary of biostratigraphic distribution of key dinocyst datums identified for Southern Ocean reference sections and correlation to the Southern Ocean calcareous nannofossil zonation of Wei and Wise (1990).

assemblages of less than five species [see Figure 5, this paper; Wilson, 1989; Hannah et al., in press; Mohr, 1990; Figure 6 in Mao and Mohr, 1995]. These data from the circum-Antarctic indicate that a significant decrease in dinocyst assemblage species richness occurred between the lower middle Eocene and the lower Oligocene in the Southern Ocean. Therefore, we suggest that an assemblage with high species richness may reflect a middle Eocene age, where an assemblage with low species richness may be used to identify a stratigraphic position of Oligocene or younger.

THIS STUDY

Previous studies highlight the potential wealth of geologic information contained in fossiliferous erratics present in coastal moraines in McMurdo Sound [e.g. Cranwell et al., 1960; Wilson, 1967; Rowe, 1974; Stott, 1982; Feldmann and Zinsmeister, 1984]. An effort to recover more geologic data from these moraines was undertaken by a team of scientists led by Dr. David M. Harwood at the University of Nebraska. During several field seasons [1992 to 1995] over 1000 glacial erratics were collected from coastal moraines around the shores of Mount Discovery, Brown Peninsula and Minna Bluff and from moraines on the flanks of Black Island, and along the floors of Salmon Valley and Miers Valley (see Figure 1). Two initial field seasons of broad regional reconnaissance led to the identification of two areas (see Figure 1) where hundreds of fossil-rich erratics were subsequently recovered. This paper presents the results

of a palynological investigation of over 80 erratics selected from the large collection of erratics housed at the University of Nebraska.

METHODS

Samples examined in this study were chosen according to the following criteria: (1) at least one sample of each lithofacies [see Table 1 and Levy and Harwood, this volume] was examined (for a sample list see Table 4); (2) glacial facies (Mm-d and Ms-d) were selected in order to constrain the age of the initiation of coastal glacial conditions in East Antarctica; (3) rare fine grained lithologies were selected for enhanced palynomorph recovery; and (4) erratics containing other fossil phyla were also examined to provide age control. Several of the erratics examined contained either clasts of other lithologies (MTD 211A) or clasts and lenses of fine-grained sediment (MTD 153, MB 109, E 303, E 365). Samples of both the matrix and the clasts or lenses were examined. A total of eighty-five samples were analyzed in this study.

Samples with identification codes beginning with 'E' were processed at the University of Nebraska using the following methods: (1) samples were crushed into 'pea-sized' fragments; (2) fragments were then washed with filtered water to remove any fine grained surface contaminants; (3) samples were treated using standard palynological techniques for acid digestion (HCl, HF); (4) residues were examined under light microscopy (LM) in order to determine whether further treatment was need-

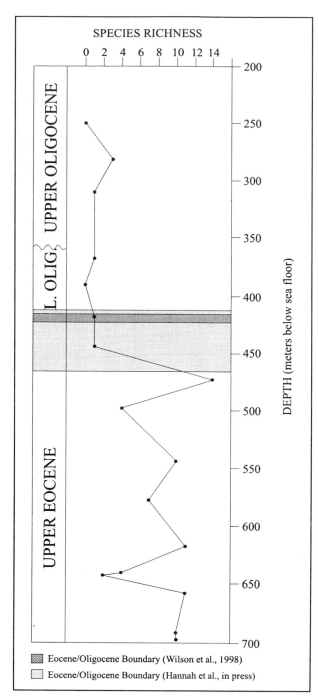

Fig. 5. Upper Eocene to lower Oligocene dinocyst assemblage species richness for the CIROS-1 drillcore. Dinocyst data is reported in Wilson (1989).

Samples with identification codes beginning with MB, MTD, SV, MV or BG were prepared by LAOLA Laboratories, Perth, Australia.

Slides were examined using an Olympus BH-1 transmitted light microscope at the Department of Geosciences, University of Nebraska-Lincoln.

Semi-quantitative abundance data for the marine component of the palynomorph assemblage are presented in Tables 2 and 3. These data were obtained by the following scale:

Very rare (X) = one specimen encountered in more than 20 fields of view

Rare (R) = one specimen encountered in 5 to 20 fields of view

Frequent (F) = one specimen encountered in every 5 fields of view

Common (C) = one specimen encountered in every field of view

Abundant (A) = more than one specimen encountered in every field of view

Identifiable fragments (fg) and identifiable opercula (op) were also recorded.

This procedure was carried out at 500X. Species identifications were confirmed at 1250X

Palynomorph Assemblage

Marine and terrestrial palynomorphs were recovered from seventy-six of the eighty-five samples (see Table 2). The marine palynomorph component comprised dinoflagellate cysts (dinocysts) as the dominant element, with lesser numbers of acritarchs, prasinophytes, scolecodonts and foraminiferal linings. The terrestrial palynomorph component consisted of pollen, spores, and fungal hyphae. Rare, freshwater, colonial chloro-coccales were also encountered. The focus of this paper is the marine palynomorphs. The terrestrial palynomorph flora is documented by Askin [this volume].

Marine palynomorphs were recovered from sixty-eight of the eighty-five samples examined. Preservation of the microflora was usually average to poor, with both complete and torn specimens present in most samples. Marine palynomorphs were common to abundant in only fifteen samples. Massive arenaceous erratics (Sm) comprise the bulk of the samples collected from McMurdo Sound and the majority of the samples examined in this study. These coarse-grained lithologies were most likely deposited in nearshore (intertidal/ subtidal) environments, which may account for the low number of samples with abundant palynomorph specimens, as marine palynomorphs (dinocysts in particular) are not usually abundant

ed; (5) if necessary, organic material was concentrated through heavy liquid separation using sodium polytungstate with a specific gravity of ~2.2; and (6) strewn slides were prepared from the organic residues and mounted with Eukitt optical adhesive.

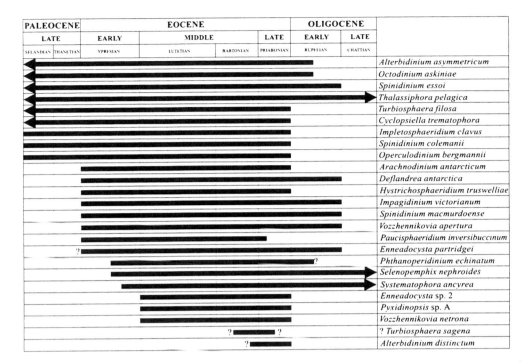

PALEOCENE		EOCENE				OLIGOCENE		
LATE		EARLY	MIDDLE		LATE	EARLY	LATE	
SELANDIAN	THANETIAN	YPRESIAN	LUTETIAN	BARTONIAN	PRIABONIAN	RUPELIAN	CHATTIAN	

Alterbidinium asymmetricum
Octodinium askiniae
Spinidinium essoi
Thalassiphora pelagica
Turbiosphaera filosa
Cyclopsiella trematophora
Impletosphaeridium clavus
Spinidinium colemanii
Operculodinium bergmannii
Arachnodinium antarcticum
Deflandrea antarctica
Hystrichosphaeridium truswelliae
Impagidinium victorianum
Spinidinium macmurdoense
Vozzhennikovia apertura
Paucisphaeridium inversibuccinum
Enneadocysta partridgei
Phthanoperidinium echinatum
Selenopemphix nephroides
Systematophora ancyrea
Enneadocysta sp. 2
Pyxidinopsis sp. A
Vozzhennikovia netrona
? *Turbiosphaera sagena*
Alterbidinium distinctum

Fig. 6. Known biostratigraphic ranges of key marine palynomorphs recovered from the McMurdo Erratics examined in this study.

in nearshore environments [Wall et al., 1977]. Furthermore, in nearshore, high-energy environments, silt-sized marine palynomorphs are likely to be winnowed from sandy deposits and transported further offshore. The paucity of marine palynomorphs in the majority of the erratics may reflect both unfavorable ecologic and hydrodynamic factors.

Lithologies in which marine palynomorph recovery was reasonable to good include: sandstone containing lenses/clasts of finer-grain size (Smc) and sandy mudstone (Mmb, Mw). As these samples are fine-grained or contain lenses of silt sized material, an increase in palynomorph abundance is not surprising. Several conglomerate samples including MTD 42 (Cmm), MB 181 (Csgc), and MB 235C (Cmc) also contained relatively abundant palynomorphs. The conglomerate samples usually contain clasts of metamorphosed sedimentary rocks that comprise basement material from Southern Victoria Land (Figure 1), but appear to contain no 'younger' Tertiary sedimentary clasts. Palynomorph assemblages in these coarse lithologies are therefore likely to be *in situ*, although reworking of older fossiliferous sediments cannot be ruled out.

Twenty-nine dinocyst genera containing forty-three species and one sub-species were encountered in the sixty-six samples (Table 3). Dinocyst assemblages are similar to both those reported from erratics studied by Wilson [1967] and those reported for the La Meseta Formation,

Seymour Island [Wrenn and Hart, 1988; Cocozza and Clark, 1992]. Six dinocyst species are proposed as new taxa.

The most commonly recovered dinocyst species are *Vozzhennikovia apertura* (in fifty-four samples) and *Enneadocysta partridgei* (in forty-eight samples). Other common taxa included *Deflandrea antarctica*, *Hystrichosphaeridium truswelliae*, *Spinidinium essoi*, *Alterbidinium asymmetricum*, and *Spinidinium macmurdoense*. The dinocyst assemblages are dominated by peridinioid cysts, with *V. apertura* as the dominant specimen in most samples. However, several samples (MTD 365(1), D1, and E 364) are dominated by specimens of the distinctive peridinioid species *Octodinium askiniae*, and one sample (MB 245) contains abundant specimens of *Alterbidinium asymmetricum*.

Species richness is generally low in most samples. Assemblages with the highest species richness (maximum twenty-eight) occur in mudstone lithofacies (Mmb and Mw) and sandstone lithofacies (Smc). Lower numbers of species usually occur in lithofacies of coarse-grain size (Sm, Ss, Sst, Ssg, Sw and Cmm). Diamictite (Dm) and mudstone with dropstones (Ms-d and Mm-d) usually have lowest species richness or are barren.

Seven acritarch genera containing fourteen species were recovered from fifty-six of the eighty-five samples examined (Table 3). Abundance is generally low, ranging

Table 1. A summary of McMurdo Erratic lithofacies.

Lithofacies		Abbreviation	Description
Sandstone			
	Massive	Sm	Well-sorted to poorly-sorted yellowish gray to greenish gray massive sandstone; scattered pebbles of various lithology may be present; invertebrate fossils are common.
	Massive with intraclasts	Smc	Moderately well-sorted to poorly-sorted yellowish grey to grayish brown massive sandstone with intraclasts of dark gray to dark grayish brown fine sandstone or mudstone. Intraclasts from sand to pebble size.
	Stratified	Ss	Moderately well-sorted olive brown sandstone with cm-scale stratification; scattered pebbles of various composition shape and size may be present.
	Stratified / trough cross-strata	Sst	Moderately well-sorted yellowish gray to olive gray sandstone with well-developed trough cross-stratification.
	Stratified / graded	Ssg	Poorly-sorted yellowish gray to dark greenish gray stratified, graded sandstone; grain size grades from basal pebbles to upper sands; beds range in thickness from less than 5mm to 4cm; both complete and fragmented fossil invertebrate shells and terrestrial organic remains may be incorporated in the coarser basal section of the graded beds.
	Weakly stratified	Sw	Moderately well-sorted to well-sorted yellowish gray to greenish gray weakly stratified sandstone; stratification is usually indicated by layers of terrestrial organic material (leaves and wood) or marine invertebrate fossils (usually molluscs).
Sandy mudstone			
	Massive / bioturbated	Mmb	Poorly-sorted dark gray sandy mudstone; dispersed pebbles and sandy lenses may be present; massive; mottled appearance indicates probable bioturbation; terrestrial macroflora (wood and leaves) and marine invertebrate macrofauna may occur.
	Stratified with dropstones	Ms-d	Poorly-sorted light olive gray sandy mudstone with moderately well-developed stratification indicated by diffuse layers of sand and mud; dispersed pebbles (?dropstones) and pelloids may be present.
	Weakly stratified	Mwb	Poorly-sorted dark yellowish brown to dark grayish brown weakly stratified sandy mudstone; mudstone pelloids may be present; stratification often masked or destroyed by bioturbation.
Mudstone			
	Massive with dropstones	Mm-d	Light olive gray massive mudstone with dispersed pebbles (?dropstones) of various lithology, shape and size; mudstone matrix may contain ostracodes and planktonic foraminifera.
Conglomerate			
	Massive - clast supported	Cmc	Unstratified poorly-sorted clast-supported conglomerate; well-rounded to sub-angular clasts range from sand to pebble in size (max 11mm); clast lithologies are varied and may possess circumgranular acicular calcite 'rinds'.
	Massive - matrix supported	Cmm	Unstratified poorly-sorted sandy matrix-supported conglomerate; well rounded to subangular clasts range up to cobble size (~90mm); clasts comprise several lithologies.
	Stratified / graded - clast supported	Csgc	Poorly-sorted stratified clast-supported conglomerate; subrounded clasts are sand to pebble size and comprise mudstone intraclasts; layers are graded and may be up to 8cm thick; terrestrial organic remains (wood and leaves) and marine invertebrates are usually incorporated within the conglomerate.
Diamictite			
	Massive / weakly stratified	Dm / Dw	Olive gray unstratified to ?weakly stratified sandy mudstone with matrix-supported clasts (between 5 and 15%) of various lithology, shape and size.

Table 2. Abundance data for palynomorphs recovered from the McMurdo Erratics. X = very rare, R = rare, F = frequent, C = common, A = abundant.

Sample No.	Dinocysts	Acritarchs	Prasinophytes	Foram Linings	Scolecodonts	Pollen / Spores	Fungal Hyphae	Sample No.	Dinocysts	Acritarchs	Prasinophytes	Foram Linings	Scolecodonts	Pollen / Spores	Fungal Hyphae
MTD 1	C	X	X	-	-	R	R	E 194	F	-	-	-	-	X	X
MTD 42	A	R	-	-	-	F	F	E 200	X	-	-	-	-	-	X
MTD 56	X	-	-	-	-	X	-	E 202	X	-	-	-	-	X	X
MTD 153(1)	A	X	-	-	-	F	F	E 203	X	-	-	-	-	-	-
MTD 153(2)	C	X	X	-	-	R	R	E 207	X	X	-	-	-	X	-
MTD 154	F	-	-	-	-	F	X	E 208	X	X	-	-	-	X	-
MTD 174A	-	-	-	-	-	X	-	E 214	C	X	-	X	-	R	R
MTD 189	X	-	-	-	-	R	-	E 215	F	R	R	-	-	R	R
MTD 190	F	-	-	-	-	F	X	E 216	X	-	-	X	-	X	-
MTD 211A(1)	-	X	-	-	-	X	-	E 219	C	X	-	X	-	F	F
MTD 211A(2)	-	-	-	-	-	-	-	E 240	-	-	-	-	-	-	-
MB 80	F	-	-	-	-	R	-	E 242D	X	-	-	-	-	X	-
MB 103(2)	X	X	-	-	-	X	-	E 243	X	X	-	-	X	R	-
MB 109(1)	A	R	-	X	-	F	F	E 244	-	-	-	-	-	X	X
MB 109(2)	A	X	-	-	-	R	F	E	R	-	-	-	-	X	X
MB 181(2)	A	R	-	-	X	F	F	E	R	-	-	X	-	-	X
MB 188B	X	X	X	-	-	X	-	E 313	X	X	-	-	-	X	X
MB 188G	F	-	-	-	-	X	-	E 317	X	X	-	-	-	X	X
MB 202	-	-	-	-	-	X	-	E 323	X	X	-	-	-	X	-
MB 212K	X	-	-	-	-	X	-	E 331	-	-	-	-	-	-	-
MB 217A	X	-	-	-	-	X	-	E 345	A	F	-	-	-	F	C
MB 235A	R	-	-	-	-	R	-	E 346	-	-	-	-	-	-	X
MB 235C	F	-	-	-	-	R	X	E 347	X	-	-	-	-	X	-
MB 244C	-	-	-	-	-	-	-	E 350	R	X	-	X	-	R	R
MB 245	C	-	-	-	-	F	F	E 351	-	-	-	-	-	X	-
MB 290G	X	-	-	-	-	-	-	E 355	X	-	-	-	-	-	X
MB 299	X	-	-	-	-	-	-	E 356	X	-	-	-	-	-	X
E 100	X	-	-	-	-	-	X	E 357	R	-	-	-	-	X	X
E 115	X	-	-	-	-	-	-	E 360	-	-	-	-	-	-	-
E 145	F	-	-	-	-	X	X	E 363	X	-	-	-	-	-	-
E 153	F	-	-	-	-	R	R	E 364	A	F	-	X	X	F	R
E 155	C	-	-	-	-	R	R	E	C	X	-	-	-	X	R
E 163	R	-	-	-	-	X	X	E	X	-	-	-	-	X	-
E 165	X	-	-	-	-	-	-	E 372	-	-	-	-	-	-	-
E 168	R	-	-	-	-	-	X	E 381	X	X	-	-	-	X	-
E 169	X	-	-	-	-	X	-	SV 3	-	-	-	-	-	X	-
E 171	R	X	-	-	-	X	-	SV 12	-	-	-	-	-	X	X
E 181	-	X	-	-	-	-	-	SIM 1	-	-	-	-	-	X	X
E 184	F	X	-	-	-	X	X	SIM 5	-	-	-	-	-	-	-
E 185	-	-	-	-	-	-	-	SIM 11	R	-	-	-	-	X	X
E 189	F	X	-	-	-	X	X	BG 1	-	-	-	-	-	-	-
E 191	X	-	-	-	-	X	-	D1	A	X	X	-	-	F	-
E 192	X	X	-	-	-	X	-								

Table 3. Marine palynomorph abundance data. X = very rare, R = rare, F = frequent, C = common, A = abundant, op = opercula, fg = fragments

DINOCYSTS	MTD 1	MTD 42	MTD 56	MTD 153(1)	MTD 153(2)	MTD 154	MTD 174A	MTD 189	MTD 190	MTD 211A(1)	MTD 211A(2)	MB 80	MB 103(2)	MB 109(1)	MB 109(2)	MB 181(2)	MB 188B	MB 188G	MB 202	MB 212K	MB 217A	MB 235A	MB 235C	MB 244C	MB 245	MB 290G	MB 299	E 100	E 115
?Alisocysta sp.	-	-	-	-	-	-	-	-	-	-	-	-	X	-	-	-	-	-	-	-	-	-	-	-	-	-	-	-	-
Alterbidinium ?distinctum	-	X	-	-	-	-	-	-	-	-	-	-	-	-	-	-	-	-	-	-	-	-	-	-	-	-	-	X	-
Alterbidinium asymmetricum	X	X	-	-	-	-	-	X	-	X	-	X	X	R	-	X	-	-	-	X	X	-	R	-	-	X	-	-	-
Arachnodinium antarcticum	X	X	-	X	-	-	-	-	-	-	-	-	X	R	X	-	X	-	-	-	-	X	-	X	-	-	-	-	-
Cerodinium cf. *C. dartmoorium*	-	-	-	-	-	-	-	-	-	-	-	-	-	X	-	-	-	-	-	-	-	-	-	-	-	-	-	-	-
Cribroperidinium giuseppi	X	X	-	X	-	-	-	-	-	-	-	-	-	X	-	-	-	-	-	-	-	-	-	-	-	-	-	-	-
Deflandrea antarctica	R	R	fg	R	X	X	-	fg	op	-	X	fg	X	R	R	op	X	-	-	fg	X	-	-	-	X	X	X	-	-
Deflandrea cf. *D. cygniformis*	-	X	-	-	-	-	-	-	-	-	-	-	-	-	-	-	-	-	-	-	-	-	-	-	-	-	-	-	-
Deflandrea cf. *D. flounderensis*	-	-	-	R	-	-	-	-	-	-	-	X	-	F	R	R	-	-	-	-	-	-	-	-	-	-	-	-	-
Deflandrea cf. *D. phosphoritica*	X	X	-	X	X	-	-	-	-	-	-	-	-	-	-	-	-	-	-	-	-	X	-	R	-	-	-	-	-
cf. *Eisenackia scrobiculata*	-	-	-	-	-	-	-	-	-	-	-	-	-	-	-	-	-	-	-	-	-	-	-	-	-	-	-	-	-
Enneadocysta partridgei	R	X	X	C	R	X	-	op	fg	-	X	-	F	F	F	op	R	-	fg	-	X	X	-	R	X	fg	-	-	-
Enneadocysta sp. 1	R	X	-	F	X	-	-	X	-	-	-	-	R	X	X	-	-	-	-	-	-	-	-	-	-	-	-	-	-
Enneadocysta sp. 2	X	-	-	X	X	-	-	-	-	-	-	-	-	R	-	X	-	X	-	-	-	-	-	-	-	-	-	-	-
cf. *Eocladopyxis peniculata*	-	X	-	-	-	-	-	-	-	-	-	-	-	-	-	-	-	-	-	-	-	-	-	-	-	-	-	-	-
Glaphyrocysta radiata	-	-	-	X	-	-	-	-	-	-	-	-	X	-	X	-	-	-	-	-	-	X	-	-	-	-	-	-	-
Hystrichosphaeridium truswelliae	F	X	-	X	X	op	-	X	-	X	-	-	R	X	X	-	-	-	X	-	X	-	X	-	-	X	-	-	-
H. cf. *H. tubiferum* subsp. *brevispinum*	X	X	-	-	-	-	-	-	-	-	-	X	-	X	-	-	-	-	-	-	-	-	-	-	-	-	-	-	-
Hystrichosphaeridium sp.	-	-	-	-	-	-	-	-	-	-	-	-	X	-	-	-	-	-	-	-	-	-	X	-	-	-	-	-	-
Impagidinium victorianum	X	X	-	X	X	-	-	-	-	-	-	-	-	X	X	-	-	-	-	-	-	-	-	-	-	-	-	-	-
Impletosphaeridium clavus	X	R	-	X	-	-	-	X	-	-	X	-	X	-	R	-	-	-	-	-	-	-	X	-	R	-	-	-	-
Impletosphaeridium sp.	-	-	-	-	-	-	-	-	-	-	-	-	-	-	-	-	-	-	-	-	-	-	-	-	-	-	-	-	-
Lejeunecysta hyalina	-	-	-	-	-	-	-	-	-	-	-	-	-	-	-	-	-	-	-	-	-	-	-	-	-	-	-	-	-
Microdinium sp.	X	-	X	-	-	X	-	-	X	-	-	-	X	-	-	X	-	-	-	-	-	-	X	-	-	-	-	-	-
Octodinium askiniae	X	X	-	-	-	-	-	-	-	-	-	-	F	-	X	-	-	-	-	X	X	-	-	-	-	-	-	-	X
Operculodinium bergmannii	X	X	-	X	-	-	-	-	-	-	-	-	X	-	X	-	-	-	-	-	-	X	-	-	-	-	-	-	-
Paucisphaeridium inversibuccinum	-	-	-	-	-	-	-	-	-	-	-	-	-	-	-	-	-	-	-	-	-	-	-	-	-	-	-	-	-
Phelodinium harringtonii	-	X	-	-	-	-	-	-	-	-	-	-	-	-	-	-	-	-	-	-	fg	-	-	-	-	-	-	-	-
Phthanoperidinium echinatum	-	X	-	X	-	-	X	-	-	-	-	X	X	X	-	-	-	-	-	-	-	-	-	-	-	-	-	-	-
Pyxidinopsis sp. A	X	X	-	-	op	-	-	-	fg	-	-	-	X	-	fg	-	-	-	-	-	-	-	X	-	-	-	-	-	-
Selenopemphix nephroides	-	-	-	-	-	-	-	-	-	-	-	-	-	-	-	-	-	-	-	-	-	-	X	-	-	-	-	-	-
Selenopemphix prionota	-	-	-	-	-	-	-	-	-	-	-	-	X	-	-	-	-	-	-	-	-	-	-	-	-	-	-	-	-
Spinidinium colemanii	-	-	-	-	-	X	-	-	X	X	-	-	F	-	X	-	-	-	X	X	-	-	-	-	-	-	-	-	-
Spinidinium essoi	R	X	-	X	X	X	-	X	-	-	-	F	R	F	-	-	-	-	-	X	-	R	-	X	-	-	-	-	-
Spinidinium macmurdoense	X	F	-	F	X	-	X	-	-	-	R	F	X	X	-	-	-	-	X	-	-	-	-	-	-	-	-	-	-
Spiniferites ramosus subsp. *reticulatus*	X	X	-	-	-	-	-	-	-	-	-	X	-	X	-	-	-	-	-	-	-	-	-	-	-	-	-	-	-
Spiniferites ramosus	R	X	-	X	X	fg	-	-	X	-	R	X	X	X	-	-	-	X	-	X	-	-	-	-	-	-	-	-	-
Systematophora ancyrea	-	X	-	-	-	-	-	-	-	-	-	X	-	X	-	-	-	-	-	-	-	X	-	-	-	-	-	-	-
Thalassiphora pelagica	X	X	-	X	-	-	-	-	-	-	-	X	X	X	R	-	-	-	-	-	X	-	-	-	-	-	-	-	-
Turbiosphaera filosa	X	-	X	-	-	X	-	X	-	X	-	-	X	-	-	X	-	-	-	R	-	-	-	-	-	-	-	-	-
Turbiosphaera sagena	-	-	-	-	-	-	-	-	-	-	-	-	-	-	-	-	-	-	-	-	-	-	-	-	-	-	-	-	-
Vozzhennikovia apertura	F	C	X	A	F	R	-	X	F	-	-	R	X	F	F	F	X	R	-	fg	X	R	-	F	-	X	-	-	-
Vozzhennikovia netrona	-	-	-	-	-	-	-	-	-	-	-	-	-	-	-	-	-	-	-	-	-	-	X	-	-	-	-	-	-
Dinocyst sp. A	X	X	-	-	-	-	X	-	-	X	X	X	-	-	-	X	-	-	-	-	-	-	-	-	-	-	-	-	-
Species Richness	25	26	3	22	12	6	0	6	11	0	0	12	3	28	15	25	7	13	0	1	1	8	20	0	14	1	2	6	2
ACRITARCHS																													
Michrystridium sp. 1	X	X	-	X	X	-	X	-	-	X	-	X	X	R	-	-	-	X	X	-	-	-	-	-	-	-	-	-	-
Michrystridium sp. 2	-	X	-	-	X	X	-	X	-	-	-	-	-	X	-	-	-	-	-	-	-	-	-	-	-	-	-	X	-
Cyclopsiella trematophora	X	X	-	X	-	-	-	-	-	-	X	X	-	-	-	-	-	X	-	-	-	-	-	-	-	-	-	-	-
Cyclopsiella sp. 1	-	-	-	-	-	-	-	X	-	-	-	-	-	-	-	-	-	-	-	-	-	-	-	-	-	-	-	-	-
Cyclopsiella sp. 2	-	-	-	-	-	-	-	-	-	-	-	-	-	-	-	-	-	-	-	-	-	-	-	-	-	-	-	X	-
Cyclopsiella sp. 3	X	-	-	-	-	-	-	-	-	-	-	-	X	X	-	-	-	-	-	-	-	-	-	-	-	-	-	-	-
Cyclopsiella sp. 4	-	-	-	-	-	-	-	-	-	-	-	-	-	-	-	-	-	-	-	-	-	-	-	-	-	-	-	-	-
Cyclopsiella sp. 5	-	-	-	-	X	-	-	-	-	-	-	-	-	-	-	-	-	-	-	-	-	-	-	-	-	-	-	-	-
Dichotisphaera sp.	-	-	-	-	-	-	X	-	X	-	X	-	-	-	-	-	-	-	-	X	X	-	-	-	-	-	-	-	-
Paralacaniella indentata	X	X	X	-	-	X	-	-	X	-	-	-	X	-	-	-	-	-	-	-	-	-	-	-	R	-	X	-	-
Scuticabolus lapidaris	-	-	-	-	-	-	-	-	-	-	-	-	-	-	-	-	-	-	-	-	-	-	-	-	-	-	-	-	-
Veryhachium sp. 1	-	-	-	-	-	-	-	-	-	-	-	-	-	-	-	-	-	-	-	-	-	-	-	-	X	-	-	-	-
Veryhachium sp. 2	-	-	-	-	-	-	-	-	-	-	-	-	-	-	-	-	-	-	-	-	-	-	-	-	-	-	-	-	-
Acritarch sp. A	-	-	X	X	X	-	-	-	-	-	-	X	-	X	-	-	-	-	-	-	-	-	-	-	-	-	-	-	-
Species Richness	5	4	1	3	3	4	0	2	4	1	0	2	1	3	2	4	2	0	0	0	0	2	2	0	4	0	0	3	0
PRASINOPHYTES																													
Cymatiosphaera sp.	.	-	-	-	-	-	-	-	-	-	-	-	-	-	-	-	-	-	-	-	-	-	-	-	-	-	-	-	-
Pterospermella sp. 1	X	-	-	-	-	-	-	-	-	-	-	-	X	-	-	-	-	-	-	-	-	-	-	-	-	-	-	-	-
Pterospermella sp. 2	-	-	-	X	-	-	-	-	-	-	-	-	-	-	-	-	-	-	-	-	-	-	-	-	-	-	-	-	-
CHLOROCOCCALES																													
Palambages sp.	X	-	-	X	-	-	-	-	-	-	-	-	-	-	-	-	-	-	-	-	-	-	-	-	-	-	-	-	-

Table 3. Marine palynomorph abundance data. X = very rare, R = rare, F = frequent, C = common, A = abundant, op =opercula, fg = fragments

DINOCYSTS	E 145	E 153	E 155	E 163	E 165	E 168	E 169	E 171	E 181	E 184	E 185	E 189	E 191	E 192	E 194	E 200	E 202	E 203	E 207	E 208	E 214	E 215	E 216	E 219	E 240	E 242D	E 243	E 244
?Alisocysta sp.	-	-	-	-	-	-	-	-	-	-	-	-	-	-	-	-	-	-	-	-	-	-	-	-	-	-	-	-
Alterbidinium ?distinctum	-	-	-	-	-	-	-	-	-	-	-	-	-	-	-	-	-	-	-	-	-	-	-	-	-	-	-	-
Alterbidinium asymmetricum	X	X	X	-	X	-	-	-	-	-	-	R	-	-	X	X	X	X	-	X	X	X	-	X	-	-	-	-
Arachnodinium antarcticum	X	-	-	-	-	-	-	-	X	-	X	-	-	X	-	-	X	-	-	-	-	X	-	-	-	-	-	-
Cerodinium cf. *C. dartmoorium*	-	-	-	-	-	-	-	-	-	-	-	-	-	-	-	-	-	-	-	-	-	-	-	-	-	-	-	-
Cribroperidinium giuseppi	-	-	-	-	-	-	-	-	-	-	-	-	-	-	-	-	-	-	-	-	-	-	-	-	-	-	-	-
Deflandrea antarctica	R	X	X	X	-	X	-	-	-	-	-	X	-	-	X	op	-	-	-	-	R	op	X	X	-	-	-	-
Deflandrea cf. *D. cygniformis*	-	-	-	-	-	-	-	-	-	-	-	-	-	-	-	-	-	-	-	-	-	X	-	-	-	-	-	-
Deflandrea cf. *D. flounderensis*	-	-	X	X	-	-	-	-	-	-	-	-	-	-	X	-	-	-	-	-	-	X	-	-	-	-	-	-
Deflandrea cf. *D. phosphoritica*	-	-	X	-	-	-	-	-	-	-	-	R	-	-	X	-	-	-	-	-	-	-	X	-	-	-	-	-
cf. *Eisenackia scrobiculata*	-	-	-	-	-	-	-	-	-	-	-	-	-	-	-	-	-	-	-	-	-	-	-	-	-	-	-	-
Enneadocysta partridgei	R/I	fg	X	R	-	op	op	-	X	-	X	X	-	X	-	-	-	-	X	fg	fg	R	-	fg	-	-	-	-
Enneadocysta sp. 1	R	-	-	-	-	-	-	-	-	-	-	-	-	-	-	-	-	-	-	-	-	-	-	-	-	-	-	-
Enneadocysta sp. 2	-	-	-	-	-	-	-	-	-	-	-	-	-	-	-	-	-	-	-	-	-	-	-	-	-	-	-	-
cf. *Eocladopyxis peniculata*	-	-	-	-	-	-	-	-	-	-	-	-	-	-	-	-	-	-	-	-	-	-	-	-	-	-	-	-
Glaphyrocysta radiata	X	-	-	-	-	-	-	-	-	-	-	-	-	-	-	-	-	-	-	-	-	-	-	-	-	-	-	-
Hystrichosphaeridium truswelliae	X	X	X	-	X	X	op	X	-	X	-	X	fg	-	X	-	-	-	-	X	X	-	X	-	-	-	-	-
H. cf. *H. tubiferum* subsp. *brevispinum*	-	-	-	-	-	-	-	-	-	-	-	-	-	-	-	-	-	-	-	-	-	-	-	-	-	-	-	-
Hystrichosphaeridium sp.	-	X	-	-	-	-	-	-	-	-	-	-	-	-	-	-	-	-	-	-	-	-	-	-	-	-	-	-
Impagidinium victorianum	-	-	-	X	-	-	-	-	fg	-	-	-	-	X	-	-	-	-	X	-	X	-	-	-	-	-	-	-
Impletosphaeridium clavus	-	-	R	-	-	X	X	-	X	-	-	-	-	-	-	-	-	-	X	X	-	R	-	-	-	-	-	-
Impletosphaeridium sp.	-	-	X	-	-	-	-	-	-	-	-	-	-	-	-	-	-	-	-	X	-	-	-	-	-	-	-	-
Lejeunecysta hyalina	X	-	-	-	-	-	-	-	-	-	-	-	-	-	-	-	-	-	X	-	-	-	-	-	-	-	-	-
Microdinium sp.	-	X	-	-	-	-	-	-	-	-	-	-	-	X	-	-	-	X	-	-	-	-	-	-	-	-	-	-
Octodinium askiniae	-	-	X	X	-	X	-	-	-	X	-	R	-	-	X	-	-	-	-	X	X	X	F	-	-	-	-	-
Operculodinium bergmannii	X	-	X	-	-	-	-	-	-	-	-	-	-	X	-	-	-	X	-	-	-	-	-	-	-	-	-	-
Paucisphaeridium inversibuccinum	-	-	X	-	-	X	-	-	-	-	-	-	-	-	-	-	-	X	-	-	-	-	-	-	-	-	-	-
Phelodinium harringtonii	-	-	-	-	-	-	-	-	-	-	-	-	-	-	-	-	-	-	-	X	-	-	-	-	-	-	-	-
Phthanoperidinium echinatum	-	-	-	-	-	-	-	-	-	X	-	-	-	-	-	-	-	fg	R	-	-	-	-	-	-	-	-	-
Pyxidinopsis sp. A	-	X	-	-	-	-	-	-	op	-	-	-	-	-	-	-	-	X	-	X	-	-	-	-	-	-	-	-
Selenopemphix nephroides	-	-	X	-	-	-	-	-	-	-	-	-	-	-	-	-	-	X	-	X	-	-	-	-	-	-	-	-
Selenopemphix prionota	X	-	X	-	-	-	-	X	-	X	-	-	-	-	-	-	-	X	-	-	-	-	-	-	-	-	-	-
Spinidinium colemanii	-	-	-	-	-	-	-	-	-	-	-	-	-	-	-	-	-	-	-	-	X	-	-	-	-	-	-	-
Spinidinium essoi	-	R	R	X	X	-	X	-	X	-	X	X	-	-	-	-	X	X	-	X	-	-	-	-	-	-	-	-
Spinidinium macmurdoense	X	X	R	-	-	X	-	-	X	-	X	-	-	X	-	-	-	R	X	-	X	-	-	-	-	-	-	-
Spiniferites ramosus subsp. *reticulatus*	X	-	X	-	-	-	-	-	-	-	-	-	-	-	X	-	-	-	-	-	-	-	-	-	-	-	-	-
Spiniferites ramosus	-	X	X	-	X	-	-	-	X	-	-	-	-	X	-	-	-	X	-	-	-	-	-	-	-	-	-	-
Systematophora ancyrea	-	-	X	fg	-	X	-	-	-	-	-	-	-	-	-	-	-	X	-	-	-	-	-	-	-	-	-	-
Thalassiphora pelagica	X	-	X	-	X	-	-	-	-	-	-	-	-	X	-	-	-	X	-	-	-	-	-	-	-	-	-	-
Turbiosphaera filosa	X	-	X	-	-	-	-	-	X	-	-	-	-	X	-	-	-	X	X	-	X	-	-	-	-	-	-	-
Turbiosphaera sagena	-	-	-	-	-	-	-	-	-	-	-	-	-	-	-	-	-	-	-	-	-	-	-	-	-	-	-	-
Vozzhennikovia apertura	R	X	F	R	-	X	-	X	-	F	-	R	X	-	X	X	X	X	-	X	F	X	X	F	-	-	X	-
Vozzhennikovia netrona	-	-	-	-	-	-	-	-	-	-	-	-	-	-	-	-	-	-	-	-	-	-	-	-	-	-	-	-
Dinocyst sp. A	-	-	X	-	-	-	-	-	-	-	-	-	-	-	-	-	-	-	-	-	X	X	-	-	-	-	-	-
Species Richenss	15	11	23	11	5	10	3	3	0	14	0	10	5	0	17	3	2	2	0	5	22	11	4	21	0	1	1	0
ACRITARCHS																												
Micrhystridium sp. 1	-	-	R	-	-	-	-	-	-	-	-	-	X	-	-	-	-	-	X	X	-	X	-	-	-	X	-	
Micrhystridium sp. 2	-	X	-	X	-	-	-	-	-	-	-	X	X	-	-	X	-	-	X	-	-	-	-	-	-	-	-	-
Cyclopsiella trematophora	-	-	X	-	X	-	-	-	-	-	-	X	-	-	-	-	-	-	X	X	-	-	-	-	-	-	-	-
Cyclopsiella sp. 1	-	-	-	-	X	-	-	-	-	-	-	X	-	-	-	-	-	-	-	-	-	-	-	-	-	-	-	-
Cyclopsiella sp. 2	-	-	-	-	-	-	-	-	-	-	-	-	-	X	-	-	X	X	-	-	-	-	-	-	-	-	-	-
Cyclopsiella sp. 3	-	-	-	-	-	-	-	-	-	-	-	-	-	-	-	-	-	-	-	-	-	X	-	-	-	-	-	-
Cyclopsiella sp. 4	-	X	-	-	-	-	-	-	-	-	-	-	-	X	-	-	-	-	-	-	-	-	-	-	-	-	-	-
Cyclopsiella sp. 5	-	X	-	X	-	-	-	-	-	-	-	-	-	-	-	-	-	-	-	-	-	-	-	-	-	-	-	-
Dichotisphaera sp.	-	-	-	-	-	-	-	-	-	-	-	-	-	-	-	-	-	-	-	-	-	-	-	-	-	-	-	-
Paralacaniella indentata	X	X	R	-	X	X	X	X	X	-	-	X	-	X	X	-	-	X	X	X	X	R	-	X	-	-	-	-
Scuticabolus lapidaris	-	-	-	-	-	-	X	-	-	-	-	X	-	-	-	-	-	X	X	-	-	-	-	-	-	-	-	-
Veryhachium sp. 1	-	-	X	-	-	-	-	-	-	-	-	-	-	-	-	-	-	-	-	-	-	-	-	-	-	-	-	-
Veryhachium sp. 2	-	-	-	-	-	-	-	X?	-	-	-	-	-	-	-	-	X?	-	X	-	-	-	-	-	-	-	-	-
Acritarch sp. A	-	-	-	-	-	-	-	X	-	-	-	X	-	-	X	-	-	-	X	-	X	-	X	-	-	-	-	-
Species Richness	1	4	5	1	3	3	1	3	1	2	0	2	3	1	3	1	3	2	3	3	6	5	0	5	0	0	1	0
PRASINOPHYTES																												
Cymatiosphaera sp.	-																											
Pterospermella sp. 1	-	-	-	-	-	-	-	-	-	-	-	-	-	-	-	-	-	-	-	-	-	X	-	-	-	-	-	-
Pterospermella sp. 2	-	-	-	-	-	-	-	-	-	-	-	-	-	-	-	-	-	-	-	-	-	R	-	-	-	-	-	-
CHLOROCOCCALES																												
Palambages sp.	-	-	-	-	-	-	-	-	-	-	-	-	-	-	-	-	-	-	-	-	-	-	-	-	-	-	-	-

Table 3. Marine palynomorph abundance data. X=very rare, R=rare, F=frequent, C=common, A=abundant, op=opercula, fg=fragments

DINOCYSTS	E 303(1)	E 303(2)	E 313	E 317	E 323	E 331	E 345	E 346	E 347	E 350	E 351	E 355	E 356	E 357	E 360	E 363	E 364	E 365(1)	E 365(2)	E 372	E 381	SV 3	SV 12	SIM 1	SIM 5	SIM 11	BG 1	D1
?Alisocysta sp.	-	-	-	-	-	-	-	-	-	-	-	-	-	-	-	-	-	-	-	-	-	-	-	-	-	-	-	-
Alterbidinium ?distinctum	-	-	-	-	-	-	-	-	-	-	-	-	-	-	-	-	-	-	-	-	-	-	-	-	-	-	-	-
Alterbidinium asymmetricum	X	X	-	X	-	-	C	-	-	X	-	-	-	X	-	-	R	X	-	-	X	-	-	-	-	-	-	R
Arachnodinium antarcticum	-	-	-	-	-	-	-	X	-	-	-	-	-	-	-	-	R	X	-	-	-	-	-	-	-	-	-	X
Cerodinium cf. *C. dartmoorium*	-	X	-	-	-	-	-	-	-	-	-	-	-	-	-	-	-	-	-	-	-	-	-	-	-	-	-	-
Cribroperidinium giuseppi	-	-	-	-	-	-	-	-	-	-	-	-	-	-	-	-	X	-	-	-	-	-	-	-	-	-	-	-
Deflandrea antarctica	X	X	X	X	-	-	R	-	-	X	-	-	-	X	-	-	X	F	X	-	X	-	-	-	-	X	-	X
Deflandrea cf. *D. cygniformis*	-	-	-	-	-	-	-	-	-	-	-	-	-	-	-	-	-	-	-	-	-	-	-	-	-	X	-	-
Deflandrea cf. *D. flounderensis*	-	-	-	-	-	-	-	-	-	-	-	-	-	-	-	-	X	X	-	-	-	-	-	-	-	-	-	X
Deflandrea cf. *D. phosphoritica*	-	-	-	-	-	-	-	-	-	X	-	-	-	-	-	-	-	-	-	-	-	-	-	-	-	-	-	X
cf. *Eisenackia scrobiculata*	-	-	-	-	-	-	-	-	-	-	-	-	-	-	-	-	-	-	-	-	-	-	-	-	-	-	-	-
Enneadocysta partridgei	X	op	op	op	-	-	R	-	-	X	-	-	-	X	-	-	F	R	X	-	op	-	-	-	-	X	-	F
Enneadocysta sp. 1	-	-	-	-	-	-	-	-	-	-	-	-	-	-	-	-	-	-	-	-	-	-	-	-	-	-	-	-
Enneadocysta sp. 2	-	-	-	-	-	-	-	-	-	-	-	-	-	-	-	-	-	-	-	-	-	-	-	-	-	-	-	X
cf. *Eocladopyxis peniculata*	-	-	-	-	-	-	-	-	-	-	-	-	-	-	-	-	X	-	-	-	-	-	-	-	-	-	-	X
Glaphyrocysta radiata	X	-	-	-	-	-	-	-	-	-	-	-	-	-	-	-	-	-	-	-	-	-	-	-	-	-	-	X
Hystrichosphaeridium truswelliae	X	X	-	op	-	-	X	-	X	X	-	-	X	-	-	X	-	fg	X	X	-	-	-	-	-	-	-	X
H. cf. *H. tubiferum* subsp. *brevispinum*	X	-	-	-	-	-	-	-	-	-	-	-	-	-	-	-	-	-	-	-	-	-	-	-	-	-	-	-
Hystrichosphaeridium sp.	-	-	-	-	-	-	X	-	-	X	-	-	X	-	-	-	-	-	-	-	-	-	-	-	-	-	-	-
Impagidinium victorianum	X	-	-	-	-	-	X	-	-	X	-	-	-	-	-	-	-	-	-	-	-	-	-	-	-	-	-	-
Impletosphaeridium clavus	X	-	-	X	-	-	X	-	-	X	-	-	-	-	-	-	X	X	-	-	-	-	-	-	-	-	-	X
Impletosphaeridium sp.	-	-	-	-	-	-	-	-	-	-	-	-	-	-	-	-	-	-	-	-	-	-	-	-	-	-	-	-
Lejeunecysta hyalina	-	X	-	-	-	-	-	-	-	-	-	-	-	-	-	-	-	-	-	-	-	-	-	-	-	-	-	X
Microdinium sp.	X	-	-	-	-	-	-	-	-	-	-	-	-	-	-	-	X	-	-	-	-	-	-	-	-	X	-	-
Octodinium askiniae	-	-	-	X	-	-	F	-	-	X	-	-	-	-	-	-	C	F	X	-	-	-	-	-	-	-	-	F
Operculodinium bergmannii	X	-	-	-	-	-	-	-	-	-	-	-	X	-	-	-	-	-	-	-	-	-	-	-	-	-	-	X
Paucisphaeridium inversibuccinum	X	X	-	-	-	-	-	-	-	-	-	-	-	-	-	-	X	X	-	-	-	-	-	-	-	-	-	X
Phelodinium harringtonii	-	-	-	-	-	-	-	-	-	-	-	-	-	-	-	-	X	-	-	-	-	-	-	-	-	X	-	-
Phthanoperidinium echinatum	-	X	-	-	-	-	-	-	-	-	-	-	-	-	-	-	R	X	-	-	-	-	-	-	-	-	-	R
Pyxidinopsis sp. A	-	-	-	-	-	-	-	-	-	-	-	-	-	-	fg	-	-	-	-	-	-	-	-	-	-	-	-	X
Selenopemphix nephroides	-	-	X	X	-	-	-	-	-	-	-	-	-	-	-	-	X	-	-	-	-	-	-	-	-	-	-	X
Selenopemphix prionota	X	-	-	-	-	-	X	-	-	X	-	-	-	-	-	-	X	X	-	-	-	-	-	-	-	X	-	X
Spinidinium colemanii	-	X	-	-	-	-	-	-	-	-	-	-	-	-	-	-	X	-	-	-	-	-	-	-	-	-	-	-
Spinidinium essoi	X	X	X	X	-	-	R	-	-	-	-	-	-	X	-	-	F	X	-	X	-	-	-	-	-	X	-	R
Spinidinium macmurdoense	X	X	-	X	-	-	-	-	-	X	-	-	-	-	-	-	X	X	-	X	-	-	-	-	-	X	-	R
Spiniferites ramosus subsp. *reticulatus*	-	X	-	-	-	-	X	-	-	-	-	-	-	-	-	-	X	-	-	-	-	-	-	-	-	-	-	-
Spiniferites ramosus	X	X	-	-	-	-	X	-	-	X	-	-	-	-	-	-	X	X	-	-	-	-	-	-	-	X	-	X
Systematophora ancyrea	X	X	-	-	-	-	-	-	-	-	-	-	-	-	-	-	X	-	-	-	-	-	-	-	-	-	-	-
Thalassiphora pelagica	-	X	-	fg	-	-	-	-	-	X	-	-	-	-	-	-	X	X	-	-	-	-	-	-	-	-	-	X
Turbiosphaera filosa	-	-	-	-	-	-	-	-	-	-	-	-	-	X	-	-	-	-	-	-	-	-	-	-	-	-	-	X
Turbiosphaera sagena	-	-	-	-	-	-	-	-	-	-	-	-	-	-	-	-	-	-	-	-	-	-	-	-	-	-	-	-
Vozzhennikovia apertura	X	R	-	X	X	-	C	-	-	X	-	-	fg	X	X	-	F	R	X	-	X	-	-	-	-	R	-	F
Vozzhennikovia netrona	X	-	-	-	-	-	X	-	-	-	-	-	-	-	-	-	X	-	-	-	-	-	-	-	-	-	-	-
Dinocyst sp. A	X	-	-	-	-	-	X	-	-	X	-	-	-	-	-	-	-	-	-	-	-	-	-	-	-	-	-	-
Species Richness	19	16	4	11	1	0	13	0	1	16	0	2	1	9	0	1	18	22	5	0	6	0	0	0	0	0	10	25
ACRITARCHS																												
Micrhystridium sp. 1	X	-	-	-	-	-	F	-	-	X	-	-	-	-	-	-	R	X	-	-	-	-	-	-	-	-	-	X
Micrhystridium sp. 2	-	-	-	-	X	-	-	-	-	-	-	-	-	-	-	-	-	X	-	-	-	-	-	-	-	-	-	-
Cyclopsiella trematophora	X	-	-	-	-	-	-	-	-	-	-	-	-	-	-	-	X	-	-	-	-	-	-	-	-	X	-	-
Cyclopsiella sp. 1	-	-	-	-	-	-	-	-	-	-	-	-	-	-	-	-	-	-	-	-	-	-	-	-	-	-	-	-
Cyclopsiella sp. 2	-	-	-	-	-	-	-	-	-	-	-	-	-	-	-	-	-	-	-	-	-	-	-	-	-	X	-	-
Cyclopsiella sp. 3	-	-	-	-	-	-	-	-	-	-	-	-	-	-	-	-	-	-	-	-	-	-	-	-	-	-	-	-
Cyclopsiella sp. 4	X	-	-	-	-	-	-	-	-	-	-	-	-	-	-	-	-	-	-	-	-	-	-	-	-	-	-	-
Cyclopsiella sp. 5	-	-	-	-	-	-	-	-	-	-	-	-	-	-	-	-	-	-	-	-	-	-	-	-	-	-	-	-
Dichotisphaera sp.	-	-	-	X	-	-	-	-	-	-	-	-	-	-	-	-	-	-	-	-	-	-	-	-	-	-	-	X
Paralacaniella indentata	X	X	X	X	-	-	X	-	-	X	-	-	-	R	-	-	X	-	-	-	-	-	-	-	-	-	-	X
Scuticabolus lapidaris	-	-	-	-	-	-	-	-	-	X	-	-	-	-	-	-	-	-	-	-	-	-	-	-	-	-	-	X
Veryhachium sp. 1	-	-	-	-	-	-	-	-	-	-	-	-	-	-	-	-	-	-	-	-	-	-	-	-	-	-	-	-
Veryhachium sp. 2	-	-	-	-	-	-	-	-	-	-	-	-	-	-	-	-	-	X	-	-	-	-	-	-	-	-	-	X
Acritarch sp. A	-	X	-	X	-	-	X	-	-	-	-	-	-	-	-	-	X	-	-	X	-	-	-	-	-	-	-	-
Species Richness	4	3	1	3	1	0	3	0	0	3	0	0	0	2	0	0	3	4	0	0	1	0	0	0	0	0	2	4
PRASINOPHYTES																												
Cymatiosphaera sp.	?	-	-	-	-	-	-	-	-	-	-	-	-	-	-	-	-	-	-	-	-	-	-	-	-	-	-	-
Pterospermella sp. 1	-	-	-	-	-	-	-	-	-	-	-	-	-	-	-	-	-	-	-	-	-	-	-	-	-	-	-	X
Pterospermella sp. 2	-	-	-	-	-	-	-	-	-	-	-	-	-	-	-	-	-	-	-	-	-	-	-	-	-	-	-	-
CHLOROCOCCALES																												
Palambages sp.	-	-	-	-	-	-	-	-	-	-	-	-	-	-	-	-	-	-	-	-	-	-	-	-	-	-	-	-

Table 4. Sample list and age data. Samples Collected 1992/93: MTD = Mount Discovery, MB = Minna Bluff, BG = Blue Glacier, SIM 1 = Sea Ice Moraine. Samples Collected 1993/94, 95/96: E 100 - E 381 from Minna Bluff and NW flank of Mount Discovery. For explanation of lithofacies abbreviations see Table 1. Key for Siliceous microfossils column: x = examined for siliceous microfossils, sample either barren or contained rare fragments, ● = Middle to upper Eocene, ■ = Oligocene/Lower Miocene, (for siliceous microfossil data see Harwood and Bohaty, this volume; Bohaty and Harwood, this volume).

Sample	Age based upon dinocyst taxon range data	Species richness	Lithofacies	Age based upon biostratigraphy + species richness	Siliceous microfossils
MTD 1	middle to upper Eocene	24	Sm	middle to upper Eocene	x
MTD 42	middle to upper Eocene	26	Cmm	middle to upper Eocene	
MTD 56	middle Eocene to lower Oligocene	3	Sm	?lower Oligocene	
MTD 153(1)	middle to upper Eocene	22	Smc	middle to upper Eocene	x
MTD 153(2)	middle to upper Eocene	12	Smc	middle to upper Eocene	
MTD 154	middle Eocene to lower Oligocene	6	Sm	middle to upper Eocene	
MTD 174A	Barren	0	Quartzite	?Paleozoic/Mesozoic	
MTD 189	middle to upper Eocene	6	Cmm	middle to upper Eocene	
MTD 190	middle to upper Eocene	11	Sm	middle to upper Eocene	
MTD 211A(1)	Barren	0	Ms-d	post-Eocene	
MTD 211A(2)	Barren	0	Quartzite	?Paleozoic/Mesozoic	
MB 80	middle to upper Eocene	12	Ss	middle to upper Eocene	
MB 103(2)	middle to upper Eocene	3	Sst	middle to upper Eocene	
MB 109(1)	middle to upper Eocene	28	Smc	middle to upper Eocene	x
MB 109(2)	middle to upper Eocene	15	Smc	middle to upper Eocene	
MB 181(2)	middle to upper Eocene	25	Ssg/Csgc	middle to upper Eocene	●
MB 188B	middle Eocene to lower Oligocene	7	Sm	middle to upper Eocene	
MB 188G	middle to upper Eocene	13	Cmm	middle to upper Eocene	
MB 202	Barren	0	Volcanic	???	
MB 212K	middle Eocene to upper Oligocene	1	Mm-d	post-Eocene	x
MB 217A	middle Eocene to upper Oligocene	1	Mm-d	post-Eocene	
MB 235A	middle to upper Eocene	8	Dm	middle to upper Eocene (reworked)	■
MB 235C	middle to upper Eocene	20	Cmc	middle to upper Eocene	
MB 244C	Barren	0	Mm-d	post-Eocene	■
MB 245	middle to upper Eocene	14	Mmb	middle to upper Eocene	x
MB 290G	middle Eocene to upper Oligocene	1	Mm-d	post-Eocene	x
MB 299	middle Eocene to upper Oligocene	2	Dm	post-Eocene	x
E 100	middle to upper Eocene	6	Sm	middle to upper Eocene	x
E 115	middle Eocene to upper Oligocene	2	Ms-d	post-Eocene	x
E 145	middle to upper Eocene	15	Sm	middle to upper Eocene	x
E 153	middle to upper Eocene	11	Sm	middle to upper Eocene	
E 155	middle to upper Eocene	22	Sm	middle to upper Eocene	x
E 163	middle to upper Eocene	11	Sm	middle to upper Eocene	x
E 165	middle to upper Eocene	5	Sm	middle to upper Eocene	
E 168	middle to upper Eocene	10	Sm	middle to upper Eocene	x
E 169	middle to upper Eocene	3	Sm	middle to upper Eocene	x
E 171	middle to upper Eocene	3	Sm	middle to upper Eocene	x
E 181	Barren	0	Sm	???	x
E 184	middle to upper Eocene	14	Smc	middle to upper Eocene	x
E 185	Barren	0	Sm	???	x
E 189	middle to upper Eocene	10	Sm	middle to upper Eocene	x
E 191	middle to upper Eocene	5	Sm	middle to upper Eocene	x
E 192	Barren	0	Sm	???	x
E 194	middle to upper Eocene	17	Sm	middle to upper Eocene	x
E 200	middle Eocene to lower Oligocene	3	Sm	?lower Oligocene	x
E 202	middle Eocene to lower Oligocene	2	Sm	?lower Oligocene	x
E 203	middle Eocene to lower Oligocene	2	Sm	?lower Oligocene	x

Table 4. Sample list and age data (continued).

Sample	Age based upon dinocyst taxon range data	Species richness	Lithofacies	Age based upon biostratigraphy + species richness	Siliceous microfossils
E 207	Barren	0	Sm	???	x
E 208	middle to upper Eocene	5	Sm	middle to upper Eocene	x
E 214	middle to upper Eocene	22	Ms-d	middle to upper Eocene	x
E 215	middle to upper Eocene	11	Sw	middle to upper Eocene	
E 216	middle Eocene to lower Oligocene	4	Ms-d	post-Eocene	x
E 219	middle to upper Eocene	21	Mmb	middle to upper Eocene	x
E 240	Barren	0	Ms-d	???	x
E 242D	middle Eocene to upper Oligocene	1	Dm	post-Eocene	x
E 243	middle Eocene to upper Oligocene	1	Dm/Dw	post-Eocene	x
E 244	Barren	0	Mm-d	???	x
E 303(1)	middle to upper Eocene	19	Sm	middle to upper Eocene	x
E 303(2)	middle to upper Eocene	16	Mmb	middle to upper Eocene	x
E 313	middle Eocene to lower Oligocene	4	Smc	?lower Oligocene	x
E 317	middle Eocene to upper Eocene	11	Sm	middle to upper Eocene	x
E 323	middle Eocene to upper Eocene	1	meta	???	x
E 331	Barren	0	Sm	???	x
E 345	middle to upper Eocene	13	Sm	middle to upper Eocene	●
E 346	Barren	0	Dm	post-Eocene	■
E 347	middle to upper Eocene	1	Dm	post-Eocene	■
E 350	middle to upper Eocene	16	Mmb	middle to upper Eocene	●
E 351	Barren	0	Dm	post-Eocene	■
E 355	middle to upper Eocene	2	meta-sed	middle to upper Eocene	x
E 356	middle Eocene to lower Oligocene	1	Sm	?lower Oligocene	x
E 357	middle to upper Eocene	9	Smc	middle to upper Eocene	x
E 360	Barren	0	Mm-d	???	x
E 363	middle to upper Eocene	1	Mm-d	post-Eocene	x
E 364	middle to upper Eocene	18	Mw	middle to upper Eocene	●
E 365(1)	middle to upper Eocene	22	Mmb	middle to upper Eocene	x
E 365(2)	middle to upper Eocene	5	Sm	middle to upper Eocene	x
E 372	Barren	0	Sm	???	
E 381	middle Eocene to lower Oligocene	6	Sm	middle to upper Eocene	x
SV 3	Barren	0	Quartzite	Paleozoic/Mesozoic	
SV 12	Barren	0	Cmc	???	
SIM 1	Barren	0	Cmm	???	
SIM 5	Barren	0	Sm	???	
SIM 11	middle Eocene to lower Oligocene	10	Sm	middle to upper Eocene	
BG 1	Barren	0	Cmm	???	
D1	middle to upper Eocene	25	Mw	middle to upper Eocene	●

between rare and frequent. Species most commonly present include *Paralecaniella indentata*, *Micrhystridium* sp. 1, and *Cyclopsiella* spp. Species richness is greatest in sample E 214 (Mw) from which six species were recovered, although species richness is generally low for all of the lithofacies.

Prasinophytes comprise a minor part of the marine palynomorph component. Two genera containing three species were recovered from five samples (see Table 3).

AGE OF THE MCMURDO ERRATICS

Biostratigraphic procedures

In studying the McMurdo Erratics, we face the following biostratigraphic limitations that restrict our ability to assign ages: (1) superposition cannot be used to determine relative stratigraphic relationships; (2) continuity of biostratigraphic ranges cannot be used to bracket the age of barren erratics or erratics that only contain long ranging taxa; (3) difficulties exist in distinguishing between spatial and temporal controls on assemblage composition. Nevertheless, relatively broad age assignments can be determined for 'suites' of erratics based on the following criteria.

A maximum age of middle Eocene may be inferred based on the absence of *Ennigmadinium cylindrifloriferum* and/or presence of one or more of the following species: *Vozzhennikovia netrona* n. sp., *Enneadocysta* sp. 2, and *Pyxidinopsis* sp. A. The following species have last appearance datums near the Eocene/Oligocene boundary: *Arachnodinium antarcticum*, *Turbiosphaera filosa*, *Operculodinium bergmannii*, *Hystricho-sphaeridium truswelliae*, *Impletosphaeridium clavus*, *Alterbidinium ?distinctum*, *Vozzhennikovia netrona*, *Pyxidinopsis* sp. A, *Enneadocysta* sp. 2, and *Cyclopsiella trematophora*. Erratics that contain one or more of these taxa are no younger than upper Eocene. Age assignments become difficult when these Eocene taxa are not present, as the absence of these species may not be due to extinction, but instead may be due to ecologic and biogeographic factors that restrict their spatial distribution.

Diversity trends within high-latitude dinoflagellate cyst assemblages show a gradual decrease in species richness throughout the Eocene and a significant decrease in species richness across the Eocene / Oligocene boundary [Mao and Mohr, 1995; Damassa and Williams, 1996]. This decrease in species richness provides a potential means to distinguish Eocene assemblages from younger assemblages. Middle Eocene dinocyst assemblages from the circum-Antarctic region commonly comprise over forty species. Upper Eocene dinocyst assemblages usually contain more than five species, and lower Oligocene assemblages usually contain less than five species [see Figure 5; Wilson, 1989; Mao and Mohr, 1995]. Erratics with moderate species richness (<40 and Δ5 species) are likely to have been deposited during the middle to upper Eocene. Erratics that comprise dinocyst assemblages that lack species with upper Eocene LADs and have low species richness (<5 species) are likely Oligocene or younger. Such inferences must, however, be made with caution, as assemblage diversity is also influenced by proximity to shoreline [e.g. Wall et al., 1977]. For example, low species richness in coarse grained lithofacies may reflect deposition within a nearshore environment and not a post-Eocene age. Lithology as well as biostratigraphic and paleoecologic data provided by other fossils (diatoms, ebridians, silicoflagellates and molluscs), must be considered in assigning ages to the McMurdo Erratics.

Age Assignments

The McMurdo Erratics reported here are divided into three groups based upon general lithologic criteria. The first, and largest, group of erratics comprise fossiliferous erratics of sandstone (Sm, Smc, Ss, Sst, Ssg, and Sw), mudstone (Mmb, Mwb) and conglomerate (Cmc, Cmm, Cscg) lithologies [Table 1; Levy and Harwood, this volume]. These facies commonly contain reasonably diverse marine palynomorph assemblages and a macroflora and fauna presented in Stilwell and Feldmann [this volume]. The second group of erratics includes mudstone (Ms-d, Mm-d) and diamictite (Dm) lithologies, that are either barren of palynomorphs or contain a poor marine palynomorph assemblage. The final group of erratics includes miscellaneous lithologies that are either barren of palynomorphs or contain only terrestrial palynomorphs.

Fossiliferous sandstone, mudstone and conglomerate. The known biostratigraphic ranges of key marine palynomorphs recovered in the fossiliferous erratics are presented in Figure 6. A maximum middle Eocene age is assigned to all erratics from which marine palynomorphs were recovered based upon the absence of *Ennigmadinium cylindrifloriferum* and maximum species richness of < 30. Erratics containing dinocyst species with biostratigraphic LADs near the Eocene/Oligocene boundary (Figures 6 and 7) are considered no younger than Eocene. We identify a middle to upper Eocene suite of erratics on these criteria (Figure 7). Biostratigraphic data furnished by terrestrial palynomorphs, diatoms, silicoflagellates, and ebridians support these age assignments [Table 4; Bohaty and Harwood, this volume; Harwood and Bohaty, this volume].

Many of the arenaceous erratics (lithofacies Sm, see Levy and Harwood, [this volume] contain assemblages that include *Vozzhennikovia apertura*, *Spinidinium macmurdoense*, *Deflandrea antarctica*, and *Enneadocysta partridgei*, all of which have lower Oligocene LADs (Figures 6 and 7). Based exclusively on taxon range data, the age assignment for these erratics is middle Eocene to early Oligocene. Species richness data enables us to improve this age assignment with those erratics with assemblages of five or more dinocyst taxa being restricted to the Eocene and assemblages of less than five dinocyst taxa being restricted to the lower Oligocene (Table 4; Figure 7). Such age assignments must, however, be verified with biostratigraphic data from other fossil taxa, including terrestrial palynomorphs [Askin, this volume] and/or invertebrate macrofossils [Stilwell, this volume], in order to rule out spatial controls on assemblage diversity.

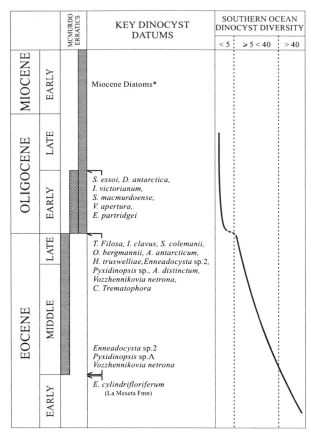

Fig. 7. Summary of key marine palynomorph datums and Southern Ocean diversity trends used to assign the McMurdo Erratics to three chronostratigraphic groups: middle to upper Eocene; ?lower Oligocene; post-Eocene (* for diatom data see Harwood and Bohaty, this volume).

Mudstone and diamictite. One goal for the present study was to determine whether pre-late Eocene glacial-marine records exist in East Antarctica. None of the fossiliferous sandstone, mudstone and conglomerate erratics reported above exhibit sedimentological characteristics typical of glacial deposits, however, the mudstone and diamictite erratics discussed below are considered to be glacial facies [Levy and Harwood, this volume]. Thirteen 'glacial' lithofacies were prepared for palynological investigation and biostratigraphic analysis. In general, dinocyst recovery from these samples was poor. Rare dinocysts were recovered from one erratic (MB 235A), although species richness was low (eight). Rare dinocysts specimens were recovered from six of the remaining twelve samples. The 'assemblages' recovered from these six samples comprised either very rare speci-

mens of a single dinocyst species or dinocyst fragments. One sample contained very rare acritarchs and pollen. The remaining five samples were either barren or contained only rare terrestrial palynomorphs.

Taphonomic processes that may account for the poor recovery of dinocysts from the glacial facies include chemical degradation and mechanical abrasion. This does not seem to be the case as these processes should alter all organic microfossils uniformly, yet terrestrial organic material including pollen, spores, fungi, leaf cuticle and wood are not affected. This suggests that taphonomic processes were not significant in modifying the marine palynomorphs and are therefore unlikely to account for the poor recovery of dinocysts in erratics of glacial facies.

Glacial mudstones and diamictites of upper Eocene age occur within the basal 270m of the CIROS-1 drillcore [Hambrey et al., 1989]. Dinoflagellate cyst species richness values in this interval of CIROS-1 range between fourteen and two, with an average value of eight taxa [Wilson, 1989]. Although dinocyst abundance decreases significantly in the basal diamictites, species richness appears to be unaffected (M. Hannah, personal communication to R. Levy, 1997). A significant decrease in dinocyst species richness is noted across the Eocene/Oligocene boundary [Figure 5; Wilson, 1989]. If glacial mudstone and diamictite facies reported here are either older than, or contemporaneous with, upper Eocene glacial sediments at the base of CIROS-1, then the erratic samples should contain dinocyst assemblages of moderate species richness (Δ5 species). However, as most of these erratics contain poor dinocyst assemblages, an Oligocene or younger age may be inferred for these mudstone and diamictite erratics. This age assignment is supported by diatom and ebridian biostratigraphy [Table 4; Harwood and Bohaty, this volume; Bohaty and Harwood, this volume].

Miscellaneous erratics. Ten miscellaneous erratics examined in this study are either barren of microfossils or contain no age- diagnostic organic-walled microfossils. Erratics BG 1 and SIM 5 are barren of fossils. Erratics E 185, E 331 and E 372 contain no microfossils, but they do contain macro-invertebrate fossils that have affinities with taxa from the Eocene La Meseta Formation on Seymour Island [Stilwell, this volume]. Erratic SV 3 is a quartzite that was most likely derived from the Beacon Supergroup (Devonian to Triassic). Erratics SIM 1, MTD 174, MB 202, E 351 contain only pollen and spores, and are discussed by Askin [this volume].

Age Summary

The McMurdo Erratics studied herein are separated into three chronostratigraphic groups based on their marine palynomorph component: middle to upper Eocene; ?lower Oligocene; and post-Eocene (Table 4; Figure 7). Middle to upper Eocene erratics are identified by a combination of marine palynomorph taxon range data and assemblage species richness greater than or equal to five. These erratics include most of the fossiliferous sandstone, mudstone and conglomerate lithologies [Table. 1; Levy and Harwood, this volume].

Many sandstone lithologies (Sm, Smc, Sst) contain low numbers of species (< 5 species) comprising palynomorph taxa with LADs in the lower Oligocene. These samples are assigned an ?early Oligocene age based on low species richness. However, these age assignments are tenuous and must be verified with biostratigraphic data from fossils other than marine palynomorphs.

Erratics assigned a post-Eocene age are either barren or commonly contain poor dinocyst assemblages consisting of *Vozzhennikovia apertura* and fragments of other taxa. The majority of the glacial facies (Ms-d, Mm-d and Dm) are included in this group.

CONCLUSION

Most of the McMurdo Erratics examined in this study appear to be from a middle to upper Eocene sequence that is hidden beneath the present East Antarctic ice sheet. The McMurdo Erratics provide a valuable source of geologic data for East Antarctica, from a time period in Earth history where little direct evidence is available. Marine palynomorph assemblages recovered from these erratics provide age control without which key paleoenvironmental and paleogeographic questions could not be approached.

Herein, we document and illustrate marine palynomorph assemblages recovered from the McMurdo Erratics. Key marine palynomorph biostratigraphic datums and assemblage diversity trends allow us to assign sub-Epoch age control to the erratics. Middle to upper Eocene erratics contain fossil marine microflora (marine palynomorphs, diatoms, ebridians, and silicoflagellates), marine macrofauna (molluscs, brachiopods, bryozoans, arthropods, and fish), terrestrial microflora (pollen and spores), terrestrial macroflora (wood and leaves), and terrestrial vertebrates. The McMurdo Erratics represent the most extensive Paleogene paleontological database presently available for East Antarctica.

These fossil organisms provide a view of marine and terrestrial communities that inhabited coastal environments before: (a) the onset of significant climatic cooling during the early Oligocene [e.g. Shackleton and Opdyke, 1973; Matthews and Poore, 1980; Miller et al., 1987]., and (b) geographic isolation of Antarctica due to the formation of deep marine basins between East Antarctica and Australia [e.g. Veevers et al., 1991 and references therein; Lawver et al., 1992].

Erratics of post-Eocene age are commonly mudstone (Ms-d, Mm-d) and diamictite (Dm) lithologies that were deposited in a glacio-marine environment [Table 1; Levy and Harwood, this volume]. These erratics usually contain poor marine palynomorph assemblages and well preserved diatom assemblages [Table 4; Harwood and Bohaty, this volume]. These assemblages post-date upper Eocene glacial lithofacies recovered from CIROS-1 in McMurdo Sound [Hambrey et al., 1989; Wilson et al., 1998].

Until drilling provides a better record of Tertiary stratigraphic sequences, the McMurdo Erratics serve as a significant paleobiotic and paleoenvironmental resource to reconstruct history of southern high latitudes. As biostratigraphic schemes continue to develop, future age refinement of dinocyst assemblages reported here will enable better age resolution and dating of individual erratics.

Where stratigraphic drilling provides improved temporal control, the McMurdo Erratics provide a wealth of spatial data that may not be recovered in a drillcore. In a land covered by ice where Cenozoic strata are poorly exposed [Webb, 1990; 1991], glacial reworking and stratigraphic drilling together will provide a means to reconstruct the history of Antarctica.

Systematic Paleontology

The taxonomic classification scheme of Fensome and others [1993] is followed. The reader should consult Wrenn and Hart [1988] for a comprehensive review of the stratigraphic occurrences of many of the species recovered in this study. Holotypes and paratypes of new species described here are housed at the University of Nebraska State Museum (UNSM).

Division DINOFLAGELLATA (Bütschli, 1885) Fensome et al., 1993
Class DINOPHYCEAE Pascher, 1914
Subclass PERIDINIPHYCIDAE Fensome et al., 1993
Order PERIDINIALES Haeckel, 1894
Family PERIDINIACEAE Ehrenberg, 1831

Genus *Alterbidinium* Lentin & Williams, 1985; emend. Khowaja-Ateequzzaman and Jain, 1991

Alterbidinium ?distinctum (Wilson, 1967) Lentin & Williams, 1985

Plate 1, figs. e and f

Deflandrea distincta Wilson, 1967, p.63-64, Figures 9-10.
Alterbia distincta (Wilson) Lentin & Williams, 1976, p.49.
Alterbia ?distincta (Wilson) Stover & Evitt, 1978, p. 93.
Alterbidinium ??distinctum (Wilson) Lentin & Williams, 1985, p. 14.

Comments. Rare specimens of *Alterbidinium ?distinctum* were recovered from erratics E 100 and MTD 42. These specimens fit the description and compare well with illustrations of the type material.

Stratigraphic Range. A. ?distinctum has been reported from the upper Eocene and lower Oligocene of DSDP Leg 71, Hole 511 (Goodman and Ford, 1983); the upper Eocene of DSDP Leg 29, holes 280A and 281 (Crouch and Hollis, 1996); and the upper Eocene of CIROS-1 in McMurdo Sound (Wilson, 1989). *A. ?distinctum* also occurs in the upper Oligocene of CIROS-1 but is probably reworked. The species is absent from middle Eocene sediments of DSDP Leg 29, Hole 280A and middle Eocene sediments on the Bruce Bank, Scotia Sea (Mao and Mohr, 1995). The species may prove to be a late Eocene marker for the circum-Antarctic region.

Alterbidinium asymmetricum (Wilson, 1967) comb. nov.

Plate 1, figs. g and h

Deflandrea asymmetrica Wilson, 1967, p. 62-63, Figures 17-21.
Alterbia asymmetrica (Wilson) Lentin & Williams, 1976, p. 48.
Senegalinium ?asymmetricum (Wilson) Stover & Evitt,

1978, p. 123.

Comments. Specimens recovered in this study fit the descriptions of the type material. Stover and Evitt (1978) provisionally placed this species in the genus *Senegalinium*. However, the generic description for *Senegalinium* indicates that species attributed to this genus must have antapical horns of equal length. Specimens consistently have antapical horns of unequal length. *Alterbidinium* differs from *Senegalinium* in having antapical horns of unequal length (Lentin and Williams, 1985). Accordingly we transfer this species to the genus *Alterbidinium*.

Stratigraphic Range. The stratigraphic range for *Alterbidinium asymmetricum* is Maastrichtian to Eocene (Wrenn and Hart, 1988; and references therein). Cretaceous occurrences are reported from the Northern Hemisphere. *A. asymmetricum* is reported from the Eocene of Argentina (Archangelsky, 1969b); Eocene on Seymour Island (Wrenn and Hart, 1988; Cocozza and Clarke, 1992); middle to upper Eocene from DSDP Leg 29, holes 280A and 281 (Crouch and Hollis, 1996); middle Eocene to lower Oligocene in the Scotia Sea (see fig. 3 herein; Mohr, 1990); and upper Eocene to lower Oligocene in McMurdo Sound (Wilson, 1989). *A. asymmetricum* occurs above the Eocene/Oligocene boundary in several Southern Ocean Deep Sea Drilling Sites (figs. 3 and 4), suggesting a total range for *A. asymmetricum* from Cretaceous to lower Oligocene. However it appears that in the Southern Ocean the species is restricted to Eocene to lower Oligocene.

Genus *Cerodinium* Vozzhennikovia, 1963; emend. Lentin & Williams, 1987

Cerodinium cf. *C. dartmoorium* (Cookson & Eisenack, 1965b) Lentin & Williams, 1987

Plate 2, figs. a and b

Plate 1

(Scale bar = 20μ)

Figs. a-d. *?Alisocysta* sp. MB 109(1), slide 1: (a) and (b) oblique view, orientation indet, two focal levels; (c) oblique view, orientation indet, high focus; (d) oblique view, high focus illustrating the spongy periphragm that extends across several paraplates.
Figs. e-f. *Alterbidinium ?distinctum* (Wilson 1967) Lentin and Williams 1985. MTD 42, slide 2: (e) ventral view, ventral surface; (f) ventral view, dorsal surface.
Figs. g-h. *Alterbidinium asymmetricum* (Wilson 1967) comb. nov. E 345, slide 3: (g) dorsal view, dorsal surface; (h) dorsal view, ventral surface.
Figs. i-k. *Arachnodinium antarcticum* Wilson and Clowes 1982. MTD 1B, slide 1: (i) orientation indet., optical view; (j) apical view, high focus. MTD 153(1), slide 1: (k) orientation indet., high focus.

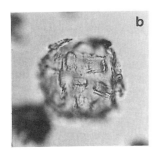

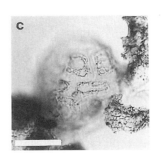

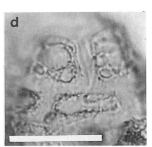

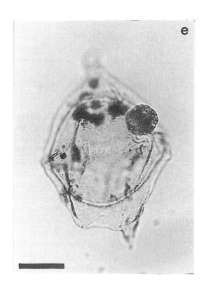

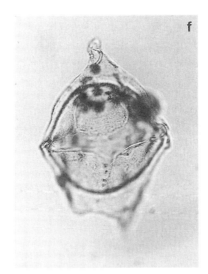

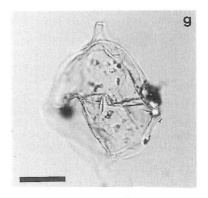

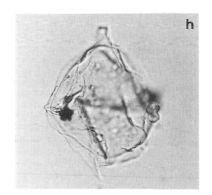

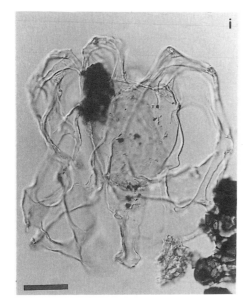

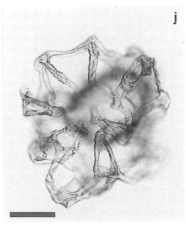

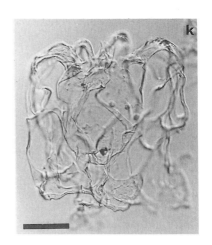

Deflandrea dartmooria Cookson & Eisenack, 1965b, p. 133-134, Plate 16, Figures 1-2, text-fig. 1.

Ceratiopsis dartmooria (Cookson & Eisenack) Lentin & Williams, 1981, p.38.

Deflandrea dartmooria Cookson & Eisenack 1965b, Lindgren, 1984, p.154.

Cerodinium dartmoorium (Cookson & Eisenack) Lentin & Williams, 1987, p.114.

Description.

Shape: A bicavate dinoflagellate cyst. The pericyst is peridinioid in form, whereas the endocyst is spherical to sub-pentagonal. The apical region of the pericyst consists of a long apical horn. Two well-developed antapical horns of equal size comprise the posterior margin of the pericyst.

Phragma: The granular endophragm is 2μ thick. Ornamentation on the thin periphragm consists of intratabular grana and coni that are separated by smooth intratabular regions up to 7μ wide. Parasutures delineate the boundaries of the paraplates.

Paratabulation: Peridinioid paratabulation is indicated by the intratabular spines and parasutures. The epicyst exhibits an ortho hexa paraplate arrangement.

Paracingulum: Laevorotatory with an offset of 6μ, delineated by parasutural ridges (2μ high) that commonly possess capitate spines along their distal margin. Paraplates are not clearly indicated within the paracingulum.

Parasulcus: Consists of a depression posterior of the cingulum in the mid-ventral region of the pericyst. Nontabular spines occur within the sulcal depression. The parasutural ridges of the paracingulum are discontinuous where the parasulcus intercepts the paracingulum. A complex flagellar scar occurs within the hypocystal region of the parasulcus. The flagellar scar is covered by a sulcul paraplate that projects laterally to the left.

Archeopyle: Endoarcheopyle consists of a broad hexa 2a archeopyle (I/I), AR = 0.48. Periarcheopyle margin is often broken and difficult to delineate. However, it appears to have a longer "a" axis. Detached opercula present within samples containing *C.* cf. *C. dartmoorium* support this observation.

Dimensions. Observed range (two specimens): pericyst length - 117 to 120μ, pericyst width - 76 to 81μ; endocyst length - 66 to 64μ, endocyst width - 76μ; apical horn: length - 33μ; antapical horn: length - 24μ.

Comments. The McMurdo Erratic specimens conform reasonably well with the description and size of the type material. Paratabulation is well delineated by parasutural lines and intratabular grana or coni, an important diagnostic feature for the species according to Stover (1973). The type material and other illustrated specimens attributed to *C. dartmoorium* (e.g. Wrenn and Hart, 1988, plate 16, figures 1-2; Stover, 1973, plate 3, figure 4) exhibit a more angular (polygonal) endocyst and a pericyst with longer antapical horns than the specimens from the McMurdo Erratics. Specimens attributed herein to *C.* cf. *C. dartmoorium* are similar in general appearance to specimens of *Deflandrea antarctica* (Type III, see below). However, specimens of *C.* cf. *C. dartmoorium* are distinguished by the following characteristics: (1) the presence of well-developed, long antapical horns, and (2) well-delineated paratabulation consisting of parasutural lines and intratabular grana or coni. It is possible that *C.* cf. *C. dartmoorium* is an ecophenotype or an evolutionary offshoot of *D. antarctica*.

Genus *Deflandrea* Eisenack, 1938; emend. Williams & Downie, 1966; emend. Lentin & Williams 1976

Deflandrea antarctica Wilson, 1967
Plate 2, figs. c-e; Plate 3, figs. a-c

Deflandrea antarctica Wilson, 1967, p. 58, 60, Figures 23, 24, 26, 27.

Plate 2

(Scale bar = 20μ)

Figs. a-c. *Cerodinium* cf. *C. dartmoorium* (Cookson and Eisenack 1965) Lentin and Williams 1987. MB 181(2), slide 1: (a) ventral view, ventral surface; (b) ventral view, dorsal surface; (c) epicystal region of the ventral surface illustrating parasutural lines and intratabular grana.

Figs. d-e. **Deflandrea antarctica** Wilson 1967. Type II specimen. MTD 42, slide 2: (d) dorsal view, dorsal surface, note non-tabular grana and spines on the pericyst; (e) dorsal view, ventral surface.

Figs. f-i. *Cribroperidinium giuseppi* (Morgenroth) Helenes 1984. MTD 1B, slide 1: (f) ventral view, ventral surface; (h) oblique view, dorsal surface (i) oblique view, ventral surface. MB 109(1), slide 1: (g) oblique view, ventral surface, note apical horn.

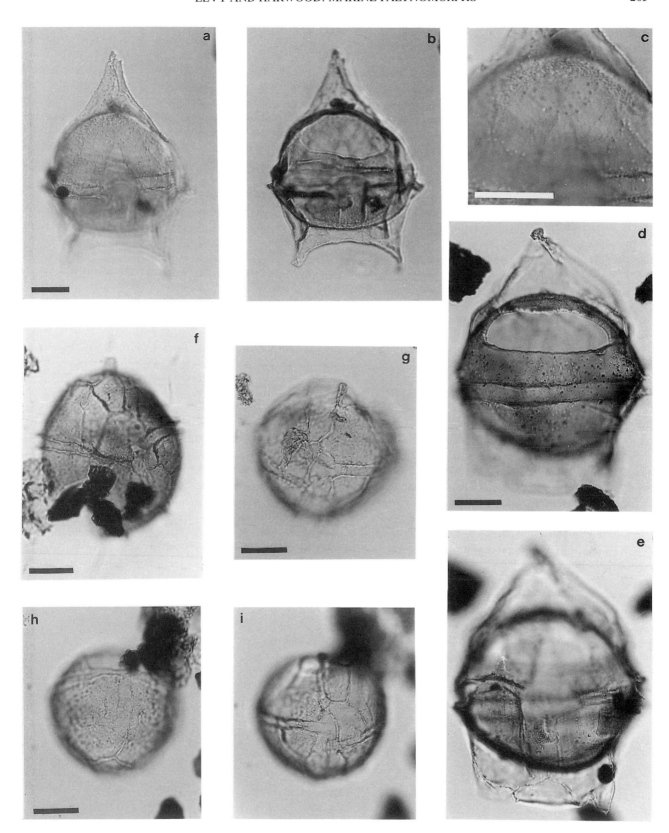

Comments. This is a common species in the McMurdo Sound material. Specimens attributed to this species exhibit variability in pericyst shape, extent of surface sculpture and degree in which the sculpture reflects paratabulation. Wilson (1967) indicated that the outer cyst of *D. antarctica* is usually covered with granules that typically form an atabulate dotted pattern. However, polygonal clusters of granules indicative of tabulation were noted. Wrenn and Hart (1988) recognized two types, distinguished on variations in pericystal sculpture, in their study of Seymour Island specimens. Type I specimens of *D. antarctica* are typically atabulate, although they may possess fine parasutural lines. Type II specimens bear paratabular sculpture that varies from granules to acicular spines.

Both Type I and Type II of *D. antarctica* were observed in this study, as well as another variation of *D. antarctica* herein designated as *D. antarctica* Type III. Paratabulation in Type III specimens is delineated clearly by the presence of both intratabular grana or coni and acicular parasutural spines (see Plate 3, fig. c).

Dimensions. Observed range (ten specimens): pericyst length - 96 to 144μ (mean 121μ), pericyst width - 78 to 95μ (mean 85μ); endocyst length - 62 to 79μ (mean 69μ), endocyst width - 70 to 84μ (mean 79μ).

Stratigraphic Range. Eocene to lower Oligocene (Wrenn and Hart, 1988)

Deflandrea cf. *D. cygniformis* Pothe de Baldis, 1966
Plate 3, figs. d and e

Deflandrea cygniformis Pothe de Baldis, 1966, p. 221-222, Plate 2, Figure c.

Comments. *D. cygniformis* has a characteristic pericystal outline consisting of a long epipericyst and relatively short hypopericyst with broad lateral paracingular projections. Specimens of *D.* cf. *D. cygniformis* recovered from the McMurdo Erratics exhibit this general shape but have a shorter antapical horns. As a result, they are significantly shorter than the holotype.

Wrenn and Hart (1988) suggest that an evolutionary relationship may exist between *D. cygniformis* and *D. antarctica*, with *D. cygniformis* being an evolutionary offshoot of *D. antarctica*. It is possible that the forms recovered from the McMurdo Erratics represent an intermediate stage in this proposed evolutionary lineage.

Dimensions. Observed range (six specimens): pericyst length - 124 to 144μ (mean 131μ) pericyst width - 77 to 86μ (mean 81μ); endocyst length - 60 to 68μ (mean 63μ), endocyst width - 62 to 76μ (mean 68μ).

Deflandrea cf. *D. flounderensis* Stover, 1973
Plate 3, figs. f and g

Deflandrea flounderensis Stover, 1973, p. 174-175, Plate 3, Figures 1 and 2.

Comments. *Deflandrea flounderensis* is characterized by a smooth, nearly circular, granulate or coarsely vermiculate endocyst, and the absence of linear markings at plate boundaries on the periphragm. These features are evident in the McMurdo Erratic specimens. Specimens from the McMurdo Erratics are smaller than the type material, being both shorter and narrower. The pericystal outline of specimens recovered from the McMurdo Erratics is similar to the pericystal outline of *D. antarctica*. However, *D.* cf. *D. flounderensis* has a smooth periphragm and thin-walled endocyst, whereas *D. antarctica* typically possesses an ornamented periphragm and a thick-walled endocyst. *D.* cf. *D. flounderensis* is otherwise similar to *D. antarctica* and may be conspecific.

Plate 3

(Scale bar = 20μ)

Figs. a-c. *Deflandrea antarctica* Wilson 1967. Type III specimen. MTD 1B, slide 1: (a) oblique view, dorsal suface; (b) oblique view, ventral surface; (c) ventral view, ventral surface, detail of parasutural acicular spines and well defined intratabular grana and spines that are characteristic of D. antartica Type III.

Figs. d-e. *Deflandrea* cf. *D. cygniformis* Pothe de Baldis 1966. MTD 42, slide 2: (d) ventral view, ventral surface; (e) ventral view, dorsal surface.

Figs. f-g. *Deflandrea* cf. *D. flounderensis* Stover 1973. MB 109(1), slide 1: (f) dorsal view, dorsal surface; (g) dorsal view, ventral surface.

Figs. h-i. *Deflandrea* cf. *D. phosphoritica* Eisenack 1938. MTD 42, slide 1: (h) optical view, note wide lateral projections in the pericyst. E 155, slide 1: (i) optical view.

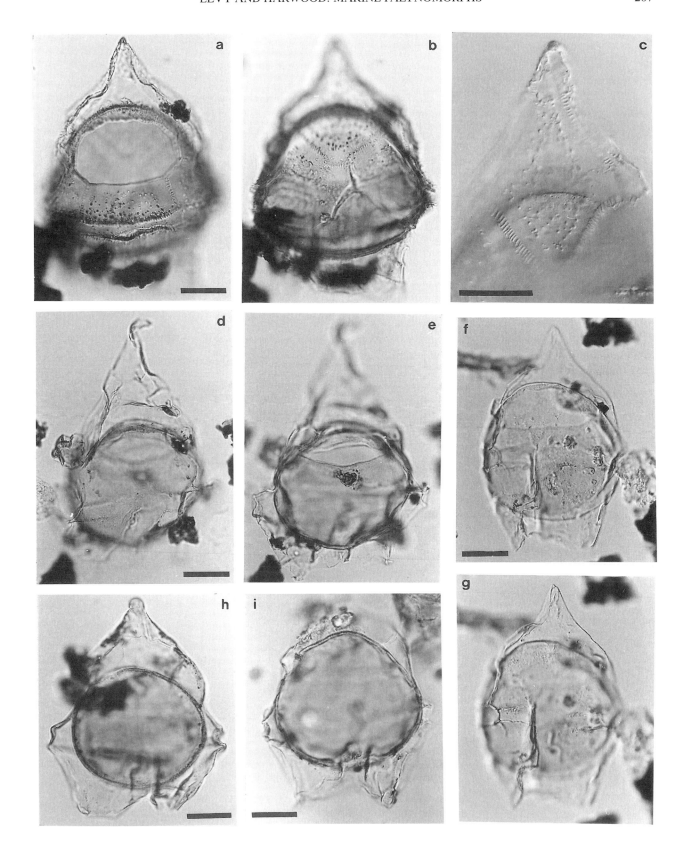

Dimensions. Observed range (five specimens): pericyst length - 102 to 118µ (mean 106µ), pericyst width - 69 to 83µ (mean 74µ); endocyst length - 56 to 66µ (mean 61µ), endocyst width - 62 to 80µ (mean 71µ).

Deflandrea cf. *D. phosphoritica* Eisenack 1938
Plate 3, figs. h and i

Deflandrea phosphoritica Eisenack, 1938, p. 187, text-figure 6.
Deflandrea cf. *D. phosphoritica* Eisenack, 1938, Wrenn and Hart, 1988, Figure 22.1

Comments. There is considerable variability in the morphology of published illustrations of *D. phosphoritica*. Specimens recovered from the McMurdo Erratics are most similar to *D.* cf. *D. phosphoritica* reported from Seymour Island (Wrenn and Hart, 1988). Both differ from the type material in that the lateral margins of the paracingular area protrude less and the margins of the hypocyst do not converge as much posteriorly. Broad lateral shoulders on the epipericyst are common.

Dimensions. Observed range (five specimens): pericyst length - 100 to 103µ (mean 101µ), pericyst width - 72 to 81µ (mean 77µ); endocyst length - 54 to 62µ (mean 57µ), endocyst width - 58 to 70µ (mean 63µ).

Genus *Phthanoperidinium* Drugg & Loeblich Jr. 1967; emend. Edwards & Bebout 1981; emend. Islam 1982
Phthanoperidinium echinatum Eaton 1976
Plate 8, figs. g-i

Phthanoperidinium echinatum Eaton, 1976, p. 298-299, Plate 17, figures 8-9, 12, text-figure 23B.
Phthanoperidinium pseudoechinatum Bujak, 1980, p. 75-76, Plate 19, figure 20, text-figure 20C.

Comments. Specimens recovered from the McMurdo Erratics are similar to the type material from England. Parasutural lines or ridges delineate paratabulation. The parasutures are bordered by penitabular spines that possess characteristic bulbous spherical (clavate) distal terminations. Several specimens have a combination IP archeopyle observed in specimens of *P. pseudoechinatum* by Bujak (1980). *P. pseudoechinatum* is considered to be a taxonomic junior synonym of *P. echinatum* by Islam (1982). Specimens recovered from the McMurdo Erratics are larger than the type species.

Dimensions. Observed range (seven specimens): pericyst length - 58 to 45µ (mean 52µ), pericyst width - 54 to 37µ (mean 44µ)

Stratigraphic Range. Damassa and others (1994) reported a middle Eocene (NP 12 to NP 17) global distribution for *P. echinatum*. This taxon appears to have a longer stratigraphic range in the Southern Ocean. *P. echinatum* has a reported Eocene range on Seymour Island (Wrenn and Hart, 1988) and an Eocene to lower Oligocene range is given for this species by Raine and others (1997).

Genus *Spinidinium* Cookson & Eisenack, 1962; emend. Lentin & Williams 1976.

Spinidinium colemanii Wrenn & Hart, 1988
Plate 9, fig. a

Deflandrea macmurdoensis Wilson, sensu Kemp, 1975, Plate 2, Figures 1-3.
Spinidinium colemanii Wrenn & Hart, 1988, p. 366-367, Figures 36.1-2, 39.2.

Comments. The McMurdo Erratic specimens differ from Seymour Island described by Wrenn and Hart (1988) in being larger.

Plate 4

(Scale bar = 20µ)

Figs. a-c. cf. *Eiseneckia scrobiculata* Morgenroth 1966. MTD 154, slide 1: (a) broken specimen, orientation indet; (c) partially disarticulated apical paraplate series. E219, slide 1: (b) broken specimen, orientation indet.
Figs. d-g. *Enneadocysta* sp. 2. MB 109(1), slide 1: (d) oblique view, optical section, note characteristic short processes, (e) oblique view, apical surface. MB 181 (2), slide 1: (f) left lateral view, high focus, (g) left lateral view, low focus.
Fig. h. *Enneadocysta partridgei* Stover and Williams 1995. MB 181(2), slide 1: (h) left lateral view, left lateral surface.
Figs. i-m. *Enneadocysta* sp. 1. MTD 153(1), slide 1: (i) orientation indet., high focus, note characteristic licrate process terminations; (j) orientation indet., low focus; (k) antapical view, antapical surface, note apical operculum in top left of figure; (l) antapical view, optical section; (m) antapical view, apical surface illustrating apical archeopyle.

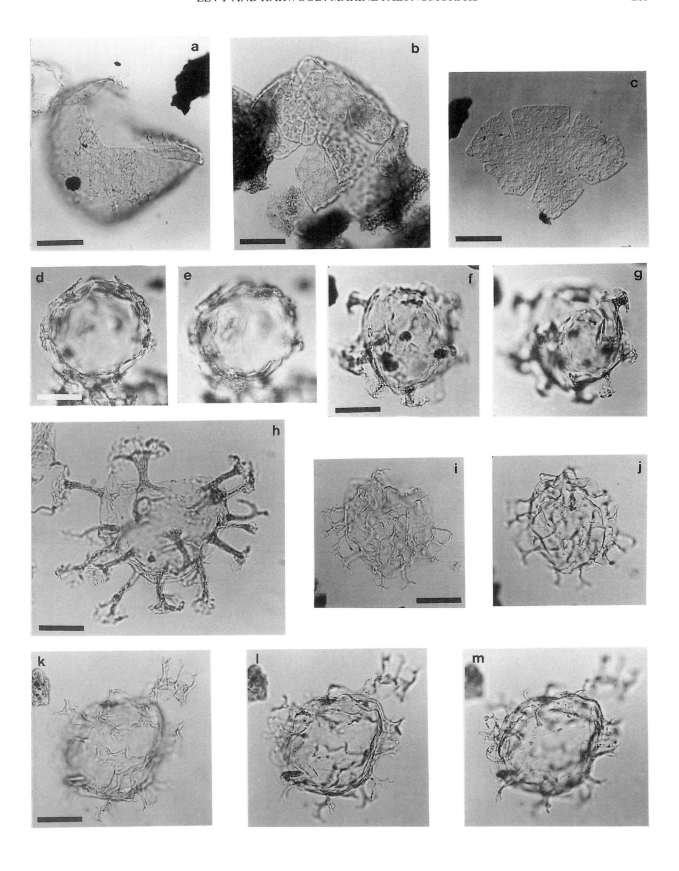

Dimensions. Observed range (four specimens): pericyst length - 78 to 57μ (mean 66μ), pericyst width - 56 to 42μ (mean 52μ).

Stratigraphic Range. Previously, *S. colemanii* has been reported from Seymour Island where it occurs in the upper Paleocene and Eocene (Wrenn and Hart, 1988).

Spinidinium essoi Cookson & Eisenack, 1967
Plate 9, fig. b

Spinidinium essoi Cookson & Eisenack, 1967, p.135, Plate 19, Figures 1-8.

Stratigraphic Range. Upper Cretaceous to lower Oligocene (Wrenn and Hart, 1988 and references therein)

Spinidinium macmurdoense (Wilson 1967) Lentin & Williams, 1976
Plate 9, figs. c and d

Deflandrea macmurdoensis Wilson, 1967, p. 60 and 62, Figures 2a and 11-16.
Spinidinium macmurdoense (Wilson, 1967) Lentin & Williams, 1976, p. 64.

Dimensions. Observed range (five specimens): pericyst length - 97 to 83μ (mean 90μ), pericyst width - 68 to 60μ (mean 63μ)

Stratigraphic Range: Eocene to lower Oligocene (Wrenn and Hart, 1988 and references therein).

Genus *Vozzhennikovia* Lentin & Williams, 1976

Vozzhennikovia apertura (Wilson, 1967) Lentin & Williams, 1976
Plate 11, figs. h-j

Spinidinium aperturum Wilson, 1967, p. 64-65, Figures 3-5, 8.
Spinidinium rotundum Wilson, 1967, p. 65-66, Figures 6-7.

Vozzhennikovia rotunda (Wilson) Lentin & Williams, 1976, p. 67.
Vozzhennikovia apertura (Wilson) Lentin & Williams, 1976, p. 65; Wrenn and Hart, 1988, p. 371-372, Figures 37.6-9, 43.3-4.

Comments. Wilson (1967) originally described two similar dinocyst species, *Vozzhennikovia apertura* (as *Spinidinium apertura*) and *Vozzhennikovia rotunda* (as *Spinidinium rotunda*), from erratics discovered on Black Island in McMurdo Sound. These two species were synonymized by Wrenn and Hart (1988), who considered each form to be an end member of a morphologic continuum of one taxon *Vozzhennikovia apertura*. Both end members of the morphologic continuum were observed in this study.

Dimensions. Observed range (eight specimens): pericyst length - 49 to 30μ (mean 41μ), pericyst width - 43 to 30μ (mean 36μ)

Stratigraphic Range. Eocene to lower Oligocene (see references in Wrenn and Hart, 1988).

Vozzhennikovia netrona n. sp.
Plate 11, figs. f and g

Vozzhennikovia apertura (Wilson) sensu Mohr, 1990, Plate 6, Figure 10.
Vozzhennikovia sp. a Crouch & Hollis, 1996, Plate 7, Figure 9.

Derivation of name. Greek, *netron*, spindle; with reference to the spindle-like shape.

Holotype. Plate 11, Figures f-g. UNSM PB99-09 Sample E 303(1), slide 4, middle to upper Eocene erratic, McMurdo Sound, Antarctica.

Description.

Shape: Pericyst sub-polygonal with long apical and left antapical horns. The lateral margins of the epipericyst and hypopericyst are typically straight to slightly convex.

Phragma: A smooth endophragm is closely appressed to the periphragm except in the apical and antapical regions.

Plate 5

(Scale bar = 20μ)

Figs. a-d. cf. *Eocladopyxis peniculata* Morgenroth 1966. E 155, slide 1: (a) ?apical view, apical surface; (b) ?apical view, optical section; (c) ?apical view, antapical surface; (d) disarticulated, spine covered plates.
Figs. e-h. *Glaphyrocysta radiata* n. sp. MTD 153(1), slide 2: (e) holotype, oblique view, dorsal surface, note the absence of both processes and fenestrate ectophram in the dorsal region. MTD 153(1), slide 2: (f) apical view, apical surface, note the solid precingular processes; (g) paratype, ventral view, ventral surface, ?sulcal depression is apparent; (h) paratype, ventral view, optical section.

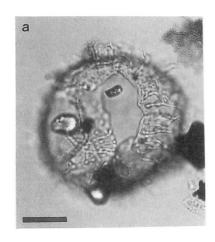

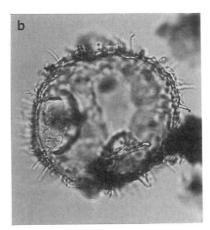

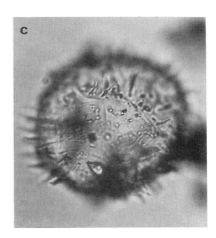

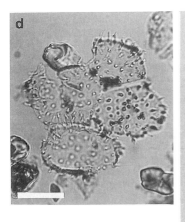

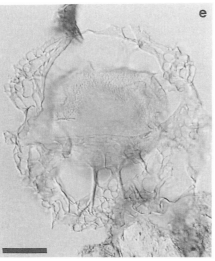

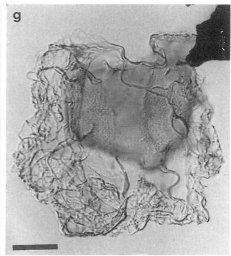

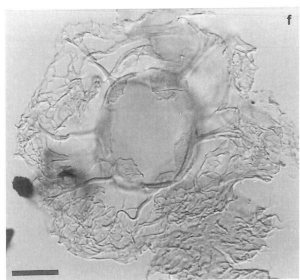

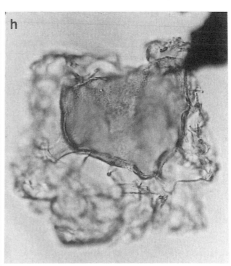

The surface of the periphragm is covered with short (2μ) capitate? spines which have a non-tabular distribution.

Paratabulation: Peridinioid, indicated by the intercalary archeopyle.

Paracingulum: Indicated by parasutural ridges that encircle the cyst in the adcingular region. The distal margins of the parasutural ridges possess capitate spines. Non-tabular spines occur within the paracingulum. The paracingulum is laevorotatory, offset by 2μ.

Parasulcus: Indicated by a break in the paracingulum. Posterior of the paracingulum, a depression in the surface of the cyst delineates the parasulcus. This depression has reduced surface ornamentation.

Archeopyle: Type I, hexa deltaform, formed by the complete removal of the 2a paraplate.

Dimensions. Observed range (six specimens): pericyst length - 61 to 85μ (mean 72μ), pericyst width - 44 to 31μ (mean 40μ), apical horn length - 14 to 20μ (mean 17μ), antapical horn length - 12 to 20μ (mean 16μ).

Comments/comparison. Possession of long apical and left antapical horns and a polygonal pericystal outline characterize this taxon. Otherwise, the species is similar to *Vozzhennikovia apertura*. This species was illustrated by Crouch and Hollis (1996), however, no description was given. Mohr (1990) recovered this taxon from ODP Hole 696 but assigned it to *Vozzhennikovia apertura*.

Stratigraphic Range. The species occurs in the middle to upper Eocene of DSDP Leg 29, Site 281 (Crouch and Hollis, 1996) and the middle to upper Eocene of ODP Hole 696B (Mohr, 1990).

Family CONGRUENTIDIACEAE Schiller, 1935

Genus *Lejeunecysta* (Gerlach, 1961) Artzner & Dörhöfer 1978; emend. Bujak in Bujak et al., 1980

***Lejeunecysta* cf. *L. hyalina* (Gerlach, 1961) Artzner & Dörhöfer, 1978**
Plate 7, fig. h

Lejeunia hyalina Gerlach, 1961, p. 169-171, Plate 26, Figures 10 and 11.
Lejeunia hyalina (Gerlach) emend. Kjellström, 1972
Lejeunecysta hyalina (Gerlach, emend. Kjellström, 1972) emend. Artzner & Dörhöfer, 1978
Lejeunecysta hyalina (Gerlach, 1961, emend. Kjellström, 1972) emend. Artzner & Dörhöfer, 1978; emend. Sarjeant, 1984.
Lejeunecysta cf. *L. hyalina* (Gerlach, 1961, emend. Kjellström, 1972) sensu Crouch and Hollis, 1996, Plate 5, figure 12.

Comments. Rare specimens recovered from the McMurdo Erratics resemble the holotype. Surficial longitudinal lines or folds that are characteristic of *L. hyalina* are present on these specimens. However, the McMurdo Erratic specimens differ from the type material in that the lateral margins of the epicyst are less convex and the paracingulum is broader. Specimens of *L.* cf. *L. hyalina* recovered in this study are similar to specimens of *L.* cf. *L. hyalina* reported from DSDP Leg 29, Hole 280A (Crouch and Hollis, 1996).

Dimensions. One well-preserved specimen: pericyst length - 120μ, pericyst width, 120μ.

Stratigraphic Range. *L.* cf. *L. hyalina* recorded from DSDP Leg 2, Hole 280A has a middle Eocene range (Crouch and Hollis, 1996).

Genus *Phelodinium* Stover & Evitt, 1978; emend. Mao & Norris, 1988

***Phelodinium harringtonii* n. sp.**
Plate 8, figs. a-d

Plate 6

(Scale bar = 20μ)

Figs. a-c. *Hystrichosphaeridium truswelliae* Wrenn and Hart 1988. MB 181(2), slide 1. (a) ventral view, ventral surface; (b) ventral view, dorsal surface; (c) detail showing well developed fenestrate polygonal platforms that occur at the distal terminations of processes.

Figs. d-f. *Hystrichosphaeridium* cf. *H. tubiferum* subsp. brevispinum (Davey and Williams 1966) Lentin and Williams 1973. MB 109(1), slide 1: (d) dorsal view, dorsal surface; (e) dorsal view, optical section; (f) detail of polygonal distal process terminations.

Figs. g-i. *Hystrichosphaeridium* sp. E 350, slide 4: (h) right lateral view, high focus showing long dorsal processes with large, fenestrate, polygonal platforms that occur at their distal terminations; (i) right lateral view, optical section showing short ventral processes. E 145, slide 3: (j) right lateral view, optical section, note long dorsal processes and relatively short ventral processes.

Figs. j-l. *Impagidinium victorianum* (Cookson and Eisenack 1965) Stover and Evitt 1978. MTD 42, slide 1: (j) ventral view, ventral surface; (k) ventral view, optical section; (l) ventral view, dorsal surface.

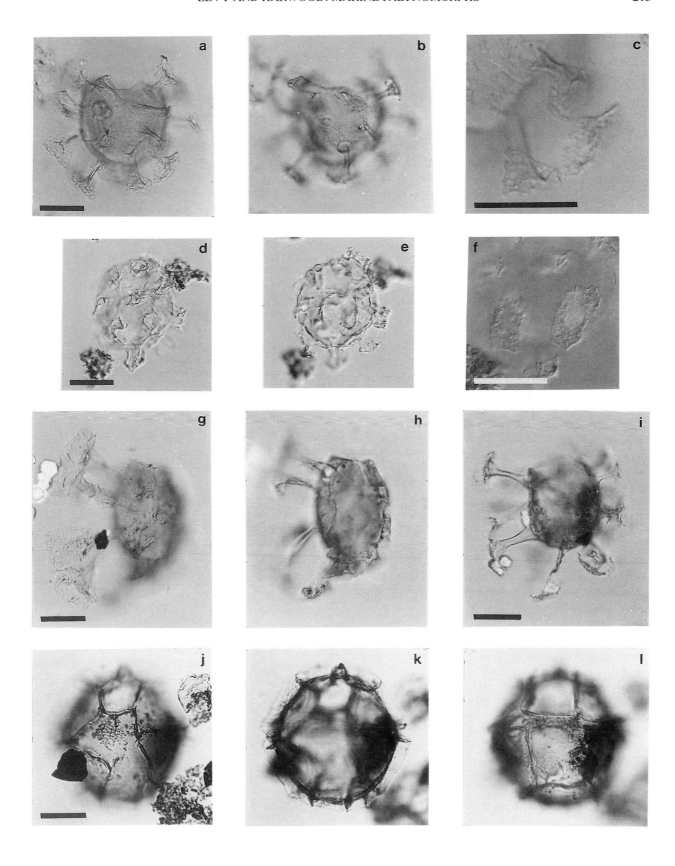

Derivation of name. Named in recognition of the contributions made to the study of Antarctic geology by H.J. Harrington.

Holotype. Plate 8, Figure a. UNSM PB99-03: Sample MTD 42, slide 2, middle to upper Eocene erratic, McMurdo Sound, Antarctica.

Diagnosis. A species of *Phelodinium* characterized by a hypocyst with concave lateral and posterior margins and a short round apical horn on the pericyst and endocyst. The cyst is usually longer than it is wide. The wall layers are smooth and are commonly closely appressed.

Description.

Shape. A cornucavate peridinioid dinoflagellate cyst. The lateral margins of the epicyst are typically straight to convex, whereas the lateral margins of the hypocyst are concave. The epicyst and hypocyst are joined across a broad equatorial region. A short, rounded apical horn is present on both the pericyst and endocyst. Two antapical horns are present and may be pointed. The posterior margin is usually concave.

Phragma. The wall layers consist of an endophragm and periphragm that are thin and smooth. The two wall layers are typically closely appressed. Separation between the periphragm and endophragm is only evident in the region of the apical and antapical horns.

Paratabulation. Peridinioid paratabulation is indicated only by the presence of an intercalary archeopyle.

Paracingulum. The position of the paracingulum is delineated by straight longitudinal lateral margins in the equatorial region. On several specimens the paracingulum is indicated by faint paracingular folds that occur in the periphragm.

Parasulcus. In specimens that possess paracingular folds the parasulcus is indicated by a longitudinal break in the folds. Otherwise, the parasulcus is not indicated.

Archeopyle. The intercalary archeopyle is formed by the complete removal of the broad hexa 2a paraplate. The AR ranges from 0.5 to 1.2 (mean 0.7).

Dimensions. Observed range (five specimens): pericyst length - 60 to 93μ (mean 80μ), pericyst width - 57 to 77μ (mean 68μ).

Comments/comparison. *Phelodinium harringtonii* has a similar shape to *P. magnificum* (Stanley) Stover and Evitt, 1978, but is smaller and lacks the frilled paracingular borders. *P. harringtonii* differs from *P. boldii* Wrenn and Hart (1988), *P. nigericum* Biffi and Grignani (1983) and *P. africanum* Biffi and Grignani (1983) by having a hypocyst with concave lateral margins, a length to width ratio greater than one, and poorly developed paracingular folds or ridges. *P. harringtonii* superficially resembles *Lejeunecysta hyalina* in specimens where the pericyst and endocyst are closely appressed. However, most specimens are cornucavate (Plate 8, fig. d).

Stratigraphic Occurrence. Middle to upper Eocene erratics, McMurdo Sound, Antarctica.

Genus *Selenopemphix* Benedek, 1972; emend. Bujak in Bujak et al., 1980.

***Selenopemphix nephroides* Benedek, 1972; emend. Bujak in Bujak et al., 1980; emend. Benedek & Sarjeant, 1981.**
Plate 8, fig. j

Plate 7

(Scale bar = 20μ)

Figs. a-b. *Impletosphaeridium clavus* Wrenn and Hart 1988. E 219, slide 1: (a) orientation indet., note nail-like processes; (b) orientation indet.

Figs. c-d. *Impletosphaeridium* sp. E 219, slide 1: (c) orientation indet., high focus, note short soild processes with either bifurcate or trifurcate distal terminations; (d) orientation indet., optical section.

Fig. e. *Microdinium* sp. MB 109(1), slide 1: orientation indet.

Figs. f-g. *Octodinium askiniae* Wrenn and Hart 1988. E 219, slide 1: (f) dorsal view, dorsal surface. MTD 42, slide 2: (g) dorsal view, dorsal surface showing octohedral archeopyle.

Fig. h. *Lejeunecysta* cf. *L. hyalina* (Gerlach 1961) Artzner and Dörhöfer 1978. E 303(2), slide 3: ventral view, ventral surface.

Figs. i-j. *Operculodinium bergmannii* (Archangelsky 1969) Stover and Evitt 1978. MTD 1B, slide 1: (i) ventral view ventral surface; (j) ventral view, dorsal surface.

Figs. k-n. *Paucisphaeridium inversibuccinum* (Davey and Williams 1966) Bujak et al. 1980; emend. Bujak et al. 1980. MTD 1B, slide 1: (k) oblique view, apical surface; (l) oblique view, antapical surface. D1, slide 2: (m) apical view, optical section. E 155, slide 1: (n) lateral view.

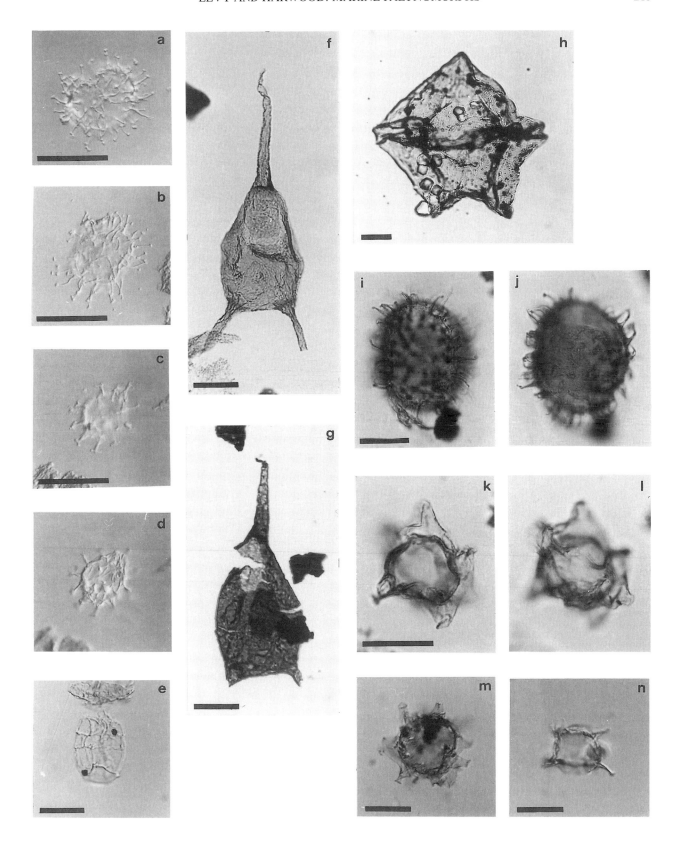

Selenopemphix nephroides Benedek, 1972, p. 47-48, Plate 11, Figure 13 and Plate 16, Figures 1-4.

Selenopemphix nephroides Benedek, 1972; emend. Bujak, 1980, p. 84.

Selenopemphix nephroides Benedek, 1972; emend. Bujak, 1980; emend. Benedek & Sarjeant, 1981, p. 333-334, 336.

Stratigraphic Range: Mohr (1990) reported *S. nephroides* from the upper middle Eocene (CP 14a) to Oligocene of ODP Hole 696B. De Coninck (1977) recorded *S. nephroides* from Ypresian strata in Belgium. Williams & Bujak (1985) reported a total global range equivalent to mid middle Eocene to upper Miocene (NP16 to NN12).

Selenopemphix prionota n. sp.
Plate 8, figs. k and l

Derivation of name. Greek, *prionota*, jagged, serrated, in reference to the crenulate to echinate paracingular frill.

Holotype. Plate 8, Figure k. UNSM PB99-05: Sample E 214, slide 5, middle to upper Eocene erratic, McMurdo Sound, Antarctica.

Description.

Shape. A peridinioid dinoflagellate cyst that is compressed along its polar axis. In polar view *Selenopemphix prionota* has an elliptical shape (because of compression along the polar axis, specimens are typically observed in polar view). In equatorial view the cyst has a polygonal shape. The lateral margins of the epicyst are straight and converge toward the apex where a short, blunt apical horn occurs. The hypocyst consists of two rounded to pointed antapical projections.

Paratabulation. Indicated by an intercalary archeopyle and paracingular frill.

Phragma. Smooth thin autophragm.

Paracingulum. Indicated by a broad equatorial region and a paracingular frill. The paracingular region is circumscribed with a broad frill that possesses echinate to crenellate sculpture along its distal margin.

Parasulcus. Indicated in polar view by a concave section in the equatorial margin of the hypocyst and in equatorial view by a longitudinal break in the paracingulum.

Archeopyle. The archeopyle appears to be an intercalary type Ia. However, in many specimens the thin autophragm is broken or torn, and as a result, the position of the archeopyle is obscured.

Dimensions. One specimen measured: dorso-ventral width - 30μ, lateral width - 36μ.

Comments/comparison. This species is similar to *Selenopemphix nephroides* but differs in that it possesses a crenulate to echinate paracingular frill. *Selenopemphix prionota* is similar to *Selenopemphix* sp. 5 of Head and Norris (1989) and may be conspecific. *S. prionota* differs from *S. coronata* (Bujak in Bujak et al., 1980) and *S. brevispinosa* (Head et al., 1989) by lacking distal bifurcation or expansion on the spines that surround the paracingulum. *Selenopemphix prionota* is a rare taxon in the McMurdo Erratics.

Stratigraphic Occurrence. Middle to upper Eocene erratics, McMurdo Sound, Antarctica.

Family uncertain

Genus *Octodinium* Wrenn & Hart, 1988

Octodinium askiniae Wrenn & Hart, 1988
Plate 7, figs. f and g

Plate 8

(Scale bar = 20μ)

Figs. a-d. *Phelodinium harringtonii* n. sp. MTD 42, slide 2: (a) holotype, dorsal view, dorsal surface; (b) paratype, dorsal view, dorsal surface; (c) dorsal view, dorsal surface, note separation of wall layers along the left lateral margin of the epicyst; (d) holotype, detail of epicystal region showing separation of wall layers beneath the apical horn.

Figs. e-f. *Pyxidinopsis* sp. A of Cocozza and Clarke (1992). MTD 1B, slide 1: (e) oblique view, right lateral suface; (f) oblique view, left lateral surface.

Figs. g-i. *Phthanoperidinium echinatum* Eaton 1976. MB 181(2), slide 1: (g) left lateral view, note intercalary archeopyle. MTD 42, slide 2: (h) dorsal view, dorsal surface; (i) dorsal view, dorsal surface, note intercalary archeopyle and characteristic rows of penitabular clavate spines.

Fig. j. *Selenopemphix nephroides* Benedek 1972; emend. Bujak 1980; emend. Benedek and Sarjeant 1981. E 219, slide 1: (j) apical polar view, optical section, dorsal surface towards the top of figure, note apical horn (dark area in center of specimen).

Figs. k-l. Selenopemphix prionota n. sp. E 214, slide 5: (k) holotype, optical section, note antapical horns (dark patches in center of specimen). E 345, slide 3: (l) paratype, ventral view, optical section.

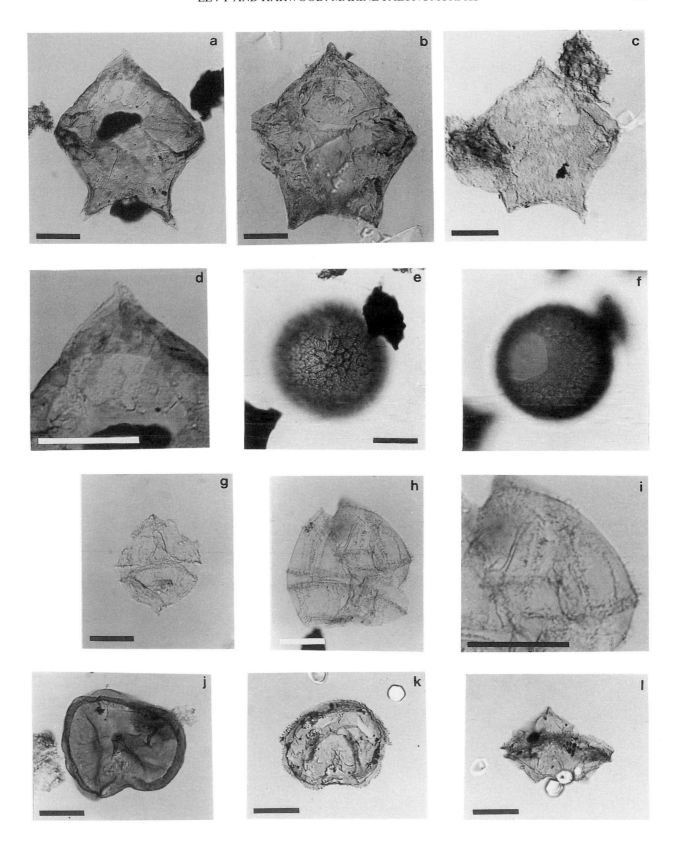

Octodinium askiniae Wrenn & Hart, 1988, p. 359-360,
   Figures 28.1-2 and 4, 29.1-7

Comments. Specimens of *Octodinium askiniae* are
abundant in several erratics. Many specimens exhibit a
characteristic octagonal archeopyle (Plate 7, figs. f and
g). Both apical and antapical horns are typically broken
and often only the central body is preserved.

***Stratigraphic Range*:** The species has been record-
ed only from the Southern Ocean. Wrenn and Hart
(1988) recovered *O. askiniae* from the Eocene La Meseta
Formation but also noted that Askin (1988) recovered *O.
askiniae* (as *Phelodinium* sp.) from the Maastrichtian on
Seymour Island. In the Weddell Sea, *O. askiniae* is
reported from the middle Eocene of the Bruce Bank,
Scotia Sea (Mao and Mohr, 1995) and middle Eocene to
?lower Oligocene of ODP Hole 696B (figs. 3 and 4, this
paper; Mohr, 1990). Crouch and Hollis (1996) recorded
O. askiniae from the early middle to upper Eocene of
DSDP Leg 29, Site 280A. Notably *O. askiniae* is not
reported from Paleocene strata on Seymour Island. The
total range for *O. askiniae* is herein considered to be
Maastrichtian to ?lower Oligocene.

Order GONYAULACALES Taylor, 1980
Family GONIODOMACEAE Lindemann, 1928

Genus *Alisocysta* Stover & Evitt, 1978

***?Alisocysta* sp.**
Plate 1, figs. a-d

Comments. Two reasonably well-preserved specimens attributable to the genus *Alisocysta* were observed
in Erratic MB 109(1). Paratabulation is indicated by penitabular septa that expand distally to form spongy plat-

forms. On one specimen, the spongy periphragm extends
across an entire paraplate (Plate 1, figs. c and d). This
type of paraplate ornamentation is characteristic of the
genus *Eisenackia*. The McMurdo Erratic specimens
probably represent a new taxon, although an insufficient
number of specimens prevent the formal erection of a
new species.

Dimensions. Observed range (two specimens): pericyst length - 39 to 46μ, pericyst width - 39 to 40μ.

**Genus *Eisenackia* Deflandrea & Cookson, 1955;
emend. Sarjeant, 1966; emend. McLean, 1973.**

cf. *Eisenackia scrobiculata* Morgenroth, 1966
Plate 4, figs. a-c

Eisenackia scrobiculata Morgenroth, 1966, p. 12-13,
   Plate 2, fig. 12, Plate 3, fig. 1.

Comments. Complete specimens of cf. *Eisenackia
scrobiculata* are rarely preserved. Commonly, paraplates
separate along the parasutures and the cyst becomes disarticulated making taxonomic determinations difficult.
As a result, the McMurdo Erratics specimens are designated cf. *E. scrobiculata*. One complete specimen and several fragments were observed.

**Genus *Eocladopyxis* Morgenroth, 1966; emend.
Stover & Evitt, 1978**

cf. *Eocladopyxis peniculata* Morgenroth, 1966
Plate 5, figs. a-d

Eocladopyxis peniculata Morgenroth 1966, p. 7-8, Plate 3,
   figs. 2-3.

Plate 9

(Scale bar = 20μ)

Fig. a. *Spinidinium colemanii* Wrenn and Hart 1988. MB 181(2), slide 1: ventral view, optical section.
Fig. b. *Spinidinium essoi* Cookson and Eisenack 1967. MB 181(2), slide 1: dorsal view, optical section.
Figs. c-d. *Spinidinium macmurdoensis* (Wilson 1967) Lentin and Williams 1976. MTD 153(1), slide 1: (c) dorsal view, optical sec-
tion. MTD 42, slide 2: (d) dorsal view, optical section.
Figs. e-g. *Spiniferites ramosus* cf. subsp. reticulatus (Davey and Williams 1966) Lentin and Williams 1973. MTD 42, slide 2: (e)
oblique view, left lateral epicyst; (f) oblique view, optical section; (g) oblique view, right lateral hypocyst.
Figs. h-i. *Spiniferites ramosus* (Ehrenberg 1838) Mantell 1854. MB 181(2), slide 1: (h-i) orientation indet, two focal levels.
Figs. j-k. *Systematophora ancyrea* Cookson and Eisenack 1965. MB 188G, slide 1: (a) oblique view, antapical surface; (b) oblique
view, apical surface. MTD 42, slide 1: (l) orientation indet.

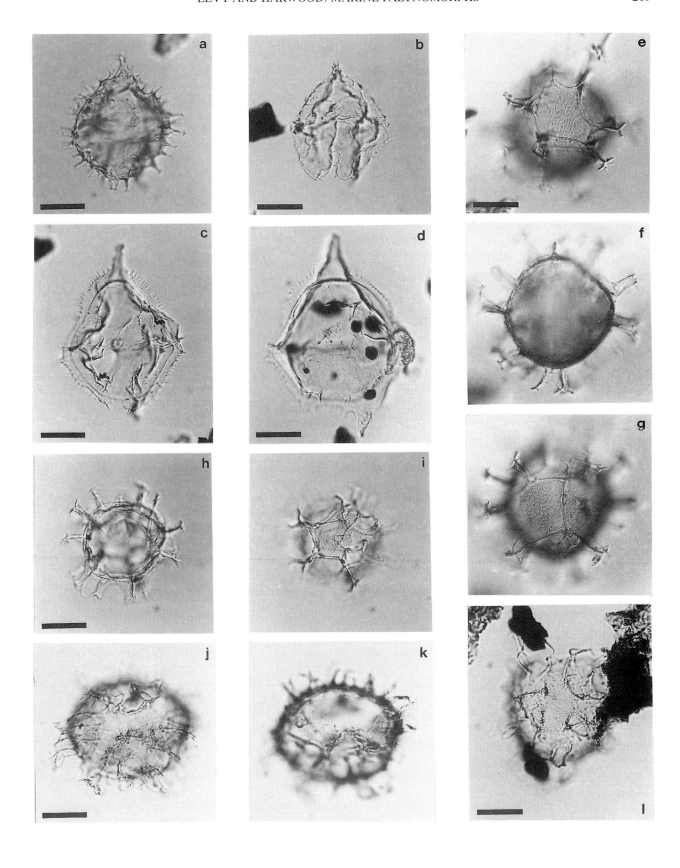

Comments. Spiny plates that may be attributed to *E. peniculata* are observed in several erratics. Complete specimens are rare and as a result, the McMurdo Erratic specimens are considered to be cf. *E. peniculata*.

Genus *Hystrichosphaeridium* Deflandre, 1937; emend. Davey & Williams, 1966b.

Hystrichosphaeridium truswelliae Wrenn & Hart, 1988
Plate 6, figs. a-c

Hystrichosphaeridium tubiferum (Ehrenberg) Deflandrea, 1937, sensu Wilson, 1967, Figure 40.
Hystrichosphaeridium truswelliae Wrenn & Hart, 1988, p. 355, Figures 25.1-4, 39.1.
Hystrichosphaeridium sp. a, Crouch and Hollis, 1996, Plate 4, figures 7 and 10.

Stratigraphic Range. Wrenn and Hart (1988) reported an uppermost lower Eocene range for this taxon. Cocozza and Clarke (1992) reported a range of lower to ?upper Eocene. We consider that *Hystrichosphaeridium* sp. a, recovered from the middle Eocene of DSDP Leg 29, Site 280A (Crouch and Hollis, 1996), is probably con-specific with *H. truswelliae*. The total stratigraphic range recognized herein is Eocene.

Hystrichosphaeridium sp. cf. *H. tubiferum* subsp. *brevispinum* (Davey & Williams, 1966b) Lentin & Williams, 1973
Plate 6, figs. d-f

Hystrichosphaeridium tubiferum var. *brevispinum* Davey & Williams, 1966b, p. 58, Plate 10, Figure 10.
Hystrichosphaeridium tubiferum subsp. *Brevispinum* (Davey & Williams) Lentin & Williams, 1973, p. 80.

Comments. This species is similar to *H. tubiferum* subsp. *brevispinum* (Davey and Williams) Lentin and Williams, in that the length of the processes are approx-imately one-third the diameter of the central body. However, the distal regions of the processes in the erratic specimens consist of broad fenestrate platforms, which are not apparent in the type material. The specimens from the McMurdo Erratics may be a subspecies of *H. truswelliae*.

Hystrichosphaeridium sp.
Plate 6, figs. h-j

Description.

Shape. Subspherical chordate dinoflagellate cysts with intratabular processes of varying length. The processes in the ventral region of the cyst are generally less than 1/3 the length of processes in the lateral and dorsal regions of the cyst. No apical paraplates were observed, therefore the length of the apical processes are unknown.

Phragma. The periphragm and endophragm are thin and smooth. The periphragm gives rise to hollow intratabular processes that possess broad fenestrate platforms at their distal terminations. Platforms range in width from 14 to 23μ (Plate 6, fig. h). The periphragm and endophragm are closely appressed, except at the base of each process.

Paratabulation. Paratabulation determined from the intratabular processes indicates a gonyaulacoid paraplate distribution consisting of ?4', 6'', 6c, 1p, 5''', 1'. Apical paratabulation is uncertain as no apical opercula were observed.

Paracingulum. Indicated by six, intratabular paracingular processes.

Parasulcus. The position of the parasulcus is indicated by the sulcal notch present in the epicyst. Sulcal paratabulation was indeterminate.

Archeopyle. Formed by complete removal of presumably four apical paraplates.

Dimensions. Observed range (three specimens): central body length - 42 to 51μ (mean 46μ, central body width - 33 to 42μ (mean 36μ); process length - 3 to 33μ.

Plate 10

(Scale bar = 20μ)

Figs. a-c. *Thalassiphora pelagica* (Eisenack 1954) Eisenack and Gocht 1960; emend. Benedek and Gocht 1981. MTD 42, slide 2: (a) ventral view, ventral surface. E 155, slide 1: (b) ventral view, ventral surface; (c) ventral view, dorsal surface.
Figs. d-i. *Turbiosphaera filosa* (Wilson 1967) Archangelsky 1968. Series of figures illustrating the varied morphology of this taxon. MTD 1B, slide 1: (d) ventral view, ventral surface, the morphology exhibited by this specimen is similar to the type material; (e) ventral view, optical section; MB 245, slide 1: (f) orientation uncertain, optical section; (g) right lateral view, high focus, specimen with short processes; (h) right lateral view, optical section; (i) left lateral view, optical section.

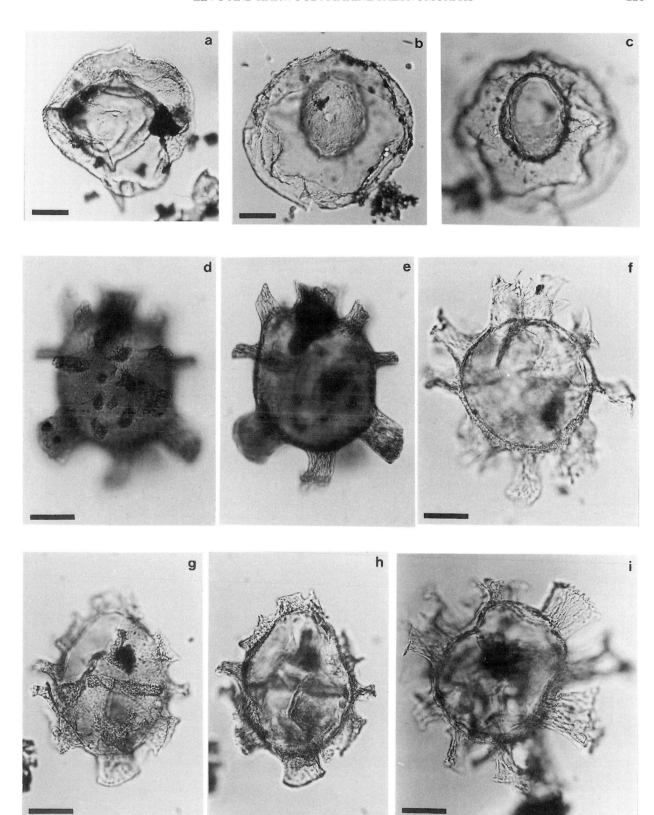

Comments/comparison. This rare form occurs in Erratics E 350 and E 145. This species is similar to *H. truswelliae*. However, whereas the apical/epicystal processes of *H. truswelliae* are shorter than the hypocystal processes, the shortest processes of *Hystrichosphaeridium* sp. occur in the ventral region of the cyst. No formal species is erected, as only three specimens of *Hystrichosphaeridium* sp. were observed.

Family GONYAULACACEAE Lindemann, 1928

Genus *Arachnodinium* Wilson & Clowes, 1982

Arachnodinium antarcticum Wilson & Clowes, 1982
Plate 1, figs. i-k

Aiora fenestrata (Deflandre & Cookson) Cookson & Eisenack, 1960, sensu Wilson, 1967, p. 69, Figures 2c, 37, 38.
Arachnodinium antarcticum Wilson & Clowes, 1982, p. 97-102, Plates 1-2, Text Figure 2.
Stratigraphic Range. Eocene (Wrenn and Hart, 1988 and references therein)

Genus *Cribroperidinium* Neale & Sarjeant, 1962; emend. Davey, 1969; emend. Sarjeant, 1982; emend. Helenes, 1984.

Cribroperidinium giuseppi (Morgenroth, 1966a)
Helenes, 1984.
Plate 2, fig. f-i

Gonyaulax giuseppi Morgenroth, 1966, p. 5-6, Plate 2, Figures 3-4.
Gonyaulacysta giuseppi (Morgenroth) Sarjeant in Davey et al., 1969, p. 9.

Millioudodinium? giuseppi (Morgenroth) Stover and Evitt, 1978, p. 159.
Cribroperidinium giuseppi (Morgenroth) Helenes, 1984, p. 121, Plate 2, Figure 6-11, Plate 4, Figure 8-13, Text-Figure 6 G-I.

Dimensions. Observed range (six specimens): length (not including apical horn) - 59 to 77μ (mean 66μ), width - 58 to 65μ (mean 62μ); length of apical horn - 7 to 12μ; height of septa - 1 to 5μ.
Stratigraphic Occurrence. Williams and Bujak report a global range of lower to upper Eocene (NP11 to NP19). Other Southern Ocean occurrences include the Eocene Rio Turbio Formation of southern Argentina (Archangelsky, 1969b as *Leptodinium* sp. , p. 194-196, Plate 2, Figures 5-6).

Genus *Enneadocysta* Stover & Williams, 1995

Enneadocysta partridgei Stover & Williams, 1995
Plate 4, fig. h
Hystrichosphaeridium sp. Cranwell et al. 1960. p. 701, Figure 1.
Cordosphaeridium diktyoplokus (Klump) Eisenack, 1963, sensu Cranwell, 1964, p. 398-404, Figures 2, 3a-3c.
Areosphaeridium diktyoplokus (Klump) Eaton, 1971, sensu Haskell & Wilson, 1975, p. 724, Plate 1, figure 1.
Areosphaeridium sp. cf. *Areosphaeridium diktyoplokus* (Klump) Eaton, 1971, sensu Goodman & Ford, 1983, p. 865, Plate 8, figure 4.
Areosphaeridium diktyoplokus (Klump) Eaton, 1971, sensu Wrenn & Hart, 1988, p. 346-347, Figure 15, no. 6.
Enneadocysta partridgei Stover & Williams, 1995, p.113-114, Plate 4, figures 4-5, Plate 5, figures 1-5.

Plate 11

(Scale bar = 20μ)

Figs. a-c. *Turbiosphaera sagena* n. sp. MB 181(2), slide 1: (a-c) holotype, ventral view, three focal levels, note distal network of strands that join the processes; (d) holotype, optical section, detail of paracingular processes illustrating distal connection; (e) paratype, dorsal view, dorsal surface, note vitrinite clasts enclosed between distally connected paracingular processes.
Figs. f-g. V*ozzhennikovia netrona* n. sp. E 303(1), slide 4: (f-g) holotype, (f) ventral view, ventral surface; (g) ventral view, dorsal surface.
Figs. h-j. *Vozzhennikovia apertura* (Wilson 1967) Lentin and Williams 1976. MB 109(1), slide 1: (h) oblique view, antapical surface. MTD 1B, slide 1: (i) ventral view, ventral surface; (j) ventral view, dorsal surface.
Figs. k-n. *Dinocyst* sp. A. MTD 42, slide 2: (k) oblique antapical view, antapical surface; (l) oblique antapical view, optical section; (k) oblique antapical view, apical surface; (n) apical view.

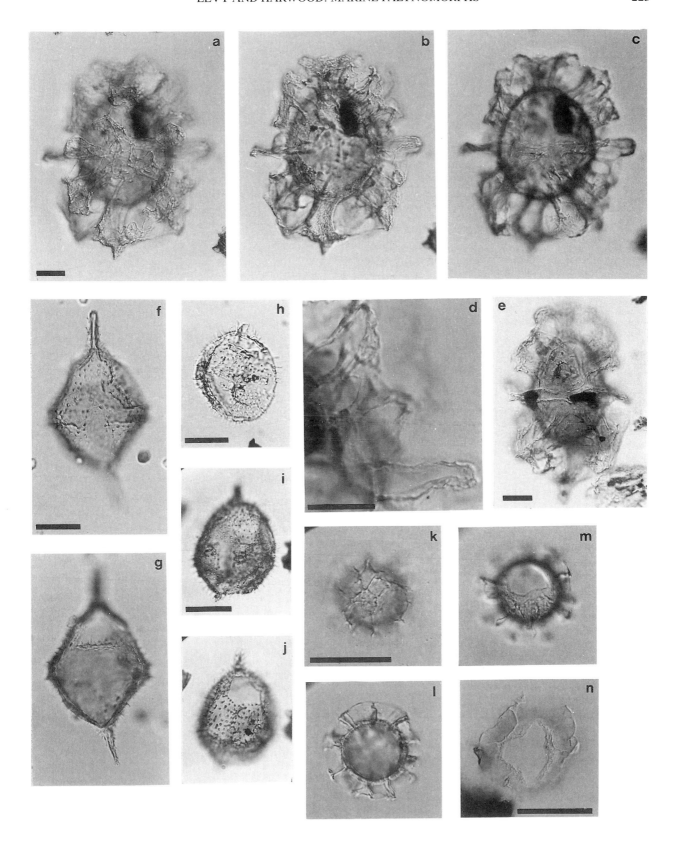

Stratigraphic Range. Stover and Williams (1995) report a range of upper middle Eocene (Bartonian) to lower Oligocene (Rupelian). On Seymour Island, *E. partridgei* occurs in the lower section of the Eocene La Meseta Formation and therefore has a lower Eocene first occurrence (Raine et al., 1997).

Enneadcysta sp. 1
Plate 4, figs. i-m

Description.

Shape. Lenticular, chordate dinoflagellate cysts with solid fibrous processes that typically possess licrate distal terminations.

Phragma. A thin, psilate to shagreenate autophragm gives rise to solid, fibrous, intratabular processes. Process shape varies within the cyst. Epicystal processes range in form from broad (6μ) processes with accessory strands (Plate 4, fig. m) to narrow processes (1μ) that bifurcate where they join the central body. Within the hypocyst, licrate processes with single stems are most common. However, some processes bifurcate distally and individual processes may be joined distally by fine strands or trabeculae.

Paratabulation. Indicated by 27+ intratabular processes (4 or 5 apical processes on operculum).

Paracingulum. Indicated by 6 or 7 intratabular processes.

Parasulcus. Indicated by a parasulcal notch in the epicyst and short parasulcal intratabular processes.

Archeopyle. An apical archeopyle (type tA) is formed by the complete removal of the apical paraplate series.

Comments/comparison. This species is characterized by fibrous intratabular processes with licrate distal terminations. Specimens are most commonly recovered in Erratic MTD 153(1). *Enneadocysta* sp. 1 possesses processes with licrate distal terminations whereas *E. partridgei* possesses processes with ragged fenestrate distal terminations. The process lengths on *Enneadocysta*. sp. 1

are usually shorter than *E. partridgei*. *E. harrisii* (Stover and Williams, 1995) has similar licrate processes but does not possess paracingular processes.

Dimensions. Observed range (five specimens): Central body length - 35 to 46μ (mean 42μ), central body width - 46 to 62μ (mean 53μ), process length - 7 to 20μ.

Enneadcysta sp. 2
Plate 4, figs. d-g

Areosphaeridium sp. A Cocozza & Clarke, 1992, p. 361-362, Figure 4b.

Enneadocysta sp. a Crouch & Hollis, 1996, Plate 4, no. 2.

Comments. This species is similar to *Enneadocysta partridgei* except it has short processes. Rare specimens were recovered in the McMurdo Erratics.

Stratigraphic Range. *Enneadocysta* sp. 2 (as *Enneadocysta* sp. a) occurs in DSDP Hole 280A, core 16, section 1 to core 14, section 1. Cores 22-17 are no older than middle Eocene and cores 12-10 are no younger than upper Eocene (Crouch and Hollis, 1996). Cocozza and Clarke (1992) report a probable middle to late Eocene age for their dinocyst associations 2 and 3 within which *Enneadocysta* sp. 2 (as *Areosphaeridium* sp. A) occurs.

Genus *Impagidinium* Stover & Evitt, 1978

Impagidinium victorianum (Cookson & Eisenack, 1965a) Stover & Evitt, 1978
Plate 6, figs. k-m

Leptodinium victorianum Cookson & Eisenack, 1965a, p.123, Plate 12, Figures 8-9.

Impagidinium victorianum (Cookson & Eisenack) Stover & Evitt, 1978.

Plate 12

(Scale bar = 20μ except figs. j-k where scale bar = 10μ)

Figs. a-b. *Cyclopsiella trematophora* (Cookson and Eisenack 1967) Lentin and Williams 1977. MTD 153(1), slide 1.
Fig. c. *Cyclopsiella* sp. 1. E 165, slide 2.
Fig. d. *Cyclopsiella* sp. 2. MTD 1B, slide 1.
Figs. e-f. *Cyclopsiella* sp. 3. MB 181(2), slide 1: (e) high focus, note spongy nature of the autophragm. E 219, slide 1: (f).
Fig. g. *Cyclopsiella* sp. 4. E 303(1), slide 4.
Figs. h-i. *?Cyclopsiella* sp. 5. MTD 154, slide 1.
Figs. j-k. *Dichotisphaera* sp. E 219, slide 1

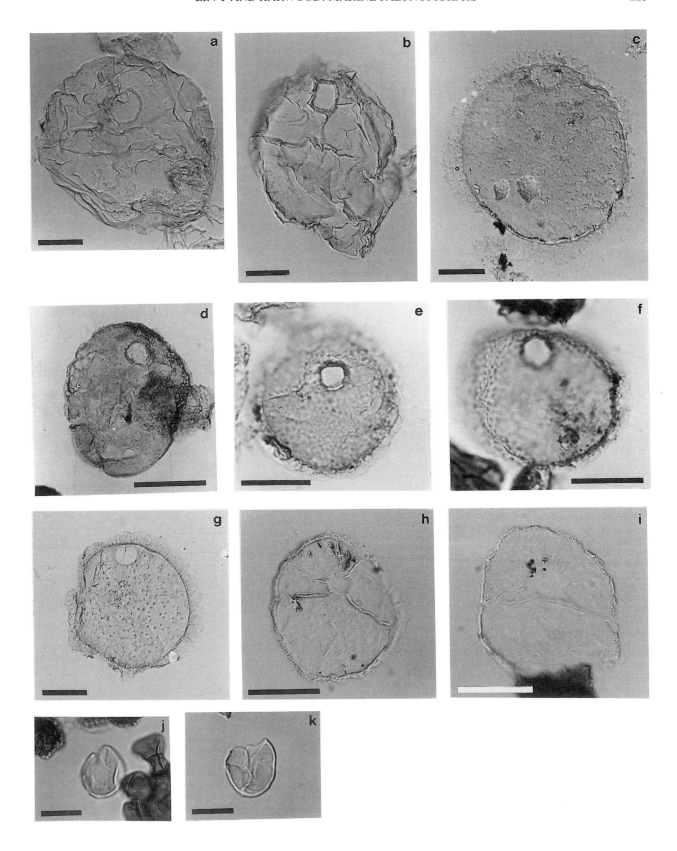

Comments. Specimens recovered from the erratics are smaller than the Australian type material, but otherwise conform with the type description.

Dimensions. Observed range (four specimens): Central body length - 54 to 60μ (mean 58μ), central body width - 56 to 60μ (mean 58μ), height of septa - 6 to 8μ. *Stratigraphic Range.* Eocene to lower Oligocene (Wrenn and Hart, 1988 and references therein).

Genus *Operculodinium* Wall, 1967

Operculodinium bergmannii (Archangelsky, 1969a) Stover & Evitt, 1978
Plate 7, figs. i and j

Cleistosphaeridium bergmannii Archangelsky, 1969a, p. 414-415, Plate 11, Figures 8, 11.
Operculodinium bergmannii (Archangelsky) Stover & Evitt, 1978, p.178.

Stratigraphic Range. The type specimen was recovered from Eocene sediments of the Rio Turbio Formation in southern Argentina (Archangelsky, 1969a). The following occurrences and corresponding ranges are reported for the Southern Ocean. Wrenn and Hart (1988) reported a lower upper Paleocene to Eocene range for specimens of *O. bergmannii* recovered from Seymour Island. The species is reported from middle Eocene sediments in the Scotia Sea (Mao and Mohr, 1995) and middle to upper Eocene sediments in the Weddell Sea (Mohr, 1990). The total range reported by Wrenn and Hart (1988) is used herein.

Genus *Pyxidinopsis* Habib, 1976

Pyxidinopsis sp. A of Cocozza & Clarke (1992)
Plate 8, figs. e and f

Pyxidinopsis sp. A Cocozza & Clarke, 1992, p. 362, Figure 5i-k.
Pyxidinopsis sp. a Crouch & Hollis, 1996, Plate 6, Figures 4-6.

Comments. *Pyxidinopsis* sp. A, described and illustrated by Cocozza and Clarke (1992) from the La Meseta Formation on Seymour Island, appears to be conspecific with specimens attributed to the genus *Pyxidinopsis* recovered from the McMurdo Erratics. They are of similar size and also have thickenings between the lumina. Specimens recovered from DSDP Hole 280A, which are illustrated by Crouch and Hollis (1996) as *Pyxidinopsis* sp. a, appear to be conspecific to both the Seymour Island material and McMurdo Erratic specimens.

Dimensions. Observed range (four specimens): Length - 53 to 61μ (mean 56μ), width - 46 to 61μ (mean 54μ).

Stratigraphic Range. The stratigraphic range for *Pyxidinopsis* sp. A on Seymour Island is middle to upper Eocene (Cocozza and Clark, 1992). Crouch and Hollis (1996) report an early middle to ?upper Eocene range for *Pyxidinopsis* sp. a on the South Tasman Rise. The range for *Pyxidinopsis* sp. A is considered herein to be middle to upper Eocene.

Genus *Spiniferites* Mantell, 1850; emend. Sarjeant, 1970

Spiniferites ramosus subsp. *reticulatus* (Davey & Williams, 1966a) Lentin & Williams, 1973
Plate 9, figs. e-g

Hystrichosphaera ramosa var. *reticulata* Davey & Williams, 1966a, p. 38, Plate 1, Figures 2-3.
Spiniferites ramosus var. *reticulatus* (Davey & Williams) Davey & Verdier, 1971

Plate 13

(Scale bar = 20μ)

Fig. a. *Micrhystridium* sp. 1. MB 80, slide 1.
Figs. b-d. Micrhystridium sp. 2. MTD 190, slide 1: (b-c) orientation indet., two focal levels. E 155, slide 1: (d) orientation indet.
Figs. e-g. *Paralecaniella indentata* (Deflandrea and Cookson 1955) Cookson and Eisenack 1970; emend. Elsik 1977. E 303(1), slide 4: (e-f) two focal levels. MTD 1B, slide 1: (g).
Figs. h-i. *Scuticabolis lapidaris* (O. Wetzel 1933) Loeblich 1967. E 219, slide 1: (h). E 200, slide 1: (i).
Figs. j-k. *Veryhachium* sp. 1. E 219, slide 1: (j). E 364, slide 1: (k).
Figs. l-m. *Veryhachium* sp. 2. D1, slide 2: (l) high focus; (m) low focus.
Figs. n-r. *Acritarch* sp. A. E 219, slide 1: (n) high focus showing foeveolate autophragm; (o) optical section; (p-r) three focal levels.

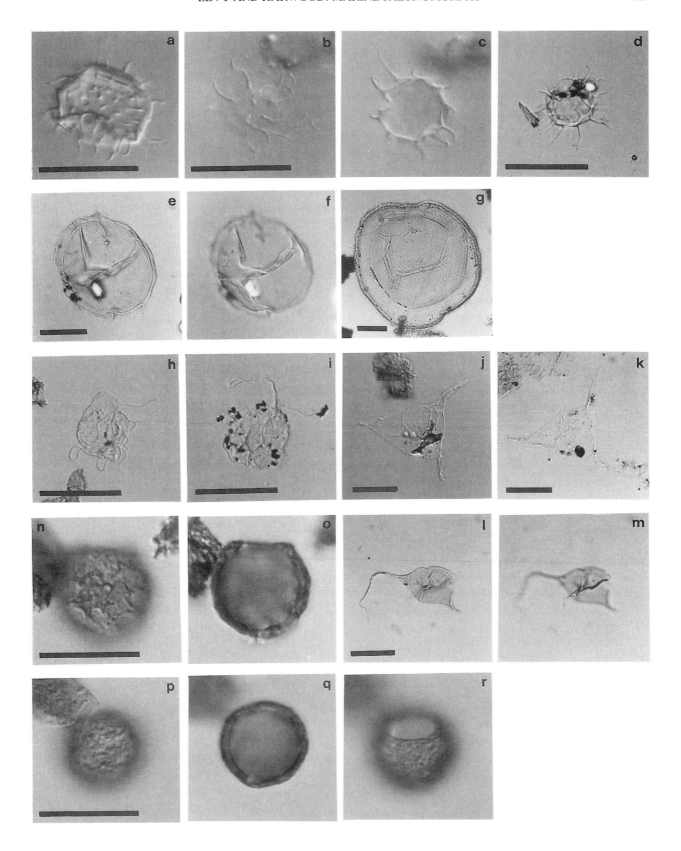

Spiniferites ramosus subsp. *reticulatus* (Davey & Williams) Lentin & Williams, 1973

Comments. The specimens encountered are within the size range of the type material. Specimens recovered from the erratics have a relatively thick (1.5 to 2.5μ) foveolate to reticulate periphragm. The endophragm is thin and appears to be smooth. Illustrations of specimens attributed to *H. ramosa* by Kemp (1975, Plate 4, Figures 4-7) are similar to the erratic specimens and are herein considered conspecific.

Stratigraphic Range. Northern Hemisphere occurrences in the Cretaceous to Paleocene. Wrenn and Hart (1988) recorded *S. ramosus* subsp. *reticulatus* from the early late Paleocene and Eocene on Seymour Island.

Spiniferites ramosus (Ehrenberg, 1838) Mantell, 1854
Plate 9, figs. h and i

Xanthidium ramosus Ehrenberg, 1838, Plate 1, Figures 1, 2 and 5.
Spiniferites ramosus Mantell, 1854, text-figure 17, nos. 4 and 5.
Ovum hispidum ramosum Lohmann, 1904, p. 21, 25
Hystrichosphaera ramosa (Ehrenberg) Wetzel, 1933b, p.144.
Bion ramosa (Ehrenberg) Eisenack, 1938, p. 243
Spiniferites ramosus (Ehrenberg) Loeblich & Loeblich, 1966, p. 56-57.

Stratigraphic Range. According to Williams and Bujak (1985), *S. ramosus* has a worldwide distribution from lower Cretaceous to Recent.

Genus *Systematophora* Klement, 1960; emend. Brenner, 1988; emend. Stancliffe & Sarjeant, 1990

Systematophora ancyrea Cookson & Eisenack, 1965a
Plate 9, figs. j-l

Systematophora ancyrea Cookson & Eisenack, 1965a, p. 126, Plate 14, Figures 1-3.

Stratigraphic Range. The type specimen was recovered from the upper Eocene Browns Creek Clays, South West Victoria, Australia. In New Zealand the species has a reported upper lower Eocene to ?lower Oligocene range (Wilson, 1982, 1984). The species is reported from the upper lower Eocene to middle Miocene for the Norwegian-Greenland Sea, DSDP Hole 338 (Manum, 1976).

Genus *Thalassiphora* Eisenack & Gocht, 1960; emend. Gocht, 1968; emend. Benedek & Gocht, 1981

Thalassiphora pelagica (Eisenack, 1954) Eisenack & Gocht, 1960; emend. Benedek & Gocht, 1981
Plate 10, figs. a-c

Pterospermopsis pelagica Eisenack, 1954, p. 71-72, Plate 12, Figures 17-18.
Pterocystidiopsis velata Deflandre & Cookson, 1955, p. 291, Plate 8, Figure 8.
Thalassiphora pelagica (Eisenack) Eisenack & Gocht, 1960, p. 513-514.
Disphaera pelagica (Eisenack) Norvick, 1973, p. 46.
Thalassiphora pelagica (Eisenack) Eisenack & Gocht 1960; emend. Benedek & Gocht, 1981.

Stratigraphic Range. Maastrichtian to lowest Miocene (Damassa et al., 1994).

Genus *Turbiosphaera* Archangelsky, 1969a

Turbiosphaera filosa (Wilson, 1967) Archangelsky, 1969a
Plate 10, figs. d-i

Cordosphaeridium filosum Wilson, 1967, p. 66, text-figure 2b, Figures 31, 32, 34.
Turbiosphaera filosa (Wilson) Archangelsky, 1969a, p. 408, 410-411, Plate 1, Figures 1-4.

Comments. Specimens identical in appearance to the type material are commonly encountered (Plate 10 figs. d and e). However, morphologic variation between

---

Plate 14

(Scale bar = 20μ)

Figs. a-c. *Cymatiosphaera* sp. E 188(B), slide 1: (a-c) three focal levels.
Figs. d-e. *Pterospermella* sp. 1. MB 109(1), slide 1.
Figs. f-h. *Pterospermella* sp. 2. E 215, slide 2. (f-g) two focal levels
Figs. i-k. *Palambages morulosa* O. Wetzel 1961. MB 181(2), slide 2: (i-j) two focal levels.

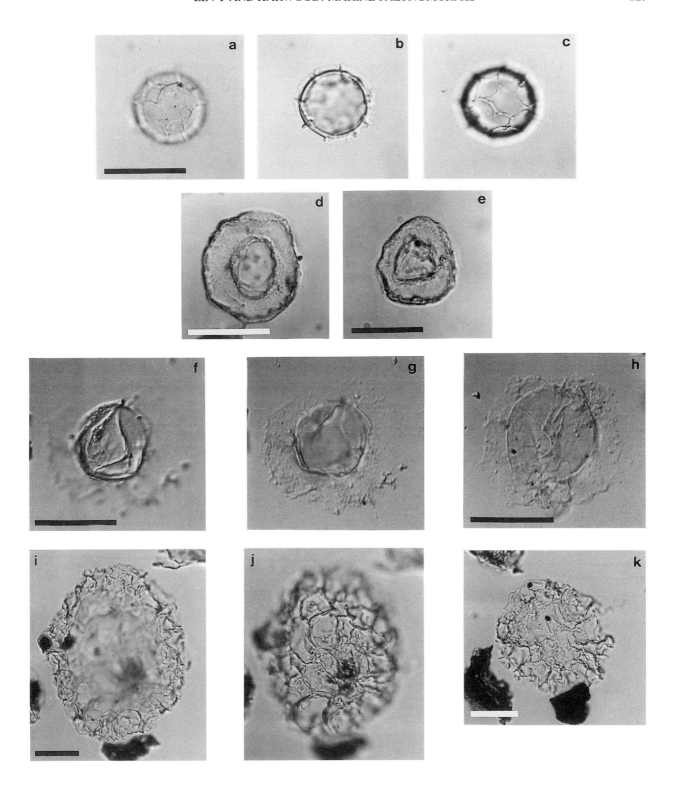

specimens of *T. filosa* recovered from the McMurdo Erratics is common. The shape of the central body varies from elliptical to sub-spherical and the length of the processes range from short (4μ) to long (21μ). These morphologic variations may prove to be biostratigraphically or paleoecologically useful.

Dimensions. Observed range (three specimens): central body length - 43 to 46μ (mean 43μ), central body width - 34 to 42μ (mean 38μ), process length - 4 to 21μ.

Stratigraphic Range. Williams and Bujak (1985) reported a global distribution of Maastrichtian to middle Eocene. Reported New Zealand occurrences range from upper Paleocene (Wilson, 1988) to lower upper Eocene (as *T.* cf. *filosa*, Wilson, 1982; 1984). Occurrences of the taxon in strata that is younger than Eocene in age are restricted to one southern hemisphere location (Kemp, 1975) and northern hemisphere sites (e.g. Evitt and Pierce, 1975; Jan du Chéne, 1977). *T. filosa* occurs below the first appearance of the calcareous nannofossil *Isthmolithus recurvus* (an upper Eocene biomarker) in all but two of the Southern Ocean drill holes utilized as biostratigraphic reference sections in this study (see biostratigraphy, this paper). Based on correlation to the calcareous nannofossil biostratigraphy, *T. filosa* is considered herein to have a regional range of Maastrichtian to upper Eocene.

Turbiosphaera sagena n. sp.
Plate 11, figs. a-e

Turbiosphaera filosa sensu Kemp, 1975, Plate 3, Figures 1 and 2.
Turbiosphaera sp. a Crouch & Hollis, 1996, Plate 7, Figure 6.

Derivation of name. Latin, *sagena*, fish-net, with reference to the fine network of fibrous strands that connect the processes.

Holotype. Plate 11, Figures a-c. UNSM PB99-07: Sample MB 181(2), slide 1, middle to upper Eocene erratic, McMurdo Sound, Antarctica.

Description.

Shape: A chordate dinoflagellate cyst with solid to fibrous intratabular processes. Central body is elongate to sub-spherical.

Phragma: The central body consists of a thick walled, fibrous autophragm. Processes are usually distally connected by fibrous, discontinuous strands or trabeculae. However, cingular processes are only rarely connected to precingular processes which separates the network of fine strands in the epicyst from the network in the hypocyst.

Paratabulation: Indicated by the intratabular processes and process complexes. Paratabulation appears to be gonyaulacacean hexiform, 4', 5-6(?)", 6c, 6(?)''', 1''''

Paracingulum: Parasutural fibrous processes indicate the position of the paracingulum. The processes are joined distally creating a hollow enclosed space across the paracingulum. These connected processes form the shelf-like projections typical of *Turbiosphaera*. The paracingulum is laevorotatory, offset by 10μ.

Parasulcus: The surface ornamentation within the parasulcus appears less fibrous than the rest of the cyst. Projections consist of low fibrous ridges and short fine processes. A break in the cingulum occurs at the anterior margin of the parasulcus.

Archeopyle: The archeopyle is formed by the complete removal of the 3" precingular paraplate. Type P (3")

Dimensions. Observed range (three specimens): Central body length - 75 to 81μ (mean 78μ), central body width - 53 to 66μ (mean 61μ); total length - 130 to 147μ (mean 139μ), total width - 83 to 118μ (mean 97μ).

Comments/comparison. The epicystal processes and hypocystal processes on specimens of *Turbiosphaera sagena* are joined distally by a fine network of fibrous strands. The paracingular processes are often hollow, formed by fine penitabular septa that are joined distally. The fibrous ectophragm distinguishes *Turbiosphaera sagena* from *T. filosa*. Species of *Araneosphaera* have processes that are joined distally, however, these processes are restricted to the hypocyst.

Stratigraphic Range. *Turbiosphaera sagena* is reported from the lower upper Eocene of DSDP Hole 281 by Crouch and Hollis (1996) (as *Turbiosphaera.* sp. a).

Family AREOLIGERACEAE Evitt, 1963

Genus *Glaphyrocysta* Stover & Evitt, 1978

Glaphyrocysta radiata n. sp.
Plate 5, figs. e-h

Derivation of name. Latin, *radiatus*, in reference to the radial nature of the solid precingular intratabular processes.

Holotype. Plate 5, Figure e. UNSM PB99-01: Sample MTD 153(1), slide 2, middle to upper Eocene erratic, McMurdo Sound, Antarctica.

Description.

Shape: A skolochordate dinoflagellate cyst with a subspherical central body. Solid intratabular processes and process complexes are joined distally by a fenestrate ectophragm. Processes are not present in the paracingular region.

Phragma: The autophragm is shagreenate to granular. Solid, fibrous, intratabular processes extend from the autophragm and are joined distally by a fenestrate ectophragm which surrounds the lateral margin of the cyst but is absent from both the mid ventral and mid dorsal regions

Paratabulation: The precingular paraplates are clearly indicated by 6 intratabular processes. Hypocystal plate arrangement is difficult to determine due to the occurrence of several accessory processes and occasional process complexes.

Paracingulum: Indicated by the absence of processes in the paracingular region of the cyst.

Parasulcus: Not clearly indicated.

Archeopyle: Forms by complete detachment of the apical plates.

Dimensions. Observed range (four specimens): Central body length - 51 to 53μ (mean 52μ), central body width - 46 to 62μ (mean 55μ); total length - 88 to 104μ (mean 95μ), total width - 83 to 117μ (mean 100μ).

Comments. The presence of intratabular pre-cingular processes perhaps indicates an affinity to *Eatonicysta* Stover and Evitt (1978). However, where *Eatonicysta* is characterized by a complete ectophragm and the presence of paracingular processes, the ventral and dorsal regions of the specimens from the McMurdo Erratics possess neither processes nor an ectophragm. Furthermore the form described here has on offset sulcal notch. In this respect the McMurdo Erratic material fits within the generic description for *Glaphyrocysta*. *G. radiata* differs from *G. semitecta* (Bujak in Bujak et al., 1980) by having simple solid processes projecting from each of the precingular paraplates and by an absence of a fenestrate ectophram in both the mid-dorsal and mid-ventral region.

Stratigraphic Occurrence. Middle to upper Eocene erratics, McMurdo Sound, Antarctica.

Family CLADOPYXIACEAE Stein, 1883

Genus *Microdinium* Cookson & Eisenack, 1960; emend. Sarjeant, 1966; emend. Stover & Evitt, 1978.

Microdinium sp.
Plate 7, fig. e

Comments. Specimens of this small taxon were rarely encountered.

Dimensions. Observed range (two specimens): Length - 23 to 30μ, width - 21 to 23μ.

Family uncertain

Genus *Impletosphaeridium* Morgenroth, 1966

***Impletosphaeridium clavus* Wrenn & Hart, 1988**
Plate 7, figs. a and b

Impletosphaeridium clavus Wrenn & Hart, 1988, p. 356-357, Figure 27.10-11, 13.

Stratigraphic Range. On Seymour Island, Antarctic Peninsula, the reported range for the species is from the lower upper Paleocene to Eocene (Wrenn and Hart, 1988; Cocozza and Clark, 1992).

Impletosphaeridium sp.
Plate 7, figs. c and d

Comments. This rare species possesses solid spines with distal process ends that bifurcate or trifurcate. This taxon has fewer processes than *Impletosphaeridium clavus*.

Dimensions. One specimen: central body diameter - 15μ; total diameter - 23μ.

Genus *Paucisphaeridium* Bujak et al., 1980

***Paucisphaeridium inversibuccinum* (Davey & Williams, 1966b) Bujak et al., 1980; emend. Bujak et al., 1980**
Plate 7, figs. k-n
?Litosphaeridium inversibuccinum Davey & Williams, 1966b, p. 82, Plate 12, Figure 3.
Paucisphaeridium inversibuccinum Bujak et al., 1980, p. 32, Plate 2, Figures 4-5.

Stratigraphic Range. Previous reported occurrences of *P. inversibuccinum* for the southern ocean include the upper lower Eocene to upper Eocene of Seymour Island (Cocozza and Clarke, 1992) and the middle Eocene of the Bruce Bank, Scotia Sea (Mao and Mohr, 1995). Williams and Bujak (1985) report a global distribution for *P. inversibuccinum* from the lower Eocene (NP 10) to lower upper Eocene (NP 18).

Dinocyst sp. A
Plate 11, figs. k-n

Description.

Shape: Murochordate dinoflagellate cyst. Central body is spherical. Surface ornamentation consists of thin septa with an average height of 0.2 x the diameter of the central body.

Phragma: A thin, smooth endophragm and fibrous, reticulate periphragm. Septa appear to be attached to endophragm at points where septa intercept.

Paratabulation: Gonyaulacoid paratabulation is indicated by apical paraplate series consisting of four paraplates. The paratabulation pattern for other paraplate series can not be determined.

Paracingulum: Not apparent.

Parasulcus: Not apparent.

Archeopyle: The archeopyle is formed by the complete removal of four apical paraplates.

Dimensions. Observed range (five specimens): Overall cyst diameter - 28 to 36μ (mean 31μ); central body diameter - 19 to 25μ (mean 22μ); septa height - 4 to 5μ.

Comments. This small murochordate species is characterized by a fibrous, reticulate periphragm that forms a network of intersecting septa. An apical archeopyle is usually apparent. The species resembles taxa attributed to *Valensiella* Eisenack (1963); emend. Courtinat (1989). However, species attributed to *Valensiella* possess an ectophragm. Specimens from the McMurdo Erratics differ from species of *Ellipsoidictyum* Klement (1960), in that a paracingulum is not obvious in the former. The McMurdo Erratic specimens are similar to *Labyrinthodinium truncatum* Piasecki (1980), however, *L. truncatum* does not have septa.

Informal group ACRITARCHA Evitt, 1963
Genus *Micrhystridium* Deflandre, 1937; emend. Downie & Sarjeant, 1963

Micrhystridium sp. 1
Plate 13, fig. a

Synopsis. A small subspherical to elliptical acritarch with short conical solid spines.

Dimensions. Observed range (four specimens): Central body diameter - 18 to 22μ (mean 21μ), process length - 5 to 6μ.

Micrhystridium sp. 2
Plate 13, figs. b-d

Comments. A small subspherical acritarch with long, slender accuminate solid spines. This species differs from *Micrhystridium* sp. 1 in that it is generally smaller and possesses fewer processes.

Dimensions. Observed range (three specimens): Central body diameter - 11 to 15μ (mean 14μ), process length - 5 to 10μ.

Genus *Cyclopsiella* Drugg & Loeblich, 1967

Cyclopsiella trematophora (Cookson & Eisenack, 1967) Lentin & Williams, 1977
Plate 12, figs. a and b

Leiosphaeridia trematophora Cookson & Eisenack, 1967, p. 136, Plate 19, Figure 13.
Cyclopsiella trematophora (Cookson & Eisenack) Lentin & Williams, 1977, p.39.

Stratigraphic Range. Wrenn and Hart (1988) report a worldwide range of Paleocene to Eocene

Cyclopsiella sp. 1
Plate 12, fig. c

Comments. This species is characterized by its relatively large size, broad flange, and pylome in an apical position.

Description.

Shape: An elliptical cyst with wide flange that surrounds the lateral margin. The cyst is compressed dorso-ventrally.

Phragma: Consists of a generally smooth, thin autophragm. Ornamentation may be present on the dorsal surface. If present, surface ornamentation consists of short, dispersed spines. The lateral margin of the cyst is surrounded by a wide (8-9μ) fibrous flange.

Paratabulation: None indicated.

Paracingulum: None indicated.

Parasulcus None indicated.

Archeopyle/Pylome: The pylome consists of a circular to elliptical opening that occurs at the apex of the cyst. The pylome opens to the apical margin.

Dimensions. Observed range (eight specimens): length (not including flange) - 79 to 101μ (mean 87μ), width (not including flange) - 68 to 77μ (mean 74μ); flange width - 8 to 9μ; wall thickness $\leq$ 1μ; pylome diameter: 10 to 14μ.

Cyclopsiella sp. 2
Plate 12, fig. d

Synopsis. A species of *Cyclopsiella* with a smooth autophragm and a reticulate ?ectophragm that partially surrounds the entire cyst.

Description.

Shape: Elliptical to subelliptical cysts that are compressed along the dorso-ventral axis.

Phragma: Consists of a thin-walled autophragm and reticulate ectophragm. The ectophragm is not continuous around the complete cyst and is particularly obvious near the lateral margins.

Paratabulation: Not indicated.

Paracingulum: Lateral indentations in the adcingular region may or may not be present.

Parasulcus: May be indicated by a longitudinal break in the ectophragm.

Pylome: A spherical opening (diameter - 4μ) posterior of the apex. The pylome is often surrounded by a thickening of the wall.

Dimensions. Observed range (two specimens): Length - 53 to 66μ, width - 46 to 52μ.

Cyclopsiella sp. 3
Plate 12, figs. e and f

Comments. This species of *Cyclopsiella* is distinguished by its thick spongy autophragm.

Description.

Shape: Elliptical to subelliptical cysts with thick, spongy autophragm. Dorso-ventrally compressed.

Phragma: Consists of a thick-walled autophragm. The surface is ornamented with pits which are circular to subcircular (diameter: 1.5μ) and/or spines.

Paratabulation: Not indicated.

Paracingulum: Lateral indentations in the adcingular region may or may not be present.

Parasulcus: Not indicated.

Pylome: A spherical opening (diameter - 6 to 9μ) that occurs 2 to 10μ posterior of the apex. The pylome is often surrounded by a thickening of the wall.

Dimensions. Observed range (nine specimens): Length (not including spongy layer) - 39 to 62μ (mean 44μ, width (not including spongy layer) - 36 to 48μ (mean 42μ); thickness of spongy layer: 2 to 9μ.

Cyclopsiella sp. 4
Plate 12, fig. g

Comments. This rare species is characterized by a striated, broad lateral flange. This species is similar to

Ascostomocystis sp. I of Manum (1976, Plate 6, figure 5).

Description.

Shape: Elliptical to subeliptical cyst with a broad, striated flange.

Phragma: Consist of a thin autophragm. Ventral surface is smooth. Dorsal surface is usually smooth or may have small dispersed spines. Broad flange (4 to 7μ wide) possesses striae that extend normal to the cyst wall.

Paratabulation: None indicated.

Paracingulum: Lateral indentations in the adcingular region may or may not be present.

Parasulcus: Not indicated.

Archeopyle/pylome: Located immediately posterior of the apex. Consists of a circular to elliptical opening (diameter: 9 to 10μ).

Dimensions. Observed range (three species): Length (not including flange) - 46 to 61μ (mean 52μ), width (not including flange) - 37 to 54μ (mean 45μ); flange: 4 to 6μ.

?Cyclopsiella sp. 5
Plate 12, figs. h and i

Description.

Shape: An elliptical acritarch with a thick (1.5 to 2μ) lateral margin and narrow, spongy flange.

Phragma: The thin autophragm is smooth on the ventral surface and has a verrucate dorsal surface. The lateral margin of the acritarch typically has a narrow spongy flange, although this is often not continuous around the entire margin.

Paratabulation: Not indicated.

Paracingulum:. An equatorial fold occurs in the mid dorso-ventral region of the acritarch. This fold may be indicative of the paracingulum. If so, then this taxon would be re-attributed to a dinoflagellate genus.

Parasulcus: Not indicated.

Archeopyle/pylome: A circular to elliptical opening occurs in the central region of the acritarch anterior of the mid dorso-ventral fold. The distance from the apex of the acritarch to the center of the pylome ranges from 13 to 19μ.

Dimensions. Observed range (four specimens): Length - 42 to 49μ (mean 45μ), width - 37 to 48μ (mean 40μ).

Comments. This form is distinguished by the presence of a fold in the middle region of the acritarch (paracingulum?) and a verrucate dorsal surface. We consider that this species may be a dinoflagellate cyst due to the presence of a ?paracingular fold, therefore, we questionably assign this taxon to *Cyclopsiella*.

Genus *Dichotisphaera* Turner, 1984

Dichotisphaera sp.
Plate 12, figs. j and k

Comments. This species is similar to *Leiosphaeridia* sp. reported by Kemp (1975, Plate 6, figures 1-3).

Genus *Paralecaniella* Cookson & Eisenack, 1970; emend. Elsik, 1977

Paralecaniella indentata (Deflandrea & Cookson, 1955) Cookson & Eisenack, 1970; emend. Elsik, 1977
Plate 13, figs. e-g

Epicephalopyxis indentata Deflandre & Cookson, 1955, p. 292, text-fig. 56, Plate 9, Figures 5-7.
Paralecaniella indentata (Deflandre & Cookson) Cookson & Eisenack, 1970, p. 323.
Paralecaniella indentata (Deflandre & Cookson) Cookson & Eisenack, 1970, emend. Elsik, 1977, p. 96, Plate 1, Figures 1-15, Plate 2, Figures 1-11.

Stratigraphic Range: Global distribution during the upper Cretaceous and Tertiary.

Genus *Scuticabolus* Loeblich, 1967

Scuticabolus lapidaris (O. Wetzel, 1933a) Loeblich, 1967
Plate 13, figs. h and i
Ophiobolus lapidaris O. Wetzel, 1933a, p. 176-179, text-fig. 5-7, Plate 2, Figures 30-34.
Scuticabolis lapidaris (O. Wetzel) Loeblich, 1967, p. 68.

Stratigraphic Range: Reported occurrences in the Northern Hemisphere range from Santonian to Danian (Wrenn and Hart 1988 and references therein). Wrenn and Hart (1988) report *S. lapidaris* (as *O. lapidaris*) from the lower upper Paleocene and indicate that it is reworked into Eocene deposits. The occurrence of *S. lapidaris* in middle to upper Eocene erratics indicates that the total range for *S. lapidaris* in the Southern Ocean may extend into the middle to upper Eocene and that the Seymour Island specimens were likely recovered in place.

Genus *Veryhachium* Denuff, 1954; emend. Downie & Sarjeant, 1963; emend. Turner, 1984

Veryhachium sp. 1
Plate 13, figs. j and k

Veryhachium sp. indet., Wrenn and Hart, 1988, p. 373.

Comments. Wrenn and Hart (1988) reported (as *Veryhachium*. sp. indet.) triangular forms with one side of the triangular body being shorter than the others. Rare specimens encountered in the erratics fit this description and are possibly conspecific.
Stratigraphic Range: The specimens on Seymour Island occur in the upper lower Eocene.

Veryhachium sp. 2
Plate 13, figs. l and m

Comments. This species possesses a polygonal central body with four processes.

Acritarch sp. A
Plate 13, figs. n-r

Comments. This small spherical acritarch is similar in general appearance to specimens of the dinocyst *Cerebrocysta bartonensis* Bujak (1980) recovered from Seymour Island (Wrenn and Hart, 1988, figures 17.6 and 17.8). However, the McMurdo Erratic specimens are smaller than the Seymour Island material. Furthermore, the opening in the cysts are usually circular and do not resemble the precingular archeopyle characteristic of *C. bartonensis*.
Description.
Shape: A small, subspherical to spherical acritarch with a circular pylome.
Phragma: The cyst consists of a thick walled (2μ) autophragm. The surface of the autophragm is foeveolate with sub circular to polygonal pits. Pits range from <1 to 3μ in diameter.
Paratabulation: None indicated.
Paracingulum: None indicated.
Parasulcus: None indicated.
Pylome: A circular pylome occurs in the ?apical region of the acritarch.
Dimensions. Observed range (four specimens): diameter - 18 to 20μ.

Division PRASINOPHYTA Round, 1971
Order PTEROSPERMATALES Schiller, 1925
Family CYMATIOSPHAERACEAE Mädler, 1963

Genus *Cymatiosphaera* (Wetzel 1933b) Deflandre, 1954

Cymatiosphaera sp.
Plate 14, figs. a-c

Comments. Rare specimens were encountered in one erratic (MB 188B) only.
Dimensions. Observed range (two specimens): Diameter - 16 to 25μ, height of septa - 2-3μ.

Family PTEROSPERMELLACEAE Eisenack, 1972

Genus *Pterospermella* Eisenack, 1972

Pterospermella sp. 1
Plate 14, figs. d and e

Comments. This rare species has a smooth spherical to elliptical central body that is surrounded by a fibrous disc-shaped 'outer body'. The central body appears to have three or four ?processes protruding from one of the polar regions.
Dimensions. Observed range (two specimens): Central body diameter - 13 to 18μ, 'outer body' diameter - 25 to 31μ.

Pterospermella sp. 2
Plate 14, figs. f-h

Comments. This rare species consists of a smooth spherical central body that is encircled with a thin-walled flange or 'outer body'.
Dimensions. Observed range (two specimens): Central body diameter - 28 to 46μ, 'outer body' diameter - 48 to 50μ.

Division CHLOROPHYTA Pascher, 1914
Class CHLOROPHYCEAE Kützing, 1843
Order CHLOROCOCCALES Marchand, 1895 orth. mut. Pascher, 1915
Family CHLOROCOCCACEAE Blackman & Tansley, 1902

Genus *Palambages* O. Wetzel, 1961

Palambages sp.
Plate 14, figs. i-k

Acknowledgements. RHL thanks Dr. Rosemary Askin and Dr. Graeme Wilson for comments and advice on dinoflagellate cyst taxonomy. Thanks also go to Dr. John Wrenn for providing type material from Seymour Island and key literature from the library at CENEX, Louisiana State University. Thanks to Dr. Margret Bolick who provided technical advice and laboratory space and John Kaser who helped with sample processing. Thanks also to Dr. Graeme Wilson and Dr. Graham Williams for their reviews of the manuscript. This work was supported in part by grants from the National Science Foundation Office of Polar Programs to D.M. Harwood (OPP-9317901 and OPP-9158075), grants from the Geological Society of America to R.H. Levy and generous donations from the geology Alumni of the University of Nebraska.

REFERENCES

Archangelsky, S.
1969a   Sobre el paleomicroplancton del Terciario inferior de Rio Turbio, Provincia Santa Cruz, *Ameghiniana, v. 5(10):* 406-416.
1969b   Estudio del paleomicroplancton de la Formacion Rio Turbio (Eoceno), Provincia de Santa Cruz, *Ameghiniana, v. 6(3):* 181-218.
Archangelsky, S. and A. Fasola
1971   Algunos elementos del paleomicroplancton del Terciario inferior de Patagonia (Argentina y Chile), *Revista del Museo de la Plata, v. 6*: 1-18.
Artzner, D.G. and G. Dörhöfer.
1978   Taxonomic note: *Lejeunecysta nom. nov. pro Lejeunia* Gerlach 1961 emend. Lentin and Williams 1976 - dinoflagellate cyst genus. *Canadian Journal of Botany, v. 56*: 1381-1382.
Askin, R.A.
1988   Campanian to Paleocene palynological succession of Seymour and adjacent islands, northeastern Antarctic Peninsula. In R.M. Feldmann and M.O. Woodburne (eds.), *Geology and Paleontology of Seymour Island, Antarctic Peninsula. Geological Society of America Memoir 169*: 131-153.
1997   Eocene - ?Earliest Oligocene terrestrial palynology of Seymour Island, Antarctica. In Ricci, C.A., (ed.), *The Antarctic Region: Geological Evolution and Processes.* Sienna, Italy: 993-996.
(this vol.)   Terrestrial palynomorphs. In J.D. Stilwell and R.M. Feldmann (eds.), *Paleobiology and Paleoenvironments of Eocene Rocks, McMurdo Sound, East Antarctica.* American Geophysical Union, Antarctic Research Series.
Barrett, P.J., (ed.).
1989   Antarctic Cenozoic history from the CIROS-1 drillhole, McMurdo Sound. *DSIR Bulletin* 245. 254p.
Barron, J.A., J.G. Baldauf, E. Barrera, J-P., Caulet, B.T. Huber, B.H., Keating, D. Lazarus, H. Sakai, H.R. Thierstein, and W. Wei.

1989   Biochronologic and magnetochronologic syn-
           thesis of ODP Leg 119 sediments from the
           Kerguelen Plateau and Prydz Bay, Antarctica. In
           J.A. Barron, B.L. Larsen et al., *Proceedings of
           the Ocean Drilling Program, Scientific Results,
           v. 119:* 813-847.

Barron, J.A., B. Larsen, and J.G. Bauldauf
989   Evidence for Late Eocene to Early Oligocene
           Antarctic glaciation and observations on Late
           Neogene glacial history of Antarctica; results
           from Leg 119. In J.A. Barron, B.L. Larsen et al.,
           *Proceedings of the Ocean Drilling Program,
           Scientific Results, v. 119:* 869-891.

Benedek, P.N.
1972   Phytoplankton aus dem Mittel und Oberoligozän
           von Tönisberg (Niederrheingebiet). *Palaeonto-
           graphica, Abt. B, v. 137:* 1-71.

Benedek, P.N. and H. Gocht.
1981   *Thalassiphora pelagica* (Dinoflagellata,
           Tertiär):   electronenmikroskopische
           Untersuchung und Gedanken zur Paläobiologie.
           Palaeontographica, Abt. B, v. 180: 39-64.

Benedek, P.N. and W.A.S. Sarjeant
1981   Dinoflagellate cysts from the Middle and Upper
           Oligocene of Tönisberg (Niederrheingebiet): a
           morphological and taxonomic restudy. *Nova
           Hedwigia, v. 35:* 313-356.

Berggren, W.A., D.V. Kent, C.C. Swisher, III, and M-P. Aubry.
1995   A revised Cenozoic geochronology and
           chronostratigraphy. In W.A. Berggren and J.
           Hardenbol, (eds.), *Geochronology, Time Scales
           and Global Stratigraphic Correlation. SEPM
           Special Publication No. 54.* p. 129-212.

Bohaty, S., and D.M. Harwood.
(this vol.) Ebridians and silicoflagellates from McMurdo
           Sound glacial erratics and the southern
           Kerguelen Plateau. In J.D. Stilwell and R.M.
           Feldmann (eds.), *Paleobiology and
           Paleoenvironments of Eocene Rocks, McMurdo
           Sound, East Antarctica.* American Geophysical
           Union, Antarctic Research Series.

Brenner, W.
1988   Dinoflagellaten aus dem Unteren Malm (Oberer
           Jura) von Süddeutschland: Morphologie,
           Ökologie,   Stratigraphie. *Tübinger
           Mikropaläontologische Mitteilungen, no. 6:*
           1-116.

Bujak, J.P.
1980   Dinoflagellate cysts and acritarchs from the
           Eocene Barton Beds of southern England. In
           Bujak, J.P., Downie C., Eaton, G.L., and
           Williams, G.L. 1980. Dinoflagellate Cysts and
           Acritarchs from the Eocene of southern
           England. *The Palaeontological Society, Special
           Papers in Palaeontology no. 24:* 36-91.

Bujak, J.P., C. Downie, G.L. Eaton, G.L., and Williams
1980   Dinoflagellate cysts and acritarchs from the
           Eocene of southern England. *The
           Palaeontological Society, Special Papers in
           Palaeontology no. 24,* 100pp.

Bujak, J.P. and E.H. Davies.
1983   Modern and fossil Peridiniineae. *American
           Association of Stratigraphic Palynologists,
           Contributions Series, v. 13:* 1-203.

Cocozza, C.D., and C.M. Clarke.
1992   Eocene microplankton from La Meseta
           Formation, northern Seymour Island. In A.M.
           Duane, D. Pirrie and J.B. Riding (eds.),
           *Palynology of the James Ross Island Area,
           Antarctic Peninsula. Antarctic Science, v. 4(3):*
           355-362.

Cookson, I.C., and L.M. Cranwell.
1967   Lower Tertiary microplankton, spores and
           pollen grains from southernmost Chile.
           Micropaleontology, v. 13: 204-216.

Cookson, I.C. and A. Eisenack.
1960   Microplankton from Australian Cretaceous sed-
           iments. *Micropaleontology, v. 6(1):* 1-18.
1962   Additional microplankton from Australian
           Cretaceous sediments. *Micropaleontology, v.
           8(4):* 485-507.
1965a   Microplankton from the Browns Creek Clays,
           SW Victoria. *Proceedings of the Royal Society
           of Victoria, v. 79:* 119-131.
1965b   Microplankton from the Dartmoor Formation
           SW Victoria. *Proceedings of the Royal Society
           of Victoria, v. 79:* 133-137.
1967   Some Early Tertiary microplankton and pollen
           grains from a deposit near Strahan, western
           Tasmania. *Proceedings of the Royal Society of
           Victoria, v. 80:* 131-140.
1970   Die Familie der Lecaniellaceae n. fam. - fossile
           Chlorophyta, Volvocales? *Neues Jahrbuch für
           Geologie und Paläontologie, Monatshefte:*
           321-325.

Courtinat, B.
1989   Les organoclasts des formations lithologiques
           du Malm dans le Jura méridional: Systematique,
           Biostratigraphie et Elements d'Interpretation
           Paleoecologique. *Documents des Laboratoires
           de Géologie de la Faculté des Sciences de Lyon,
           no. 105:* 1-361.

Cranwell, L.M.
1964   Hystrichospheres as an Aid to Antarctic dating
           with special reference to the recovery of
           Cordosphaeridium in erratics at McMurdo
           Sound. *Grana Palynologica, v. 5(3):* 397-405.
1969   Antarctic and circum-Antarctic palynological
           contributions. *Antarctic Journal of the United
           States, v. 4(4):* 197-198.

Cranwell, L.M., H.J. Harrington, and I.G. Speden
1960   Lower Tertiary microfossils from McMurdo Sound, Antarctica. *Nature, v. 184(4701):* 1782-1785.

Crouch, E.M. and C.J. Hollis
1996   Paleogene palynomorph and radiolarian biostratigraphy of DSDP Leg 29, Sites 280 and 281 South Tasman Rise. Institute of Geological and Nuclear Sciences Science Report 96/19, 46pp.

Damassa, S. P., G.L. Williams, H. Brinkhuis, H., J.P. Bujak, and A.J. Powell
1994   Short Course in Paleogene Dinoflagellate Cysts, Utrecht, June, 1994. Laboratory of Palaeobotany and Palynology, Netherlands Research School of Sedimentary Geology, Institute for Paleoenvironment and Paleoclimate, Utrecht, 391pp.

Damassa, S.P., and G.L. Williams.
1996   Species diversity patterns in North Atlantic Eocene-Oligocene dinoflagellates. In A. Moguilevsky and R. Whatley (eds.), *Microfossils and Oceanic Environments,* p. 187-203. University of Wales, Aberystwyth - Press.

Davey, R.J.
1969   Non-Calcareous microplankton from the Cenomanian of England, Northern France and North America, Part I. Bulletin of the British Museum (Natural History) *Geology, v. 17:* 103-180.

Davey, R.J. and J.-P. Verdier.
1971   An Investigation of microplankton assemblages from the Albian of the Paris Basin. *Verhandelingen der Koninklijke Nederlandsche Akademie van Wetenschappen, Afdeling Natuurkunde, Eerste Reeks, v. 26:* 1-58.

Davey, R.J. and G.L. Williams.
1966a   The genera *Hystrichosphaera* and *Acomosphaera.* In Davey, R.J., Downie, C., Sarjeant, W.A.S., and Williams, G.L., Studies on Mesozoic and Cainozoic Dinoflagellate Cysts. *Bulletin of the British Museum (Natural History) Geology, Supplement 3:* 28-52.

1966b   The genus *Hystrichosphaeridium* and its allies. In Davey, R.J., Downie, C., Sarjeant, W.A.S., and Williams, G.L., Studies on Mesozoic and Cainozoic Dinoflagellate Cysts. Bulletin of the British Museum (Natural History) *Geology, Supplement 3:* 53-105.

Deflandre, G.
1937   Microfossiles des silex crétacés. Deuxième partie. Flagellés *incertae sedis* Hystrichosphaeridés. Sarcodinés. Organismes Divers. *Annales de paléontologie, v. 26:* 51-103.

1954   Systématique des Hystrichosphaeridés: sur l'Acception du Genre *Cymatiosphaera* O.

Wetzel. *Comptes rendus de la Société géologique de France, v. 12:* 257-258.

Deflandre, G. and I.C. Cookson.
1955   Fossil microplankton from Australian Late Mesozoic and Tertiary sediments. *Australian Journal of Marine and Freshwater Research, v. 6(2):* 242-313.

Denuff, J.
1954   *Veryhachium,* genre nouveau d'Hystrichosphères du Primaire. *Compte rendu sommaire des séances de la Societe géologique de France, no. 13:* 305-306.

Downie, C. and W.A.S. Sarjeant.
1963   On the interpretation and status of some hystrichosphere genera. *Palaeontology, v. 6(1):* 83-96.

Drugg, W.S. and A.R. Loeblich, Jr.
1967   Some Eocene and Oligocene phytoplankton from the Gulf Coast, U.S.A. *Tulane Studies in Geology, v. 5:* 181-194.

Eaton, G.L.
1971   A morphogenetic series of dinoflagellate cysts from the Bracklesham Beds of the Isle of Wight, Hampshire, England. In Farinacci, Λ. (ed.), *Second Planktonic Conference, Rome, 1970, Proceedings.* Rome: Edizioni Tecnoscienza: 355-379.

1976   Dinoflagellate cysts from the Bracklesham Beds (Eocene) of the Isle of Wight, southern England. *Bulletin of the British Museum (Natural History) Geology, v. 26(6):* 332p + 21plates.

Edwards, A.R. and K. Perch-Nielsen.
1975   Calcareous nannofossils from the southern Southwest Pacific, Deep Sea Drilling Project, Leg 29. In J.P. Kennett, R.E. Houtz, et al., *Initial Reports of the Deep Sea Drilling Project, v. 29:* 469-539.

Edwards, L.E. and J.W. Bebout.
1981   Emendation of *Phthanoperidinium* Drugg and Loeblich 1967, and a description of *P. brooksii* sp. nov. from the Eocene of the Mid-Atlantic outer continental shelf. *Palynology, v. 5:* 29-41.

Ehrenberg, C.G.
1838   Über das Massenverhältniss der jetzt lebenden Kiesel-Infusorien und über ein neues Infusorien-Conglomerat als Polirschiefer von Jastraba in Ungarn. Abhandlungen der Preussischen Akademie der Wissenschaften, 1836, p. 109-135.

Eisenack, A.
1938   Die Phosphoritknollen der Bernsteinformation als Überlieferer tertiären Planktons. Schriften der Physikalisch - ökonomischen *Gesellschaft zu Königsberg, v. 70:* 181-188.

1954   Mikrofossilien aus Phosphoriten des samländischen Unteroligozäns und über die Einheitlichkeit der Hystrichosphaerideen. *Palaeontographica, Abt. A, v. 105:* 49-95.

1963   Zur Membranilarnax-Frage. Neues Jahrbuch für Geologie und Paläontologie, *Monatshefte, v. 5:* 98-103.

1972   Kritische Bemerkung zur Gattung *Pterospermopsis* (Chlorophyta, Prasinophyceae). Neues Jahrbuch für Geologie und Paläontologie, *Monatshefte, v. 10:* 596-601.

Eisenack, A. and H. Gocht.
1960   Neue Namen für einige Hystrichosphären der Bernsteinformation Ostpreussens. Neues Jahrbuch für Geologie und Paläontologie, *Monatshefte, v. 11:* 511-518.

Elsik, W.C.
1977   *Paralecaniella indentata* (Defl. & Cooks. 1955) Cookson & Eisenack 1970 and allied dinocysts. *Palynology, v. 1:* 95-102.

Evitt, W.R.
1961   Observations on the morphology of fossil dinofla-gellates. *Micropaleontology, v. 7:* 511-518.

1963   A discussion and proposals concerning fossil dinoflagellates, hystrichospheres, and acritarchs, II. *National Academy of Sciences, Washington, Proceedings, v. 49:* 298-302.

Evitt, W.R., and S.T. Pierce.
1975   Early Tertiary ages for the coastal belt of the Franciscan Complex, northern California. *Geology, v. 3(8):* 433-436.

Feldmann, R.M., and W.J. Zinsmeister.
1984   First occurrence of fossil decapod crustaceans (Callianassidae) from the McMurdo Sound region, Antarctica. *Journal of Paleontology, v. 58(4):* 1041-1045.

Fensome, R.A., F.J.R. Taylor, G. Norris, W.A.S. Sarjeant, D.I. Wharton, and G.L. Williams
1993   A Classification of Fossil and Living Dinoflagellates. *Micropaleontology, Special Publication, v. 7:* 1-351.

Gerlach, E.
1961   Mikrofossilien aus dem Oligozän und Miozän Nordwestdeutschlands, unter besonderer Berücksichtigung der Hystrichosphaeren und Dinoflagellaten. Neues Jahrbuch für Geologie und Paläontologie, *Abhandlungen, v. 112(2):* 143-228.

Gocht, H.
1968   Zur Morphologie und Ontogenie von Thalassiphora (Dinoflagellata*), Palaeontographica, Abt. A, v. 129:* 149-156.

Goodman, D.K. and L.N. Ford, Jr.
1983   Preliminary dinoflagellate biostratigraphy for the middle Eocene to lower Oligocene from the southwest Atlantic Ocean. In W.J. Ludwig, V.A. Krasheninnikov et al., *Initial Reports of the Deep Sea Drilling Project, Leg 71:* 859-877.

Habib, D.
1976   Neocomian dinoflagellate zonation in the western North Atlantic. *Micropaleontology, v. 21:* 373-392.

Hall, S.A.
1977   Cretaceous and Tertiary dinoflagellates from Seymour Island, Antarctica. *Nature, v. 267:* 239-241.

Hambrey, M.J., P.J. Barrett, and P.H. Robinson
1989   Stratigraphy. In Barrett, P.J. (ed.), Antarctic Cenozoic History from the CIROS-1 Drillhole, McMurdo Sound. *DSIR Bulletin 245:* 23-48.

Hambrey, M.J. and P.J. Barrett
1993   Cenozoic sedimentary and climatic record, Ross Sea region, Antarctica. In J.P. Kennett and D.A. Warnke (eds.), *The Antarctic Paleoenvironment: A Perspective on Global Change, Part Two.* American Geophysical Union, Antarctic Research Series, v. 60: 91-124.

Hannah, M., M. Bianca Cita, R. Coccioni, and S. Monechi
(in press)   The Eocene / Oligocene boundary at 70 degrees south, McMurdo Sound Antarctica. *Terra Antarctica.*

Harrington, H.J.
1969   Fossiliferous rocks in moraines at Minna Bluff, McMurdo Sound. *Antarctic Journal of the United States, v. 4(4):* 134-135.

Harwood, D.M., and S. Bohaty
(this vol.)   Marine diatom assemblages from Eocene erratics, McMurdo Sound, Antarctica. In J.D. Stilwell and R.M. Feldmann (eds.), *Paleobiology and Paleoenvironments of Eocene Rocks, McMurdo Sound, East Antarctica.* American Geophysical Union, Antarctic Research Series.

Haskell, T.R. and G.J. Wilson.
1975   Palynology of sites 280-284, DSDP Leg 29, off southeastern Australia and western New Zealand. In J.P. Kennett, R.E. Houtz, et al., *Initial Reports of the Deep Sea Drilling Project, v. 29:* 723-741.

Head, M.J. and G. Norris.
1989   Palynology and dinocyst stratigraphy of the Eocene and Oligocene in ODP Leg 105, Hole 647A, Labrador Sea. In S.P. Srivastava, M. Arthur, B. Clement, et al. *Proceedings of the Ocean Drilling Program, Scientific Results, v. 105:* 515-550.

Head, M.J., G. Norris, and P.J. Mudie
1989   Palynology and dinocyst stratigraphy of the Miocene in ODP Leg 105, Hole 645E, Baffin Bay. In S.P. Srivastava, M. Arthur, B. Clement, et al. *Proceedings of the Ocean Drilling Program, Scientific Results, v. 105:* 467-514.

Helenes, J.
1984   Morphological analysis of Mesozoic-Cenozoic *Cribroperidinium* (Dinophyceae), and taxonomic implications. *Palynology, v. 8:* 107-137.

Islam, M.A.

1982   Archeophyle structure in the fossil dinoflagellate *Phthanoperidinium*, Review of Palaeobotany and Palynology, v. 36: 305-316.

Jan du Chéne, R.

1977   Etude palynologique du Miocene Supérieur Andalou (Espagne). *Revista Española de Micropaleontologia, v. 9:* 97-114.

Kemp, E.M.

1975   Palynology of Leg 28 Drill Sites, Deep Sea Drilling Project. In D.F. Hayes, L.A. Frakes, et al., *Initial Reports of the Deep Sea Drilling Project, v. 28:* 599-623.

Khowaja-Ateequzzaman, Garg, R. and K.P. Jain

1991   Some observations on dinoflagellate cyst genus *Alterbidinium* Lentin and Williams 1985. *Palaeobotanist, v. 39(1):* 37-45.

Kjellström, G.

1972   Archaeopyle formation in the genus *Lejeunia* Gerlach 1961 emend. *Geologiska Föreningens i Stockholm Förhandlingar, v. 94:* 467-469.

Klement, K.W.

1960   Dinoflagellaten und Hystrichosphaerideen aus dem unteren und mittleren Malm südwestdeutschlands. *Palaeontographica, Abt. A, v. 114:* 1-104.

Lawver, L.A., L.M. Gahagan and M.F. Coffin

1992   The development of paleoseaways around Antarctica. In J.P. Kennett and D.A. Warnke (eds.), *The Antarctic Paleoenvironment: A Perspective on Global Change, Part One.* American Geophysical Union, Antarctic Research Series, v. 56: 7-30.

Lentin, J.K. and G.L. Williams.

1973   Fossil dinoflagellates: index to genera and species. Geological Survey of Canada, Paper no. 73-42: 1-176.

1976   A monograph of fossil peridinioid dinoflagellate cysts. Bedford Institute Oceanography Report BI-R-75-16, 1-237.

1977   Fossil dinoflagellates: index to genera and species, 1977 edition. Bedford Institute of Oceanography Report Series BI-R-77-8, 209p.

1981   Fossil dinoflagellates: index to genera and species, 1981 edition. Bedford Institute of Oceanography Report Series BI-R-81-12, 345p.

1985   Fossil dinoflagellates: index to genera and species, 1985 edition. Canadian Technical Report of Hydrography and Ocean Sciences No. 60, 451p.

1987   Status of the fossil dinoflagellate genera *Ceratiopsis* Vozzhennikovia 1963 and *Cerodinium* Vozzhennikovia 1963 Emend. *Palynology, 11:* 113-116.

1993   Fossil dinoflagellates: index to genera and species, 1993 edition. American Association of Stratigraphic Palynologists, Contributions Series No. 28, 856p.

Levy, R.H., and D.M. Harwood.

(this vol.)   Sedimentary lithofacies and inferred depositional environments of the McMurdo Sound erratics. In J.D. Stilwell and R.M. Feldmann (eds.), *Paleobiology and Paleoenvironments of Eocene Rocks, McMurdo Sound, East Antarctica.* American Geophysical Union, Antarctic Research Series.

Lindgren, S.

1984   Acid resistant peridinioid dinoflagellates from the Maastrichtian of Trelleborg, southern Sweden. *Stockholm Contributions in Geology, v. 39(6):* 145-201.

Loeblich, A.R. III.

1967   Nomenclatural notes in the Pyrrhophyta, Xanophyta and Euglenophyta. *Taxon, v. 16:* 68-69.

Loeblich, A.R. Jr. and A.R. Loeblich, III

1966   Index to the genera, subgenera, and sections of the Pyrrhophyta. *Studies in Tropical Oceanography, Miami, no. 3,* x + 94p.

Lohmann, H.

1904   Eier und sogenannte Cysten der Plankton-Expedition. Anhang: Cyphonautes. Ergebnisse der Plankton-Expedition der Humboldt Stiftung, v. 4: 1 62.

Mantell, G.A.

1850   *A pictorial atlas of fossil remains, consisting of coloured illustrations selected from Parkinson's "Organic remains of a former world", and Artis's "Antediluvian phytology".* Henry G. Bohn, London, xii + 207p.

1854   *The medals of creation: or, first lessons in geology and the study of organic remains.* Second edition, Henry G. Bohn, London, 2 vols., 930p.

Manum, S.B.

1976   Dinocysts in Tertiary Norwegian-Greenland sea sediments (deep Sea Drilling Project Leg 38), with observations on palynomorphs and palynodebris in relation to environment. In M. Talwani, G. Udintsev, et al., *Initial Reports of the Deep Sea Drilling Project, v. 38:* 897-919.

Mao, S. and B.A.R. Mohr.

1995   Middle Eocene dinocysts from Bruce Bank (Scotia Sea, Antarctica) and their paleoenvironmental and paleogeographic implications. *Review of Palaeobotany and Palynology, v. 86:* 235-263.

Mao, S. and G. Norris.

1988   Late Cretaceous-Early Tertiary dinoflagellates and acritarchs from the Kashi Area, Tarim Basin, Xinjiang Province, China. *Royal Ontario Museum, Life Sciences Contributions 150,* 93p.

Matthews, R.K. and R.Z. Poore.
1980　Tertiary ^{18}O record and glacio-eustatic sea-level fluctuations. *Geology, v. 8*: 501-504.

McLean, D.M.
1973　Emendation and transfer of *Eisenackia* (Pyrrhophyta) from the Microdiniaceae to the Gonyaulacaceae. *Geologiska Föreningens I Stockholm Förhandlingar, v. 95*: 261-265.

McIntyre, D.J., and G.J. Wilson.
1966　Preliminary palynology of some Antarctic Tertiary erratics. *New Zealand Journal of Botany, v. 4(3)*: 315-321.

Miller, K.G.
1992　Middle Eocene to Oligocene stable isotopes, climate, and deep-water history: The Terminal Eocene Event? In D.R. Prothero and W.A. Berggren (eds.), *Eocene-Oligocene Climatic and Biotic Evolution.* Princeton University Press. p. 160-177.

Miller, K.G., R.G. Fairbanks, and G.S. Mountain
1987　Tertiary oxygen isotope synthesis, sea level history, and continental margin erosion. *Paleoceanography, v. 2(1):* 1-19.

Miller, K.G., J.D. Wright and R.G. Fairbanks
1991　Unlocking the ice house: Oligocene-Miocene oxygen isotopes, eustacy, and margin erosion. *Journal of Geophysical Research, v. 96(B4):* 6829-6848.

Mohr, B.A.R.
1990　Eocene and Oligocene sporomorphs and dinoflagellate cysts from Leg 113 drill sites, Weddell Sea, Antarctica. In P.F. Barker, J.P. Kennett, et al., *Proceedings of the Ocean Drilling Program, Scientific Results, v. 113*: 595-612.

Morgenroth, P.
1966　Mikrofossilien und Konkretionen des Nordwesteuropäischen Untereozäns. *Palaeontographica Abt. B, v. 119:* 1-53.

Neale, J.W. and W.A.S. Sarjeant.
1962　Microplankton from the Speeton Clay of Yorkshire. *Geological Magazine, v. 99:* 439-458.

Norvick, M.S.
1973　The microplankton genus *Disphaeria* Cookson and Eisenack Emend. *Bulletin of Australian Bureau of Mineral Resources, Geology and Geophysics, v. 140:* 45-46.

Okada, H., and D. Bukry.
1980　Supplementary modification and introduction of code numbers to the low-latitude coccolith biostratigraphic zonation (Bukry, 1973; 1975). *Marine Micropaleontology, v. 5(3):* 321-5.

Partridge, A.D.
1976　The geological expression of eustacy in the Early Tertiary Gippsland Basin. *APEA Journal:* 73-79

Piasecki, S.
1980　Dinoflagellate cyst stratigraphy of the Miocene Hodde and Gram Formations, Denmark. *Geological Survey of Denmark, Bulletin, v. 29:* 53-76.

Pothe de Baldis, E.D.
1966　Microplankton del Terciaro de Tierra del Fuego. *Ameghiniana, v. 4:* 219-228.

Raine, J.J., R.A. Askin, E.M. Crouch, M.J. Hannah, R.H. Levy, and J.H. Wrenn
1997　Palynomorphs. In M.J. Hannah and J.J. Raine (eds), *Southern Ocean Late Cretaceous / Early Cenozoic Biostratigraphic Datums.* Institute of Geological and Nuclear Sciences Science Report 97/4: 25-33.

Rowe, G.H.
1974　A petrographic and paleontological study of Lower Tertiary erratics from Quaternary moraines, Black Island, Antarctica. B.Sc (Hons) Thesis Lodged in the Library, Victoria University of Wellington, New Zealand. 91p.

Sarjeant, W.A.S.
1966　Dinoflagellate cysts with *Gonyaulax*-type tabulation. In Davey, R.J., Downie, C., Sarjeant, W.A.S., and Williams, G.L., Studies on Mesozoic and Cainozoic Dinoflagellate Cysts. *Bulletin of the British Museum (Natural History) Geology, Supplement 3:* 107-156.

1969　Taxonomic changes. In Davey, R.J., Downie, C., Sarjeant, W.A.S., and Williams, G.L., Appendix to "Studies on Mesozoic and Cainozoic Dinoflagellate Cysts". *Bulletin of the British Museum (Natural History) Geology, Appendix to Supplement 3:* 7-15.

1970　The genus *Spiniferites* Mantell, 1850 (Dinophyceae). *Grana Palynologica, v. 10:* 74-78.

1982　The dinoflagellate cysts of the *Gonyaulacysta* group: a morphological and taxonomic restudy. *American Association of Stratigraphic Palynologists Contributions Series, 9:* 81p.

1984　Restudy of some dinoflagellate cysts from the Oligocene and Miocene of Germany. *Journal of Micropalaeontology, v. 3(2):* 73-94.

Shackleton, N.J. and N.D. Opdyke.
1973　Oxygen isotope and paleomagnetic stratigraphy of equatorial Pacific core V28-238: oxygen isotope temperatures and ice volumes on a 10^5 year and 10^6 year scale. *Quaternary Research, v. 3*: 39-55.

Stancliffe, R.P.W. and W.A.S. Sarjeant.
1990　The Complex Chordate Dinoflagellate Cysts of the Bathonian to Oxfordian (Jurassic): Their

Taxonomy and Stratigraphic Significance. *Micropaleontology, v. 36(3)*: 197-228.

Stilwell, J.D.
(this vol.) Eocene mollusca (Bivalvia, Gastropoda and Scaphopoda) from McMurdo Sound: systematics and paleoecologic significance. In J.D. Stilwell and R.M. Feldmann (eds.), *Paleobiology and Paleoenvironments of Eocene Rocks, McMurdo Sound, East Antarctica.* American Geophysical Union, Antarctic Research Series.

Stilwell, J.D., and R.M. Feldmann (eds.).
(this vol.) *Paleobiology and Paleoenvironments of Eocene Rocks, McMurdo Sound, East Antarctica.* American Geophysical Union, Antarctic Research Series.

Stott, L.D.
1982 A re-evaluation of the age and nature of Cenozoic erratics from McMurdo Sound, Antarctica. Unpublished B.S. Thesis, The Ohio State University, 52pp.

Stott, L.D., B.C. McKelvey, D.M. Harwood, and P.-N. Webb
1983 A revision of the ages of Cenozoic erratics at Mount Discovery and Minna Bluff, McMurdo Sound. *Antarctic Journal of the United States*, 1983 Review: 36-38.

Stover, L.E.
1973 Palaeocene and Eocene species of *Deflandrea* (Dinophyceae) in Victorian coastal and offshore basins, Australia. In J.E. Glover and G. Platford (eds.), *Mesozoic and Cenozoic Palynology: Essays in Honour of Isabel Cookson.* Special Publications of the Geological Society of Australia, v.4: 167-188.

Stover, L.E. and W.R. Evitt.
1978 Analyses of pre-Pleistocene organic-walled dinoflagellates. Stanford University Publications, *Geological Sciences, v. 15*, 300p.

Stover, L.E. and G.L. Williams.
1995 A revision of the Paleogene dinoflagellate genera *Areosphaeridium* Eaton 1971 and *Eatonicysta* Stover and Evitt 1978. *Micropaleontology, v. 41(2)*: 97-141.

Truswell, E.M.
1991 Data report: palynology of sediments from Leg 119 drill sites in Prydz Bay, East Antarctica. In J. Barron, B. Larsen, et al., *Proceedings of the Ocean Drilling Program, Scientific Results, v. 119*: 941-945.

Turner, R.E.
1984 Acritarchs from the type area of the Ordovician Caradoc Series, Shropshire, England. *Palaeontographica Abt. B, v. 190*: 87-57.

Veevers, J.J., C. McA. Powell, and S.R. Roots
1991 Review of seafloor spreading around Australia. I. Synthesis of the patterns of spreading. *Australian Journal of Earth Sciences, v. 38*: 373-389.

Vozzhennikova, T.F.
1961 K voprosu o sistematike iskopayemyke Peridiney. *Akademiya Nauk SSSR (Doklady Earth Science Sections), v. 139(6)*: 1461-1462.
1963 Pirrofitovye Vodorosli (Phylum Pyrrhophyta). In Yu. A. Orlov (ed.), *Osnovy Paleontologii (Fundamantals of Paleontology), v. 14*: 179-185.

Wall, D.
1967 Fossil microplankton in deep-sea cores from the Caribbean Sea. *Palaeontology, v. 10*: 95-103.

Wall, D., B. Dale, G.P. Lohmann, and W.K. Smith.
1977 The environmental and climatic distribution of dinoflagellate cysts in modern marine sediments from regions in the North and South Atlantic Oceans and adjacent seas. *Marine Micropaleontology, v. 2*: 121-200.

Webb, P-N.
1990 The Cenozoic history of Antarctica and its global impact. *Antarctic Science, v. 2(1)*: 3-21.
1991 A review of the Cenozoic stratigraphy and paleontology of Antarctica. In M.R.A. Thomson, J.A. Crame and J.W. Thomson, (eds.), Geological Evolution of Antarctica. Cambridge University Press: 599-607.

Wei, W.
1992 Updated nannofossil stratigraphy of the CIROS-1 core from McMurdo Sound (Ross Sea). In S.W. Wise, Jr., R. Schlich, et al., *Proceedings of the Ocean Drilling Program, Scientific Results, v. 120*: 1105-1117.

Wei, W and S. W. Wise, Jr.
1990 Middle Eocene to Pleistocene calcareous nannofossils recovered by Ocean Drilling Program Leg 113 in the Weddell Sea. In P.F. Barker, J.P. Kennett, et al., *Proceedings of the Ocean Drilling Program, Scientific Results, v. 113*: College Station, TX, p. 639-666.

Wetzel, O.
1933a Die in organischer Substanz erhaltenen Mikrofossilien des baltischen Kreide-Feuersteins mit einem sediment-petrographischen und stratigraphischen Anhang. *Palaeontographica, Abt. A, v. 77*: 141-188.

Wetzel, O.
1933b Die in organischer Substanz erhaltenen Mikrofossilien des baltischen Kreide-Feuersteins mit einem sediment-petrographischen und stratigraphischen Anhang. *Palaeontographica, Abt. A, v. 78*: 1-110.
1961 New microfossils from Baltic Cretaceous flintstones. *Micropaleontology, v. 7(3)*: 337-350.

Williams, G.L. and J.P. Bujak.
1985 Mesozoic and Cenozoic dinoflagellates. In H.M. Bolli, J.B. Saunders and K. Perch-Nielsen (eds.), *Plankton Stratigraphy, Volume 2.* Cambridge University Press. p. 847-964.

Williams, G.L. and C. Downie.
1966 Further dinoflagellate cysts from the London Clay. In Davey, R.J., Downie, C., Sarjeant, W.A.S., and Williams, G.L., Studies on Mesozoic and Cainozoic Dinoflagellate Cysts. *Bulletin of the British Museum (Natural History) Geology, Supplement 3:* 215-235.

Wilson, G.J.
1967 Some new species of Lower Tertiary dinoflagellates from McMurdo Sound, Antarctica. *New Zealand Journal of Botany, v. 5(1):* 57-83.

1984 New Zealand Late Jurassic to Eocene dinoflagellate biostratigraphy - A summary. *Newsletters on Stratigraphy, v.13(2):* 104-117.

1987 Dinoflagellate biostratigraphy of the Cretaceous-Tertiary boundary, mid-Waipara River section, North Canterbury, New Zealand. *New Zealand Geological Survey Record 20:* 8-15.

1988 Paleocene and Eocene dinoflagellate cysts from Waipawa, Hawkes Bay, New Zealand. *New Zealand Geological Survey Paleontological Bulletin 57,* 96p.

1989 Marine palynology. In Barrett, P.J. (ed.), Antarctic Cenozoic History from the CIROS-1 Drillhole, McMurdo Sound. *DSIR Bulletin 245:* 129-133.

Wilson, G.J. and C.D. Clowes.
1982 *Arachnodinium,* A new dinoflagellate genus from the Lower Tertiary of Antarctica. *Palynology, v. 6:* 97-103.

Wilson, G.S., A.P. Roberts, K.L. Verosub, F. Florindo, and L. Sagnotti
1998 Magnetobiostratigraphic chronology of the Eocene-Oligocene transition in the CIROS-1 core, Victoria Land margin, Antarctica: implications for Antarctic glacial history. *GSA Bulletin, v. 110(1):* 35-47.

Wise, S.W., Jr.
1983 Mesozoic and Cenozoic calcareous nannofossils recovered by Deep Sea Drilling Project Leg 71 in the Falkland Plateau region, southwest Atlantic Ocean. In W.J. Ludwig, V.A. Krasheninnikov, et al., *Initial Reports of the Deep Sea Drilling Project, Leg 71:* 481-550.

Wise, S.W., Jr., Breza, J.R., Harwood, D.M., and Wei, W.
1991 Paleogene glacial history of Antarctica. In D.W. Müller, J.A. McKenzie and H. Weissert (eds.), *Controversies in Modern Geology,* Academic Press, Harcourt Brace Jovanovich. p. 133-171.

Wise, S.W. Jr., J.R. Breza, D.M. Harwood, W. Wei, and J.C. Zachos
1992 Paleogene glacial history of Antarctica in light of Leg 120 drilling results. In S.W. Wise, Jr., R. Schlich, et al., *Proceedings of the Ocean Drilling Program, Scientific Results, v. 120:* 1001-1030.

Wrenn, J.H., and S.W. Beckman.
1982 Maceral, total organic carbon, and palynological analyses of Ross Ice Shelf Project Site J9 cores. *Science, v. 216(9):* 187-189.

Wrenn, J.H., and G.F. Hart.
1988 Paleogene dinoflagellate cyst biostratigraphy of Seymour Island, Antarctica. In R.M. Feldmann and M.O. Woodburne (eds.), *Geology and Paleontology of Seymour Island, Antarctic Peninsula. Geological Society of America Memoir 169:* 321-447.

David M. Harwood, Department of Geosciences, 214 Bessey Hall, University of Nebraska - Lincoln, Lincoln, NE 68588-0340, U.S.A.

EOCENE PLANT MACROFOSSILS FROM ERRATICS, MCMURDO SOUND, ANTARCTICA

Mike Pole

Department of Botany, University of Queensland, Brisbane, QLD 4072, Australia

Bob Hill

Department of Plant Science, University of Tasmania, Hobart, TAS 7001, Australia

David Harwood

Department of Geology, University of Nebraska, Lincoln, NE 68588-0340, USA

Glacial erratics of Eocene sediments at McMurdo Sound, Antarctica, contain plant macrofossils. These include *Araucaria* leaves, at least two species of *Nothofagus* based on leaves, and three types of non-*Nothofagus* dicotyledonous leaves. One of the *Nothofagus* species has plicate vernation and was therefore deciduous. Two types of *Nothofagus* fruits are present, two-flanged and three-flanged. The three-flanged fruits indicate that they came from a four-valved cupule and thus from one of the (currently) temperate subgenera.

INTRODUCTION

Fragments of two glacial erratics containing plant macrofossils were discovered during the 1992-93 season and a further one was discovered on a return trip when the macrofossils were collected in 1993-94 [Stilwell *et al.*, 1993; See general locality map of fossil sites in Introduction]. Subsequent study of the dinoflagellates indicates their age is middle-late Eocene [Levy and Harwood, this volume]. These specimens are very significant as they are only the third find of Tertiary leaves from East Antarctica - the Antarctic mainland [Previous finds are recorded in Hill, 1989; Hill, Harwood and Webb, 1996]. The aims of this paper are to document the plant macrofossils and to compare them with previously described Eocene plant fossils from Antarctica.

METHODS

Fossiliferous erratics were each given a number prefixed with "E". Specimens were collected in the field using hammer and chisel and further preparation of macrofossils was with a compressed-air chisel in the laboratory. For ease of comparison line drawings are provided of the angiosperm leaves. These were drawn by tracing onto sheet plastic at x5 magnification using a Nikon Profile Projector, then reduced for publication using a photocopier and redrawn on architectural paper using Indian ink. To indicate preservation mode some specimens are illustrated as photographs. Extant *Nothofagus* seeds were illustrated using a JEOL JSM 820 Scanning Electron Microscope (SEM) operated at 15kV after being sputter-coated with gold. They are mounted on stubs (prefixed with 'S') and stored in the Department of Plant Science, University of Tasmania. All fossil material is catalogued in the paleobotany collection of the Smithsonian Institution. Specimen numbers are prefixed with PB94.

LOCALITIES

1.   Minna Bluff (78°25'S, 165°50'E): erratic sandstone boulder E153.

Four or five large (1-2 m long) coarse sandstone boulders were closely spaced on moraine at the edge of

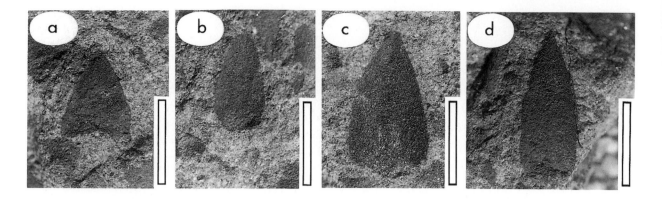

Fig. 1. *Araucaria* leaves from locality E215. a, PB94-540, b, PB94-542 c, PB94-539 d, PB94-538. Scale = 1 cm.

Minna Bluff and clearly result from fragmentation of a larger block. The boulders contain wood [Francis, this volume], molluscs [Stilwell, this volume], and leaves.

2.   Minna Bluff (78°25'S, 165°50'E): erratic siltstone boulder E219.

A single, 60 cm diameter boulder of gray calcareous siltstone was located approximately 100m away from the boulders described above. The boulder contained a fragment of Teredo-bored wood [Francis, this volume], molluscs [Stilwell, this volume], and leaves.

3.   Mount Discovery (78°17'S, 165°35'E): erratic sandstone boulder E215.

A single 40 cm length boulder of coarse sandstone was located in moraine near the foot of Mount Discovery, approximately 16 km from the Minna Bluff localities. The boulder had abundant leaf fragments but no molluscs or well-preserved wood.

RESULTS

Araucaria Leaves

Locality E215 contains several roughly triangular-shaped objects varying in length from 11-19 mm, which are interpreted as scale-leaves of *Araucaria* (Figure 1). Although no cone-scales were seen their shape compares well with other, isolated but certain araucarian scale leaves in the fossil record [e.g. Pole, 1995] and various extant species of *Araucaria*. The lack of obvious keel and broad, relatively large size, suggests comparison with sections *Columbea*, *Intermedia*, or *Bunya* rather than *Eutacta* [Wilde and Eames, 1952].

Locality E215: PB94-538 to PB94-542, PB94-553

Nothofagus Leaves

No cuticular details are preserved on the fossils and fine details are generally absent (Figure 2 illustrates a range of preservation modes). Identification of *Nothofagus* leaves is based on a comparison of gross form and venation with extant *Nothofagus*. The important characters are the ovate to elliptical shape of the lamina, the simple craspedodromous, relatively straight and evenly spaced lateral veins, and the irregularly toothed margin. These characters are not diagnostic of *Nothofagus*, and the identification must take into consideration the circumstantial evidence of associated *Nothofagus* pollen [see Levy and Harwood, this volume] and *Nothofagus* seeds (see below). The possibility that these leaves may include *Fagus* is acknowledged [e.g. Romero and Dibbern, 1985], but is considered most unlikely.

Based on two extremes of morphology, at least two species of *Nothofagus* leaf are present at locality E219.

1.   A large form which has prominent, nearly straight lateral veins, and no signs of the original vernation. This form is typified by PB94-524 (Figure 3a) which has a maximum preserved length of 45 mm, and a width of at least 65 mm. The original length of this specimen is likely to have been at least 100 mm making it one of the largest *Nothofagus* leaves, extant or fossil. The margin is not preserved on this specimen, but on an associated fragment which could belong to the same species, three evenly-spaced, sharp teeth per lateral vein commonly occur (Figure 4d). The size of this fossil leaf is comparable to the extant South American *N. alpina* (Poepp. & Endl.) Oerst. and *N. alessandri* Espinosa. Large fossil *Nothofagus* leaves include *N. cretacea* Zastawniak

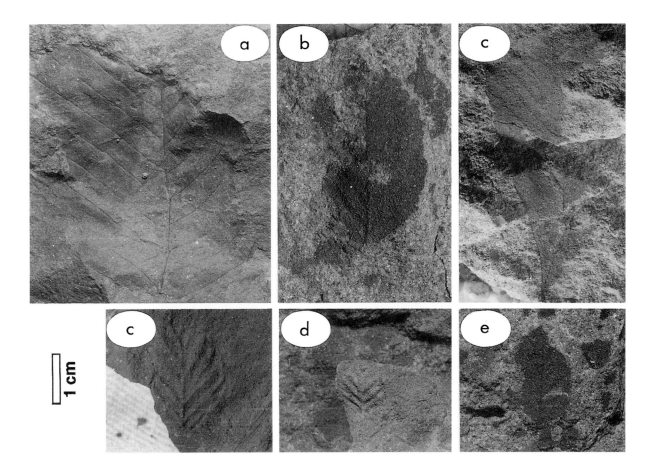

Fig. 2. Angiosperm leaves showing typical preservation mode at each locality. a, b, d-f, *Nothofagus*; c, indet angiosperm. a, PB94-524, b, PB94-545 (E215), c, PB94-518 (E153), d, PB94-519, e, PB94-520, f, PB94-543 (E215). All to same scale at lower left.

[1994] from the Late Cretaceous of the South Shetland Islands, Antarctica, up to 77 mm long and 47 mm wide; *N. ulmifolia* (Ett.) Oliver, from the Late Cretaceous of New Zealand, up to 85 mm long and 35 mm wide [Pole, 1992]; *N. azureus* Pole [1993] from the Miocene of New Zealand, up to 87 mm long, and *N. plicata* Scriven, McLoughlin & Hill [1995], up to 100 mm long. Doktor *et al.* [1996] illustrated a *Nothofagus* leaf about 68 mm wide from the Eocene of Seymour Island. The largest *Nothofagus* described by Tanai [1986] from Patagonia was *N. simplicidens* Dúsen, up to 86 mm long and 32 mm wide. The margin of *N. cretacea* is unknown, and those of the other large *Nothofagus* leaves are all different in some way, either the number of teeth per lateral vein, or in their position and shape. The new fossils appear to be a distinct species.

2.   A small form (10-20 mm long) which has distinct plicate vernation, including PB94-519, (Figure 3c),

PB94-522 (Figure 4g), PB94-520 (Figure 4h), and PB94-523 (Figure 3i). This is similar to extant *N. gunnii* (Hook. fil.) Oerst., *N. pumilio* (Poepp. & Endl.) Krasser, and the Oligocene fossil described from the CIROS-1 drillcore by Hill [1989] as aff. *N. gunnii*. The marginal details are not clear enough for closer comparison although there are hints the margin is crenate. No similar small-leaved forms have been described from Seymour Island.

The remaining specimens from locality E219 are too fragmentary for certain placing, but probably represent the larger form. Leaves from locality E215 certainly include the small form (Figure 4b), as well as larger specimens (Figure 4a, c) which may represent a third species.

Locality E219: PB94-519 - PB94-535
Locality E215: PB94-537, PB94-543-PB94-547, PB94-549- PB94-558

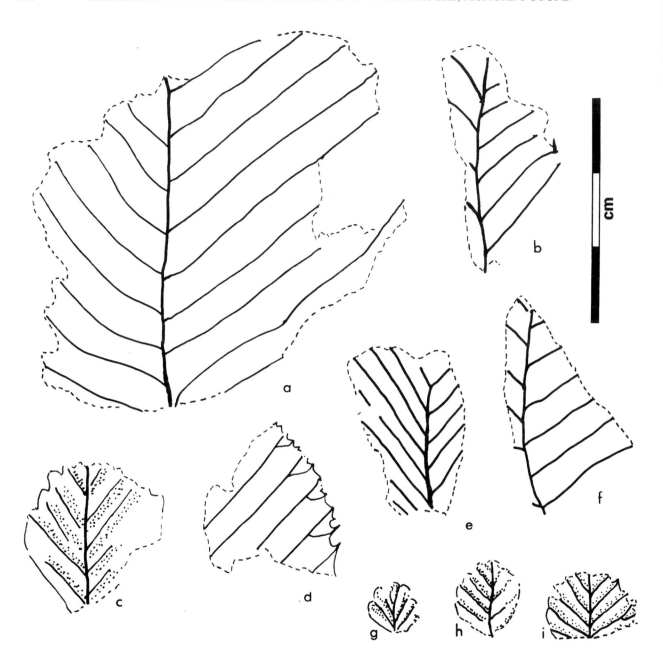

Fig. 3. *Nothofagus* leaves from locality E219. a, PB94-524 (E219), b, PB94-533, c, PB94-519 d, PB94-525 (E219) e, PB94-526 (E219) f, PB94-530 g, PB94-522 h, PB94-520 i, PB94-523. Stipple indicates vernation. All to same scale.

Nothofagus Seeds

Two structures are interpreted as *Nothofagus* fruits by direct comparison with fruits of extant species. One, PB94-548 (Figure 5b), is essentially two dimensional, measuring 8x7 mm, while the other, PB94-536 (Figures 5d, e) has an additional third flange at right-angles and measures 6.5x3 mm. Such a three-flanged fruit came from a four-valved cupule (it needs the space between the cupule valves to fit in) and thus came from one of the three temperate subgenera; *Fuscospora*, *Lophozonia*, or *Nothofagus* [Hill and Read, 1991]. The two-flanged fruits could come from any of the four subgenera.

Locality E215: PB94-548, Locality E219: PB94-536

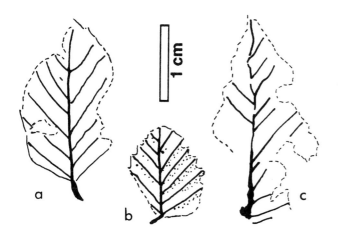

Leaves from locality E153 have the poorest preservation, but include leaves which are unlikely to be *Nothofagus*. PB94-515 (Figure 6a) is a simple, entire-margined leaf, PB94-517 (Figure 6b) is a fragment of a base and might be *Nothofagus*, PB94-516 (Figure 6c) is a fragment of mid-lamina having strong, widely-separated, curving lateral veins, PB94-518 (Figure 6d) has a toothed margin but with a very acute base which is unlike any known *Nothofagus*.

DISCUSSION

The fossils described in this paper add an entirely new plant macrofossil locality to Antarctica. Previous records of Tertiary plant macrofossils are restricted to the

Fig. 4. *Nothofagus* leaves from locality E215. a, PB94-543, b, PB94-547, c, PB94-553. All to same scale.

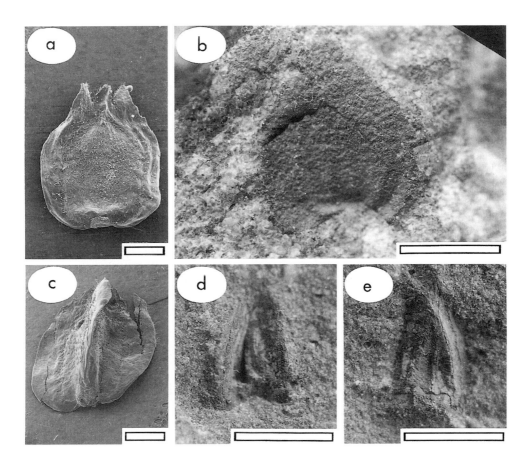

Fig. 5. *Nothofagus* seeds. a, extant *N. cunninghamii*, SEM of two-flanged fruit, S850, scale = 1 mm, b, fossil two-flanged fruit, PB94-548 from locality E215, scale = 5 mm, c, extant *N. cunninghamii*, SEM of three-flanged seed, third flange is projecting upwards, S851, scale = 1 mm, d, fossil three-flanged fruit showing third flange projecting down into sediment, PB94-536a from locality E219, scale = 5 mm, e, fossil three-flanged fruit, PB94-536b (counterpart of d), scale = 5 mm.

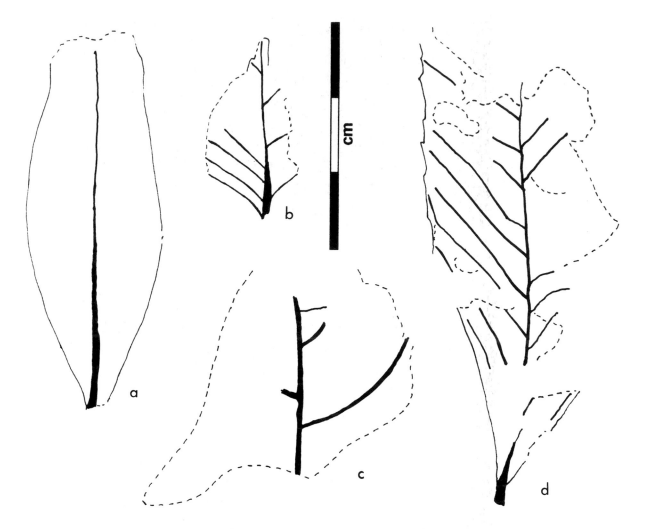

Fig. 6. Unidentified leaves from locality E153. (non-*Nothofagus*) a, PB94-515 b, PB94-517 c, PB94-516 d, PB94-518. All to same scale.

region around the Antarctic Peninsula, a single leaf from the CIROS-1 borehole in McMurdo Sound, and from the Sirius Group of the Transantarctic Mountains. The new fossils are the most southerly Eocene plant assemblages known and the oldest Tertiary plant remains from East Antarctica. The location of deposition of the sediments is unknown but the locality today was at a similar paleolatitude in which the middle-late Eocene La Meseta Formation of the Seymour Island was deposited (around 65-70°S, based on the paleogeographic reconstruction of Lawver *et al.* 1992, and the Antarctic apparent polar wander curve of DiVenere *et al.*, 1994) as well as the Late Oligocene *Nothofagus* leaf fossil from the CIROS-1 borehole [Hill, 1989]. The Sirius Formation (85°S paleolatitude) was deposited at a much higher paleolati-

tude [Hill and Truswell, 1993; Hill *et al.*, 1996]. They reinforce conclusions that Antarctica was forested, and that *Nothofagus* (the southern beech) formed a dominant portion of the forest biomass for much of the Tertiary [Hill and Scriven, 1995]. The *Nothofagus* fruits described here are the first known from Antarctica. Added to the pollen record [Levy and Harwood, this volume] they confirm that *Nothofagus* was reproducing sexually. The *Araucaria* leaves described here are the first record from East Antarctica.

Although the leaves are poorly preserved anatomically, they retain some three-dimensionality which indicates some were deciduous. This is based on the plicate vernation of the leaves (the manner of leaf folding while within the bud) [Philipson and Philipson, 1978], which is still evident on some specimens making them look 'fan-like' (Figure 3d, e). Plicate vernation within *Nothofagus* is only found within deciduous species [Mirbel, 1827; Philipson and Philipson, 1979]. These leaves were probably dropped at the start of the polar winter which would have brought several months of darkness. The other *Nothofagus*, (although plicate vernation is not evident) were likely to have been deciduous too. Some deciduous species of *Nothofagus* lose their plication as the leaves expand (i.e. the evidence of plication can only be used in a positive way - plicate leaves were definitely deciduous, but with non-plicate leaves the evidence is ambiguous).

Araucaria has a disjunct distribution today, but is found on all landmasses with *Nothofagus* except New Zealand and Tasmania. Both *Araucaria* and *Nothofagus* occur in regions with a low range of mean annual temperature [Axelrod, 1984; Veblen *et al.*, 1996]. An element of rainfall seasonality has been noted for some *Araucaria* [Webb, 1968].

South America is a promising area to discover a climate similar to those represented by the fossil assemblages. *Araucaria* grows as an emergent over deciduous *Nothofagus* in the Valdivian Andes, for instance on Vulcan Llaima in Chile, which often experiences snow in the winter. This climate is likely to have been similar to that part of Antarctica near where the Mount Discovery sediments accumulated. Further south in Patagonia lower diversity forests dominated by *Nothofagus* occur under more extreme climates. The *Nothofagus* assemblage of Minna Bluff could result from deterioration in climate compared with Mount Discovery. In the opposite sense, the assemblage from Minna Bluff, with no clear *Nothofagus* may have been deposited under more clement climate than Mount Discovery.

The three fossiliferous blocks are all unique in their combination of taxa. Locality E153 is the only one which includes angiosperm leaves which are unlikely to be *Nothofagus*, E219 contains only *Nothofagus* (but at least two species), while E215 contains both *Nothofagus* and *Araucaria*. There are at least three possibilities to account for this. The samples may represent different time intervals, thus sampling communities which reflect different climates; they may represent natural spatial patchiness of vegetation, or they may reflect taphonomic biases in different sedimentary environments. There is a clear danger of over-interpreting a small amount of data.

The three assemblages may be compared with the succession of assemblages documented from Seymour Island which suggest increasing dominance by *Nothofagus* and decreasing taxonomic diversity with time:

1. The Paleocene Cross Valley Member of the Sobral Formation is dominated by ferns and secondarily by *Nothofagus*. Podocarp and araucarian conifers, and a variety of other angiosperms are present [Case, 1988]. Case regards this as the same horizon studied by Dusen [1908].

2. The middle Eocene of the La Meseta Formation, according to Case [1988], is dominated by a large-leaved *Nothofagus* and what he regarded as two species of fern. An araucarian conifer is also present. Doktor *et al.* [1996] collected an assemblage they interpreted as dominated by *Nothofagus* and what they described as a new genus and species, *Knightiophyllum andreae*, possibly belonging to the Proteaceae. They also collected ferns, araucarians, probable podocarps and one specimen of a clearly non-*Nothofagus* dicotyledon. The importance of *Nothofagus* in this unit may be greater than either Case or Doktor *et al.* concluded. Firstly, one of us (Pole) suggests that both of Case's fern morphs from this time are a single, additional species of *Nothofagus* (*cf. N. melanoides* Pole, 1992). Secondly, Doktor *et al.* applied *Knightiophyllum* to some leaves described by Case [1988] as *Nothofagus*. It is not clear to us how *Knightiophyllum* is distinguished from *Nothofagus*,. There is also a nomenclatural problem (not addressed here) with *Knightiophyllum* which, as a genus, was first erected by Ettingshausen [1887] for New Zealand material. Berry [1916] also inadvertently used this name, as Dilcher and Mehrotra [1969] pointed out. This range of forms is not unexpected - *Nothofagus* leaf morphology today is demonstrably only a fraction of what has been attained throughout the existence of the genus.

3. The late Eocene leaf assemblages of the La Meseta Formation consist entirely of *Nothofagus* leaves [Case, 1988].

Climatically, *Nothofagus* is usually taken to indicate cool, ever-wet conditions. Specific limits for growth of living *Nothofagus* were discussed by Hill *et al.* [1996]; for instance, a minimum temperature no less than -22°C and air temperatures during the growth season of at least 5°C for several weeks. They also suggested that *Nothofagus beardmorensis* Hill, Harwood, & Webb from the Sirius Group may have been more frost tolerant. However, the presence of additional taxa, such as *Araucaria*, in the McMurdo Sound assemblages suggests the climate was far from extreme. In Patagonia today *Araucaria* grows with *Nothofagus* around latitude 39°S and extends to the tree line [Armesto, *et al.*, 1995; Veblen *et al.*, 1995]. Further south there is a broad vegetational trend towards a simple, almost pure *Nothofagus* forest near sea level at the southern tip near 55°S and at higher altitudes in low latitudes. Thus while fossil leaf assemblages containing only *Nothofagus* leaves may represent extreme conditions, simply local dominance in less extreme conditions, or preservational bias, the assemblage in locality E215 with *Nothofagus* and *Araucaria* are good evidence of less extreme conditions.

Acknowledgements. Many thanks to the staff of NSF at McMurdo Base and especially to the crews of VXE-6 Squadron. This research was funded by a NSF grant. Most work by the first author was carried out while he was a research associate at the Department of Plant Science, University of Tasmania funded by an ARC grant to R.S. Hill, then completed at the Department of Botany, University of Queensland, funded by an ARC grant to M.E. Dettmann and G. Stewart. We are grateful for the comments of J. A. Case and an anonymous referee.

REFERENCES

Armesto, J. J., C. Villagran, J. C. Aravena, C. Pérez, C. Smith-Ramirez, M. Cortés, and L. Hedin.
1995   Conifer forests of the Chilean coastal range, in Ecology of the Southern Conifers, edited by N.J. Enright and R.S. Hill, pp. 156-170, Melbourne University Press, Melbourne.

Axelrod, D. I.
1984   An interpretation of Cretaceous and Tertiary biota in polar regions. *Palaeogeogr., Palaleoclimatol., Palaeoecol. 45*, 105-147.

Berry, E. W.
1916   The lower Eocene floras of southeastern North America. *U.S. Geol. Surv. Prog. Pap., 91.*

Case, J. A.
1988   Paleogene floras from Seymour Island, Antarctic Peninsula. *Geol. Soc. Am. Mem., 169*, 523-530.

Dilcher, D. L., and B. Mehrotra
1969   A study of leaf compressions of *Knightiophyllum* from Eocene deposits of southeastern North America. *Am. J. Bot., 56*, 936-943.

DiVenere, V. J., D. V. Kent, I. W. D. Dalziel
1994   Mid-Cretaceous paleomagnetic results from Marie Byrd Land, West Antarctica: a test of post-100 Ma relative motion between East and West Antarctica. *J. Geophys. Res., 99 (B8)*, 15, 115-15, 139.

Doktor, M., A. Gazdzicki, A. Jerzmanska, S. Porebski, and E. A. Zastawniak
1996   Plant-and-fish assemblage from the Eocene La Meseta Formation of Seymour Island (Antarctic Peninsula) and its environmental implications. *Pal. Polonica 55*, 127-146.

Dusen, P.
1908   Die Tertiare Flora der Seymour-Insel. *Wissenschaftliche Ergebnisse der Schwedischen Sudpolar-Expedition 1901-1903 3*, 1-26.

Ettingshausen, C. V.
1887   Beitrage zur Kenntniss der Fossilen Flora Neuseelands. *Denkschriften der Akademie der Wissenschaften, Wien 53*, 143-194.

Francis, J. E.
This volume Fossil wood from Eocene High Latitude forests, McMurdo Sound, Antarctica.

Hill, R. S.
1989   Fossil leaf, in Antarctic Cenozoic History from the CIROS-1 Drillhole, McMurdo Sound, edited by P.J. Barrett, *DSIR Bull., 245*, 143-144.

Hill, R. S., and J. A. Read
1991   A revised infrageneric classification of *Nothofagus* (Fagaceae). *Bot. J. Linn. Soc. 105*, 37-72.

Hill, R. S., and L. J. Scriven
1995   The angiosperm-dominated woody vegetation of Antarctica: a review. *Rev. Palaeobot. Palynol., 86*, 175-198.

Hill, R. S. and E. M. Truswell
1993   *Nothofagus* fossils in the Sirius Group, Transantarctic Mountains: leaves and pollen and their climatic implications. The Antarctic paleoenvironment: a perspective on global change. *Ant. Res. Ser., 60*, 67-73

Hill, R. S., D. M. Harwood, and P. Webb
1996   *Nothofagus beardmorensis* (Nothofagaceae), a new species based on leaves from the Pliocene Sirius Group, Transantarctic mountains, Antarctica. *Rev. Palaeobot. Palynol., 94*, 11-24.

Lawver, L. A., L. M. Gahagan, and M. F. Coffin
1992   The development of paleoseaways around Antarctica, in 'The Antarctic paleoenvironment: a perspective on global change, part one', edited by J. P. Kennett and D. A. Warnke, pp. 7-30, Antarctic Research Series 56, American Geophysical Union, Washington, D.C.

Levy, R. H., and D. M. Harwood
This volume Marine playnomorph biostratigraphy and age(s) of the McMurdo Sound Erratics.

Mirbel, C. F. B.
1827   Description de quelques especies nouvelles de la familee des Amentacees. *Mém. Mus. nat. d'hist. nat. (Paris), 14*, 462-474.

Philipson, W. R., and M. N. Philipson
1978   Leaf vernation in *Nothofagus*. *N.Z. J. Bot., 17*, 417-421.

Pole, M. S.
1992   Cretaceous macrofloras of eastern Otago, New

Zealand: angiosperms. *Aust. J. Bot.,* *40,* 169-206.

Pole, M. S.
1993   Early Miocene flora of the Manuherikia Group, New Zealand. 8. Nothofagus. *J. Roy. Soc. N.Z.,* *23,* 329-344.

Pole, M. S.
1995   Late Cretaceous macrofloras of eastern Otago, New Zealand: Gymnosperms. *Aust. Syst. Bot. 8,* 1067-1106.

Romero, E. J., and M. C. Dibbern
1985   A review of the species described as *Fagus* and *Nothofagus* by Dusen. *Palaeontographica Abt. B, 197,* 123-137.

Scriven, L. J., S. McLoughlin, and R. S. Hill
1995   *Nothofagus plicata* (Nothofagaceae), a new deciduous Eocene macrofossil species, from southern continental Australia. *Rev. Palaeobot. Palynol., 86,* 199-209.

Stilwell, J. D.
This volume   Eocene mollusca (Bivalvia, Gastropoda and Scaphopoda) from McMurdo Sound: systematics and paleoecologic significance.

Stilwell, J. D., R. H. Levy, and D. M. Harwood
1993   Preliminary paleontological investigation of Tertiary glacial erratics from the McMurdo Sound region, East Antarctica, *Ant. J.,* 16-19.

Tanai, T.
1986   Phytogeographic and phylogenetic history of the genus *Nothofagus* Bl. (Fagaceae) in the southern hemisphere. *J. Fac. Sci., Hokkaido Univ., Ser. iv, 21,* 505-582.

Veblen, T. T., B. R. Burns, T. Kitzberger, A. Lara, and R. Villalaba
1995   The ecology of the conifers of southern South America, in Ecology of the Southern Conifers, edited by N.J. Enright and R.S. Hill, pp. 120-155, Melbourne University Press, Melbourne.

Veblen, T. T., R. S. Hill, and J. Read (Eds.)
1996   The Ecology and Biogeography of *Nothofagus* Forests. Yale University Press, New Haven and London.

Webb, L. J.
1968   Environmental relationships of the structural types of Australian rainforest vegetation. *Ecol. 49,* 296-311.

Wilde, M. H., and A. J. Eames
1952   The ovule and 'seed' of *Araucaria bidwilli* with discussion of the taxonomy of the genus, II. Taxonomy. *Ann. Bot. 16,* 27-47.

Zastawniak, E.
1994   Upper Cretaceous leaf flora from the Blaszyk Moraine (Zamek Formation), King George Island, South Shetland Islands, West Antarctica. *Acta Palaeobot. 34,* 119-163.

David Harwood, Department of Geology, University of Nebraska, Lincoln, NE 68588-0340, USA

Bob Hill, Department of Plant Science, University of Tasmania, Hobart, TAS 7001, Australia

Mike Pole, Department of Botany, University of Queensland, Brisbane, QLD 4072, Australia

possible to identify whether the wood was conifer or angiosperm and some distinctive features could be discerned. The following identifications have been made.

Type A Conifer wood

This wood is composed of vertical tracheids (average width 25μm) and has a distinctive arrangement of bordered pits on the radial walls of the tracheids. The pits are dominantly in biseriate rows with an alternating arrangement, giving the pits a hexagonal outline (Figure 4b). The medullary rays are notably short, typically 2-3 cells in height. Cross-field pits were hard to determine but there appear to be 5-6 small pits per cross-field.

Samples: E153f, E153dh, possibly E153g, all from Minna Bluff.

Comment: The arrangement of the bordered pits in the tracheids in biseriate rows, the alternating pattern of the pits and their hexagonal shape is characteristic of fossil conifer wood belonging to the form-genus *Araucarioxylon* Kraus.

Type B Conifer wood

This wood is composed of vertical tracheids (average width 30μm) with uniseriate rows of bordered pits (mean width 20 μm) on the radial walls. The rays are tall and uniseriate, sometimes consisting of up to 27 cells. The most distinctive feature are the single, sometimes paired, oval pores in each cross-field (Figure 4c); these pits range up to 30μm x 20μm in size. Resin canals are absent and it is not possible to determine whether parenchyma is present. Scattered tracheids are filled with an opaque mineral, probably pyrite.

Sample: E153b, Minna Bluff

Comment: The tracheid pitting is characteristic of wood of podocarpoid or cupressinoid type but the most interesting features are the large oval pits in the cross-fields. The cross-field pitting is very distinctive of wood of some types of the living Podocarpaceae, including species of *Phyllocladus* and *Dacrydium* [Greguss, 1955; Patel, 1967a, b, 1968]. As fossil wood it can be assigned to the conifer form-genus *Phyllocladoxylon* Gothan.

Type C Angiosperm wood

Sections of wood in both transverse and longitudinal directions are available for the specimen from the concretion, although very little wood is present in each and it is rather distorted. This is an angiosperm wood composed of fibres and vessels. The vessels are arranged in ring porous arrangement with maximum density of large vessels along early part of ring (Figure 4d). Vessels are either solitary or grouped in pairs in a radial direction.

In tangential section vessels of two sizes are apparent; larger ones with diameter 80μm, and small ones with diameter 20μm. The end connections between vessels are hard to determine but appear to be a mixture of both pitted and scalariform plates. The intervessels pits are circular or oval in outline, mostly opposite in arrangement (Figure 4e). The rays are dominantly uniseriate with small circular simple ray pits.

Samples E219, possibly E153d and E153e, Minna Bluff.

Comment: Features of this wood, such as the nature of the vessel distribution, vessel ends and pitting, allow it to be identified as *Nothofagoxylon* Gothan. *Nothofagoxylon* wood of Tertiary age has been described from King George Island, South Shetland Islands and from several sites in Chile [Torres, 1984a,b; Jagmin, 1987. It has also been described from Seymour Island by Gothan [1908] and Torres *et al.* [1994]. The small size and poor preservation of the Minna Bluff specimens does not permit valid comparison with the different species of *Nothofagoxylon* fossils described.

Growth rings

The wood samples are far too small to exhibit long series of growth rings. However, some rings are present in the angiosperm wood from the calcareous concretion and a single ring can be measured in one other wood section. They provide some interesting information but

Fig. 4a. Transverse thin section of small branch from the carbonate concretion (E219). This shows that only a small portion of the wood remains, most of it having been destroyed by the action of the wood-boring bivalve Toredolites which has left large circular sediment-filled cavities. Scale = 3mm.

Fig. 4b. Longitudinal thin section of specimen E153f showing alternately arranged bordered pits (P) on the radial walls of the tracheids, typical of the conifer form-genus *Araucarioxylon*. Scale = 70μm.

Fig. 4c. Longitudinal thin section showing the medullary rays cells of specimen E153b and the conspicuous large oval pits (P) in the cross-fields of the rays. This is typical of the conifer form-genus *Phyllocladoxylon*. Scale = 70μm.

Fig. 4d. Transverse thin section of specimen E219, showing a distinct growth ring and concentration of vessels (V) in the early part of the growth ring. (RB = ring boundary). This is wood of the angiosperm *Nothofagoxylon*. Scale = 5mm.

Fig. 4e. Longitudinal section of specimen E219, illustrating the vessels (V) and intra-vessel pitting (P). Scale = 60μm.

detailed climatic reconstructions cannot be made from such small samples.

In the wood from the concretion (E219) growth rings of 1mm, 0.5mm and 0.45mm were measured in transverse section (Figure 4d). These are narrow rings and indicate that the tree was growing quite slowly at this point. This may have been due to low temperatures, unfavourable site conditions or the position of the wood on the tree.

One growth ring is apparent in the section of E153b, marked by a decrease in tracheid width across the ring. It is 2.88mm in width, measured in longitudinal section, and represents reasonably favourable growth. The radial longitudinal diameter of individual cells across the ring were measured and the results plotted in Figure 5 as a simple plot of changing cell diameter and the cumulative sum of deviations from the mean cell diameter. This can give an indication of growth characteristics throughout a single growing season [e.g. Creber and Chaloner, 1984; Francis et al., 1994]. The plot indicates that during this particular growing season the tree suffered two periods of adverse conditions that caused the growth to slow down; the second drop in cell size (from about cells 42 to 63) is severe enough to suggest that a false ring has been formed, perhaps due to an intermittent water shortage during the growing season. There is some disruption to the cells at the end of the ring but it is not apparent that this is due to frost damage.

DISCUSSION

The fossils found in these sediments represent a new Eocene plant locality in Antarctica at higher latitudes than previously known. Reconstructions by Lawver et al. [1992] indicate that the paleolatitude of the McMurdo region was approximately 78°S during the Eocene. At this latitude plants would have had to tolerate several months of darkness during the winter, during which time they probably became dormant [Read and Francis, 1992], but grew continuously during the long hours of summer sunlight. Light levels and temperatures were obviously not low enough to prohibit the growth of forests at this paleolatitude .

The occurrence of plants here indicates that land adjacent to the McMurdo region was forested during the Eocene, and although the plant material has clearly been transported from its site of growth, the presence of rather delicate angiosperm leaves [Pole et al., this volume] suggests that the plant material was not transported for long distances.

The three main types of wood found (the conifers *Araucarioxylon* and *Phyllocladoxylon* and the angiosperm

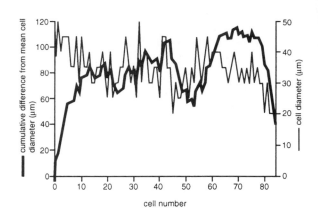

Fig. 5. Graph illustrating cell dimensions across one growth ring in specimen E153. The thin line illustrates the change in radial cell diameter through the growth ring. The thick line shows the trend of the cumulative difference in radial diameter from the mean cell diameter and highlights more clearly the two periods in which growth was adversely affected (approximately between cells 20-32 and 42- 63); the latter one is a very marked decrease in growth and may have been caused by an environmental hazard such as intermittent drought (but not freezing).

Nothofagoxylon) have structures comparable to living trees of the Araucariaceae, Podocarpaceae (genus *Phyllocladus*)and the southern beech *Nothofagus*. These taxa are present in Antarctica in Tertiary sequences from the James Ross Basin and the South Shetland Islands. The Araucariaceae are represented by wood, leaves, pollen and cuticle through the Late Cretaceous to the Eocene, though not recorded from the Oligocene, according to Askin [1992]. The Podocarpaceae and Nothofagaceae are present as fossils from the Late Cretaceous to Oligocene. Collections of fossil wood that include these three taxa have also been described by Torres et al. [1994] from La Meseta Formation on Seymour Island and from Paleogene strata on King George Island [Torres and Lemoigne, 1988] so the forest composition seems consistent across this region of Antarctica during the Early Tertiary.

The three fossil tree types represent the remains of ancestral forests that covered South America, Antarctica and Australasia when these continents were joined. Although the ecological tolerances of these trees in the past may have been different from comparable living types, some indication of the environmental conditions can be deduced from the distribution of comparable trees today. At present *Phyllocladus* is widespread in the cool wet sclerophyll forests of temperate regions of Tasmania and New Zealand [Barker, 1995] (the Cool Temperate rainforests of Webb, 1959) and occurs together with

Nothofagus. Nothofagus also is also widespread in the cool temperate forests of South America and is able to tolerate winter temperatures as low as -22°C in southern Chile [Alberdi *et al.*, 1985].

Species of the Araucariaceae tend to represent less extreme climates, being most common in New Guinea and present in northern Australia [Page and Clifford, 1981]. In South America one species occurs in the sub-tropical to temperate zones from 18-30°S with mean annual temperatures of 10 - 18°C (microthermal/ mesothermal climates of Wolfe and Upchurch, 1987) while *A. araucana* has a more southerly but limited distribution from 37-40°S, mostly at high altitudes [Armesto *et al.*, 1995; Veblen *et al.*, 1995] and occurs with various species of *Nothofagus*. It is able to tolerate a range of rainfall and temperatures, surviving in high altitudes with winter snow.

Extrapolating these ecological constraints back to the Eocene based on only three wood types is limiting, but simple comparison suggests that Eocene environments in the McMurdo region were cool temperate (approximately equivalent to the microthermal climates of Wolfe and Upchurch (1987) with mean annual temperatures less than 13°C) with perhaps some cold conditions during winter but probably without long periods of sub-zero temperatures. The limited information from growth rings in the wood show that climates were seasonal, as would be expected at these latitudes, and that the growing season may have been interrupted by adverse conditions such as drought. There is no clear evidence for cold climates with long sub-zero winters that would have permitted significant permanent ice formation, although winter snow may have been possible.

The Eocene wood in the McMurdo erratics does not therefore appear to show evidence of dramatic climate cooling, as has been suggested for the end of the Eocene from other sources. Marine isotope studies from around the paleolatitude of the Antarctic Peninsula suggest that oceanic cooling began in the middle-late Eocene, with major ice sheet development in the early Oligocene [Gazdzicki *et al.*, 1992; Zachos *et al.*, 1992; Ditchfield *et al.*, 1994]. Temperatures nearer the South Pole may have been cold enough by the late Eocene for the formation of at least mountain glaciers, if not ice at sea level [Ditchfield *et al.*, 1994]; this would not conflict with evidence from the wood.

SUMMARY

Fossil wood in erratic blocks of Eocene sediments in the McMurdo region was derived from adjacent forested land. The main wood types identified in the fossil assemblage are the conifer woods *Araucarioxylon* and *Phyllocladoxylon*, and the angiosperm *Nothofagoxylon*. They are similar to woods of living Araucariaceae, Podocarpaceae [species *Phyllocladus/Dacrydium*] and Nothofagaceae, trees which are found together today in cool temperate regions of South America, Australia and New Zealand. The presence of this wood in Eocene sediments suggests that climates in these high paleolatitudes of about 78°S were cool temperate (microthermal), possibly with some winter snow but not cold enough to allow the existence of extensive ice near sea level. Accurate dating of this material will help constrain the history of climate cooling and onset of glaciation in Antarctica.

Acknowledgements. I wish to thank D. Harwood and R. Levy (University of Nebraska) for the opportunity to work on this material and the Trans-Antarctic Association, Leeds University Academic Development Fund, NSF and the Natural Environment Research Council for funding. Mike Pole and Gary Wilson are also thanked for their help in the field.

REFERENCES

Alberdi, M., M. Romero, D. Rios, and H. Wenzel.
1985 Altitudinal gradients of seasonal frost resistance in *Nothofagus* communities of southern Chile., *Acta Oecologica-oecologica Plantarum. 6*, 21-30.

Armesto, J. J., C. Villagrán, J. C. Aravena, C. Pérez, C. Smith-Ramirez, M. Cortés, and L. Hedin.
1995 Conifer forests of the Chilean coastal range, *In* N. J. Enright, and R. S. Hill (Eds.), *Ecology of the Southern Conifers*, pp.156-170. Melbourne University Press, Melbourne.

Askin, R. A.
1992 Late Cretaceous-Early Tertiary Antarctic outcrop evidence for past vegetation and climates. *Antarctic Research Series, 56*: 61-73, figs.1-3, AGU, Washington, D.C.

Barker, P.C.J.
1995 *Phyllocladus asplenifolius*: Phenology, germination and seedling survival. *New Zealand Journal of Botany, 33*, 325-337.

Creber, G. T., and W. G. Chaloner.
1984 Influence of environmental factors on the wood structure of living and fossil trees. *Botanical Review, 50*, 357-448.

Ditchfield, P. W., J. D. Marshall, and D. Pirrie.
1994 High-latitude paleotemperature variation - new data from the Tithonian to Eocene of James Ross Island, Antarctica. *Palaeogeography Palaeoclimatology Palaeoecology 107:* 79-101.

Francis, J. E.
1986 Growth rings in Cretaceous and Tertiary wood from Antarctica and their palaeoclimatic implica-

tions. *Palaeontology, 29,* 665-684.

Francis, J. E., and R. S. Hill.
1996   Pliocene fossil plants from the Transantarctic Mountains: evidence for climate from growth rings and fossil leaves. *Palaios, 11,* 389-396.

Francis, J. E., K. J. Wolfe, M. J. Arnott, and P. J. Barrett.
1994   Permian climates of the southern margins of Pangea: evidence from fossil wood in Antarctica. *Canadian Society of Petroleum Geologists Memoir, 17,* 275-282.

Gazdzicki, A. J., M. Gruszczynski, A. Hoffman, K. Malkowski, S. Marenssi, H. Stanislaw, and A. Tatur.
1992   Stable carbon and oxygen isotope record in the Paleogene La Meseta Formation, Seymour Island, Antarctica. *Antarctic Science, 4,* 461-468.

Gothan, W.
1908   Die fossilen holzer von der Seymour und Snow Hill Insel. *Wissenschaftliche Ergebnisse Schwedischen Sudpolarexpedition.* 1901-3. *3,* 1-33.

Greguss, P.
1955   *Identification of living gymnosperms on the basis of xylotomy,* 263 pp. Akademia Kiado, Budapest.

Jagmin, N.
1987   Estudo anatômico dos troncos fósseis de Admiralty Bay, King George Island (Peninsula Antártica). *Acta Biologica Leopoldensia., 9,* 81-98.

Lawver, L.A., L. M. Gahagan, and M. F. Coffin.
1992   The development of paleoseaways around Antarctica. *Antarctic Research Series, 56*: 7-30, AGU, Washington, D.C.

Levy, R. H., J. D. Stilwell, D. M. Harwood, J. E. Francis, and M. Pole.
1995   Eocene paleobiology of East Antarctica - a report on evidence from fossiliferous glacial erratics in McMurdo Sound. *VII International Symposium on Antarctic Earth Sciences,* p. 243, University of Siena, Italy.

Page, C. N., and H. T. Clifford.
1981   Ecological biogeography of Australian conifers and ferns. *In* A. Keast (Ed.), *Ecology and Biogeography of Australia,* pp. 473-498. Junk, The Hague.

Patel, R. N.
1967a   Wood anatomy of the Podocarpaceae indigenous to New Zealand 1. *Dacrydium. New Zealand Journal of Botany, 5,* 171-184.
1967b   Wood anatomy of the Podocarpaceae indigenous to New Zealand 2. *Podocarpus. New Zealand Journal of Botany, 5,* 307-321.
1968   Wood anatomy of Podocarpaceae indigenous to New Zealand. 3. *Phyllocladus. New Zealand Journal of Botany, 6,* 3-8,

Read, J., and J. E. Francis.
1992   Responses of some southern hemisphere tree species to a prolonged dark period and their phytogeographic and palaeoecological implications for high-latitude Cretaceous and Tertiary floras. *Palaeogeography Palaeoclimatology Palaeoecology, 99,* 271-290.

Stilwell, J. D., R. H. Levy, and D. M. Harwood.
1993   Preliminary paleontological investigation of Tertiary glacial erratics from the McMurdo Sound region, East Antarctica. *Antarctic Journal of the United States, 28,* 16-19.

Torres, T.
1984a   Identificacion de madera fosil del Terciario de la Isla Rey Jorge, Islas Shetland del Sur, Antartica, *Congreso Latinoamericano de Paleontologia, Mexico,* pp. 555-565.
1984b   *Nothofagoxylon antarcticus* n. sp., madera fosil del Terciario de la isla Rey Jorge, islas Shetland del Sur, Antártica. *Series Científica INACH, 31,* 39-52.

Torres, T., and Y. LeMoigne.
1988   Maderas fósiles terciarias de la Formación Caleta Arctowski, isla Rey Jorge, Antártica. *Series Científica INACH, 37:* 69-107.

Torres, T., S. Marenssi, and S. Santillana.
1994   Maderas fosiles de la isla Seymour, Formacion La Meseta, Antártica. *Series Científica INACH, 44:* 17-38.

Truswell, E. M.
1991   Antarctica: a history of terrestrial vegetation. *In* R. Tingey (Ed.), *The Geology of Antarctica,* pp. 499-528. Oxford University Press, Oxford.

Veblen, T. T., B. R. Burns, T. Kitzberger, A. Lara, and R. Villabla.
1995   The ecology of the conifers of southern South America. *In* N. J. Enright and R. S. Hill (Eds.), *Ecology of the southern conifers.* pp.120-155, Melbourne University Press, Melbourne.

Webb, L. J.
1959   A physiognomic classification of Australian rain forests. *Journal of Ecology, 47*: 551-570.

Wolfe, J. A., and G. R. Upchurch.
1987   North American nonmarine climates and vegetation during the Late Cretaceous. *Palaeogeography Palaeoclimatology Palaeo-ecology, 61*: 33-77.

Zachos, J, C, J. R. Breza, and S. W. Wise.
1992,   Early Oligocene ice-sheet expansion on Antarctica: stable isotope and sedimentological evidence from Kerguelen Plateau, southern Indian Ocean. *Geology,* 20, 569-573.

Jane E. Francis, Earth Sciences, University of Leeds, Leeds, UK

EOCENE MOLLUSCA (BIVALVIA, GASTROPODA AND SCAPHOPODA) FROM MCMURDO SOUND: SYSTEMATICS AND PALEOECOLOGIC SIGNIFICANCE

Jeffrey D. Stilwell

School of Earth Sciences, James Cook University of North Queensland, Townsville, Queensland 4811, Australia

The Eocene Fossil Mollusca from the McMurdo Sound erratics represent the sole record of this important macroinvertebrate group of this age from East Antarctica. This study is the first to document the fauna. A total number of 65 species (at least 28 new) of bivalves, gastropods and scaphopods have been recorded from these erratics, which represent a spectrum of predominantly shallow marine environments and facies. New species proposed are *Linucula? mcmurdoensis* n. sp., *Acila?* n. sp., *Saccella eoantarctica* n. sp., *Yoldiella?* n. sp., *Pseudotindaria? levyi* n. sp., *Solemya surolongata* n. sp., *Limopsis (Limopsista?) antarctominuta* n. sp., *Brachidontes sandalius* n. sp., *Chlamys s. l.* n. sp., *Crassostrea antarctogigantea* n. sp., *Thyasira (Conchocele) antarctosulcata* n. sp., *Cardita subrectangulata* n. sp., *Nemocardium (Pratulum?) minutum* n.., *"Eurhomalea" claudiae* n. sp., *Hiatella harringtoni* n. sp., *Calliostoma s. l.* n. sp., *Falsimargarita? vieja* n. sp., *Astreaea liliputia* n. sp., *Drepanocheilus (Tulochilus) erebus* n. sp., *Struthiolarella mcmurdoensis* n. sp., *Sigapatella (Spirogalerus?) colossa* n. sp., *Taniella (Pristinacca?)* n. sp., *Euspira bohatyi* n. sp., *Pseudofax?* n. sp., *Cominella? s. l.* n. sp., *Fusinus?* n. sp., *Cylichnania?* n. sp., among probable others. Shallow suspension-feeding bivalves dominate the macrofauna (about 41.5% of the total molluscan fauna). Deposit-feeding bivalves and gastropods comprise about 24.0% of the fauna, followed by carnivores (about 23.0%) and the least abundant epifaunal grazers and/or browsers (about 9.0% of total). Nearly monotypic concentrations of molluscan remains dominate the macrofauna and these deposits are characterized by authochthonous and parauthochthonous assemblages.

INTRODUCTION

Fossiliferous erratics recently recovered from the vicinity of Mount Discovery and Minna Bluff glacial moraine deposits contain a wealth of new data on the Eocene faunal composition of Antarctica. The macrofauna is dominated by a moderately diverse assemblage of molluscs in a spectrum of marine shallow-water facies and environments. Data from McMurdo Sound provide new information on the importance of Antarctica and the high southern latitudes with regard to the origin and evolution of the modern fauna. Further, the molluscan fauna provides for the first time a means of comparing coeval

molluscs of East Antarctica with the better known and well-preserved, diverse assemblages of Seymour and Cockburn islands, Antarctic Peninsula, in order to assess paleobiogeographic links between these widely separated regions.

PREVIOUS INVESTIGATIONS ON EOCENE MOLLUSCA OF ANTARCTICA

Eocene molluscs of Seymour Island, Antarctic Peninsula (latitude 64 ^{0} 15' S, longitude 56 ^{0} 45' W), located some 100 km southeast of the tip of the Peninsula, have the honorable distinction of being the first fossils to be described and named from the Antarctic continent. These fossils were collected during a Norwegian whaling expedition on the barque *Jason* commanded by Captain C. A. Larsen, who landed on the east side of Seymour Island in mid-November 1892 in search of food, namely seals [Zinsmeister, 1988]. The crew of the *Jason* had been sealing in the ice south of the Orkneys for about twenty days in late 1892 before meeting up with the Scottish whaling barque *Balaena* in the vicinity of "Graham Land" near Seymour Island (first sighted and named Cape Seymour by Captain James Clark Ross on January 6, 1843, see account by Ross, 1847, 2: 343-345). An early account of Larsen's discovery of fossils, apparently overlooked in the literature, is portrayed by the artist of the *Balaena*, which landed on Seymour Island Christmas Day of 1892 on the 1892-93 Dundee Antarctic Expedition. The artist, W. G. Burn Murdoch [1894, p. 251], stated "It is a marvel that no scientific expedition has been sent down here since the days of Ross...Captain Larsen of the *Jason* has landed, and he tells us he found beds of fossils on the beach, shells, and tree trunks. Some of the fossil shells he showed us resemble very large cockles." The large cockles that Murdoch describes are undoubtedly the robust cucullaeid bivalve *Cucullaea raea* Zinsmeister, 1984. A more detailed account in the same work is given by *William S. Bruce in* Murdoch [1894, p. 356, 364], the naturalist of the *Balaena*, who wrote "Captain Larsen landed on the South Orkneys and Seymour Island, and in the latter he found some fossils which had fallen from a decomposing cliff. These are the first fossils ever brought from Antarctica. There are specimens among them of *Cucullaea*, *Cytherea*, and *Nataea*, and pieces of a coniferous tree. They are probably of Tertiary age, and indicate a warmer climate than now prevails in these high southern latitudes...we must investigate the nature and distribution of the rocks, which contain for the palaeontologist an entirely new fossil fauna and flora." Bruce's comment on the fossil bivalves is seemingly the first to note that Antarctica has not always been locked in a perpetual freezer and that warmer climes existed during the Tertiary. The fossils obtained by Dr. Donald and Capt. Larsen were published in two short notes by Sharman and Newton [1894, 1897], who described *Cucullaea donaldi* and *Eurhomalea antarctica* [*Venus*], respectively.

This important fossil discovery on Seymour Island prompted the organization of an expedition by one of Sweden's foremost scientists/explorers, Otto Nordenskjöld, nearly a decade later. The Swedish South Polar Expedition of 1901-03 resulted in significant discoveries of Tertiary and Cretaceous fossils, mostly molluscs, but at a cost. Nordenskjöld and his crew were stranded in

Antarctica for 2 1/2 years because his ship *Antarctic* was lost when it became trapped by ice movements. Because of this tragedy Nordenskjöld was unable to exploit what is now known to be the most diverse assemblage of Cretaceous and Tertiary fossils on the Antarctic continent but, nevertheless, a moderately diverse assemblage of 26 Eocene mollusc species from Seymour Island was described by Wilckens [1911] in the expedition reports. In the same series, Wilckens [1924] described a small, poorly preserved, molluscan faunule of Eocene age from Cockburn Island located a few kms to the north of Seymour Island at the boundary of Admiralty Sound and Erebus and Terror Gulf (Note: Cockburn Island was first visited by Captain Ross on New Year's Day of 1843, but due to ice conditions the crew could not land until January 6th, at which time they made a quick three hour survey in the morning, and erroneously surmised that the island was entirely volcanic, and thus collected a few rock samples and concentrated largely on the interesting botany of the island [Ross, 1847, 2: 333-343]). Wilckens described 10 species from Cockburn Island, but left all in open nomenclature. Several steinkerns were also figured. Nearly all of the identifiable taxa are now known to be conspecific with Seymour Island species recorded from the La Meseta Formation. Zinsmeister and Stilwell [1990] published the only account of Cockburn Island Eocene molluscs since Wilckens [1924], describing the minute ringiculid gastropod *Ringicula cockburnensis*.

Published work on molluscs did not commence again until a joint Argentine/American expedition explored Seymour Island during the 1974-75 austral summer, resulting in several significant papers by Zinsmeister [1976a-b, 1977, 1978, 1979, 1982, 1984] and Zinsmeister and Camacho [1980, 1982]. Extensive fossil collections formed by subsequent expeditions to Seymour Island during the austral summers of 1981-82, 1983-84, 1985, and 1986-87, sponsored by the Division of Polar Programs of the National Science Foundation, led to the publication of a monographic treatment of 170 species of molluscs from the La Meseta Formation, of which 123 were considered new [Stilwell and Zinsmeister, 1992].

Research on the fossils of East Antarctica began with the discovery of erratics in McMurdo Sound in 1959 by H. J. Harrington of New Zealand Geological Survey (Institute of Geological and Nuclear Sciences, Lower Hutt), who made further collections of "...several hundreds of fragments of mudstones, calcareous sandstone and conglomerate or diamictites" in 1969 [Harrington, 1969]. The only published account of molluscs resulting from these collections was by Hertlein [1969], who studied an erratic collected by R. C. Wood during the 1968-69 season which contained gastropods identified as *Struthiolarella cf. variabilis* Wilckens. Hertlein's short paper is the first report of Paleogene macrofossils from the Ross Sea area.

Renewed interest in the erratics as a potential source of important macro- and microfossil data to help bridge a major chasm in our knowledge of the Antarctic Eocene biota led D. M. Harwood of University of Nebraska-Lincoln to organise an expedition to McMurdo Sound during the 1991-92 austral summer. Further expeditions and collections made by the author and others during the 1992-93, 1993-94 and 1995-96 austral summers form the basis of this paper. A total of approximately 1250 fossiliferous erratics have been collected from McMurdo Sound in the vicinity of Mount Discovery, Minna Bluff and Black Island (see Introduction, this volume), and the number of taxa recorded has soared to more than 70 species of macroinvertebrates, predominantly molluscs (at least 65 species; see Tables 1-2). A preliminary survey and checklist of the macroinvertebrate fauna recorded in the erratics was presented by Stilwell *et al.* [1993].

AGE

The age of the molluscs described herein is difficult to pinpoint because of the marked endemic component of the fauna and flora at species level reflecting geographic and genetic isolation of Antarctica during the Eocene. Several taxa are found to be conspecific with those from the La Meseta Formation, but few are restricted to short intervals on Seymour Island. The endemic component of microfossil fauna/flora on Seymour and Cockburn islands is also strong [see Wrenn and Hart, 1988; Askin *et al.*, 1991; Askin, this volume; Lee, this volume]. Dinoflagellate cysts recovered from decapod-bearing erratics from Mount Discovery belong to a suite of many long-ranging species, poorly suited for precise biostratigraphical control and one of the most diagnostic species, *Hystrichosphaeridium truswelliae*, has a reported age range of early to late early Eocene [Stilwell *et al.*, 1997]. The overlapping ranges of other dinoflagellate species found in the erratics indicate an overall age range of late early Eocene to middle Eocene [*ibid*]. Dinoflagellates and molluscs associated with a pseudontorn bird bone discovered in the erratics points further to a middle to late Eocene age [Stilwell *et al.*, 1997].

PALEOECOLOGY

The paleoecology of Tertiary Antarctic macroinvertebrates has been little studied. The Eocene Mollusca recovered from the McMurdo Sound erratics are derived from a spectrum of environments and facies, but predominantly a sandy, shallow shelf environment. A detailed account of the sedimentary environments represented in the erratics is given by Levy *et al.* [this volume] and is not repeated herein.

In descending order of species-level diversity and importance, the molluscan fauna from the erratics encompasses four general groupings: infaunal and epifaunal suspension feeders, deposit feeders, epifaunal grazers and browsers, and carnivores. Shallow infaunal suspension feeders dominate the macrofauna and comprise such taxa as *Solemya surolongata* n. sp. [burrower], *Cucullaea* sp. *cf. C. donaldi* Sharman and Newton, *Limopsis (Limopsista?) antarctominuta* n. sp., *Saxolucina sharmani* [Wilckens], *Miltha?* sp., *Thyasira (Conchocele) antarctosulcata* n. sp., *?Anisodonta truncilla* Stilwell and Zinsmeister, *Cyclocardia* sp., *Nemocardium (Pratulum?) minutum* n. sp., *Crassatella* sp., *?Gomphina iheringi* Zinsmeister, *"Eurhomalea" claudiae* n. sp., *Cyclorismina?* n. sp.? *cf. "C." marwicki* Zinsmeister, *?Eumarcia (Atamarcia) robusta* Stilwell and Zinsmeister, Veneridae genus et species indeterminate, *Hiatella harringtoni* n. sp. [nestling bivalve], *Panopea akerlundi* Stilwell and Zinsmeister [deep burrower], *Panopea* n. sp? *cf. P. philippii* Zinsmeister [deep burrower], *Periploma* n sp.? *cf. P. topei* Zinsmeister, and *Teredo* sp. Suspension-feeding bivalves comprise about 79.5% of the bivalve fauna and some 41.5% of the total molluscan fauna. Epifaunal suspension feeders comprise mainly epibyssate taxa such as *Aulacomya* sp. *cf. A. anderssoni* Zinsmeister, *Brachidontes sandalius* n. sp., *Eburneopecten* sp., *Chlamys s. l.* n. sp., Anomiidae genus et species indeterminate, *Crassostrea antarctogigantea* n. sp. [cemented], and *Cardita subrectangulata* n. sp. Deposit feeders make up about 20.5% of the total bivalve fauna [about 24% of total molluscan fauna] and include *Linucula? mcmurdoensis* n. sp., *Leionucula nova* (Wilckens), *Acila?* n. sp., *Saccella eoantarctica* n. sp., *Yoldiella?* n. sp., and *Pseudotindaria? levyi* n. sp. Carnivores dominate the gastropod fauna at 48% [about 23%] and include *Taniella (Pristinacca?)* n. sp., *Euspira bohatyi* n. sp., *Polinices (Polinices)* sp. *cf. P. subtenuis* (von Ihering), *?Penion australocapax* Stilwell and Zinsmeister, *Pseudofax?* n. sp., *Cominella? s. l.* n. sp., *Austroconella* sp. *cf. A. verrucosa* Stilwell and Zinsmeister, *?Eobuccinella brucei* Stilwell and Zinsmeister, *Fusinus?* n. sp., Turridae genus et species indeterminate, *Acteon eoantarcticus* Stilwell and Zinsmeister, *Crenilabium suromaximum* Stilwell and Zinsmeister, *?Ringicula (Ringicula) cockburnensis* Zinsmeister and Stilwell, and *Cylichnania?* n. sp.

Deposit feeders are somewhat less abundant at about 27.5% of recorded gastropod species, including Rissoidae genus et species indeterminate, Cerithiidae genus et species indeterminate, *Colposigma euthenia* Stilwell and Zinsmeister, *Zeacolpus?* sp., *Arrhoges (Antarctohoges) diversicostata* (Wilckens), *Drepanocheilus (Tulochilus) erebus* n. sp., *Struthiolarella mcmurdoensis* n. sp., *Perissodonta* n. sp.? *cf. P. laevis* (Wilckens). Epifaunal grazers and/or browsers [20.5% of gastropods, about 9% of total] comprise *Cellana feldmanni* Stilwell and Zinsmeister, Patellacea genus et species indeterminate, *Calliostoma s. l.* n. sp. [alternatively possibly a carnivore], *Falsimargarita? vieja* n. sp., Trochidae genus et species indeterminate, and *Astraea lilliputia* n. sp. *Sigapatella (Spirogalerus?) colossa* n. sp. may have been an epifaunal suspension feeder. The Scaphopoda, represented by fragments, were probably infaunal deposits feeders and/or carnivores. It is interesting to note that some of the McMurdo Sound molluscan fauna may be chemosynbiotic. Some shallow water lucinacean bivalves, including Solemyidae and Lucinidae, are noted for being chemosynbiotic taxa and use decaying plant and animal matter in very locally hydrogen sulfide rich niches for chemosynbiosis. Further, some of these molluscs are associated with vent settings and require an energy source of methane, hydrogen sulfide and low oxygen.

Mollusca numerically dominate the macrofauna. Many of the most fossiliferous erratics are nearly monotypic concentrations of various taxa such as *Struthiolarella mcmurdoensis* n. sp., *Drepanocheilus (Tulochilus) erebus* n. sp., Turridae genus et species indeterminate, and *"Eurhomalea" claudiae* n. sp., and are of low species-level diversity. E145 from Mount Discovery has the most diverse, well-preserved molluscan assemblage. Few articulated bivalves are noted in the erratics apart from deep burrowers *Panopea akerlundi* Stilwell and Zinsmeister and *Panopea* n. sp.? *cf. P. philippii* Zinsmeister and although bedding planes in these blocks are disrupted by burrowing, these bivalves are probably in living position. Some specimens of *"Eurhomalea" claudiae* n. sp. are also articulated. Few large molluscs apart from calcified oysters are encountered in the erratics.

The skeletal concentrations such as that found in the largest oyster dominated erratics would fall under the category of mixed skeletal oyster bioherms, as other invertebrate debris is present. The large number of erratics dominated by oyster remains in the Mount Discovery moraine indicates that this horizon was probably extensive during the Eocene. These concentrations are mostly autochthonous. Most of the erratics are probably parautochthonous concentrations of slightly transported remains, as the number of entire molluscs is relatively high. Some specimens, such as those of *Struthiolarella mcmurdoensis* n. sp. and *Panopea* n. sp? *cf. P. philippii* Zinsmeister, are encrusted by serpulid worms, indicating a period of exposure on the sea floor after death. The composition of the molluscs and sedimentological evidence point to a shallow shelf setting in open marine conditions for most of the erratics.

SYSTEMATIC PALEONTOLOGY

Summarized herein is a systematic catalogue of the Eocene molluscan fauna recovered from the erratics of McMurdo Sound during four expeditions this decade. In this study, fossils were collected from key areas using traditional means (e.g. rock hammer and chisel). Fossils were prepared in the laboratory using a pneumatic air scribe and various dental tools. Fossils were coated with ammonium chloride before macrophotography and gold and/or platinum for scanning electron microscopy. Wherever possible and appropriate, previously described species are redescribed, figured and taxonomically updated.

The specimens and collections used in this study are housed in many institutions in the United States and abroad; these are: United States National Museum (USNM), Purdue University (PU), University of Nebraska-Lincoln (UNL), Institute of Polar Studies, Ohio State University (IPS), James Cook University of North Queensland, Townsville, Australia (JCU), University of Otago, Dunedin, New Zealand (UO), and Naturhistoriska Riksmuseet, Stockholm, Sweden (MO).

Systematic arrangement of the Bivalvia generally follows Vokes [1980]; Vaught [1989]; and Beu and Maxwell [1990]; that of the Gastropoda Ponder and Warén [1988]; Vaught [1989]; and Beu and Maxwell [1990]; and Scaphopoda follows Vaught [1989] and Steiner [1992]. The traditional ordinal names Archaeogastropoda, Mesogastropoda and Neogastropoda are used in this study. However, these names are included with the proviso that future work may see the abandonment of these names because these taxonomic groupings may be grades rather than clades, as advocated by Bieler [1992], among others. Superfamilial endings of "-acea" are preferred here to "-oidea", because the latter may be confused with and is close to ordinal endings of "-oida". Further, a genus name followed by the term *sensu stricto* implies that the subgenus name is the same as the genus, whereas a genus name followed by the term *sensu lato* implies that the subgenus is uncertain.

Phylum MOLLUSCA Linné, 1758
Class BIVALVIA Lié, 1758
Subclass PALEOTAXOONTA Korobkov, 1954
Order NUCULOIDA Dall, 1889
Superfamily NUCULACEA Gray, 1824
Family NUCULIDAE Gray, 1824
Family NUCULINAE Gray, 1824
Genus *Linucula* Marwick, 1931

Linucula Marwick, 1931, p. 49.

Type. (By original designation) *Nucula ruatakiensis*
Marwick, 1926.
Biogeographic element. Palaeoaustral, as interpret-
ed herein.

Linucula? mcmurdoensis, new species
Plate 1, figs. D and J

Diagnosis. Minute, subtrigonal to subovate, moderately inflated nuculid with extremely weak radial elements and ventral crenulations; distinguished from the
type, *Linucula ruatakiensis* Marwick, 1926, in having
ambiguous radial sculpture and much more posterior
beaks.
Description. Shell minute, moderately inflated,
subtrigonal to subovate; umbo prominent, moderately
opisthogyrate, located approximately a third of length of
shell from posterior margin; posterodorsal dorsal margin
short, weakly convex, merging with steeply rounded posterior margin; anterodorsal margin moderately long,
weakly convex, merging with moderately narrowly
rounded anterior margin; anterior margin more narrowly
rounded than posterior margin, yielding a slightly oblique
profile; ventral margin smooth, broadly rounded; shell
mostly smooth apart from more than 50 semi-regular,
punctuated, slightly raised, commarginal growth pauses;
surface and subsurface radial elements extremely weak;
hinge details unknown; ventral margin very finely crenulated.
Dimensions. Holotype USNM 490735 length 3.0
mm, height 2.5 mm; paratype USNM 490736 length 3.25
mm, height 3.0 mm.
Types. Holotype USNM 490735; paratype USNM 4-
90736.
Figured specimen. Holotype.
Material. One well-preserved, articulated specimen
and three other poorly preserved individuals.
Localities. E145 (type), E189.
Geographic distribution. McMurdo Sound.
Discussion. This puzzling, minute nuculid has ves-

tiges of a finely crenulate ventral margin on the paratype
precluding its placement in *Leionucula* Quenstedt; otherwise it is remarkably similar to species of *Leionucula*.
The crenulate margin and radial sculpture, albeit ambiguous and weak, is consistent with *Lincula* Marwick. The
best preserved specimen, the holotype, is articulated so
that the presence/absence of dorsal crenulations characteristic of *Linucula* cannot be determined.
The Neogene New Zealand type species of *Linucula*,
L. ruatakiensis [Marwick, 1926, p. 327, pl. 75, figs. 7, 9;
see also Fleming, 1966, p. 102, pl. 2, figs. 34-35], is easily distinguished from *Linucula? mcmurdoensis* n. sp. in
having a larger shell, more trigonal outline reflecting
more centrally situated beaks, and much stronger radial
sculpture. The outline, size, and sculpture of *Linucula?
mcmurdoensis* n. sp. is remarkably similar to the Recent
species, *Ennucula eltanini* Dell, 1990 [p. 8, figs. 6-7],
from deep waters off the western coast of Tierra del
Fuego, southern South America, but the Antarctic species
is more broadly ovate in outline.
All specimens of *Linucula? mcmurdoensis* n. sp.
were taken from medium-grained sandstone, indicating a
shallow shelf environment of deposition. This species is
the smallest of all Eocene bivalves encountered in the
erratics
Etymology. Name derived from McMurdo Sound,
the species' type locality.

Genus *Acila* H. Adams and A. Adams, 1858

Acila H. and A. Adams, 1858, p. 545.

Type species. (By subsequent designation,
Stoliczka, 1871) *Nucula divaricata* Hinds, 1843.
Biogeographic element. Indo-Pacific/Tethyan, as
interpreted herein.

Acila?, new species
Plate 1, fig. E

Dimensions. USNM 490769 length 6.5 mm, height
5.5 mm (incomplete external mold).
Figured specimen. USNM 490769.
Material. One partial external mold.
Locality. E155.
Geographic distribution. McMurdo Sound.
Discussion. The characteristic divaricate sculpture
of regular, successive, overlapping chevron-shaped ribs
in this new species is consistent with the Paleogene to
Recent nuculid bivalve *Acila* H. and A. Adams [1858],
but the incomplete nature of the only recorded specimen

of this species prevents detailed assessment here. The divaricate ribs in *Acila?* n. sp. are crossed by weak commarginal growth pauses, creating a subdued pustulose surface on the external part of the valve. The internal part of the ventral margin is mostly smooth apart from very weak crenulations. The ventral margin is moderately well-rounded and convex. I know of no other species of *Acila* recorded from Antarctic fossil record or the Recent. Although similar in having divaricate sculpture, the lucinid bivalve *Divaricella* von Martens [1880] has generally finer, less regular divaricate ribs, compared to species of *Acila*. The McMurdo Sound species has a very small shell relative to its rather robust and regular, divaricate ribs and broadly spaced growth pauses, which become much narrower adjacent to the ventral margin.

The type species of *Acila, A. divaricata* [Hinds, 1843] [see Hanley *in* Sowerby, 1860, p. 155, pl. 230, fig. 151; Keen *in* Moore, 1969, p. N231, fig. A3-8a, b; Abbott and Dance, 1983, p. 289, bottom row, middle figure], from the Recent of China and Japan has nearly identical divaricate and commarginal sculpture, compared with *Acila?* n. sp. The ventral margin of both species is equally convex but, unfortunately, hinge and dorsal margin details are not preserved in the external mold of the McMurdo Sound specimen.

Subfamily NUCULOMINAE Maxwell, 1988
Genus *Leionucula* Quenstedt, 1930

Leionucula Quenstedt, 1930, pp. 110 and 112.

Type. (By original designation) *Nucula albensis* d'Orbigny, 1844.

Biogeographic element. Cosmopolitan [Keen *in* Moore, 1969, p. N231].

Discussion. Stilwell [1993, pp. 362-363] reviewed the controversy surrounding the relationship between *Ennucula* Iredale, 1931, and *Leionucula* Qunstedt, 1930. Evidence presented therein indicates that the separation of these genera, as several authors have done in the past, is more than questionable, given the remarkable similarity in both external and internal morphology. *Leionucula* is Recent. The range in Antarctica is Cretaceous to Pliocene.

Leionucula nova (Wilckens, 1911)
Plate 1, figs. a, b, and c

Nucula nova Wilckens, 1911, p. 5, pl. 1, figs. 4a, 4b, and 5.
Nucula sp., Wilckens, 1924, pp. 11-12, pl. 1, figs. 16-17.
Nucula (Leionucula) nova Wilckens; Zinsmeister, 1984, p. 1501, figs. 3C-3E; Stilwell and Zinsmeister, 1992, p. 47, pl. 1, figs. c-e.
Leionucula nova (Wilckens); Stilwell, 1993, p. 364.

Plate 1

Figs. A, B, and C. *Leionucula nova* (Wilckens). (A) USNM 490737, E145, length = 15.0 mm, x3. (B) USNM 490738, E145, length = 9.0 mm, x3. (C) USNM 490739, E155, length = 8.5 mm, x3.

Figs. D and J. *Linucula? mcmurdoensis* n. sp. (D) Holotype USNM 490735, E145, length = 3.0 mm, x14. (J) Holotype, USNM 490735, E145, length = 3.0 mm, x100 (SEM of microsculpture of ventral margin).

Fig. E. *Acila?* n. sp. (E) USNM 490769, E155, length = 6.5 mm, x3.

Fig. F. *Leionucula* n. sp.? (F) USNM 490793, E240, length = 27.5 mm, x3.

Figs. G and K. *Saccella eoantarctica* n. sp. (G) Holotype USNM 490740, E372, length = 6.0 mm, x3.5. (K) Paratype USNM 490794, E147, length = 4.5 mm, x3.

Figs. H and L. *Pseudotindaria? levyi* n. sp. (H) Holotype USNM USNM 490741, E184, length = 6.0 mm, x3. (L) Holotype USNM 490741, E184, length = 6.0 mm, x9.5.

Fig. I. *Yoldiella?* n. sp. (I) USNM 490795, E145, length = 4.0 mm, x15.5.

Figs. M and N. *Neilo beui* Stilwell and Zinsmeister. (M) USNM 490743, E145, length = 39.5 mm, x1. (N) USNM 490742, E155, length = 43.5 mm, x1.

Figs. O and Q. *Solemya surolongata* n. sp. (O) Holotype USNM 490744, E184, length = 38.5 mm, x1. (Q) Paratype USNM 490745, E184, length = 27.0 mm, x1.

Figs. P and R. *Cucullaea sp. cf. C. donaldi* Sharman and Newton. (P) USNM 490747, E359, length of partial hinge = 34.5 mm, x1.3. (R) USNM 490746, E359, length = 46.5 mm, x1.3.

Discussion. This species is reminiscent of, but seemingly not closely linked to, extant Austral species of *Yoldiella* and *Pseudotindaria*. Of the many species of *Yoldiella* described from Antarctica, *Pseudotindaria? levyi* n. sp. from McMurdo Sound most resembles *Y. profundorum* [Melvill and Standen, 1912] [see Dell, 1990, pp. 14-15, figs. 21-22], but the rostrum is more developed and truncated in new McMurdo Sound species. This new group is comparable to the type of *Yoldiella*, *Y. lucida* Lovén [1846] [see Hanley *in* Sowerby, 1860, p. 145, pl. 227, figs. 23-25], but is still more elongate with a more developed rostrum. Another species with a similar outline and sculpture is the late Eocene New Zealand neilonellid *Pseudotindaria delli* Maxwell [1992] [pp. 59-60, pl. 1, figs. e-g], but the McMurdo Sound species is, as above with other comparable species, more elongate with a better developed rostrum. However, placement of the McMurdo Sound species in *Pseudotindaria* Sanders and Allen [1977] seems more appropriate than *Yoldiella*, until internal details are known.

Etymology. Species named in honor of Richard. H. Levy of University of Nebraska-Lincoln for his major contributions to the study of the McMurdo Sound erratics.

Family MALLETIIDAE H. and A. Adams, 1857
Genus *Neilo* A. Adams, 1854

Neilo A. Adams, 1854, p. 93.

Type species. (By monotypy) *Neilo cumingii* A. Adams, 1854.

Biogeographic element. Cosmopolitan (McAlester *in* Moore, 1969, p. N233).

Neilo beui Stilwell and Zinsmeister, 1992
Plate 1, figs. M and N

Neilo beui Stilwell and Zinsmeister, 1992, p. 52, pl. 1, figs. m, p and q.

Dimensions. USNM 490742 length 43.5 mm, height 23.0 mm; USNM 490743 length 39.5 mm, height 19.5 mm.

Figured specimens. USNM 490742-490743.

Material. One well-preserved specimen and other fragments.

Localities. E145, E155.

Geographic distribution. Seymour Island and McMurdo Sound.

Discussion. Specimens of a species of *Neilo* from McMurdo Sound, especially the beautifully preserved individual USNM 49042, match perfectly *Neilo beui* Stilwell and Zinsmeister [1992], a biostratigraphically useful species known only from a restricted interval in the upper part of Unit V in the La Meseta Formation of Seymour Island [see Stilwell and Zinsmeister, 1992, p. 31, Fig. 40, of biostratigraphic plot of bivalve taxa in La Meseta Formation; also see this work for comments on relationships with other taxa]. The presence of *Neilo beui* in Seymour Island and McMurdo Sound deposits indicates that the species had long-ranging, planktotrophic (teleplanic) larval capabilities.

Marshall [1978, p. 425] stated that species of *Neilo* can be important depth and climate indicators and are often encountered in "fine-grained, poorly oxygenated muds rich in anaerobic bacteria". *Neilo beui*, an inferred deposit feeder of varying degrees of mobility, has been recorded only in medium- to coarse-grained sandstones in Antarctica. These specimens of *Neilo beui* may have been transported from their original environments to shallower waters, but the well-preserved nature of the material and the rather thin shell indicates only minimal transport.

Subclass CRYPTODONTA Neumayr, 1884
Order SOLEMYOIDA Dall, 1889
Superfamily SOLEMYACEA Gray, 1840
Family SOLEMYIDAE Gray, 1840
Genus *Solemya* Lamarck, 1818

Solemya Lamarck, 1818, p. 488.

Type species. (By subsequent designation, Children, 1823) *Solemya mediterranea* (= *Tellina togata* Poli, 1795).

Biogeographic element. Cosmopolitan [Cox *in* Moore, 1969, p. N241].

Solemya surolongata, new species
Plate 1, figs. O and Q

Diagnosis. Quite narrow, subrectangular to elongate-ovate *Solemya* with 1.5 mm broad radiating ribs; distinguished from coeval Antarctic species from Seymour Island, *S. peteri* Zinsmeister, 1984, in having a smaller, much narrower and elongate shell with more abundant, narrower radiating ribs.

Description. Shell medium-sized, thin-shelled, very narrowly elongate-ovate to subrectangular, compressed; umbones small, beak seemingly slightly sunken, obsolete, more posterior to mid point, about a quarter of the length of shell from posterior margin; anterodorsal and posterodorsal margins nearly straight; anterior and posterior margins very narrowly rounded; ventral margin very long, mostly straight; shell ornamented with strongly radiating, moderately strong and prominent, subequally spaced, flat-topped ribs about 1.5 mm wide, becoming perpendicular posteriorly and oblique anteriorly.

Dimensions. Holotype USNM 490744 length 38.5 mm, height 13.5 mm; paratype USNM 490745 length 27.0 mm, height 7.5 mm.

Types. Holotype USNM 490744; paratype USNM 490745.

Figured specimens. USNM 490744, USNM 490745.

Type locality. E184.

Material. Two specimens.

Geographic distribution. McMurdo Sound.

Discussion. Both species and specimens of *Solemya* are rare in Tertiary deposits of the southern circum-Pacific, as noted by Zinsmeister [1984]. Described species are: *Soleyma antarctica* Philippi [1887] of the Miocene of Chile, *Solemya* sp. of Marwick [1931] from the Miocene of New Zealand, an undescribed species from the middle Eocene of New Zealand [JDS, unpublished data], *Solemya peteri* Zinsmeister [1984] of the Eocene of Seymour Island, and *Solemya surolongata* n. sp. from the Eocene of McMurdo Sound.

Solemya surolongata n. sp. is closely related to *S. peteri* of Seymour Island, differing only in having narrower radiating ribs and a smaller, more elongate shell. The umbones are more sunken and the ventral margin much straighter in *S. surolongata* n. sp., compared with *S. antarctica* [Philippi, 1887] [p. 179, pl. 42, fig. 5], suggesting a more distant relationship. The generally conservative morphology of species of *Solemya* makes the deduction of a putative ancestor for *S. surolongata* n. sp. difficult, but a strong candidate is a new species from the latest Cretaceous of Northland, New Zealand, described by Stilwell [1994].

Etymology. Species derived from the Spanish "sur" (equivalent to "south") and for its narrowly elongate shell outline.

Subclass PTERIOMORPHA Beurlen, 1944
Order ARCOIDA Stoliczka, 1871
Superfamily ARCACEA Lamarck, 1809
Family CUCULLAEIDAE Stewart, 1930
Genus *Cucullaea* Lamarck, 1801

Cucullaea Lamarck, 1801, p. 116.

Type species. (By subsequent designation, Children, 1823) *Cucullaea auriculifera* Lamarck, 1801.

Biogeographic element. Indo-Pacific/Tethyan, as interpreted herein.

Cucullaea sp. *cf. C. donaldi* Sharman and Newton, 1894
Plate 1, figs. P and R

cf. Cucullaea donaldi Sharman and Newton, 1894, pp. 50 and 51, fig. 1 (not fig. 2); Wilckens, 1911, pp. 6-11, figs. 8 and 9 (not 6 or 7); Zinsmeister, 1984, pp. 1505 and 1506, figs. 4E-4G; Stilwell and Zinsmeister, 1992, p. 53, pl. 1, figs. s, t, and w.

Dimensions. USNM 490746 length 46.5 mm, height 38.5 mm; USNM 490747 length of partial hinge 34.5 mm.

Figured specimens. USNM 490746-490747.

Material. Two poorly preserved specimens.

Locality. E359.

Discussion. This small *Cucullaea* is probably conspecific with *C. donaldi* Sharman and Newton [1894], recorded only from the uppermost units VI and VII in the La Meseta Formation of Seymour Island. Specimen USNM 490746, a poorly preserved leached example, is 46.5 mm long, quite comparable to the neotype of *C. donaldi* [see Stilwell and Zinsmeister, 1992, pl. 1, figs. s, t, and w], which is 52 mm long. The subrhomboid outline of the McMurdo Sound species is also consistent with *C. donaldi*. A partial external mold of the hinge of a second specimen from McMurdo Sound, USNM 490747, has a rather broad irregular ligament that degenerates into irregular crenulations in the central part of the hingeplate, as found by Zinsmeister [1984] in *C. donaldi*.

Cucullaea cf. C. donaldi has been recognized only from one erratic, E359, associated with an abundance of large oyster remains. Species-level diversity is low in this block, but also includes a large turritellid gastropod as steinkerns and external molds similar in size to *Colposigma capitanea* Stilwell and Zinsmeister [1992], known only from units V and VI of the La Meseta Formation.

Superfamily LIMOPSACEA Dall, 1895
Family LIMOPSIDAE Dall, 1895
Genus *Limopsis* Sassi, 1827

Limopsis Sassi, 1827, p. 476.

Type species. (By original designation) *Arca aurita* Brocchi, 1814.

Subgenus *Limopsista* Finlay and Marwick, 1937

Limopsista Finlay and Marwick, 1937, p. 24.

Type species. (By original designation) *Limopsis (Limopsista) microps* Finlay and Marwick, 1937.

Biogeographic element. Originally thought to be endemic to New Zealand, this subgenus may be paleo-austral, if the McMurdo Sound species proves to be allocated to this group.

Limopsis (Limopsista?) antarctominuta, new species
Plate 2, figs. G and I

Diagnosis. Subequivalved and somewhat equilateral species with less than 10 mostly equally spaced growth pauses and very weak commarginal threads; distinguished from the type species, *Limopsis (Limopsista) microps* Finlay and Marwick, 1937, in having more growth pauses and weaker interstitial commarginal threads.

Description. Shell minute for family, but average-sized for subgenus, moderately thick, moderately to greatly inflated, mostly equilateral and subequivalve, subcircular to only slightly subtrigonally ovate; umbones rounded, moderately large and inflated; anterodorsal and posterodorsal margins weakly declivous and mostly straight merging with well-rounded anterior and posteri-or margins; ventral margin moderately broad, but well rounded; shell mostly smooth apart from less than 10 subequally spaced growth pauses and very fine commarginal threads; hinge details unknown apart from a fraction of the taxodont hinge on one specimen.

Dimensions. Holotype USNM 490748 length 5.0 mm, height 4.5 mm; paratype USNM 490749 length 3.5 mm, height 3.5 mm, width of paired valves 2.0 mm; paratype USNM 490750 length of fragment 4.0 mm.

Types. Holotype USNM 490748; paratypes USNM 490749-490750.

Figured specimens. USNM 490748-490749.

Localities. E183; E203 (type), E375.

Material. Five specimens.

Geographic distribution. McMurdo Sound.

Discussion. Although full hinge details of this minute taxodont species are not available, the outline and sculpture are very close, indeed, to the late Early Paleocene limopsid species, *Limopsis (Limopsista) microps* Finlay and Marwick [1937] [pp. 24-25, pl. 1, figs. 12-13; see also Fleming, 1966, p. 120, pl. 11, figs. 119-120; Stilwell, 1994, pl. 50, figs. 5-9], recorded from Wangaloan rocks of South Island, New Zealand. With regard to external features, the only detectable differences between the McMurdo Sound and New Zealand species relate to number of growth pauses and strength of interstitial threads, which are weaker in the Antarctic species. Until hinge details are better known in the McMurdo Sound species, the subgenus-level assigned can only be deemed tentative.

Plate 2

Figs. A and F. *Eburneopecten* sp. (A) USNM 490756, E373, length of block with specimens = 250 mm. (F) USNM 490857, E376, length = 21.5 mm, x1.

Figs. B-E. *Brachidontes sandalius* n. sp. (B) Paratype USNM 490752, length = 10.5 mm, x3. (C) Paratype USNM 490754, E207, length = 21.0 mm, x3. (D) Paratype USNM 490753, E207, length of fragment = 6.5 mm, x15 (SEM). (E) Holotype USNM 490751, E207, length = 10.5 mm, x3.

Figs. G and I. *Limopsis (Limopsista?) antarctominuta* n. sp. (G) Holotype USNM 490748, E203, length = 5.0 mm, x3.5. (I) Paratype USNM 490749, length = 3.5 mm, x15.

Fig. H. *Chlamys s. l.* n. sp. (H) USNM 490757, E327, length = 14.5 mm, x1.

Fig. J. Anomiidae genus et species indeterminate. (J), USNM 490763, E189, length = 8.0 mm, x4,

Figs. K-O. *Crassostrea antarctogigantea* n. sp. (K) Holotype USNM 490758, E359, length = 108 mm, x0.5. (L) Paratype USNM 490760, E336, length = 112 mm, x0.5. (M) Paratype USNM 490761, E359, length = 73 mm, x0.9. (N) Paratype USNM 490762, E359, length of hinge =69.0 mm, x1. (O) Paratype USNM 490760, E359, length = 112 mm, x0.5.

Fig. P. *Aulacomya sp. cf. A. anderssoni* Zinsmeister. (P) USNM 490755, E379, length = 60.5 mm, x0.75.

Darragh [1994, p. 83] noted that *Limopsis microps* "looks almost like a species of *Glycymeris*", but gave no further comment. *Limopsis (Notolimopsis)* Maxwell [1969] of the Limopsidae is closely related to *Limopsis (Limopsista)*, but has twice as many hinge teeth.

Etymology. Species named for its occurrence in Antarctica and for its small size.

Order MYTILOIDA Férussac, 1822
Superfamily MYTILACEA Rafinesque, 1815
Family MYTILIDAE Rafinesque, 1815
Subfamily MYTILINAE Rafinesque, 1815
Genus *Aulacomya* Mörch, 1853

Type species. (By subsequent designation, von Ihering, 1900) *Mytilus magellanicus* Chemnitz, 1785 (= *Mytilus ater* Molina, 1782).

Biogeographic element. Indo-Pacific/Tethyan, as interpreted herein.

Aulacomya sp. *cf. A. anderssoni* Zinsmeister, 1984
Plate 2, fig. P

cf. Aulacomya anderssoni Zinsmeister, 1984, pp. 1507 and 1508, fig. 5A; Stilwell and Zinsmeister, 1992, pp. 56 and 58, pl. 2, figs. e and g.

Dimensions. USNM 490755 length 60.5 mm, height 26.5 mm.

Figured specimen. USNM 490755.

Material. One articulated, poorly preserved specimen.

Locality. E379.

Discussion. A poorly preserved mytilid from Mount Discovery is probably conspecific with the Eocene Seymour Island species *Aulacomya anderssoni* Zinsmeister [1984], recorded from Units III-V. This specimen, USNM 490755, is largely decorticated, but vestiges of ornament including punctuated growth pauses and the overall outline match *A. anderssoni* very well, apart from the seemingly straighter anterodorsal margin of the McMurdo Sound specimen. This could be a preservational artifact. No further comment on this species is warranted until further material is recovered.

Genus *Brachidontes* Swainson, 1840

Brachidontes Swainson, 1840, p. 384.

Type species. (By monotypy) *Modiola sulcata* Lamarck, 1819 (*non* 1805) (= *Mytilus citrinus* Röding, 1798, = *Arca modiolus* Linné, 1767).

Biogeographic element. Cosmopolitan (Soot-Ryen *in* Moore, 1969, p. N273).

Brachidontes sandalius, new species
Plate 2, figs. B-E

Diagnosis. Average-sized to somewhat small *Brachidontes* with a foot- or sandal-shaped outline and moderately strong to strong umbonal ridge, punctuated growth pauses, and more than 40 irregularly developed radial riblets, strongest at periphery of shell; distinguished from the Recent type, *B. modiolus* (Linné, 1758), in having a shorter more squat ovate shell, a more broadly convex posterodorsal margin, and much weaker radial ornamentation.

Description. Shell small- to medium-sized for genus, thin-shelled, moderately inflated, transversely ovate (slightly modioliform); beaks subterminal; anterodorsal margin broadly convex, mostly straight from beaks to a point midway of entire length of shell; posterodorsal margin short, steep, narrowly rounded; posterior margin narrowly rounded; umbonal ridge moderately to strongly developed, weakening slightly towards posterior margin; surface mostly smooth apart from punctuated commarginal growth pauses and numerous threads, and more than 40, mostly weak, irregularly and variably developed and wavy, radial riblets, strongest near margins of shell; hinge details unknown.

Dimensions. Holotype USNM 490751 length 10.5 mm, height 4.0 mm; paratype USNM 490752 length 10.5 mm, height 4.5 mm; paratype USNM 490753 length of posterior fragment 6.5 mm; paratype USNM 490754 length 21.0 mm, height 7.0 mm.

Types. Holotype USNM 490751; paratypes USNM 490752-490754.

Figured specimens. USNM 490751-490754.

Localities. E184, E206, E207 (type).

Material. Nine specimens and many fragments.

Geographic distribution. McMurdo Sound.

Discussion. *Brachidontes sandalius* n. sp. from McMurdo Sound is closely allied with *Arcuatula sootryeni* Stilwell and Zinsmeister [1992] [p. 58, pl. 3, figs. c, e, and g] from the uppermost units VI and VII of the La Meseta Formation of Seymour Island, and is most likely congeneric. These taxa can be assigned to *Brachidontes* Swainson [1840] because they have subterminal umbones and characteristic radial sculpture. However, a closely related group *Hormomya* Mörch [1853] can be difficult to distinguish from *Brachidontes* [see comments by Soot-Ryen, 1955, pp. 43-44], but *Hormomya* usually has more terminal umbones. The McMurdo Sound species differs from *B. sootryeni* in being consistently much smaller with a less convex posterodorsal margin and an overall more sandal-shaped shell. These two taxa were probably derived from the same parent stock.

The outline of the Recent North and South American species *Brachidontes exustus* [Linné, 1758] [see Abbott and Dance, 1983, fig. on p. 298] is very similar in outline to *B. sandalius* n. sp., but is more arcuate than the McMurdo Sound species. *Brachidontes sandalius* n. sp. is one of the more common bivalves recovered from the erratics.

Etymology. Species name derived from the Latin "sandalium" (equivalent to "slipper" or "sandal") for its sandal-shaped outline.

Order PTERIOIDA Newell, 1965
Suborder PTERIINA Newell, 1965
Superfamily PECTINACEA Rafinesque, 1815
Family PECTINIDAE Rafinesque, 1815
Genus *Eburneopecten* Conrad, 1865

Eburneopecten Conrad, 1865, p. 140.

Type species. (By original designation) *Pecten scintillatus* Conrad, 1865.
Biogeographic element. Indo-Pacific/Tethyan, as interpreted here.

Eburneopecten sp.
Plate 2, figs. a and f

Dimensions. USNM 490756 length of block with multiple specimens 250 mm; specimens range from 30-40 mm in length and height; USNM 490857 length 21.5 mm.
Figured specimens. USNM 490756, USNM 490857.
Material. Five poorly preserved specimens.
Localities. E373, E376.
Geographic distribution. McMurdo Sound.
Discussion. Pectinid bivalves are generally scarce in the erratics, but in one block, USNM 490756, *Eburneopecten* sp. is the most common bivalve in a matrix of medium- to coarse-grained sandstone. Associated with the pectens are brachiopod and oyster and lucinid bivalve remains. The outline, auricle morphology and vestiges of commarginal sculpture in the McMurdo pectens are consistent with the latest Cretaceous to Paleogene genus *Eburneopecten* Conrad [1865], but the poor preservation of the material at hand prevents any meaningful comparisons. The type species, *E. scintillatus* [Conrad, 1865]

[see Hertlein *in* Moore, 1969, p. N352, fig. C75-1a, b; Toulmin, 1977, p. 316, pl. 56, figs. 4-5], from the Eocene of North America has the same outline as *Eburneopecten* sp. from Antarctica, but the shell is much smaller and the auricles in the right valve are longer and less inclined in the North American species. The oldest species recorded is a new taxon from the latest Cretaceous of Chatham Islands, South Pacific [Stilwell, 1998], indicating that the origin of this group was in the high southern latitudes.

Genus *Chlamys* Röding, 1798

Chlamys Röding, 1798, p. 161.

Type species. (By subsequent designation, Hermannsen, 1847) *Pecten islandicus* Müller, 1776.
Biogeographic element. Cosmopolitan (Herlein *in* Moore, 1969, p. N355; Kauffman, 1973, p. 359).

Chlamys s. l., new species
Plate 2, fig. H

Dimensions. USNM 490757 length 14.5 mm incomplete, height 25.0 mm incomplete.
Figured specimen. USNM 490757.
Material. One incomplete specimen.
Locality. E327.
Geographic distribution. McMurdo Sound.
Discussion. This new pectinid is distinct from described pectinids from the La Meseta Formation of Seymour Island [see Stilwell and Zinsmeister, 1992, pp. 60 and 61, pl. 4, figs. a-d and f], but the subgenus is uncertain due to the incomplete nature of the sole recorded specimen. The posterior auricle in this specimen is moderately long, horizontal and square cut. The sculpture is of about 15 subequally spaced and nearly flat-topped radial ribs and very weak commarginal growth striae.

The sculpture of *Chlamys s. l.* n sp. is very close to a New Zealand late Eocene species identified as *Chlamys (s. l.)* n. sp. by Maxwell [1992, p. 65, pl. 3, figs. g-i] and also to a Patagonian Tertiary species *Amussium cossmanni* von Ihering [1907, p. 260, pl. 9, fig. 59 a-b], but the McMurdo Sound species has fewer, more robust radial ribs.

Superfamily ANOMIACEA Rafinesque, 1815
Family ANOMIIDAE Rafinesque, 1815

Anomiidae genus et species indeterminate
Pl. 2, fig. J

Dimensions. USNM 490763 length 8.0 mm, height 10.5 mm; USNM 490764 length 7.5 mm, height 10.5 mm.

Figured specimen. USNM 490763.

Material. Three poorly preserved specimens and fragments.

Localities. E189, E373.

Geographic distribution. McMurdo Sound.

Discussion. These very thin, subcircular bivalves are consistent with taxa within the variable family Anomiidae, but are too poorly preserved for in-depth comment on relationships. The petite shell of the McMurdo Sound species is reminiscent of the early Tertiary Patagonian species, *Pododesmus juliensis* von Ihering [1907] [pp. 267 and 268, pl. 10, fig. 65 a-b], but the Patagonian species is a bit larger with a less irregular shell. No coeval anomiid bivalves have been recorded from the La Meseta Formation of Seymour and Cockburn islands, Antarctic Peninsula.

Order OSTREOIDA Férussac, 1822
Suborder OSTREINA Férussac, 1822
Superfamily OSTREACEA Rafinesque, 1815
Family OSTREIDAE Rafinesque, 1815
Subfamily CRASSOSTREINAE Torigoe, 1981
Genus *Crassostrea* Sacco, 1897

Crassostrea Sacco, 1897, p. 15.

Type species. (By original designation) *Ostrea virginica* Gmelin, 1791.

Biogeographic element. Cosmopolitan (Stenzel *in* Moore, 1971, p. N1129; Kauffman, 1973, p. 359).

Crassostrea antarctogigantea, new species
Plate 2, figs. k-o; Plate 3, figs. a and b

Diagnosis. Immense, thick and robust elongate-ovate to obliquely-ovate *Crassostrea* with very inflated (60 mm or greater) left valve, semi-regularly foliose shell with about 5 mm layers, and large ligament pit that is nearly twice as high as wide; differs from *Crassostrea ingens* [Zittel, 1864] of the Late Miocene of New Zealand in being somewhat smaller with a less arcuate, more oblique shell, less developed ridges and grooves adjacent to ligament pit, which is more oblique.

Description. Shell very large (some specimens more than 160 mm high), variable in outline, but mostly narrowly elongate and oblique to elongate-ovate, shell very thick and robust, calcitic; shape of shell variable in some instances due to xenomorphic sculpture reflecting contours of substrate of attachment; length of shell usually less than half of height; left valve very inflated, 60 mm or more in most specimens; right valve flatter, not as foliose as left valve; all specimens slightly curved; dorsal margin in both valves irregular, moderately narrow, becoming slightly more expanded ventrally; some specimens with a gently convex dorsal margin; ventral margin well-rounded; exterior of shell semi-regularly foliose in rather thick about 5 mm layers; ligament area quite large, in most specimens higher than wide, depressed subcentrally with elongate pit, which is nearly twice as high as long; pit with adjacent obliquely trending ridges and grooves; vestiges of poorly developed chomata in one specimen, just ventral of hinge; adductor scar large, subcentral, steeply inclined; internal ventral margin bounded by broad groove about 35 mm wide in holotype, extending from margin of adductor muscle scar to ventral margin.

Dimensions. Holotype USNM 490758 length 108 mm, height 162 mm; paratype USNM 490759 length 100 mm, height 140 mm; paratype USNM 490760 length 112 mm, height 120 mm; USNM 490761 length 73 mm,

---

Plate 3

Figs. A and B. *Crassostrea antarctogigantea* n. sp. (A) Paratype USNM 490759, E359, length = 100 mm, x0.5. (B) Paratype USNM 490759, E359, length = 100 mm, x0.5.

Figs. C, D, F and H. *Saxolucina sharmani* (Wilckens). (C) USNM 490766, E145, length = 18.5 mm, x1. (D) USNM 490767, E155, length = 21.0 mm, x1.1. (F) USNM 490768, E145, length = 19.0 mm, x3.5. (H) USNM 490765, E185, length = 18.5 mm, x3.

Figs. E, L, and M. *?Anisodonta truncilla* Stilwell and Zinsmeister. (E) USNM 490796, E207, length = 3.5 mm, x3. (L) ?USNM 490796, E207, length = 3.5 mm, x28 (SEM). (M) USNM 490797, E207, length = 3.5 mm, x23 (SEM).

Figs. G, I, and K. *Miltha?* sp. (G) USNM 490772, E330, length = 31.0 mm, x1. (I) USNM 490773, E330, length = 26.0 mm, x1. (K) USNM 490771, E375, length = 29.5 mm, x1.3.

Fig. J. *Thyasira (Conchocele) australosulcata* n. sp. (J) Holotype USNM 490770, E194, length = 73.0 mm, x0.8.

height 105 mm nearly complete; paratype USNM 490762 length of hinge 69 mm.

Types. Holotype USNM 490758; paratypes USNM 490759-490762.

Figured specimens. USNM 490758-490762.

Localities. E336, E359 (type).

Material. Four specimens and many incomplete specimens.

Geographic distribution. McMurdo Sound.

Discussion. This massive crassostreine oyster is related closely to Tertiary species of *Crassostrea* of New Zealand and South America. It is interesting to note that coeval Antarctic species described from Seymour Island, *Ostrea antarctica* Zinsmeister [1984] [pp. 1510 and 1511, fig. 6A; see also Stilwell and Zinsmeister, 1992, p. 62, pl. 4, fig. e] and *O. seymourensis* Zinsmeister [1984] [p. 1511, fig. 6B; see also Stilwell and Zinsmeister, 1992, p. 64, pl. 4, fig. h], are not linked closely with *Crassostrea antarctogigantea* n. sp. from McMurdo Sound.

The closest relative of *Crassostrea antarctogigantea* n. sp. seems to lie with *C. ingens* [Zittel, 1864] [p. 54, pl. 13, fig. 3; see also Beu and Maxwell, 1990, p. 280, pl. 32, fig. f] from the Late Miocene of New Zealand. After Zittel's description of *Ostrea ingens*, Ortmann [1902, pp. 99-110, pls. 15, 16 (double-page), 17, 18, and 19, fig. 1a-e] described the same species from many Tertiary Patagonian localities, but stated that this gargantuan species is not known from older Tertiary localities. In the same work, Ortmann listed many synonymies, which will not be repeated herein. Indeed, very little variation can be seen between the Patagonian and New Zealand material, apart from the more diffuse ridges and grooves bordering the shallower ligament pit in the New Zealand material.

Crassostrea antarctogigantea n. sp. is generally smaller with a more curved and oblique shell, and a more oblique ligament pit, compared to *C. ingens*. The Recent type species, *C. virginica* Gmelin [1791] [see Wood and Hanley, 1856, p. 62, pl. 11, fig. 68; *Stenzel in* Moore, 1971, pp. N1128-N1131, figs. J101 and J102; Abbott and Dance, 1983, unnumbered fig., p. 318], is a much more gracile species with a very narrow, elongate shell and more irregular external growth squamae, compared with *C. antarctogigantea* n. sp. This new species is the largest of all Eocene East Antarctic bivalves. Based on the large erratics thus far recovered including *C. antarctogigantea* n. sp., this species originally formed immense bioherms along the shallow shelf. Associated with the oysters are other bivalves and gastropods that indicate fully marine conditions of normal salinity.

Etymology. Species name derived from its occurrence in Antarctica and its large size.

Subclass HETERODONTA Neumayr, 1884
Order VENEROIDA H. and A. Adams, 1856
Superfamily LUCINACEA Fleming, 1828
Family LUCINIDAE Fleming, 1828
Subfamily LUCININAE Fleming, 1828
Genus *Saxolucina* Stewart, 1930

Saxolucina Stewart, 1930, p. 184.

Type species. (By original designation) *Lucina saxorum* Lamarck, 1806.

Biogeographic element. Indo-Pacific/Tethyan, as interpreted herein.

Saxolucina sharmani (Wilckens, 1911)
Plate 3, figs. C, D, F, and H

Phacoides sharmani Wilckens, 1911, p. 12, pl. 1, fig. 11.
Saxolucina sharmani (Wilckens), Zinsmeister, 1984b, p. 1513, figs. 7M and 7N; Stilwell and Zinsmeister, 1992, p. 64, pl. 4, figs. j and n.

Dimensions. USNM 490765 length 18.5 mm, height 18.5 mm; USNM 490766 length 19.0 mm, height 19.0 mm (specimen articulated, but partially opened); USNM 490767 length 21.0 mm, height 21.0 mm; USNM 490768 length 19.0 mm, height 15.5 mm incomplete.

Figured specimens. USNM 490765-490768.

Material. 10 specimens and many partial specimens.

Localities. E145, E155, E184, E185, E207.

Geographic distribution. Seymour Island and McMurdo Sound.

Discussion. The abundant material at hand of this lucinid species is virtually indistinguishable from the long-ranging species *Saxolucina sharmani* (Wilckens, 1911) of the La Meseta Formation, Seymour Island. Some of the McMurdo Sound specimens are slightly more inflated than others and the growth pauses slightly more developed in a few specimens, revealing minor intraspecific variability. Zinsmeister [1984] noted that the *Lucina promaucana* Philippi [1887] [p. 175, pl. 24, fig. 6; see Ortmann, 1902, pp. 130 and 131, pl. 27, fig. 4a-b], a widespread species in Tertiary rocks of southern South America, is rather similar in outline and ornamentation, but *S. sharmani* lacks a lateral tooth in the hinge. In addition, the anterodorsal margin of *L. promaucana* is longer than in *S. sharmani*.

Von Ihering [1907] [p. 288, pl. 13, fig. 90a-b] described *Phacoides promaucana crucialis* as a subspecies common in the Tertiary of Argentina, but the differences between this species and *L. promaucana* are probably of species-level. The beaks of *P. crucialis* are more prosogyrous and the anterodorsal margin is more concave, compared with *L. promaucana*. The outline of *P. crucialis* is reminiscent of Tertiary species of *Gonimyrtea* Marwick [1929], but the hinge details cannot be ascertained in von Ihering's description and figure.

Subfamily MILTHINAE Chavan *in* Moore, 1969
Genus *Miltha* H. and A. Adams, 1857

Miltha H. and A. Adams, 1857, p. 468.

Type species. (By original designation) *Lucina childreni* Gray, 1825.
Biogeographic element. Indo-Pacific/Tethyan (*cf.* Fleming, 1967, p. 115).

Miltha? sp.
Plate 3, figs. G, I, and K

Dimensions. USNM 490771 length 29.5 mm, height 29.0 mm; USNM 490772 length 31.0 mm, height 27.5 mm incomplete; USNM 490773 length 26.0 mm, height 28.5 mm.
Figured specimens. USNM 490771-490773.
Material. Three specimens.
Localities. E330, E375.
Geographic distribution. McMurdo Sound.
Discussion. These large subcircular bivalves may be congeneric with the Paleocene to Recent lucinid *Miltha* H. and A. Adams [1857] and probably represent a new species. The outline of this species is similar to *Saxolucina sharmani* [Wilckens, 1911], discussed above, but *Miltha?* sp. is much larger with a more compressed shell. The outline and sculpture is also reminiscent of *Miltha agilis* [Finlay and Marwick, 1937] [pp. 27-28, pl. 3, figs. 3, 6, and 10; Fleming, 1966, p. 158, pl. 30, figs. 303-305; Stilwell, 1994, pp. 776-779, pl. 53, figs. 1-5], from the late Early Paleocene of New Zealand, but *Miltha?* sp. has a smaller shell, weaker commarginal sculpture and slightly less prosogyrous beaks.

The type species of *Miltha, M. childreni* (Gray, 1825) [see Abbott and Dance, 1983, p. 322, unnumbered fig.], from the Recent of South America has a more circular shell, more elevated umbones, and a much more obliquely truncated posterodorsal margin, compared with *Miltha?* sp.

Family THYASIRIDAE Dall, 1901
Genus *Thyasira* Leach *in* Lamarck, 1818

Thyasira Leach *in* Lamarck, 1818, p. 195.

Type species. (By original designation) *Amphidesma flexuosa* Lamarck, 1818.

Subgenus *Conchocele* Gabb, 1866

Conchocele Gabb, 1866, p. 27.

Type species. (By original designation) *Conchocele disjuncta* Gabb, 1866.
Biogeographic element. Indo-Pacific/Tethyan, as interpreted herein.
Discussion. Chavan *in* Moore [1969, p. N508] stated that the range of *Thyasira (Conchocele)* is Oligocene to Recent, but recent work in the southern hemisphere extends this range into the Cretaceous. *Thyasira (Conchocele)* has been recorded in Campanian to Maastrichtian rocks of New Zealand, New Caledonia, South America, and Antarctica [Stilwell, 1994].

The autecology of large thyasirid bivalves has been studied by many researchers, and this group of bivalves has been observed to be deep-burrowing, suspension-feeding, slow-moving, and is most characteristic of low-diversity faunal assemblages, sometimes populating areas of oxygen-poor, hydrogen sulfide-rich waters and nutrient-poor habitats [see comments by Kauffman, 1967; Kauffman *in* Moore, 1969; McKerrow, ed., 1978; Freneix, 1980, 1981; Hickman, 1984; Evans, 1985; Reid and Brand, 1986; Macellari, 1988; and Beu and Maxwell, 1990].

Thyasira (Conchocele) australosulcata, new species
Plate 3, fig. J

Diagnosis. Shell quite large for *Thyasira (Conchocele)*, somewhat ovate to subtrigonal, moderately oblique; ratio of length to height marginally less than 1:1; beaks small, very prosogyrate; ventral margin broadly convex; sulcus deep, characterized by a wide 10.5 mm band extending from umbones to subangular intersection of posterodorsal and posterior margin; shell mostly smooth apart from commarginal growth pauses adjacent to ventral margin that broaden posteriorly; distinguished from the closely related latest Cretaceous Antarctic and South American species, *Thyasira (Conchocele) townsendi* (White, 1887), in having a bigger shell, broader, more convex ventral margin, more prosogyrous beaks, a slightly shallower sulcus, and weaker commarginal sculpture.

Description. Shell very large for subgenus, robust, moderately thick, moderately inflated, moderately declivous from subcentral (more dorsal) part of disc to ventral margin, subovate to subtrigonal, moderately oblique; ratio of length to height just under 1:1; beaks small, very prosogyrate, nearly flush with anterior margin; umbones marked, well-rounded; anterodorsal margin short, very steeply inclined, nearly straight, merging towards moderately narrowly rounded anterior margin; posterodorsal margin long, gently sloping, broadly convex; posterior margin bluntly rounded, only slightly sharpened; intersection of posterodorsal and posterior margins subangular, reflecting pronounced sulcus; ventral margin long, broadly convex; sulcus marked, deep, extending the entire length of posterodorsal margin by a broad 10.5 mm band; lunule poorly developed; surface of shell mostly smooth apart from bunched growth pauses adjacent to ventral margin, becoming more spaced towards posterior margin, and weak subsurface radial elements; internal details unknown.

Dimensions. Holotype USNM 490770 length 73.0 mm, height 72.0 mm, width of single right valve 19.0 mm.

Type. Holotype USNM 490770.

Figured specimen. Holotype.

Type locality. E194.

Material. Holotype. A fragmentary specimen from E194 may represent this species.

Geographic distribution. McMurdo Sound.

Discussion. A lineal relationship between the Maastrichtian South American and Antarctic species, *Thyasira townsendi* White, 1887, p. 14, pl. 3, figs. 1-2; see also Weller, 1903, p. 415, pl. 1, figs. 2-3; Wilckens, 1910, p. 53, pl. 2, fig. 31a-c, pl. 3, fig. 1; Zinsmeister and Macellari, 1988, pp. 273 and 276, fig. 9-7, 8], and the McMurdo Sound species, *Thyasira (Conchocele) australosulcata* n. sp., seems probable, given the relatively minor morphological disparities between these taxa. *Thyasira townsendi* is herein assigned to *Thyasira (Conchocele)*, as its morphology is consistent with this group. *Thyasira (Conchocele) australosulcata* n. sp. has a larger shell, a broader ventral margin, more prosogyrous beaks, and weaker commarginal ornamentation, compared to *T. (C.) townsendi*. It is surprising that no Paleogene species of this subgenus have been recorded from Seymour or Cockburn islands given the remarkable fossil record, but the rare occurrence of this group in McMurdo Sound, represented by the single right valve, indicates further that this specimen may have been trans-

ported from its original habitat into much shallower water. The McMurdo Sound specimen is quite robust and large and somewhat worn apart from the ventral margin indicating some transport. *Thyasira (Conchocele) townsendi* in the Lopez de Bertodano Formation of Seymour Island is interpreted to have lived in a middle shelf environment in a siltstone facies, similar to a new species recorded from Maastrichtian middle to lower shelf deposits of Kaipara, Northland, New Zealand [Stilwell, 1994].

Etymology. Species name derived from its Austral occurrence and its pronounced sulcus.

Superfamily CYAMIACEA Philippi, 1845
Family SPORTELLIDAE Dall, 1895
Genus *Anisodonta* Deshayes, 1858

Anisodonta Deshayes, 1858, p. 542.

Type species. (By monotypy) *Anisodonta complanata* Deshayes, 1858.

Biogeographic element. Cosmopolitan (Keen *in* Moore, 1969, p. N540).

?Anisodonta truncilla Stilwell and Zinsmeister, 1992
Plate 3 , figs. E, L and M

Anisodonta truncilla Stilwell and Zinsmeister, 1992, pp. 68 and 69, pl. 5, figs. i-k.

Dimensions. USNM 490796 length 3.5 mm, height 3.0 mm; USNM 490797 length 3.5 mm, height 2.5 mm.

Figured specimens. USNM 490796-490797.

Material. Two specimens.

Locality. E207.

Geographic distribution. McMurdo Sound and Seymour Island, if conspecific.

Discussion. These subquadrate specimens with a moderately developed medioposterior carina are very similar to the sportellid bivalve, *Anisodonta truncilla* Stilwell and Zinsmeister [1992], described from Units II and III of the La Meseta Formation, the only detectable difference being the smaller size and apparent absence of radial threads in the McMurdo Sound material. The latter difference may be a reflection of preservation. The specimens also resemble *Cyamiomactra laminifera* [Lamy, 1906] [see recent work on this species by Dell, 1990, pp. 50 and 51, fig. 100] from subantarctic islands and also off the coast of the Antarctic continent from 15 to 1281 m depth.

Superfamily CARDITACEA Fleming, 1820
Family CARDITIDAE Fleming, 1820
Subfamily CARDITINAE Fleming, 1820
Genus *Cardita* Brugière, 1792

Cardita Brugière, 1792, p. 401.

Type species. (By subsequent designation, Gray, 1847) *Chama calyculata* Linné, 1758.
Biogeographic element. Indo-Pacific/Tethyan as interpreted herein.

Cardita subrectangulata, new species
Plate 4, fig. a

Diagnosis. Relatively small *Cardita* with subrectangular to subquadrate outline; shell sculptured with about 12 strong, equal spaced radial riblets; distinguished from Recent type species, *C. calyculata* [Linné, 1758] in being smaller, having fewer riblets, and having a more quadrate outline.

Description. Shell rather small for genus, moderately thin, inflated, subquadrate to subrectangular, expanded posteriorly; shell greatly inflated in central portion of disc, becoming steeply declivous ventrally and dorsally; beaks nearly flush with anterior margin; anterodorsal margin short, steep, only slightly truncated, slightly concave, merging towards narrowly rounded anterior margin; posterodorsal margin moderately long, mostly straight, merging towards moderately narrowly rounded posterior margin; ventral margin mostly straight, long; shell ornamentation of about 12 pronounced, equally spaced radial riblets; hinge and internal details unknown.

Dimensions. Holotype USNM 490774 length 11.5 mm, height 10.0 mm.
Type. Holotype USNM 490774.
Figured specimen. Holotype.
Material. Holotype.
Type locality. E373.
Geographic distribution. McMurdo Sound.
Discussion. *Cardita subrectangulata* n. sp. is the only record of this genus in the Eocene of Antarctica and is not too far removed from the Recent type, *Chama calyculata* Linné [1758] [see Wood and Hanley, 1856, p. 53, pl. 9, fig 10; *Chavan in* Moore, 1969, p. N548, fig. E48-1], differing in being smaller with a more quadrate outline and fewer radial riblets. Otherwise, these two taxa have many overall common features, characteristic of this widespread group.

Family CARDIIDAE Lamarck, 1809
Subfamily VENERICARDIINAE Chavan *in* Moore,

1969
Genus *Cyclocardia* Conrad, 1867

Cyclocardia Conrad, 1867, p. 191.

Type species. (By subsequent designation, Stoliczka, 1871) *Cardita borealis* Conrad, 1831.
Biogeographic element. Cosmopolitan (Chavan *in* Moore, 1969, p. N551).
Discussion. *Cyclocardia* Conrad [1867] has had a long history in Antarctica with species spanning the mid Eocene to Recent [Dell, 1990; Stilwell and Zinsmeister, 1992].

Cyclocardia sp.
Plate 4, figs. B-D

Dimensions. USNM 490777 length of fragment 5.0 mm, height 4.5 mm; USNM 490778 length 4.5 mm, height 5.0 mm.
Figured specimens. USNM 490777-490778.
Material. Two poorly preserved specimens.
Localities. E200; E374.
Distribution. McMurdo Sound.
Discussion. These poorly preserved fragments compare well with species of the long-ranging Cenozoic genus *Cyclocardia* Conrad [1867], including the coeval Seymour Island species *Cyclocardia mesembria* Stilwell and Zinsmeister [1992] [pp. 69-70, pl. 5, figs. q-r] and much younger northern late Tertiary species such as *Venericardia granulata* Say [see Conrad, 1838, pp. 12-13, pl. 7, fig. 1], which has nearly identical granulose radial rib development. The trend of the ventral margin in the McMurdo Sound species is much more convex and, hence, different from the broad ventral margin of the Recent Antarctic species, *C. astartoides* [Martens, 1878] [see synonym *C. antarctica* Smith, 1907, p. 2, pl. 2, figs. 15-15a; Dell, 1990, pp. 59-60, figs. 98-99], suggesting a distant relationship between these taxa.

Subfamily PROTOCARDIINAE Keen, 1951
Genus *Nemocardium* Meek, 1876

Nemocardium Meek, 1876, p. 167.

Type species. (By subsequent designation, Sacco, 1899) *Cardium semiasperum* Deshayes, 1858.
Biogeographic element. Indo-Pacific/Tethyan as interpreted here.

Subgenus *Pratulum* Iredale, 1924
Pratulum Iredale, 1924, pp. 182, 207.

Type species. (By original designation) *Cardium thetidis* Hedley, 1902.

Biogeographic element. Indo-Pacific/Tethyan (*cf.* Kauffman, 1973, p. 372; *cf.* Piccoli *et al.*, 1986).

Nemocardium (Pratulum?) minutum n. sp.
Plate 4, figs. E and I

Diagnosis. Small *Nemocardium (Pratulum?)* with gently declivous antero- and posterodorsal margins and about 35 subequal radial riblets that become wider and rounded ventrally; riblets slightly pustulose due to crossing commarginal growth lines; distinguished from Recent Australian type species *N. (P.) thetidis* (Hedley, 1902) in being much smaller with slightly broader radial riblets ventrally and absence of posterior scales.

Description. Shell small for genus and subgenus, subcircular, thin, polished, only slightly inflated, nearly equilateral; length to height ratio about 1:1; umbones slightly inflated, close to center; antero- and posterodorsal margins equally short and gently sloping merging with well-rounded anterior and posterior margins; ventral margin convex; surface sculptured with approximately 35 subequal radial riblets that broaden and become more rounded ventrally; riblets partially beaded reflecting equally spaced growth lines; interspaces smooth apart from weak commarginal growth lines; internal details unknown; ventral margin weakly crenulate.

Dimensions. Holotype USNM 490801 length 3.0 mm, height 3.0 mm; paratype USNM 490779 length 3.0 mm, height 2.5 mm.

Types. Holotype USNM 490801; paratype USNM 490779.

Figured specimens. USNM 490801; USNM 490779.

Material. One complete specimen and one external mold.

Locality. E145 (type); E374.

Geographic distribution. McMurdo Sound.

Discussion. This new species is likely to be closely related to a suite of Australian and New Zealand species, *Nemocardium (Pratulum) thetidis* [Hedley, 1902] [see Macpherson and Gabriel, 1962, p. 339, fig. 387] and *N. (P.) modicum* [Marwick, 1944] [p. 265, pl. 36, figs. 17-18; see also Fleming, 1966, p. 168, pl. 35, figs. 368-370; Stilwell, 1994, pl. 54, figs. 9-17], respectively, but the absence of posterior scales on the McMurdo Sound material forces a tentative subgenus-level assignment herein. *Nemocardium (Pratulum?) minutum* n. sp. is distinguished from *N. (P.) thetidis* in being much smaller with broader and more rounded radial riblets ventrally without posterior scales. *Nemocardium (Pratulum) modicum* from the late Early Paleocene of New Zealand has been placed in this subgenus by several authors although it, too, lacks posterior scales unlike the type [see Marwick, 1944; Fleming, 1966; Beu and Maxwell, 1990]. This species is

---

Plate 4

Fig. A. *Cardita subrectangulata* n. sp. (A) Holotype USNM 490774, E373, length = 11.5 mm, x1.5.

Figs. B-D. *Cyclocardia* sp. (B) USNM 490778, E200, length = 4.5 mm, x3.8. (C) USNM 490777, E374, length = 5.0 mm, xc.25. (D) USNM 490777, E374, length = 5.0 mm, x4.

Figs. E and I. *Nemocardium (Pratulum?) minutum* n. sp. (E) Holotype USNM 490801, E145, length = 3.0 mm, x3.5. (I) Paratype USNM 490779, E374?, length = 3.0 mm, x4.

Fig. F. *?Gomphina iheringi* Zinsmeister. (F) USNM 490798, E168, length = 8.5 mm, x4.

Figs. G and J. *Crassatella* sp. (G) USNM 490776, E207, length = 37.0 mm, x1. (J)USNM 490775, E145, length = 62.5 mm, x1.

Figs. H, K, M, and P. *"Eurhomalea" claudiae* n. sp (H) Paratype USNM 490782, E348, length (hinge) = 10.0 mm, x3.5. (K) Holotype USNM 490780, length = 27.0 mm, x3. (M) Holotype USNM 490780, length = 27.0 mm, x3. (P) Paratype USNM 490781, length (hinge) = 6.5 mm, x2.

Figs. L and O. *Cyclorismina?* n. sp. *cf. "C". marwicki* Zinsmeister. (L) USNM 490783, length = 30.0 mm, x1. (O) USNM 490784, E325, length = 43.0 mm, x0.8.

Figs. N and Q. *?Eumarcia (Atamarcia) robusta* Stilwell and Zinsmeister. (N) USNM 490785, Locality MA80 036.2 (Peter Webb, Ohio State University), length = 78.5 mm, x0.8. (Q) USNM 490785, Locality MA80 036.2, length = 78.5 mm, x0.8.

Biogeographic element. Paleoaustral as inferred herein.

Cyclorismina?, new species, *cf. "C." marwicki* Zinsmeister, 1984

Plate 4, figs. L and O

cf. *"Cyclorismina" marwicki* Zinsmeister, 1984, p. 1522, figs. 10A-10C; Stilwell and Zinsmeister, 1992, p. 82, pl. 8, figs. l-n.

Dimensions. USNM 490783 length 30.0 mm, height 28.5 mm, width of paired valves 17.0 mm; USNM 490784 length 43.0 mm, height 41.5 mm, width of single valve 18.5 mm.

Figured specimens. USNM 490783-490784.

Material. Two specimens and two probable fragments.

Localities. E185, E208, E325, unnumbered E.

Geographic distribution. McMurdo Sound.

Discussion. Circular venerid bivalves are extensively common in Tertiary deposits worldwide and without hinge details genus-level placement of these taxa is difficult at best. Moderately sized, nearly circular bivalves collected from the erratics are nearly identical to *"Cyclorismina" marwicki* Zinsmeister [1984] of Seymour Island, except that the McMurdo Sound specimens are more inflated and marginally more circular. Hinge details of the McMurdo Sound specimens remain unknown.

Of note, Zinsmeister [1984] based his assignment of the Seymour Island species on what was known of *Cyclorismina* at the time of his writing. Additional study by me of the details of the hinge of the type specimens of *Cyclorismina woodsi* Marwick [1927], from the latest Cretaceous type species of New Zealand, indicates that the diverging cardinal teeth of the Antarctic species is consistent with *C. marwicki* and that they are congeners. *Cyclorismina* can be added to the growing list of taxa present in New Zealand in the Late Cretaceous and after the K-T event found in the Paleogene of Antarctica.

Genus **Eumarcia** Iredale, 1924

Eumarcia Iredale, 1924, pp. 182 and 211.

Type species. (By original designation) *Venus fumigata* Sowerby, 1853.

Subgenus **Atamarcia** Marwick, 1927

Atamarcia Marwick, 1927, p. 622.

Type species. (By original designation) *Eumarcia sulcifera* Marwick, 1927.

Biogeographic element. Paleoaustral as inferred herein.

?Eumarcia (Atamarcia) robusta Stilwell and Zinsmeister, 1992

Plate 4, figs. N and Q

?Eumarcia (Atamarcia) robusta Stilwell and Zinsmeister, 1992, pp. 83 and 84, pl. 9, figs. a, c, and f.

Dimensions. USNM 490785 length 78.5 mm incomplete, height 99.0 mm, width of paired valves 45.5 mm.

Figured specimen. USNM 490785.

Material. One poorly preserved specimen and probable fragments.

Locality. MA80 036.2 (Peter Webb locality, Ohio State University).

Geographic distribution. McMurdo Sound and possibly Seymour Island.

Discussion. This large, robust, articulated bivalve bears a striking resemblance to *Eumarcia (Atamarcia) robusta* Stilwell and Zinsmeister [1992] [pp. 83 and 84, pl. 9, figs. a, c, and f] from coeval deposits of Seymour Island. The long narrow escutcheon, nymph, large ovate adductor scars, and vestiges of moderately spaced commarginal grooves and riblets in the McMurdo Sound species are consistent with *E. (A.) robusta*, but details of the pallial line and sinus of *E. (A). robusta* are wanting in the Seymour Island material. These details are preserved in the McMurdo Sound specimen (USNM 490786), which reveals a very deep, very narrowly rounded to finger-like pallial sinus. The shell of this specimen is very thick at nearly 6.5 mm. The anterior and posterior parts of the shell in the McMurdo Sound specimen are incomplete, so that an accurate comparison with other taxa is not possible. The rather pointed and deep pallial sinus details of this individual are also consistent with other Tapetinae, including *Amiantis* Carpenter [1864].

Veneridae genus et species indeterminate
Plate 5, fig. F

Dimensions. USNM 490786 length 58.0 mm, height 45.0 mm, width of paired valves about 28.5 mm.

Figured specimen. USNM 490786.

Material. One steinkern with vestiges of shell material preserved.

Locality. E359.

Geographic distribution. McMurdo Sound.

Distribution. McMurdo Sound.

Discussion. The partial hinge preserved and vestiges of ornament in this poorly preserved steinkern are comparable with the Tapetinae. A positive latex cast was made of the hinge, revealing a rather large grooved cardinal tooth 3b in the right valve and also an inclined, shallowly grooved cardinal 1. Cardinal tooth 3a may also be present, nearly flush with the lunular margin, but this part of the hinge did not preserve well. The ornament is of equally spaced commarginal riblets. The outline and sculpture of this specimen is similar to the coeval Seymour Island tapetine species, *Eumarcia (Eumarcia) australissa* Stilwell and Zinsmeister [1992] [pp. 82 and 83, pl. 8, figs. r and s], represented only by a single right valve.

Order MYOIDA Stoliczka, 1870
Suborder MYINA Stoliczka, 1870
Superfamily HIATELLACEA Gray, 1824
Family HIATELLIDAE Gray, 1824
Genus *Hiatella* Daudin, in Bosc, 1801

Hiatella Daudin, in Bosc, 1801, p. 120.

Type species. (By subsequent designation, Winckworth, 1932) *Hiatella monoperta* (equivalent to *Mya arctica* Linné, 1767).

Biogeographic element. Indo-Pacific/Tethyan, as interpreted herein.

Hiatella harringtoni, new species
Plate 5, figs. a and b

Diagnosis. Moderately small subtrapezoidal *Hiatella* with inflated umbones, descending antero- and posterodorsal margins, and very fine sculpture of bunched commarginal threads separated by subequally spaced growth pauses; distinguished from *H. tenuis* (Wilckens, 1911) in having a much smaller and more delicate shell, sloping dorsal margins, a more convex ventral margin, and much finer commarginal sculpture of threads.

Description. Shell small- to medium-sized for genus, thin, compressed, subtrapezoid; beaks small, prosogyrous, just anterior of center; umbones moderately inflated; umbonal ridge strongly developed, equally strong from beak to intersection of ventral and posterior margins; anterodorsal margin moderately short, mostly straight, very gently declivous, merging with narrowly rounded anterior margin; posterodorsal margin moderately long, very gently convex, more steeply sloping, merging with a short, obliquely truncated posterior margin; ventral margin broadly convex; shell finely sculptured with many equally spaced commarginal threads in 0.5 mm wide bunches, separated by subequal growth pauses; hinge plate apparently narrow; internal details unknown.

Dimensions. Holotype USNM 490787 length 11.0 mm, height 7.0 mm; paratype USNM 490788 length 9.5 mm, height 5.5 mm.

Types. Holotype USNM 490787; paratype USNM 490788.

Figured specimens. USNM 490787-490788.

Material. Four specimens.

Localities. E145 (type), E155, E189.

Geographic distribution. McMurdo Sound.

Discussion. The morphological differences between *Hiatella harringtoni* n. sp. and *H. tenuis* [Wilckens, 1911] [pp. 19 and 20, pl. 1, fig. 22; see also Zinsmeister, 1984, p. 1525, figs. 10D and 10E] from the uppermost units of the La Meseta Formation are marked, indicating that these taxa probably were derived from different stock. *Hiatella harringtoni* n. sp. has a much smaller, more delicate shell, a sloping posterodorsal margin, a broadly convex ventral margin, and quite fine ornament, compared with *H. tenuis* which has a much larger and more robust shell, a nearly straight posterodorsal margin, a nearly straight ventral margin, and undulating commarginal ribs. The umbones of *H. harringtoni* n. sp. are also much more inflated. The intraspecific variation of *H. tenuis* and *H. harringtoni* n. sp. is strong, reflecting the probable nestling habit of these taxa, consistent with extant forms.

Etymology. Species named in honor of H. J. Harrington for his discovery of the erratics in McMurdo Sound.

Genus *Panopea* Ménard de la Groye, 1807

Panopea Ménard de la Groye, 1807, p. 135.

Type species. (By subsequent designation, Children, 1823) *Panopea aldrovani* Ménard de la Groye, 1807 (= *Mya glycymeris* Born, 1778) (I.C.Z.N. Opinion 1414, 1986) (Darragh and Kendrick, 1991, p. 88).

Biogeographic element. Indo-Pacific/Tethyan (*cf.* Kauffman, 1973, p. 372).

Panopea akerlundi Stilwell and Zinsmeister, 1992
Plate 5, figs. D, E, and H

Panopea akerlundi Stilwell and Zinsmeister, 1992, pp. 88 and 89, pl.10, figs. a and c.

Dimensions. USNM 490789 length 80.5 mm, height 43.0 mm, width of paired valves 32.0 mm; USNM 490790 length 75.0 mm, height 40.5 mm, width of paired valves about 22.5 mm.

Figured specimens. USNM 490789-49790.

Material. Four specimens and fragments.

Localities. E100, E330, E336.

Geographic distribution. Seymour Island and Mc-Murdo Sound.

Discussion. These well-preserved specimens of *Panopea* are virtually indistinguishable from *P. akerlundi* Stilwell and Zinsmeister [1992] [pp. 88 and 89, pl. 10, figs. a and c] from Unit III of the La Meseta Formation, Seymour Island, and are deemed conspecific. Unit III is considered to be a lateral equivalent, in part, to Unit II, which has been dated by microfossils to be of early? to middle Eocene age [R. Askin, personal communication, 1988, *in* Stilwell and Zinsmeister, 1992, p. 20]. Thus, the presence of *Panopea akerlundi* in McMurdo Sound may indicate an older age for some of the erratics, if the restricted age range of this species on Seymour Island is taken into account.

Panopea, new species?, *cf. P. philippii* Zinsmeister, 1984

Plate 5, fig. C

Dimensions. USNM 490791 length 85.0 mm incomplete, height 60.5 mm.

Figured specimen. USNM 490791.

Material. One specimen.

Locality. E153.

Geographic distribution. McMurdo Sound and possibly Seymour Island.

Discussion. This possible new species is remarkably similar to *Panopea philippii* Zinsmeister [1984] [p. 1525, figs. 10I and 10K; see also Stilwell and Zinsmeister, 1992, p. 88, pl. 10, figs. d and f] from the upper units V-VII of the La Meseta Formation, especially with respect to the very broad shell reflecting a nearly 1:1 length to height ratio and broad commarginal growth undulations. The estimated shell length of USNM 490791 from

---

Plate 5

Figs. A and B. *Hiatella harringtoni* n. sp. (A) Holotype USNM 490787, E145, length = 11.0 mm, x3.4. (B) Paratype USNM 490788, E145, length = 9.5 mm, x3.3.

Fig. C. *Panopea* n. sp.? *cf. P. philippii* Zinsmeister. (C) USNM 490791, E153, length = 85.0 mm, x1.

Figs. D, E, and H. *Panopea akerlundi* Stilwell and Zinsmeister. (D) USNM 490789, E336, length = 80.5 mm, x0.75. (E) USNM 490790, E330, length = 75.0 mm, x1. (H) USNM 490790, E330, length = 75.0 mm, x1.

Fig. F. Veneridae genus et species indeterminate. (F) USNM 490786, E359, length = 58.0 mm, x0.8.

Fig. G. *Periploma* n. sp.? *cf. P. topei* Zinsmeister. (G) USNM 490792, E374, length = 15.0 mm, x3.

Figs. I and J. Bivalvia genus et species indeterminate. (I) USNM 490799, E352, length = 16.5 mm, x3.3. (J) USNM 490800, E352, length = 14.5 mm, x3.3.

Figs. K and L. *Cellana feldmanni* Stilwell and Zinsmeister. (K) USNM 490802, E168, length = 20.5 mm, x1. (L) USNM 490803, E168, length = 21.5 mm, x1.

Figs. M and N. Patellacea genus et species indeterminate. (M) USNM 490804, E173, length = 6.5 mm, x3.3. (N) USNM 490805, E184, length = 5.5 mm, x3.

Figs. O, P, and S. *Calliostoma s. l.* n. sp. (O), USNM 490812, Peter Webb locality (Ohio State University), height = 4.0 mm, x3.5. (P) USNM 490812, Peter Webb locality, height = 4.0 mm, x3.5. (S) USNM figs. Q, 490812, Peter Webb locality, height = 4.0 mm, x8.5.

Figs. Q, R, and T. *Falsimargarita? vieja* n. sp. (Q) Holotype USNM 490806, E378, height = 5.5 mm, x7.5. (R)USNM 490806, E378, height = 5.5 mm, x3. (T) USNM 490806, E378, height 5.5 mm, x3.

Fig. U. Trochidae genus et species indeterminate. (U) 490807, E168, height = 6.5 mm, x3.

McMurdo Sound is about 85.0 mm, some 35 mm longer than the holotype of *P. philippii*. *Panopea zelandica* Quoy and Gaimard [1835] [see Valenciennes, 1839, pp. 19 and 20, pl. 3, fig. 2a and 2b; also Powell, 1979, p. 428, pl. 78, fig. 11 and fig. 113-1 through 3] from the Late Miocene to Recent of New Zealand is apparently related to the McMurdo Sound species, but the length to height ratio of *P. zelandica* is slightly less.

Superfamily ANOMALODESMATA Dall, 1889
Order PHOLADOMYOIDA Newell, 1965
Superfamily PHOLADOMYACEA Newell, 1965
Family PERIPLOMATIDAE Dall, 1895
Genus *Periploma* Schumacher, 1817

Periploma Schumacher, 1817, p. 115.

Type species. (By monotypy) *Corbula margaritacea* Lamarck, 1801.
Biogeographic element. Cosmopolitan (Keen *in* Moore, 1969, p. N849).

Periploma, new species?, *cf. P. topei* Zinsmeister, 1984
Plate 5, fig. G

Dimensions. USNM 490792 length 15.0 mm, height 9.5 mm.
Figured specimen. USNM 490792.
Material. Single left valve.
Locality. E374.
Geographic distribution. McMurdo Sound and possibly Seymour Island.
Discussion. The single periplomatid specimen recognized during fossil preparation of E374 is virtually indistinguishable from *Periploma topei* Zinsmeister [1984] [pp. 1525 and 1526, figs. 10F and 10G; see also Stilwell and Zinsmeister, 1992, p. 89, pl. 10, figs. 10F and 10G], present in Units II-V (middle? Eocene) of the La Meseta Formation. The McMurdo Sound specimen, which has a 2.25 mm boring just below the anterodorsal margin, is one third of the size of adult individuals of *P. topei* and may be a juvenile. More material is required to confirm a conspecific relationship with *P. topei*.

Bivalvia genus et species indeterminate
Plate 5, figs. I and J

Dimensions. USNM 490799 length 16.5 mm, height 8.0 mm; USNM 490800 length 14.5 mm, height 7.0 mm.
Figured specimens. USNM 490799-490800.
Material. Two poorly preserved specimens.

Locality. E352.
Geographic distribution. McMurdo Sound.
Discussion. These elongated and narrow shells are similar to several families of bivalves including Mactridae, Donacidae, Psammobiidae, and others. Only faint broad commarginal growth pauses are preserved on the specimens. The antero- and posterodorsal margins are gently sloping, the posterior margin is narrowly rounded and the ventral margin is broadly convex. More material is needed of this interesting species, which is unknown from Seymour Island.

Class GASTROPODA Cuvier, 1797
Subclass PROSOBRANCHIA Milne Edwards, 1848
Order PATELLOIDA von Ihering, 1876
Suborder PATELLINA von Ihering, 1876
Superfamily PATELLACEA Rafinesque, 1815
Family NACELLIDAE Thiele, 1929
Genus *Cellana* H. Adams, 1869

Cellana H. Adams, 1869, p. 273.

Type species. (By original designation) *Nacella cernica* H. Adams, 1869.
Biogeographic element. Indo-Pacific/Tethyan (*cf.* Powell, 1979).

Cellana feldmanni Stilwell and Zinsmeister, 1992
Plate 5, figs. K and L

Cellana feldmanni Stilwell and Zinsmeister, 1992, p. 90, pl. 11, fig. a.

Dimensions. USNM 490802 length 20.5 mm, width 16.0 mm; USNM 490803 length 21.5 mm, width 17.0 mm.
Figured specimens. USNM 490802-490803.
Material. Four specimens.
Localities. E155, E168, E169.
Geographic distribution. Seymour Island and McMurdo Sound.
Discussion. The geographic range of *Cellana feldmanni* Stilwell and Zinsmeister [1992] is expanded herein to include East Antarctica. The original description of this species was based on the single, moderately preserved holotype from the uppermost Unit VII of the La Meseta Formation. Four specimens were collected from the Mount Discovery erratics and one of these, a nearly perfect external mold, has sufficient detail to expand our knowledge of the morphology of this species. Although the Mount Discovery specimens are smaller than the

holotype, the sculpture configuration and length to width ratio are nearly identical, suggesting they are merely immature individuals of this species. Specimen USNM 490803 has regularly spaced radial ribs that are spaced at a single rib per about 0.75 mm and each fourth to fifth rib is slightly stronger. The concentric growth pauses are irregularly spaced, but strongest close to the periphery which is very weakly crenulated. This low-profiled shell is about 5 mm high. The apex is subcentrally located almost exactly one-third of the length of the shell from the periphery. See Stilwell and Zinsmeister [1992, p. 90] for comments on possible relationships.

Patellacea genus et species indeterminate
Plate 5, figs. M and N

Dimensions. USNM 490804 length 6.5 mm, width 4.5 mm; USNM 490805 length 5.5 mm, width 4.0 mm.
Figured specimens. USNM 490804-490805.
Material. Six poorly preserved specimens.
Localities. E135, E173, E184.
Geographic distribution. McMurdo Sound.
Discussion. These poorly preserved, minute, cap-shaped limpets are recorded from medium- to coarse-grained sandstone, probably from near-shore high-energy facies. Identification of these patellacean limpets is uncertain due to preservation, but the very deep shell, nearly smooth periphery, broad concentric growth rings, and moderately broad radial ribs (preserved in part on one specimen), suggest a possible affinity to the Acmaeidae. Internally, the shell is mostly smooth apart from very weak crenulations.

Suborder TROCHINA Cox and Knight, 1960
Superfamily TROCHACEA Rafinesque, 1815
Family TROCHIDAE Rafinesque, 1815
Subfamily CALLIOSTOMATINAE Thiele, 1924
Genus *Calliostoma* Swainson, 1840

Calliostoma Swainson, 1840, p. 351.

Type species. (By subsequent designation, Hermannsen, 1846) *Trochus conulus* Linné, 1758.
Biogeographic element. Indo-Pacific/Tethyan as interpreted herein.

Calliostoma s. l., new species
Plate 5, figs. O, P, and S

Dimensions. USNM 490812 height 4.0 mm, diameter of last whorl 6.5 mm.

Figured specimen. USNM 490812.
Material. One moderately preserved specimen.
Locality. Peter Noel-Webb, Ohio State University, Mount Discovery locality.
Geographic distribution. McMurdo Sound.
Discussion. This new species is represented by a sole specimen discovered during the course of this investigation from an erratic collected last decade by Peter Noel-Webb, Ohio State University. This calliostomatine species has remnants of sculpture, but insufficient detail to assign it to the plethora of *Calliostoma* Swainson [1840] subgenera. Four gradate whorls are preserved. The spire angle of this specimen, USNM 490812, is about 90° and whorl inflation is very rapid with a capacious, compressed last whorl that reveals a biangulate profile of a very distinct and sharp peripheral keel and more subdued keel adapically. The sculpture is of more-or-less moderately strong equally spaced spiral cords that are slightly beaded obliquely, reflecting prosocline growth lines. The flattened base is sculptured with wavy spiral riblets and the basal constriction is very rapid. Part of the base is incomplete, so whether or not there is an umbilicus is unknown. The protoconch is also wanting. The whorl profile and sculpture of *Calliostoma s. l.* n. sp. is reminiscent of the Recent New Zealand species *Maurea turnerarum* Powell, 1964 [see Powell, 1979, p. 62, pl. 19, fig. 9]. Also similar is *Trochus fricki* Philippi [1887] [pp. 95 and 96, pl. 12, fig. 7] from the Tertiary of Chile, but this species is slightly larger with more flush whorls.

Genus *Falsimargarita* Powell, 1951

Falsimargarita Powell, 1951, p. 93.

Type species. (By original designation) *Margarites gemma* Smith, 1915.
Biogeographic element. Endemic.
Discussion. *Falsimargarita* Powell [1951] is a rare calliostomatine genus that has been recorded solely in the Antarctic. Powell [1951, p. 93] erected *Falsimargarita* because it can be differentiated from *Calliostoma* Swainson [1840] in having a deep and open umbilicus, a rather thin externally iridescent shell, and distinct radular features. Further, the four recorded species of *Falsimargarita* have fine, sharp spiral riblets and axial threads. A moderately preserved specimen from Mount Discovery, USNM 490806, fits comfortably in *Falsimargarita* and is probably an early representative of the genus. The presence of four Recent species, all in deep water [see Dell, 1990], and the morphological

diversity displayed with respect to sculpture configuration of these taxa suggest that the origin of the genus does, indeed, extend back into the Tertiary.

Falsimargarita? vieja, new species
Plate 5, figs. Q, R, and T

Diagnosis. Relatively small *Falsimargarita* with a spire angle of about 105°, about 30 sharp spiral riblets, and a relatively steep base; separated from the type species, *F. gemma* (Smith, 1915) in being much smaller with straighter less wavy spiral riblets and a steeper base.

Description. Shell small for genus, thin and fragile, depressed turbinate; spire low, apparently paucispiral, of slightly inflated, convex whorls; spire angle approximately 105°; protoconch unknown; umbilicus deep, narrow; suture slightly impressed; growth lines orthocline at periphery of last whorl, not preserved elsewhere; last whorl moderately compressed, but well-inflated; peripheral angulation weak; shell sculptured with many fine spiral riblets; last whorl with about 30 closely spaced, sharp riblets, becoming weaker at onset of basal constriction below angulation; aperture large, subcircular.

Dimensions. Holotype USNM 490806 height 5.5 mm incomplete, diameter of last whorl 8.0 mm.

Type. Holotype USNM 490806.

Figured specimen. USNM 490806.

Material. Holotype.

Locality. E378 (type).

Geographic distribution. McMurdo Sound.

Discussion. As most of the spire is missing with the only specimen available of this species, the genus-level assignment is tentative, but seemingly appropriate given the remarkable similarity of this species with Recent species of *Falsimargarita* from Antarctica. *Falsimargarita? vieja* n. sp. is probably an early member of the group and is very close indeed to the type species, *F. gemma* [Smith, 1915] [p. 62, pl. 1, fig. 1; see also discussion and figures by Dell, 1990, p. 93, figs. 148-152], recorded from deep water down to 2525 m off the coast of the Antarctic continent and also near the South Shetland islands. *Falsimargarita? vieja* n. sp. is nearly half as small as the holotype of *F. gemma*, has a slightly steeper base, and has less wavy spirals that are slightly more spaced, but these differences are minor.

Etymology. Species name derived from the Spanish "viejo" (equivalent to "old") for its early record in Antarctica.

Trochidae genus et species indeterminate

Plate 5, fig. U

Dimensions. USNM 490807 height 6.5 mm, diameter of last whorl 10.5 mm.

Figured specimen. USNM 490807.

Material. One specimen.

Locality. E168.

Geographic distribution. McMurdo Sound; possibly Seymour Island.

Discussion. Only vestiges of shell are preserved on this specimen, but the whorl profile suggests that it belongs to the Trochidae. The whorl profile is similar to *Falsimargarita? vieja* n. sp., but the preserved shell fragments show no sculpture, only possible weak growth lines. The depressed turbinate whorl profile of USNM 490807 is also similar to *"Antisolarium" abstrusum* Stilwell and Zinsmeister [1992] [p. 94, pl. 11, figs. j-l], recorded from Unit VI of the La Meseta Formation, but no more meaningful comment can be made.

Family TURBINIDAE Rafinesque, 1815
Subfamily TURBININAE Rafinesque, 1815
Genus *Astraea* Röding, 1798

Astraea Röding, 1798, p. 69.

Type species. (By subsequent designation, Suter, 1913) *Trochus imperialis* Gmelin, 1791 (=*T. heliotropium* Martyn, 1784).

Biogeographic element. Indo-Pacific/Tethyan as interpreted herein.

Astraea lilliputia, new species
Plate 6, figs. a-e, and g

Diagnosis. Minute *Astraea* that has a spire angle of about 98°, a mammilate protoconch of 2 1/2 smooth whorls, rudimentary development of about 12 peripheral spines, four to five strong spirals and axials, and narrow, deep umbilicus with thickened labial callus; distinguished from the type species, *A. heliotropium* [Martyn, 1784], in being much smaller with more flush whorls and much reduced spines.

Description. Shell quite small for genus (height about 2.5 mm), moderately solid, low-spired trochiform; spire conic, slightly gradate, low, of some four steeply sided gently convex whorls; whorl inflation moderately rapid; spire angle approximately 98°; protoconch large, paucispiral, moderately mammillate, of 2 1/2 smooth rounded whorls; sutures slightly impressed on early juvenile spire whorls, becoming more flush on mature whorls; whorls somewhat clasping; last whorl moderately inflated, uniangulate with a well-developed, sharpened basal keel; antepenultimate through last whorl steep sided, nearly straight apart from sculpture; growth lines prosocline; shell highly ornamented with strong axial ribs and nearly equally strong spirals, creating rather weak nodes; peripheral keel wavy with about 12 blunt spines; last whorl sculptured adapical of keel with four moderately pustulose spirals and about 16 strong axials subparallel with axials on previous whorls; penultimate whorl similarly sculptured with five strong wavy spirals; base mostly flat, heavily sculptured with weak basal keel and five strong heavily beaded spirals; interstices between peripheral keel and basal keel marked by single weak spiral cord; umbilicus narrow and deep with marked thickened callus (probable immature individual); aperture moderately large, subovate.

Dimensions. Holotype USNM 490808 height about 2.5 mm, diameter of last whorl 3.5 mm.

Type. Holotype USNM 490808.

Figured specimen. USNM 490808.

Material. Holotype.

Locality. E189 (type).

Geographic distribution. McMurdo Sound.

Discussion. This beautiful and minute species is represented by the single exquisitely preserved holotype. *Astraea lilliputia* n. sp. is the sole representative of the Turbininae in the Antarctic Recent and fossil record and greatly extends the geographic and temporal range of the group to include the highest southern latitudes. The holotype, USNM 490808, is most likely an immature specimen, as the umbilicus is open, but very narrow, reflecting probably early development of the labial callus. *Astraea lilliputia* is one of the earliest members of the genus and bears little resemblance to other congeneric taxa from the Tertiary and Recent because of the rudimentary development of peripheral spines in the Antarctic species. The type species, *A. heliotropium* [Martyn, 1784] [see Montfort, 1810, p. 199 for text and single large reversed woodcut of sinistral figure on preceding page 198; Suter, 1913, pp. 166 and 167, 1915, pl. 41, fig. 1; Powell, 1979, pp. 66 and 67, pl. 11, figs. 2 and 3; Abbott and Dance, 1983, p. 49, second row figures], collected during Captain Cook's voyages to New Zealand, is a Miocene to Recent Austral species that is very large and spinose, compared with *A. lilliputia* n. sp. and is not closely related. I can find no New Zealand and Australian nominal taxa that come close to *A. lilliputia*, morphologically. However, a Tertiary Chilean species described by Philippi [1887] [p. 96, pl. 12, fig. 3] as *Trochus araucanus* has a whorl profile and quite blunt spines on the

periphery very similar to *Astraea lilliputia* n. sp. and may be congeneric. No sculpture is visible on the figure of *A.? araucana* given by Philippi.

Etymology. Species named for its diminutive size.

Superfamily RISSOACEA Gray, 1847
Family RISSOIDAE Gray, 1847

Rissoidae genus et species indeterminate
Plate 6, fig. F

Dimensions. USNM 490856 height 2.5 mm, diameter of last whorl 2.0 mm.

Figured specimen. USNM 490856.

Material. One specimen.

Locality. E181.

Geographic distribution. McMurdo Sound.

Discussion. The strongly convex whorl profile, minute shell, impressed sutures and ornamentation of strong axials and much weaker spiral sculpture of specimen USNM 490856 is consistent with the Rissoidae. No details of the aperture or protoconch are available for comment, but the above characteristics are reminiscent of Recent New Zealand species assignable to *Alvinia* Monterosato [1884] *s. l.*

Superfamily CERITHIACEA Fleming, 1822
Family CERITHIIDAE Fleming, 1822

Cerithiidae genus et species indeterminate
Plate 6, figs. H and I

Dimensions. USNM 490853 height 10.5 mm, diameter of last whorl 4.5 mm.

Figured specimen. USNM 490853-490854.

Material. One specimen. USNM 490854 may also be conspecific.

Locality. E145.

Geographic distribution. McMurdo Sound.

Discussion. This poorly preserved specimen, USNM 490852, has a high-spired cerithiiform outline with an acute spire angle of about 17° and strong ornament of marked and spaced opisthocyrt axials that extend from suture to suture and broad spirals that create a reticulate, pustulose network of intersecting spirals and axials. There are secondary spirals in the interspaces between the primaries. The sutures are impressed and the whorl profiles are gently convex. All of these features are common to the Cerithiidae, but the genus remains uncertain. No similar taxon has been described from the La Meseta Formation.

Family TURRITELLIDAE Woodward, 1851
Subfamily TURRITELLINAE Woodward, 1851
Genus *Colposigma* Finlay and Marwick, 1937

Colposigma Finlay and Marwick, 1937, pp. 39 and 40.

Type species. (By original designation) *Colposigma mesalia* Finlay and Marwick, 1937.

Biogeographic element. Paleoaustral as inferred herein.

Discussion. *Colposigma* Finlay and Marwick [1937] is one of the most diverse Paleocene to Eocene gastropods in the southern hemisphere with eight recorded species from New Zealand (*Colposigma mesalia* Finlay and Marwick, 1937, late Early Paleocene; *C. gairi* Marwick, 1960, middle Eocene; *C. imparcincta* Finlay and Marwick, 1937, middle Eocene; *C. plebeia* Marwick, 1960, middle Eocene), Australia (*Colposigma uniangulata* Darragh, 1997, mid Paleocene), Antarctica (*C. euthenia* Stilwell and Zinsmeister, 1992, late early? to late Eocene; *C. capitanea* Stilwell and Zinsmeister, 1992, p. middle to late Eocene), and southern South America (*C. exigua* (Ortmann, 1899), middle? late? Eocene). *Colposigma euthenia* Stilwell and Zinsmeister [1992] has been recorded in Units I-VI of the La Meseta Formation and its disappearance at the top of Unit VI corresponds to a major faunal change, in particular, a sharp decrease in species-level diversity which may reflect a facies change or mark a climatic change such as declining sea-surface temperatures at the close of the Eocene [see Stilwell and Zinsmeister, 1992, pp. xi, 43]. *Colposigma* became extinct at the end of the Eocene.

Allmon *et al.* [1990, p. 597] suggested that all extinct and extant Turritellidae genus- and subgenus-level groups be assigned to *Turritella s. l.* because the taxonomy is still unresolved. Finlay and Marwick [1937] reluctantly erected *Colposigma* for turritellids that have a deeper and lower sinus, distinct from other closely relat-

ed taxa such as *Zeacolpus* [Finlay, 1926], not *Mesalia* Gray [1840]. The sinus of *Colposigma* is considered to be parasigmoid (apex about the adapical third), the protoconch is multispiral cyrtoconoid, the shell is rather small, and the sculptural configuration is of simple regular ribs with the abapical ribs being the strongest [*cf.* Finlay and Marwick, 1937, p. 39; Marwick, 1957, p. 20; Marwick, 1971, p. 10]. Because of these believed important differences, compared with other turritellids, *Colposigma* is retained herein.

Colposigma euthenia Stilwell and Zinsmeister, 1992
Plate 6, fig. c; Plate 7, figs. a-b

Colposigma euthenia Stilwell and Zinsmeister, 1992, pp. 94 and 95, pl. 11, figs. r-t; Stilwell, 1994, pp. 840 and 841.

Dimensions. USNM 490809 height of fragment 4.5 mm; USNM 490810 height 7.0 mm nearly complete, diameter of last whorl 3.0 mm; USNM 490851 height 5.5 mm, diameter of last whorl 2.0 mm.

Figured specimens. USNM 490809-490810, 490851.

Material. Five specimens, mostly incomplete.

Localities. E155, E207, E333.

Geographic distribution. McMurdo Sound; Seymour and Cockburn islands.

Discussion. *Colposigma euthenia* Stilwell and Zinsmeister [1992] [pp. 94 and 95, pl. 11, figs. r-t] is one of the most abundant and widespread of all Eocene Antarctic gastropods and is recognized herein in McMurdo Sound for the first time. Previously, the species was recognized in all units apart from Unit VII of the La Meseta Formation on Seymour Island and Units I and/or II of the formation on Cockburn Island. Stilwell [1994, p. 845] indicated that *C. mesalia* Finlay and Marwick [1937] [p. 40, pl. 5, figs. 7-8; see also Fleming,

---

Plate 6

Figs. A, B, D, E, and G. *Astraea lilliputia* n. sp. (A) Holotype USNM 490808, E189, height = about 2.5 mm, x75. (B) Holotype USNM 490808, E189, height = about 2.5 mm, x17. (D) Holotype USNM 490808, E189, height = about 2.5 mm, x17. (E) Holotype USNM 490808, E189, height about 2.5 mm, x50. (G) Holotype USNM 490808, E189, height = about 2.5 mm, x25.

Fig. C. *Colposigma euthenia* Stilwell and Zinsmeister. (C) USNM 490851, E333, height = 5.5 mm, x3.

Fig. F. Rissoidae genus et species indeterminate. (F) USNM 490856, E181, height = 2.5 mm, x18.

Figs. H and I. Cerithiidae genus et species indeterminate. (H) USNM 490853, E145, height 10.5 mm, x4. (I) USNM 490854, E376, height = 10.0 mm, x7.5.

1966, p. 238, pl. 70, figs. 824, 827-829] from the late Early Paleocene of New Zealand may be the ancestor of *C. euthenia*. *Colposigma mesalia* has more subdued sculpture and more convex whorls, compared with *C. euthenia*.

Genus *Zeacolpus* Finlay, 1926

Zeacolpus Finlay, 1926, p. 388.

   Type species. (By original designation) *Turritella vittata* Hutton, 1873.
   Biogeographic element. Paleoaustral, if indeed in McMurdo Sound; otherwise endemic in New Zealand.

Zeacolpus? species
Plate 7, figs. C and I

   Dimensions. USNM 490811 height of fragment 11.0 mm; USNM 490852 height of fragment 50.5 mm.
   Figured specimens. USNM 490811, USNM 490852.
   Material. Three fragments.
   Locality. E358, E359.
   Geographic distribution. McMurdo Sound.
   Discussion. A positive latex cast made from an external mold of a turritellid gastropod indicates a possible relationship with *Zeacolpus* Finlay [1926], but only two whorls are preserved with little sculptural detail. The whorl profile of this species is concave to somewhat campanulate. The sculptural configuration,according to the Marwick [1957] system is equally strong spirals AB with A being directly flush with the slightly impressed suture adaperturally and B situated about 1.5 mm abaperturally from suture. There are about four secondary spirals in the interstices between A and B. This species is undoubtedly new, but more material is needed.

Superfamily STROMBACEA Rafinesque, 1815
Family APORRHAIDAE Gray, 1850
Subfamily ARRHOGINAE Popenoe, 1983
Genus *Arrhoges* Gabb, 1868

Arrhoges Gabb, 1868, p. 145.

   Type species. (By monotypy) *Chenopus occidentale* Beck, 1847.
   Biogeographic element. Indo-Pacific/Tethyan, as interpreted herein.

Subgenus *Antarctohoges* Stilwell and Zinsmeister, 1992

Antarctohoges Stilwell and Zinsmeister, 1992, p. 103.

Plate 7

Figs. A and B. *Colposigma euthenia* Stilwell and Zinsmeister. (A) USNM 490810, E155, height = 7.0 mm. (B) USNM 490809, E207, height = 4.5 mm, x14.

Figs. C and I. *Zeacolpus?* sp. (C). USNM 490811, E359, height = 11.0 mm, x3. (I) USNM 490852, E358, height 50.5 mm, x0.6.

Figs. D, F, H, and J. *Drepanocheilus (Tulochilus) erebus* n. sp. (D) Paratype USNM 490816, E189, length (wing) = 6.5 mm, x3. (F) Holotype USNM 490814, E189, heigt = 13.5 mm, x3.5. (H) Paratype USNM 490817, E189, height = 13.5 mm, x3. (J) Paratype USNM 490815, E189, height = 18.5 mm, x3.2.

Fig. E. *Arrhoges (Antarctohoges) diversicostata* (Wilckens). (E) USNM 490813, Peter Webb Locality (Ohio State University), height = 25.0 mm, x3.2.

Figs. G, K, L, M, N, and O. *Struthiolarella mcmurdoensis* n. sp. (G) Paratype USNM 490819, E191, height = 31.0 mm, x1. (K) Holotype USNM 490818, E191, height = 29.0 mm, x1. (L) Paratype USNM 490819, E191, height = 31.0 mm, x1. (M) Paratype USNM 490821, height = 18.5 mm, x3.2. (N) Paratype USNM 490821, height = 18.5 mm, x3.2. (O) Holotype USNM 490818, E191, height = 29.0 mm, x1.

Figs. P-R. *Perissodonta* n. sp.? *cf. P. laevis* (Wilckens). (P) USNM 490826, height = 11.0 mm, x3.2. (Q) USNM 490822, E333, height = 11.0 mm, x3.2. ® USNM 490823, E145, height = 11.5 mm, x3.5.

Figs. S and T. *Sigapatella (Spirogalerus?) colossa* n. sp. (S) Holotype USNM 490824, E359, diameter = 35.5 mm, x1. (T) Paratype USNM 490825, E359, diameter = 35.0 mm, x1.

Figs. U-Y. *Taniella (Pristinacca?)* n. sp. (U) USNM 490832, E181, height = 4.25 mm, x2.8. (V) USNM 490832, E181, height = 4.25 mm, x3.3. (W) USNM 490831, E189, height 3.5 mm, x3.4. (X) USNM 490831, E189, height 3.5 mm, x3.4. (Y) USNM 490832, E181, height = 4.25 mm, x8.2.

Type species. (By original designation) *Chrysodomus? diversicostata* Wilckens, 1911.

Biogeographic element. Endemic.

Discussion. Stilwell and Zinsmeister [1992] erected the aporrhaid subgenus *Arrhoges (Antarctohoges)* as it lacks the terminal, upturned, blunt lobe of *Arrhoges s. s.* and, in addition, the Antarctic species have gently arcuate and thickened outer lips, compared with *Arrhoges s. s. Arrhoges (Antarctohoges)* further has a last whorl that encroaches to a degree on the penultimate whorl, a short siphonal canal and a subovate aperture. The subgenus is known only from the Eocene of Antarctica and is considered endemic.

Arrhoges (Antarctohoges) diversicostata (Wilckens, 1911)
Plate 7, fig. E

Chrysodomus? diversicostata Wilckens, 1911, pp. 32 and 42, pl. 1, fig. 35.

Arrhoges (Antarctohoges) diversicostata (Wilckens); Stilwell and Zinsmeister, 1992, pp. 103 and 104, pl. 12, figs. v and w, pl. 13, fig. I.

Dimensions. USNM 490813 height 25.0 mm, diameter of last whorl 10.0 mm.

Figured specimen. USNM 490813.

Material. One specimen.

Locality. Mount Discovery, Peter Noel-Webb locality, Ohio State University.

Geographic distribution. McMurdo Sound; Seymour Island.

Discussion. The strong axially sinuous ribs of the only recorded specimen, USNM 490813, are more characteristic of *Arrhoges (Antarctohoges) diversicostata* [Wilckens, 1911] than to *A. (A.) arcuacheilus* Stilwell and Zinsmeister [1992] from the La Meseta Formation. The neotype of *A. (A.) diversicostata*, USNM 441683, is 28.5 mm in height, comparable to the McMurdo Sound specimen. On Seymour Island, *A. (A.) diversicostata* has been recorded predominantly from the upper units VI-VII of the La Meseta Formation and only possibly from Units I-III. The presence of this species in McMurdo Sound greatly expands its geographic range. No other *Arrhoges (Antarctohoges)*-like gastropods have been recorded from older Paleocene or Cretaceous deposits in the James Ross Basin of the Antarctic Peninsula or elsewhere around the Antarctic continent.

Genus *Drepanocheilus* Meek, 1864

Drepanocheilus Meek, 1864, p. 35.

Type species. (By original designation) *Rostellaria americana* Evans and Shumard, 1857 *non* d'Orbigny (= *Drepanocheilus evansi* Cossmann, 1904).

Biogeographic element. Indo-Pacific/Tethyan, as interpreted herein.

Subgenus *Tulochilus* Finlay and Marwick, 1937

Tulochilus Finlay and Marwick, 1937, p. 63.

Type species. (By original designation) *Drepanocheilus (Tulochilus) bensoni* Finlay and Marwick, 1937.

Biogeographic element. Paleoaustral.

Discussion. Stilwell [1994] [pp. 855-861] expanded the morphology of *Drepanocheilus (Tulochilus) bensoni* Finlay and Marwick [1937] as a result of the discovery of a nearly perfect specimen of this species in Paleocene rocks at Wangaloa, South Island, New Zealand. Previously, the species was represented by the single, rather battered, holotype from Boulder Hill and a few fragments. The subgenus, *Drepanocheilus (Tulochilus)* Finlay and Marwick [1937] can be differentiated from the closely related *Drepanocheilus s. s.* Meek [1864] as the former has strongly angled rather than convex spire whorls, in having the outer lip extending further up the spire, in having strong tubercles, and in having a less strongly produced columella [Finlay and Marwick, 1937; Wenz, 1940; Beu and Maxwell, 1990]. The broadly conical protoconch of *D. (T.) bensoni* is large and long in proportion to its height and size (about 17.5% of height) and consists of three large whorls, including the relatively large, compressed, strongly convex nucleus [Stilwell, 1994, p. 857]. The discovery of *Drepanocheilus (Tulochilus)* in the Eocene of East Antarctica yields additional information on the subgenus and also adds to the growing list of common taxa found both in the Paleocene of New Zealand and Eocene of Antarctica. These specimens have a smaller, narrower protoconch, indicating that this character is variable in the subgenus.

Drepanocheilus (Tulochilus) erebus, new species
Plate 7, figs. D, F, H, and J

Diagnosis. Large for subgenus, but still relatively small for family at about 15-20 mm in height; small spire angle of about 15°; protoconch apparently small, polygyrate; very adapically pointed wing; 10-15 more-or-less equally spaced spiral riblets and more than 15 strong tubercles at periphery; differentiated from type species, *Drepanocheilus (Tulochilus) bensoni* Finlay and

Marwick (1937), in having a much larger shell, more spire whorls, smaller protoconch, and more spiral rib elements.

Description. Shell quite small for family (height about 15-20 mm), thin, delicate, alate; spire angle approximately 15°; growth lines weak, opisthocline; spire high with more than 6 angulate, subquadrate whorls; protoconch incomplete, but apparently polygyrate with small nuclear whorls; last whorl biangulate, with moderately strong subcentral, tubercle-bearing keel and poorly developed peribasal keel just adaperturally; spire whorls with axially elongated, opisthocline, moderately blunt tubercles and about 10-15 weak, mostly equally-spaced, straight to slightly wavy, spiral riblets; whorl inflation slow and constant; last whorl slightly to moderately inflated, ornamented with more than 15 strong tubercles at periphery and approximately 30 spirals (some incised) that diverge on wing and become deflected adapically; tubercles end on wing; wing well-developed with adapically projecting, slightly concave spike; wing with weak, orthocline growth lines; inner surface of wing with deep primary groove adjacent to spiral ridge in line with tubercles; neck apparently reduced to small spike; aperture elongated, sublenticular with narrow notch; inner lip with moderately narrow callus.

Dimensions. Holotype USNM 490814 height 13.5 mm, diameter of last whorl 11.0 mm; paratype USNM 490815 height 18.5 mm; paratype USNM 490816 length of wing 6.5 mm; paratype USNM 490817 height of fragment 13.5 mm.

Types. USNM 490814-490817.

Figured specimens. USNM 490814-490817.

Material. 12 incomplete specimens.

Localities. E189 (type).

Geographic distribution. McMurdo Sound.

Discussion. Erratic E189, collected from Mount Discovery, contains a nearly monotypic concentration of a new aporrhaid gastropod, *Drepanocheilus (Tulochilus) erebus*, described herein, and a small accumulation of cirriped plates, also described in this volume [see *Buckeridge* paper, this volume]. This is the first record of the genus and subgenus in Antarctica, previously known from only the New Zealand Paleocene [Finlay and Marwick, 1937; Stilwell, 1994]. The shell of *Drepanocheilus erebus* n. sp. is very fragile and thin; thus, most specimens are decorticated and broken. These specimens are remarkably similar to the type species, *Drepanocheilus (Tulochilus) bensoni* Finlay and Marwick [1937] [p. 63, pl. 8, figs. 3-4; Beu and Maxwell, 1990, p. 81, pl. 2, fig. e; Stilwell, 1994, pp. 855-861, pl.

59, figs. 15, 17-18], from the late Early Paleocene of South Island, New Zealand, suggesting a descendant-ancestor relationship. *Drepanocheilus (Tulochilus) erebus* n. sp. can be distinguished from *D. (T.) bensoni* in having a larger shell, more spire whorls, a smaller protoconch, and more spiral riblets. These differences are relatively minor. The type species of *Drepanocheilus s. s.*, *D. (D.) americanus* [Evans and Shumard, 1857] [see Meek, 1876, pp. 325 and 326, pl. 32, figs. 8a and b; Wenz, 1940, p. 913, fig. 2683], from the latest Cretaceous of North America, is easily separated from *D. (T.) erebus* in having much more convex, less angulate whorls, more axially elongated tubercles, and a more adapically pointed wing.

Etymology. Species named after the active, majestic volcano Mount Erebus on Ross Island, which can be seen from virtually all Mount Discovery fossil sites.

Family STRUTHIOLARIIDAE Fischer, 1887
Subfamily STRUTHIOLARIINAE Zinsmeister and Camacho, 1980
Genus *Struthiolarella* Steinmann and Wilckens, 1908

Struthiolarella Steinmann and Wilckens, 1908, pp. 53-60.

Type species. (By original designation) *Struthiolarella ameghinoi* von Ihering, 1897.

Biogeographic distribution. Paleoaustral.

Discussion. The struthiolariid gastropod *Struthiolarella* is one of the most speciose members of this widespread Cretaceous to Recent Austral family with roots extending back into the Paleocene of Patagonia. The oldest member of *Struthiolarella* is *S. senoniana* [Camacho and Zinsmeister, 1989] from the Early? Paleocene of southwestern Patagonia, followed by Eocene and younger Chilean and Argentine species, *S. ameghinoi* [Von Ihering, 1897], *S. hatcheri* [Ortmann, 1899], *S. ornata* [Sowerby, 1846], and *S. densestriata* [von Ihering, 1897]. In Antarctica, *Struthiolarella* is represented by the Eocene species *S. variabilis* Wilckens [1911] (Units III-VI of La Meseta Formation), *S. shackeltoni* Zinsmeister and Camacho [1980] (Unit V of La Meseta Formation), *S. steinmanni* Stilwell and Zinsmeister [1992] (Unit V of La Meseta Formation), and *S. mcmurdoensis* n. sp. from the McMurdo Sound erratics.

The distinguishing features of *Struthiolarella* have been discussed at length by Zinsmeister and Camacho [1980] and will not be repeated herein, except to state that sculpture, sinus and suture details separate this genus

Type species. (By monotypy) *Fusus dilatatus* Quoy and Gaimard, 1835.

Biogeographic element. Paleoaustral as interpreted herein.

?Penion australocapax Stilwell and Zinsmeister, 1992
Plate 8, figs. M and O

?*Penion australocapax* Stilwell and Zinsmeister, 1992, p. 128, pl. 17, figs. h-j.

Dimensions. USNM 490833 height 64.0 mm, diameter of last whorl 33.0 mm incomplete.

Figured specimen. USNM 490833.

Material. One poorly preserved specimen.

Locality. E145.

Geographic distribution. McMurdo Sound and possibly Seymour Island.

Discussion. This poorly preserved, large fusiform shell may be conspecific with *Penion australocapax* Stilwell and Zinsmeister [1992] [p. 128, pl. 17, figs. h-j] from Units II-V of the La Meseta Formation. Vestiges of spiral sculpture on specimen USNM 490833 from Mount Discovery reveal flattened, equally spaced spiral ribs very similar to *P. australocapax*.

Genus *Pseudofax* Finlay and Marwick, 1937

Pseudofax Finlay and Marwick, 1937, pp. 78-80.

Type species. (By original designation) *Phos ordinarius* Marshall, 1917.

Biogeographic element. Paleoaustral, as interpreted herein.

Pseudofax?, new species?
Plate 8, fig. N

Dimensions. USNM 490834 height 5.0 mm incomplete, diameter of last whorl 4.5 mm incomplete.

Figured specimen. USNM 490834.

Material. One poorly preserved specimen.

Locality. E373.

Geographic distribution. McMurdo Sound.

Discussion. This probable new species is known solely from a fragment of the teleoconch whorls and may be assignable to the Paleocene to Eocene paleoaustral buccinid *Pseudofax*. The sculpture of the McMurdo Sound specimen, USNM 490834, has more-or-less orthocline axial elements that are crossed by spaced spiral cords of equal strength, consistent with the variable sculptural configuration of species of *Pseudofax*, such as the type species, *P. ordinarius* [Marshall, 1917] [p. 456, pl. 35, figs. 24-25; see Finlay and Marwick, 1937, p. 80, pl. 9, figs. 16 and 18; Stilwell, 1994, pp. 973-977, pl. 69, figs. 16, 18-23, pl. 70, figs. 1-4], from the Paleocene of New Zealand.

Genus *Cominella* J. E. Gray *in* M. E. Gray, 1850

Cominella J. E. Gray *in* M. E. Gray, 1850, p. 72.

Type species. (By subsequent designation, Iredale, 1918) *Buccinum testudineum* Bruguiere, 1789 (= *Buccinum maculosum* Martyn, 1784.

Biogeographic element. Paleoaustral as interpreted herein.

Cominella? s. l., new species?
Plate 8, fig. F

Dimensions. USNM 490835 height 7.5 mm, diameter of last whorl 4.5 mm.

Figured specimen. USNM 490835.

Material. One specimen.

Locality. E (unnumbered erratic from Mount Discovery).

Geographic distribution. McMurdo Sound.

Discussion. This small, moderately robust, bucciniform species may be allied with the paleoaustral *Cominella* J. E. Gray *in* M. E. Gray [1850], but the subgenus is uncertain. A relationship with *C. (Josepha)* [Tenison-Woods, 1879] may be possible as the McMurdo Sound species, see specimen USNM 490835, has a small, narrow, moderately high-spired shell with strong, widely spaced, opisthocline axials and many closely spaced, spiral riblets, very similar to New Zealand species of *C. (Josepha)*. The axials on the penultimate whorl of *C.? s. l.* n. sp.? are orthocline.

Genus *Austrocominella* von Ihering, 1907

Austrocominella von Ihering, 1907, p. 344.

Synonym. *Zelandiella* Finlay, 1926 (Stilwell, 1994, pp. 978-981).

Type species. (By original designation) *Cominella (Austrocominella) fuegensis* von Ihering, 1907.

Biogeographic element. Paleoaustral as interpreted herein.

Austrocominella, species, *cf. A. verrucosa* (Stilwell and

Zinsmeister, 1992)
Plate 9, figs. a-c

cf. *Zelandiella verrucosa* Stilwell and Zinsmeister, 1992, pp. 135 and 136, pl. 18, figs. o-r.

Dimensions. USNM 490836 height 15.0 mm, diameter of last whorl 12.0 mm; USNM 490837 height 11.5 mm, diameter of last whorl 9.5 mm.
Figured specimens. USNM 490836-490837.
Material. Four specimens.
Localities. E330, E374, E375.
Geographic distribution. McMurdo Sound and possibly Seymour Island.
Discussion. The clasping sutures, bucciniform outline, strong axial tubercles or nodes, and spiral ornamentation of this species is strongly reminiscent of *Austrocominella verrucosa* [Stilwell and Zinsmeister, 1992] [pp. 135 and 136, pl. 18, figs. o-r] from Units I-V of the La Meseta Formation. Only incomplete specimens of this species were recovered and without details of the aperture a more concrete identification cannot be made at this time.

Genus *Eobuccinella* Stilwell and Zinsmeister, 1992

Eobuccinella Stilwell and Zinsmeister, 1992, p. 122.

Type species. (By monotypy) *Eobuccinella brucei* Stilwell and Zinsmeister, 1992.
Biogeographic element. Endemic.

?Eobuccinella brucei Stilwell and Zinsmeister, 1992
Plate 9, figs. E and G

?Eobuccinella brucei Stilwell and Zinsmeister, 1992, p. 122, pl. 15, figs. i and j.

Dimensions. USNM 490838 height 11.0 mm, diameter of last whorl 8.0 mm; USNM 490839 height 9.5 mm, diameter of last whorl 7.5 mm.
Figured specimens. USNM 490838-490839.
Material. Two specimens.
Locality. E116.
Geographic distribution. McMurdo Sound and possibly Seymour Island.
Discussion. The broadly fusiform outline and remnants of spiral sculpture of these two specimens, USNM 490838-490839, match perfectly *Eobuccinella brucei* Stilwell and Zinsmeister [1992] [p. 122, pl. 15, figs. i and j] from Units II-V of the La Meseta Formation, but

details of the columella are wanting. Hence, a tentative assignment is made herein.

Family FASCIOLARIIDAE Gray, 1853
Subfamily FUSININAE Swainson, 1840
Genus *Fusinus* Rafinesque, 1815

Fusinus Rafinesque, 1815, p. 145.

Type species. (By monotypy) *Murex colus* Linné, 1758.
Biogeographic element. Cosmopolitan (*cf.* Wenz, 1943, p. 1260).

Fusinus?, new species
Plate 9, fig. N

Dimensions. USNM 490840 height of fragment 24.5 mm.
Figured specimen. USNM 490840.
Material. One fragment.
Locality. E148.
Geographic distribution. McMurdo Sound.
Discussion. This very high-spired, narrowly fusiform shell has an acute spire angle of about 22° and sculpture of five spiral ribs, blunt tubercles and very weak prosocline growth lines. The suture is impressed. These characteristics are common to *Fusinus* Rafinesque [1815] and the spire morphology of the McMurdo Sound species is highly reminiscent of the tropical Recent type species, *F. colus* [Linné, 1758] [see Kiener, 1846, pp. 5 and 6, pl. 4, fig. 1 and also figures of synonym *F. tuberculata* Lamarck, pp. 9 and 10, pl. 7, fig. 1; Wenz, 1943, p. 1260, fig. 3589; Abbott and Dance, 1983, p. 187, colored figure], but differs in having more gradual whorl inflation reflecting slightly more expanded spire whorls, coarser spirals, and more central tubercles, compared with *F. colus*. *Fusinus suraknisos* Stilwell and Zinsmeister [1992] [pp. 131 and 132, pl. 18, figs. a-c] from uppermost Unit VII of the La Meseta Formation and *F. graciloaustralis* Stilwell and Zinsmeister [1992] [pp. 132 and 134, pl. 18, figs. c and d] from Unit III of the La Meseta Formation are both robust species with broader spires, coarser spiral sculpture, stronger tubercle development, and more angular whorl profiles, compared with *F.?* n. sp. and are not closely related to these species. *Fusinus?* n. sp. may be more closely related to coeval narrowly spired New Zealand taxa such as *Falsicolus* (= *Fusinus?) bensoni* [Allan, 1926] and *F. alta* [Marshall, 1919]. The latter two taxa may be conspecific [Beu and Maxwell, 1990].

Dimensions. USNM 490845 height 4.0 mm, diameter of last whorl 2.0 mm; USNM 490846 height 7.5 mm, diameter of last whorl 4.5 mm; USNM 490847 height 5.5 mm, diameter of last whorl 3.0 mm.

Figured specimens. USNM 490845-490847.

Material. Six specimens.

Locality. E145.

Geographic distribution. McMurdo Sound and Seymour Island.

Discussion. Several specimens of *Acteon eoantarcticus* Stilwell and Zinsmeister [1992] [p.168, pl. 24, figs. w and x], recorded previously from Units III-V of the La Meseta Formation, were prepared from erratic E145, associated with the first Paleogene vertebrate remains from East Antarctica. The preservation is exceptional in this particular block, so that detail of the sculpture is amenable to scanning electron microscopy. The prepared specimens compare well with *A. eoantarcticus*, except that the overall whorl profile is slightly less convex in the McMurdo Sound material, compared to the Seymour Island material. As species of *Acteon* can be quite variable morphologically, this variation in the Antarctic material is probably intraspecific. The outline and sculpture of *A. eoantarcticus* is very similar to New Zealand late Early Paleocene species, *A. semispiralis* Marshall [1917] and *A. wangaloa* Finlay and Marwick [1937], and a close phylogenetic link seems probable.

Genus *Crenilabium* Cossmann, 1889

Crenilabium Cossmann, 1889, pp. 306 and 307.

Type species. (By original designation) *Acteon aciculatus* Cossmann, 1889.

Biogeographic element. Indo-Pacific/Tethyan, as interpreted herein.

Crenilabium suromaximum Stilwell and Zinsmeister, 1992
Plate 9, figs. F and K

Crenilabium suromaximum Stilwell and Zinsmeister, 1992, p. 169, pl. 24, figs. a, b and j.

Dimensions. USNM 490848 height 5.0 mm, diameter of last whorl 2.0 mm.

Figured specimen. USNM 490848.

Material. One specimen.

Locality. E145.

Geographic distribution. McMurdo Sound and Seymour Island.

Discussion. *Crenilabium suromaximum* Stilwell and Zinsmeister [1992] [p. 169, pl. 24, figs. a, b, and j] is recognized in the erratics of McMurdo Sound and is represented by a single specimen. Previously, *C. suromaximum* was recorded only from Units II-V from the La Meseta Formation, where it is very rare. *Crenilabium suromaximum* is much lower spired compared with other Austral Paleogene species such as *C.* n. sp. of Stilwell [1994] [p. 1121-1125, pl. 79, figs. 1-4, 6-7, 11] from the late Early Paleocene of New Zealand and also the Paris Basin type species, *C. aciculatum* Cossmann [1889] [pp. 306-307, pl. 8, fig. 30 and above enlargement of aperture].

Family RINGICULIDAE Philippi, 1853
Genus *Ringicula* Deshayes, 1838

Ringicula Deshayes, 1838, p. 342.

Type species. (By subsequent designation, Gray, 1847) *Auricula ringens* Lamarck, 1804.

Biogeographic element. Cosmopolitan (Zinsmeister and Stilwell, 1990, p. 692).

Subgenus *Ringicula s. s.*

?Ringicula (Ringicula) cockburnensis Zinsmeister and Stilwell, 1990
Plate 9, figs. H and T

?Ringicula (Ringicula) cockburnensis Zinsmeister and Stilwell, 1990, pp. 374-375, Fig. 3.1-3.6.

Dimensions. USNM 490849 height of fragment 3.5 mm; USNM 490858 height 3.5 mm.

Figured specimen. USNM 490849; USNM 490858.

Material. Two poorly preserved specimens.

Locality. E204.

Geographic distribution. McMurdo Sound and possibly Cockburn Island.

Discussion. This probable species of *Ringicula* Deshayes [1838] may be conspecific with *Ringicula s. s. cockburnensis* Zinsmeister and Stilwell [1990] [pp. 374 and 375, fig. 3.1-3.6] from the Eocene of Cockburn Island, but the preservation of the available material is poor. Specimen USNM 490849 is a partial fragment of the spire that has a very thickened prominent varix and a low spire of only a few whorls. The sculpture of the specimen is not preserved well enough for any meaningful comment.

Superfamily PHILINACEA Gray, 1850
Family CYLICHNIDAE A. Adams, 1850

Genus *Cylichnania* Marwick, 1931

Cylichnania Marwick, 1931, p. 153.

Type species. (By original designation) *Cylichnania bartrumi* Marwick, 1931.
Biogeographic element. Paleoaustral.

Cylichnania?, new species
Plate 9, fig. D

Dimensions. USNM 490850 height 3.5 mm, diameter of last whorl 2.0 mm.
Figured specimen. USNM 490850.
Material. One specimen.
Locality. E189.
Geographic distribution. McMurdo Sound.
Discussion. This minute, slender cylindrical shell with an involute spire strongly resembles the hitherto Tertiary New Zealand and Australian genus *Cylichnania* Marwick [1931], especially the late Early Paleocene species *C. impar* Finlay and Marwick [1937] [p. 93, pl. 13, fig. 1; Fleming, 1966, p. 382, pl. 142, fig. 1690; Stilwell, 1994, pp. 1159-1160, pl. 81, figs. 1-3], but no sculpture is preserved on specimen USNM 490850, just faint orthocline growth lines.

Gastropoda genus et species indeterminate
Plate 9, fig. P

Dimensions. USNM 490855 height of fragment 3.5 mm.
Figured specimen. USNM 490855.
Material. One fragment.
Locality. E376.
Geographic distribution. McMurdo Sound.
Discussion. The affinities of this gastropod fragment are uncertain. The fragment has marked quite straight axials and weak broad spiral elements quite similar to *Turbonilla obtusa* Philippi [1887] [p. 91, pl. 11, fig. 13] from the Tertiary of Navidad, Chile.

Class SCAPHOPODA Bronn, 1862
Order DENTALIIDA Da Costa, 1776
Family DENTALIIDAE Gray, 1834

Dentaliidae genus et species indeterminate
not figured

Discussion. Poorly preserved scaphopod fragments were recovered from erratics E145 and E376. No orna-

ment is preserved on the material which is mostly decorticated and fragmentary. Accurate identification of these scaphopods is not possible at this time.

Acknowledgments. The writer would firstly like to thank D. Harwood for inviting me to participate on three expeditions to East Antarctica to collect fossils used in this report and for his enthusiasm and support of my work. R. Levy, S. Bohaty, and J. Kaser made this project a success due to their hard work in the field and fossil finding prowess. I wish to thank Rich Levy, in addition, for his help with this project and for going above and beyond the call of duty. The personnel of McMurdo Station, Ross Island, made this paper possible because of their willingness to help in any way possible while we were in the field and at the Station. L. Hally assisted me while photographing specimens. R. Henderson provided advice and valuable suggestions during the course of this investigation.

REFERENCES

Abbott, R. T., and S. P. Dance.
1983   *Compendium of Seashells.* 410 p. E. P. Dutton, Inc., New York.

Adams, A.
1850   Monograph of the family Bullidae. *In* G. B. Sowerby, *Thesaurus Conchyliorum.* London, 2:553 608.
1854   Further contributions towards the natural history of the Trochidae. *Proceedings of the Zoological Society of London,* 22:37-41.
1856   Descriptions of thirty-four species of bivalve Mollusca (*Leda, Nucula,* and *Pythina*) from the Cuming collection. *Proceedings of the Zoological Society of London,* 24:47-53.
1860   On some new genera and species of Mollusca from Japan. *Annals and Magazine of Natural History,* 3(5):405-422.

Adams, H.
1869   Descriptions of a new genus and fourteen new species of marine shells. *Proceedings of the Zoological Society of London* for 1869:272-275.

Adams, H., and A. Adams.
1853-1858 *The Genera of Recent Mollusca.* 1(1853-1854, 484 p.); 2 (1854-1858, 661 p.); 3(1858), 136 pls. London.

Allan, R. S.
1926   Fossil Mollusca from the Waihao Greensands. *Transactions of the New Zealand Institute,* 56:338-346.

Allen, J. A.
1978   Evolution of the deep-sea protobranch bivalves. *Philosophical Transactions of the Royal Society of London, 284B*:387-401.

Allmon, W. D., J. C. Nieh, and R. D. Norris
1990   Drilling and peeling of turritelline gastropods

since the Late Cretaceous. *Palaeontology, 33*(3):595-611.

Askin, R. A.
This volume   Eocene spores and pollen from Mount Discovery, East Antarctica. *In* J. D. Stilwell and R. M. Feldmann (Eds.), *Paleobiology and Paleo-environments of Eocene Rocks, McMurdo Sound, East Antarctica.*

Askin, R. A., D. H. Elliot, J. D. Stilwell, and W. J. Zinsmeister
1991   Stratigraphy and paleontology of Campanian and Eocene sediments, Cockburn Island, Antarctic Peninsula, *Journal of South American Earth Sciences, 4*(1-2):99-117.

Beck, H.
1847   Verzeichness der Naturaliensammlung, welche auf Befehl Sr. Majestät des Königs aus vershie-denen Königlichen Musâen in Kopenhagen...zur 24. Versamlung Deutscher Naturforscher und Ärzte nach Kiel gesandt war. Amtl. Ber. 24. *Vers. Deutsch. Naturf. Ärzte Keil, 1846*:109- 110.

Beu, A. G., and P. A. Maxwell.
1990   Cenozoic Mollusca of New Zealand (drawings by R. C. Brazier). *New Zealand Geological Survey Bulletin, 58*:1-518.

Bieler, R.
1992   Gastropod phylogeny and systematics. Annual Review of Ecology and Systematics, *23*:311-338.

Blainville, H.-M. Ducrotay de.
1816-1830   *Dictionaire des Sci. Naturelles.* Paris, 60 volumes.

Born, I. von.
1778   *Index Rerum Naturalum Musei Caesarei Vindo-bunensis Pars. Testacea.* Vindobonae, 458 p.

Bosc, L. A. G.
1801-1803   Histoire naturelle des coquilles...5 vols. *In* R. R. Castel (ed.) *Histoire naturelle de Buffon, classée...d'après le systême de Linné...par* R. R. Castel...nouvelle édition. Déterville, Paris, 80 vols.

Boshier, D. P.
1960   The fossil history of some New Zealand Calyp-traeidae (Gastropoda). *New Zealand Journal of Geology and Geophysics, 3*(3):390-399.

Brambilla, G.
1976   I molluschi Pliocenici di Villalvernia (Ales-sandria). I. Lamellibranchi. *Memorie Della Società Italiana di Scienze Naturali Museo Civico di Storia Naturale di Milano, 21*(3):81-128.

Brocchi, G. B.
1814   Conchiologia fossile subapennina, con osservaz-ioni geologiche sugli Apennini e sul suolo adia-cente. *Stamperia Reale,* Milan, 2 vols., 712 p.

Bronn, H. G.
1862   Malacoza Acephala. *Die Klassen und Ordnungen des Thier-Reichs, 3*(1):1-518.

Brown, T.
1849   *Illustrations of the Fossil Conchology of Great Britain and Ireland.* Smith, Elder and Co., London, 273 p.

Bruguiére, M.
1789-1816   Encyclopédie méthodique, ou par ordre de matières; par une société de gens de lettres, de savans et d'artistes...*Histoire naturelles des vers. Tome sixième [Vol. 1].* Chez Panckoucke, libraire...Liège, chez Plomteux, Imprimeur des Etats, Paris, 344 p; [1792:345-758].

Bucquoy, E., P. Dautzenberg, and G. Dollfuss.
1882-1886   Les mollusques marins du Roussillion. 1. *Gastropodes.* Paris, 570 p.

Beurlen, K.
1944   Beiträge zur Stammgeschichte der Muscheln. *Bayerische Akademie der Wissenschäftlich Sitzungsberichte,* 1-2:133-145.

Camacho, H. H., and W. J. Zinsmeister.
1989   La familia Struthiolariidae Fischer, 1884 (Mol-lusca: Gastropoda) y sus representantes del Terciario patagónico. *Actas del 40 Congreso Argentino de Paleontología y Biostratigrafia, 4*:98-110.

Carpenter, P. P.
1864   Supplementary report on the present state of our knowledge with regard to the mollusca of the West Coast of North America. *Report of the British Association for the Advancement of Science, 33*(1863):517-686.

Catlow, A., and L. Reeve.
1845   *The Chonchologist's Nomenclator. A catalogue of all the Recent species of shells, included under the subkingdom "Mollusca"*...Reeve Brothers, London, 326. P.

Chemnitz, J. H. [see Martini, F. H. W.].

Children, J. G.
1822-1824   Lamarck's genera of shells translated from the French with plates from original drawings by Miss Anna Children. *Quarterly Journal of Science, 14*:64-86 (1822); *14*:(298-322); *15*:23-52 (1823); *15*:23-52 (1823); *15*:216-258 (1823); *16*:49-79 (1823); *16*:241-264 (1824).

Clarke, A. H.
1959   New abyssal mollusks from off Bermuda collected by the Lamont Geological Observatory. *Proceedings of the Malacological Society of London, 33*:231-238.

Conrad, T. A.
1831   *American Conchology No. 3.* [see A. Catlow and L. Reeve, 1845].
1832-1835   *Fossil shells of the Tertiary formations of North America.* Illustrated by figures drawn from stone from nature. Pt. 1:1-20, pls. 1-6 [1832]; 2:21-26, pls. 7-14 [1832]; 3:29-38 [1833]; 4:39-46 [1833]

[2nd edition pp. 29-56, pls. 15-1, 1835]. J. Dobson, Philadelphia.

1838-1845 *Fossils of the Tertiary formations of the United States.* Illustrated by figures, drawn from nature. Pt. 1:V-XVI, 1 bis 32, pls. 1-17 [1838]; 2:33-56, pls. 18-29 [1840]; 3:57-89, pls. 30 bis 49 [1845]. J. Dobson, Philadelphia.

1865 Descriptions of new Eocene shells of the United States. *American Journal of Conchology, 1*:142-149.

1867 Descriptions of new genera and species of Miocene shells, with notes on other fossil and recent species. *American Journal of Conchology, 2*:257-270.

Cossmann, M.

1886-1913 Contribution illustré des coquilles fossiles de l'Eocène des environs de Paris...*Annales de la Société Royale Malacologique de Belgique* [see Wenz, 1938-1944 for collation of this complex work].

1904 Réctifications de Nomenclature. *Essais critiques de paléozoologie,* (1904):165-167.

1920 Supplément aux mollusques éocéniques de la Loire-Inférieure. *Bulletin de la Société Sciences naturelles Ouest France, 5*(3):53-141.

Costa, E. M. Da.

1776 *Elements of Conchology: or, an introduction to the knowledge of shells.* B. White, London, 318 p.

Cox, L. R , and J. B. Knight

1960 Suborders of the Archaeogastropoda. *Proceedings of the Malacological Society of London, 33*(6):262-264.

Cuvier, G.

1797 Tableau Élémentaire de l'Histoire Naturelle des Animaux. Paris, 710 p.

Dall, W. H.

1889 On the hinge of pelecypods and its development, with an attempt toward a better division of the group. *American Journal of Science, 38*:445-462.

1890-1903 Contributions to the Tertiary fauna of Florida with special reference to the Miocene Silex beds of Tampa and the Pliocene beds of Caloosahatchie River. *Transactions of the Wagner Free Institute of Science,* Philadelphia [see Keen, 1958, p. 564, for parts].

Darragh, T. A.

1994 Paleocene bivalves from the Pebble Point Formation, Victoria, Australia. *Proceedings of the Royal Society of Victoria, 106*:71-103.

1997 Gastropoda, Scaphopoda, Cephalopoda and new Bivalvia of the Paleocene Pebble Point Formation, Victoria, Australia. *Proceedings of the Royal Society of Victoria, 109*(1):57-108.

Darragh, T. A., and G. W. Kendrick.

1980 Eocene bivalves from the Pallinup Siltstone near Walpole, Western Australia. *Journal of the Royal Society of Western Australia, 63*(1):5-20.

1991 Maastrichtian Bivalvia (excluding Inoceramidae) from the Miria Formation, Carnarvon Basin, north Western Australia. *Records of the Western Australian Museum Supplement, 36*:1-102.

Dell, R. K.

1990 Antarctic Mollusca. *Royal Society of New Zealand Bulletin, 27*:1-311.

Dell, R. K., and C. A. Fleming.

1975 Oligocene-Miocene bivalve Mollusca and other macrofossils from Sites 270 and 272 (Ross Sea), DSDP Leg 28. *Initial Reports of the Deep Sea Drilling Project,* Washington, *28*:693-703.

Deshayes, G. P.

1830-1832 Encyclopédie méthodique ou par ordre de matiéres...*Histoire naturelle des Vers et Mollusques.* Paris.

1838 Traité elementaire de Conchyliologie avec l'application de cette *Science a la Géologie.* Paris, 2 vols.

1856-1866 Description des animaux sans vertebres decouverts dans le bassin de Paris pour servir de supplément à la description des coquilles fossiles des environs de Paris, comprenant une revue generale de toutes les espèces actuellement connues. J. B. Baille, Paris, 3 vols. and 2 atlases.

Evans, J., and B. F. Shumard.

1857 On some new species of fossils from the Cretaceous formation of Nebraska Territory. *Transactions of the Academy of Sciences of St. Louis, 1*:38-42.

Evans, R. B.

1985 Geology and paleoecology of Cretaceous rocks in northern Kaipara, New Zealand. *New Zealand Journal of Geology and Geophysics, 28*:609-622.

Férussac, A. E. de

1822 *Tableaux systématiques des animaux mollusques.* Paris and London, 111 p.

Finlay, H. J.

1924 New shells from New Zealand Tertiary beds. *Transactions of the New Zealand Institute, 55*:450-479.

1926 A further commentary on New Zealand molluscan systematics. *Transactions of the New Zealand Institute, 57*:320-485.

Finlay, H. J., and J. Marwick.

1937 The Wangaloan and associated faunas of Kaitangata-Green Island Subdivision. *New Zealand Geological Survey Bulletin, 15*:1-140.

Fischer, P.

1880-1887 *Manuel de conchyliologie et de paléontologie conchyliologique.* F. Savy, Paris, 1369 p.

Fleming, C. A.

1962 New Zealand biogeography: a paleontologist's

approach. *Tuatara, 10*(2):53-108.

1966   Marwick's Illustrations of New Zealand shells with a checklist of New Zealand Cenozoic Mollusca. *New Zealand Department of Scientific and Industrial Research Bulletin, 173*:1-456.

1967   Cenozoic history of Indo-Pacific and other warm-water elements in the marine Mollusca of New Zealand. *Venus, 25*(3-4):105-117.

1979   The Geological History of New Zealand and its Life. *Auckland University Press,* Oxford University Press, 141 p.

Fleming, J.
1820   *Mollusca. New Edinburgh Encyclopedia.* American Edition 13, Philadelphia.

1822   *The Philosophy of Zoology, or a general view of the structure, functions, and classifications of animals.* London, vol. 1 (432 p.); vol. 2 (618 p.).

1828   *A History of British Animals, exhibiting the descriptive characters and systematic arrangement etc.* Bell and Bradfute, Edinburgh, 565 p.

Forbes, E.
1838   *Malacologia Monensis. A catalogue of the Mollusca inhabiting the Isle of Man and the neighbouring sea.* J. Carefree and Son, Edinburgh, 63 p.

Freneix, S.
1980   Bivalves Nèocrètacès de Nouvelle-Calèdonie. Signification biogèographique, biostratigraphique, palèoècologique. *Annales de Palèontologie Invertèbrès, 66*(2):67-134.

1981   Faunes de bivalves du Sènonian de Nouvelle-Calèdonie. Analyses palèobiogèographique, biostratigraphique, palèoècologique. *Annales de Palèontologie Invertèbrès, 67*(1):13-32.

Gabb, W. M.
1866-1869 Cretaceous and Tertiary fossils. Geological Survey of California. *Paleontology, 2*:1-299.

1868   An attempt at a revision of the two familes Strombidae and Aporrhaidae. American Journal of Conchology, *4*:137-149.

Gmelin, J. F.
1791   Caroli a Linné, etc., Systema naturae per regna tria naturae, etc., ed. 13. *Leipzig, 1*(6):3021-3910.

Gray, J. E.
1824   *Supplement to the Appendix, Parry's First Voyage, 1819-1820.* London, 37 p.

1825   A list and description of some species of shells not taken notice of by Lamarck. *Annals of Philosophy, 25*:134-140, 407-415.

1834   Alphabetical list of the figures of Mollusca. *In* E. Griffith and E. Pidgeon, *The Animal Kingdom, arranged in conformity with its organisation, by Baron Cuvier...with supplementary additions to each order...Vol. 12. The Mollusca and Radiata.*

Whittacker and Co., London, p. 595-601.

1840   *Synopsis of the contents of the British Museum,* 42nd edition. British Museum (London), 370 p.

1847   A list of the genera of Recent Mollusca, their synonyms and types. *Proceedings of the Zoological Society of London, 15:*129-219.

1850   Systematic arrangement of the figures. *In* M. E. Gray, *Figures of Molluscous Animals,* London, *4*:63-206.

1853   On the divisions of ctenobrancheous gasteropodous Mollusca into larger groups and families. *Proceedings of the Zoological Society of London, 21*:32-44.

Gray, M. E.
1842-1857 *Figures of Molluscous Animals.* London, 5 vols. Griffin, M., and M. A. Hünicken.

1994   Late Cretaceous-early Tertiary gastropods from southwestern Patagonia, Argentina. *Journal of Paleontology, 68*(2):257-274.

Groye, Ménard de la
1807   Mémoire sur un nouveau genre de coquille bivalve-équivalve, de la famille des Solénoides. *Museum Histoire Naturelle Annales,* Paris, 9:131-139.

Harrington, H. J.
1969   Fossiliferous rocks in moraines at Minna Bluff, McMurdo Sound. *Antarctic Journal of the United States, 4*(4):134-135.

Hedley, C.
1902   Scientific results of the trawling expedition of H. M. O. S. "Thetis". Brachiopoda and Pele-cypoda. *Memoirs of the Australian Museum, 4*(5):287-324.

Hermannsen, A. N.
1846-1849 *Indicis generum malacozoorum primordia. Nomina subgenerum, generum, familiarum, tibuum, ordinum, classum; adjectis auctoribus temporibus, locis, systematics atque literariis, etymis, synonymis.* T. Fischer, Cassel, *1*:1-637 [1846]; *2*:1-717 [1847-1849].

Hertlein, L. G.
1969   Fossiliferous boulder of early Tertiary age from Ross Island, Antarctica. *Antarctic Journal of the United States, 4*(4):199-201.

Hickman, C. S.
1984   Composition, structure, ecology, and evolution of six Cenozoic deep-water mollusk communities. *Journal of Paleontology, 58*(5):1215-1234.

Hinds, R. B.
1843   Descriptions of new shells from the collection of Captain Belcher. *Annals and Magazine of Natural History, 11*(1):16-21.

Hutton, F. W.
1873   *Catalogue of the Tertiary Mollusca and Echinodermata of New Zealand.* Government Printer

for Colonial Museum and Geological Survey Department, Wellington, 48 p.

1877   Descriptions of some new Tertiary Mollusca from Canterbury. *Transactions of the New Zealand Institute, 9*:593-598.

Ihering, H. von.

1876   Versuch eines natürlichen Systemes der Mollusken. *Jahrbücher des deutschen malakozoologischen Gesellschaft, 1876*:97-148.

1897   Os molluscos dos terrenos tertiarios da Patagonia. *Revista do Museu Paulista, 2*:217-328.

1900   On the South American species of Mytilidae. *Proceedings of the Malacological Society of London, 4*(2):84-98.

1905-1907   Les mollusques fossiles du Tertiaire et du Crétacé Supérieur de l'Argentine. *Anales del Museo Nacional de Buenos Aires,* Series III, *7*:1-611 [published in 39 parts, November 21, 1905-September 13, 1907: date of each part appears on bottom page of each part].

Iredale, T.

1918-1921   Molluscan nomenclatural problems and their solution. *Proceedings of the Malacological Society of London, 13*:28-40; *14*:198-208.

1924   Results from Roy Bell's molluscan collections. *Proceedings of the Linnean Society of New South Wales, 49*:179-278.

1931   Australian molluscan notes No. 1. *Records of the Australian Museum, 18*:201-235.

Kabat, A. R.

1991   The classification of the Naticidae (Mollusca; Gastropoda): review and analysis of the supraspecific taxa. *Bulletin of the Museum of Comparative Zoology, 152*:417-449.

Kase, T.

1984   *Early Cretaceous marine and brackish-water Gastropoda from Japan.* The National Science Museum, Tokyo, 263 p.

1990   Late Cretaceous gastopods from the Izumi Group of southwest Japan. *Journal of Paleontology, 64*(4):563-578.

Kauffman, E. G.

1967   Cretaceous *Thyasira* from the Western Interior of North America. *Smithsonian Miscellaneous Collections, 152*(1):1-159.

1973   Cretaceous Bivalvia. *In* A. Hallam (Ed.), *Atlas of Paleobiogeography,* Elsevier Scientific Publication Company, pp. 353-383.

Keen, A. M.

1951   Outline of a proposed classification of the pelecypod family Cardiidae. *Minutes of the Conchological Club of Southern California,* p. 6-8.

Kiener, L. C. (and P. Fischer).

1834-1880   *Species général et Iconographie des coquilles vivantes, comprenant la collection du Muséum d'Histoire naturelle de Paris, la collection Lamarck, celle du Prince Masséna...et les découvertes récentes des voyageurs.* Chez Rousseau, J.-B. Baillière, Paris, 11 vols. [see Wenz, 1944, p. 1551 for collation of parts].

Korobkov, I. A.

1954   *Handbook and Methodological Guide to Tertiary Mollusca.* Lamellibranchiata. Leningrad, 444 p. [in Russian]

Lamarck, J. B. de

1799   Prodrome d'une nouvelle classification des coquilles, comprenant une rédaction apropriée des caractères génériques, et l'establissement d'un grand nombre des genres nouveaux. *Mémoire de la Société d'Histoire Naturelle de Paris, 1799*:63-90.

1801   *Système des Animaux sans Vertèbres, ou Tableau Gènèral des Classes, des Ordres et des Genres de ces Animaux.* Paris, 432 p.

1802-1806   Mémoires sur les fossiles des environmens de Paris. *Annales du Muséum National d'Histoire Naturelle.* Paris [see Wenz, 1944, p. 1555 for parts].

1809   *Philosophie Zoologique. 1*:1-422; *2*:1-473. Paris.

1815-1822   *Histoire Naturelle de Animaux sans Vertèbres.* Paris, 7 vols [see Wenz, 1944, pp. 1555 and 1556 for collation of parts].

Lamy, E.

1906   Lamellibranches recuellis par l'Expedition Antarctique Française du Dr. Charcot. *Bulletin du Muséum d'histoire naturelle, 12*:44-52.

Lesson, R. P.

1830   Mollusques...Chapter 11 *In* M. L. I. Duperry, *Voyage autour du monde...sur...la Coquille pendant...1822-1825 par M. L. I. Duperry. Zoologie, 2*(1). A. Bertrand, Paris, pp. 239-448.

Levy, R. H.

This volume   Eocene dinoflagellate cysts from McMurdo Sound, East Antarctica. *In* J. D. Stilwell and R. M. Feldmann (Eds.), *Paleobiology and Paleoenvironments of Eocene Rocks, McMurdo Sound, East Antarctica,* Antarctic Research Series 76, American Geophysical Union, Washington, D.C.

Linné, C.

1758   *Systema Naturae.* Edition 10. Holmiaea, Stockholm, 823 p.

1767   *Systema Naturae.* Editio duodecima reformata. "Vermes Testacea", *1*(2):533-1327.

Lovén, D. C.

1846   Index Molluscorum litora Scandinaviae occidentalia habitantium. Ofversigt af Finska vetenskaps-societatens förhandlingar, Helsingfors, Homiae, 50 p.

Macellari, C. E.
1988 Stratigraphy, sedimentology and paleoecology of Upper Cretaceous/Paleocene shelf-deltaic sediments of Seymour Island. *In* R. M. Feldmann and M. O. Woodburne (Eds.), *Geology and Paleontology of Seymour Island, Antarctic Peninsula, Geological Society of America Memoir, 169*, pp. 25-53.

Macpherson, J. H., and C. J. Gabriel
1962 *Marine Molluscs of Victoria,* Melbourne University Press, Melbourne, 475 p.

Marincovich, L.
1977 Cenozoic Naticidae (Mollusca: Gastropoda) of the northwestern Pacific. *Bulletins of American Paleontology, 83*(294):165-494.

Marshall, B. A.
1978 The genus *Neilo* in New Zealand (Mollusca: Bivalvia). *New Zealand Journal of Zoology, 5*:425-436.

Marshall, P.
1917 The Wangaloa Beds. *Transactions of the New Zealand Institute, 49*:450-460.
1919 Some new fossil species of Mollusca. *Transactions of the New Zealand Institute, 51*:253-258.

Martens, E. von
1878 Einige conchylien aus den kälteren Meeresgegenden der südlichen Erdhalfte. *Sitzungsbericht der Gesellschaft naturforschender Freunde zu Berlin, 1878*:20-26.
1880-1894 *Conchologische Mittheilungen als Fortsetzung der Novitates conchologicae.* Cassel, 3 vols. [see Wenz and Zilch, 1959-1960, p. 767, for parts].

Martini, F. H. W., and J. H. Chemnitz
1769-1795 *Neues systematisches Conchylien-Cabinet.* Nuremberg, 11 vols.

Martyn, T.
1784-1787 The Universal Conchologist, exhibiting the figure of every known shell accurately drawn and painted after nature: with a new systematic arrangement by the author Thomas Martyn. Sold at his house, No. 16 Great Marlborough Street, London. 4 vols., 161 plates [not available in nomenclature, except for a few names validated in ICZN Opinion 479].

Marwick, J.
1924 The Tertiary and Recent Naticidae and Naricidae of New Zealand. *Transactions of the New Zealand Institute, 55*:545-579.
1926 Molluscan fauna of the Waiarekan Stage of the Oamaru Series. *Transactions of the New Zealand Institute, 56*:307-316
1927 The Veneridae of New Zealand. *Transactions of the New Zealand Institute, 57*:567-635.
1929 Tertiary molluscan fauna from Chatton, South-

land. *Transactions of the New Zealand Institute, 59:*903-934.
1931 The Tertiary Mollusca of the Gisborne District. *New Zealand Geological Survey Bulletin 13*:1-177.
1944 New Zealand fossil and Recent Cardiidae (Mollusca). *Transactions of the Royal Society of New Zealand, 73*:181-192.
1950 Notes on the southern family Struthiolariidae, *Journal de Conchyliologie, 90*:234-239.
1957 New Zealand genera of Turritellidae, and the species of *Stirocolpus. New Zealand Geological Survey Paleontological Bulletin 27*:1-54.
1960 Early Tertiary Mollusca from Otaio Gorge, South Canterbury. *New Zealand Geological Survey Paleontological Bulletin 33*:1-32.
1971 New Zealand Turritellidae related to *Zeacolpus* Finlay (Gastropoda). *New Zealand Geological Survey Paleontological Bulletin 44*:1-87.

Maxwell, P. A.
1969 Middle Tertiary Mollusca from North Otago and South Canterbury, New Zealand. *Transactions of the Royal Society of New Zealand, 6*(13):155-185.
1988 Comments on "A reclassification of the Recent genera of the subclass Protobranchia (Mollusca: Bivalvia)" by J. A. Allen and F. Hannah (1986). *Journal of Conchology, 33*:85-96.
1992 Eocene Mollusca from the vicinity of McCulloch's Bridge, Waihao River, South Canterbury, New Zealand: paleoecology and systematics. New Zealand Geological Survey Paleontological Bulletin 65:1-280.

McKerrow, W. S. (ed.).
1978 *The Ecology of Fossils.* MIT Press, Cambridge, 384 p.

Medina, F. A., and C. A. Rinaldi.
1978 Transposicion charnelar en *Bathytormus wilckensis* sp. nov. (Bivalvia). *Ameghiniana, 15*(3-4):422-428.

Meek, F. B.
1864 Checklist of the invertebrate fossils of North America. Jurassic Formation. Notes and explanations. *Smithsonian Miscellaneous Collections, 7*:27-29, 39-40.
1876 A report on the invertebrate Cretaceous and Tertiary fossils of the Upper Missouri Country. *United States Geological Survey of the Territories Reports, 9*:1-629.

Melvill, J. C., and R. Standen
1912 The marine Mollusca of the Scottish National Antarctic Expedition. Part II. *Transactions of the Royal Society of Edinburgh, 48:*333-366.

Milne-Edwards, H.
1848 Note sur la classification naturelle des mol-

lusques gastéropodes. *Annales des Sciences Naturelles Zoologiques, Série 3, 9*:102-112.

Molina, G. I.

1782   Saggio sulla storia naturale del Chile. Bologna, 367 p.

Monterosato, T. Allery de.

1884   *Nomenclature generica e specifica di alcune conchiglie mediterranee.* Palermo, 152 p.

Montfort, Denys de.

1808-1810 *Conchyliologie systématique, et classification méthodique des coquilles; offrant leurs figures, leur arrangement générique, leurs descriptions caractéristiques, leurs noms; ainsi que leur synonymie en plusieurs langues. Volume 1: Coquilles univalves, cloisonnés* [1808, 409 p.]; *Volume 2: Coquilles univalves, non cloisonnées* [1810, 676 p.]. Paris, Chez F. Schoell, Libraire.

Moore, R. C. (ed.).

1969   *Treatise on invertebrate paleontology, Part 6. Volumes 1-2. Bivalvia.* Geological Society of America and University of Kansas Press, Lawrence, 952 p.

1971   Ibid. Vol. 3. Oysters, p. N953-N1224.

Mörch, O. A. L.

1853 [1852?] *Catalogus conchyliorum quae reliquit D. Alponso D'Aguirra & Gadea, Comes de Yoldi, regis daniae cubiculariorum princeps...Fasciclus primus.* Cephalophora. Hafniae, Typis Ludovici Kleini, 170 p.

Morton, J. E.

1956   The evolution of *Perissodonta* and *Tylospira* (Struthiolaridae). *Transactions of the Royal Society of New Zealand, 83*(3):515-524.

Müller, O. F.

1776   *Zoologigae Danicae Prodromus seu Animalium Daniae et Norvegiae Indigenarum Characters,* Nomina et Synonyma Imprinis popularium. Havniae, 282 p.

Murdoch, W. G. Burn.

1894   *From Edinburgh to the Antarctic.* An artist's notes and sketches during the Dundee Antarctic Expedition of 1892-93; with a chapter by W. S. Bruce naturalist of the barque "Balaena". Longmans, Green and Co., London, 364 p.

Neumayr, R. M.

1884   *Zur morphologie des Bivalvenschlosses. Mathematischen und Naturforschungen Classe der Kaiserlichen* Akademie der Wissenschaften in Wien, Sitzungs Berichte, *88*:385-419.

Newell, N. D.

1965   Classification of the Bivalvia. *American Museum Novitates,* 2206:1-25.

Orbigny, A. d'

1839-47 Voyage dans l'Amerique meridionale pendant 1826-1833. *Zoologie,* 4-6, 758 p.

1842-44 Paléontologie Française. Description zoologique et géologique de tous les animaux mollusques et rayonnés fossiles de France. Terrains Crétacés. Mollusques. Tome Troisième. Lamellibranches. G. Masson, Paris, 807 p.

Ortmann, A. E.

1899   The fauna of the Magellanian beds of Punta Arenas, Chile. *American Journal of Science, 8*:427-432.

1902   Tertiary macroinvertebrates. *In* W. B. Scott (Ed.), *Reports of the Princeton Expedition to Patagonia 1896-1899.* Volume 4, Paleontology I, Part 2, p. 45-332. Princeton, New Jersey and E. Schweizerbart'sche Verlagshandlung, Stuttgart.

Philippi, R. A.

1844 [1845?] *Enumeratio Molluscorum Siciliae cum viventium tum in tellure tertiaria fossilium quae in itinere suo observavit.* Vol. 2. Halis Saxonum (Halle), 303 p.

1853   *Handbuch der Conchyliologie und Malakozoologie.* Halle, 547 p.

1887   *Fósiles Tertiarios y Cuartarios de Chile.* F. A. Brockhaus, Leipzig, 256 p.

Piccoli,., S. Sartori, and A. Franchino

1986   Mathematical model of the migration of Cenozoic benthic molluscs in the Tethyan belt. *Memorie degli Instituti di Geologia e Mineralogia,* Universida di Padova, *38*:207-244.

Poli, J. X.

1791-1826 *Testacea útriusque Siciliae eorumque Historia et Anatome tabulis aeneis illustrata.* Parmae, 3 vols.

Ponder, W. F., and A. Warén.

1988   Classification of the Caeonogastropoda and Heterostropha - a list of the family-group names and higher taxa. *Malacological Review,* Supplement 4:288-326.

Popenoe, W. P.

1983   Cretaceous Aporrhaidae from California: Aporrhainae and Arrhoginae. *Journal of Paleontology, 57*(4):742-765.

Powell, A. W. B.

1951   Antarctic and Subantarctic Mollusca. Pelecypoda and Gastropoda. *Discovery Reports, 26*:47-196.

1964   New Zealand molluscan systematics with descriptions of new species. Part 4. *Records of the Auckland Institute and Museum, 6*(1):11-20.

1979   New Zealand Mollusca. Marine, Land and Freshwater Shells. Collins, Auckland, 500 p.

Quenstedt, W.

1930   Die Anpassung an die grabende Lebensweise in der Geschichte der Solenmyiden und Nuculaceen. *Geologische und Palaeontologische Abhandlungen (N. F.), 18*(1):1-119.

Quoy, J. C. R., and P. Gaimard.

1832-1835 Voyage de découverts de l'Astrolabe, éxécuté par ordre du Roi pendant les années 1826-1827-1828-1829, sous le commandement de M. J. Dumont d'Urville. Mollusques. *Zoologie, Tome 2,* 686 p.[1832-1833]; *Tome 3,* 954 p. [1834-1835]. [Mollusques, p. 1-644]. *Atlas zoologique,* 200 pl. [each section numbered separately]. J. Tastu, Paris.

Rafinesque, C. S.

1815 *Analyses de la nature ou tableau de l'universe et des corps organises.* Palermo, 224 p.

Reid, R. G. B., and D. G. Brand

1986 Sulfide-oxidising symbiosis in lucinaceans: implications for bivalve evolution. *The Veliger,* 29:3-24.

Rio, C. del.

1997 Cenozoic biogeographic history of the eurythermal genus *Retrotapes,* new genus (subfamily Tapetinae) from southern South America and Antarctica. *The Nautilus, 110*(3):77-93.

Röding, P. F.

1798 *Museum Boltenianum sive Catalogus cimeliorum e tribus regnis naturae quae olim collegerat Joa. Frid. Bolten, M. D. p. d., Pars Secunda.* Johan Christi Trapii, Hamburg, 119 p.

Ross, J. C.

1847 *A Voyage of Discovery and Research in the Southern and Antarctic Regions, during the years 1839-43.* Vol. 1 (366 p.); vol. 2 (447 p.). John Murray, London.

Sacco, F.

1890-1904 *I molluschi dei terreni tertiarii de Piemonte e della Liguria.* Torino, pts. VI-XXX.

Sanders, H. L., and J. Allan.

1977 Studies on the deep-sea Protobranchia (Bivalvia), the family Tindariidae and the genus *Pseudotindaria. Bulletin of the Museum of Comparative Zoology,* Harvard University, *148*:23-59.

Sassi, A.

1827 Saggio geologico sopra il Bacino terziario di Albenga. *Giornale Ligustico di Scienze Lettere e Arti, 1*:467-484.

Schmidt, A.

1818 *Versuch über die best Einrichtung zur Aufstellung, Behandlung und Aufbewahrung der verschiedenen Naturkörper und Geganstände der Kunst...*J. Perthes, Gotha, 252 p.

1856 Der Geschlechtsapparat der Stylommatophoren in taxonomischer Hinsicht. *Abhandlungen naturwissenschaftlichen Vereines,* Halle, *1*:1-52.

Schumacher, C. F.

1817 *Essais d'un nouveau système des habitations des vers testacés.* Copenhagen, 287 p.

Sharman, G., and E. T. Newton

1894 Notes on some fossils from Seymour Island, in the Antarctic regions obtained by Dr. Donald. *Transactions of the Royal Society of Edinburgh, 37*(30):707-709.

1897 Notes on some additional fossils collected at Seymour Island, Graham Land, by Dr. Donald and Captain Larsen. *Proceedings of the Royal Society of Edinburgh, 22*(1):58-61.

Smith, E. A.

1875 Description of some new shells from Kerguelen's Island. *Annals and Magazine of Natural History, 16*(4):67-73.

1907 Lamellibranchiata. National Antarctic Expedition, 1901-1904, *Natural History, 2*:1-7.

1915 Mollusca. Pt. 1, Gastropoda, Prosobranchia, Scaphopoda and Pelecypoda. British Antarctic ("Terra Nova") Expedition, 1910. *Natural History Report, Zoology, 2*:61-112.

Soot-Ryen.

1955 A report on the family Mytilidae (Pelecypoda). *Allan Hancock Pacific Expeditions, 20*(1):1-78.

Sowerby, G. B. II.

1832-1841 *The conchological illustrations, or coloured figures of all the hitherto unfigured Recent shells* [Vol. 1, 1832-1833, parts 1-34; Vol. 2, 1833-1841, "parts" 35-200]. Sowerby, London, 116 p. and 200 pl.

1842-1887 *Thesaurus Conchyliorum, or monographs of genera of shells.* Edited by G. B. Sowerby, Jun. Sowerby, Great Russell Street, London. 5 vols., issues in 44 parts [see Woodward, 1915, p. 1981, for collation of parts, dates and authors].

1846 Description of Tertiary fossil shells from South America. *In* C. Darwin, *Geological Observations on South America* (Appendix), London, p. 249-264.

Sowerby, J. de C.

1812-1846 *The Mineral Conchology of Great Britain: or coloured figures and descriptions of those remains of testaceous animals or shells which have been preserved at various times and depths in the earth.* London, 7 vols.

Steiner, G.

1992 Phylogeny and classification of Scaphopoda. *Journal of Molluscan Studies, 58*:385-400.

Steinmann, G., and O. Wilckens

1908 Kreide und Tertiärfossilien aus den Magellansländern, gesammelt von der schwedischen Expedition 1895-1897. *Arkiv für Zoologi, 4*:1-119.

Stewart, R. B.

1930 Gabb's California Cretaceous and Tertiary type Lamellibranchs. *Academy of Natural Sciences of Philadelphia Special Publications, 3*:1-314.

Stilwell, J. D.
1992   Review of Eocene Mollusca from the vicinity of McCulloch's Bridge, Waihao River, South Canterbury, New Zealand: paleoecology and systematics (New Zealand Geological Survey Bulletin 65, 278 p.), by P. A. Maxwell. *Geological Survey of New Zealand Newsletter*, 97:49-54.
1993   New Early Paleocene Mollusca from the Wangaloa Formation of South Island, New Zealand. *Journal of Paleontology, 67*(3):360-369.
1994   Latest Cretaceous to earliest Paleogene molluscan faunas of New Zealand: changes in composition as a consequence of the break-up of Gondwana and extinction. Ph.D. dissertation, 1630 p., The University of Otago, Dunedin, New Zealand.
1998   Late Cretaceous Mollusca from the Chatham Islands, New Zealand, *Alcheringa* 22:29-85.

Stilwell, J. D., and M. Griffin.
in prep.   A monograph of the Austral Cretaceous to Recent Struthiolariidae (Mollusca: Gastropoda): systematics, paleoecology and paleobiogeography.

Stilwell, J. D., C. M. Jones, R. H. Levy, and D. M. Harwood.
in press   First fossil bird from East Antarctica. *Antarctic Journal of the United States.*

Stilwell, J. D., R. H. Levy, R. M. Feldmann, and D. M. Harwood.
1997   On the rare occurrence of Eocene callianassid decapods (Arthropoda) preserved in their burrows, Mount Discovery, East Antarctica. *Journal of Paleontology, 71*(2):284-287.

Stilwell, J. D., R. H. Levy, and D. M. Harwood.
1993   Preliminary paleontological investigation of Tertiary glacial erratics from the McMurdo Sound region, East Antarctica. *Antarctic Journal of the United States - Review* 1993:16-19.

Stilwell, J. D., and W. J. Zinsmeister.
1992   *Molluscan systematics and biostratigraphy. Lower Tertiary La Meseta Formation, Seymour Island, Antarctic Peninsula.* American Geophysical Union Antarctic Research Series 55, 192 p.

Stoliczka, F.
1870-1871   *Cretaceous fauna of Southern India. The Pelecypoda, with a review of all known genera of this class, fossil and Recent. Vol. 3.* Memoirs of the Geological Survey of India, Palaeontologia Indica, 6th series. Calcutta, 535 p.

Suter, H.
1913-1915   *Manual of the New Zealand Mollusca. With an atlas of quarto plates.* Government Printer, Wellington, 1120 p., Atlas [1915].

Swainson, W.
1840   *A Treatise on Malacology; or the natural classification of shells and shell-fish.* Lardner's Cabinet Cyclopedia, London, 419 p.

Tenison-Woods, J. E.
1879   On some new Tasmanian marine shells. *Papers and Proceedings and Report of the Royal Society of Tasmania for 1878*:32-40.

Thiele, J.
1924 [1921?] Revision des Systems der Trochacea. *Mitteilungen aus dem Zoologischen Museum in Berlin, 11*:47-74.
1929-1935   *Handbuch der Systematischen Weichtierkunde.* Jena, 1154 p.

Torigoe, K.
1981   Oysters in Japan. *Journal of Science of Hiroshima University, Series B, Division 1 (Zoology), 29*(2):291-419.

Toulmin, L. D.
1977   *Stratigraphic distribution of Paleocene and Eocene fossils in the Eastern Gulf Region.* Vols. 1-2. Geological Survey of Alabama monograph *13*, 602 p., 72 pls.

Troschel, F. H.
1848   Bericht über die Leistungen in der Naturgeschichte der Miollusken 1846. *Archiv für Naturgeschichte, 13*(2):151-208.

Valenciennes, M. A.
1839   Description de l'animal de la *Panopée australe*, et recherches sur les autres espèces vivantes ou fossiles de ce genre. *Archives du Muséum d'histoire naturelle,* Paris, 38 p.

Vaught, K. C.
1989   *A classification of the living Mollusca.* R. T. Abbott and K. J. Boss (eds.). American Malacologists, Inc., Melbourne, Florida, 195 p.

Verrill, A. E., and K. J. Bush
1897   Revision of the genera of Ledidae and Nuculidae of the Atlantic Coast of the United States. *American Journal of Science,* ser. 4, *3*(13):51-63.

Vokes, H. E.
1980   *Genera of the Bivalvia: a systematic and bibliographic catalogue* (revised and updated). Paleontological Research Institution, Ithaca, New York, 307 p.

Weller, S.
1903   The Stokes collection of Antarctic fossils. *Contributions from the Walker Museum, 1*(5):65-71.

Wenz, W.
1938-1944   Gastropoda. *In* G. H. Schinderwolf (Ed.), *Handbuch der Paläozoologie.* Borntraeger, Berlin, 1639 p.

Wenz, W., and A. Zilch.
1959-1960   *Gastropoda. Teil 2. Euthyneura.* Gebrüder Borntraeger, Berlin-Nikolasse, 834 p.

White, C. A.
1887   *Contributions to the Paleontology of Brazil, Mus. Nac.* Rio de Janeiro Archivos, 7:1-273, 28 plates.

Wilckens, O.
1910   Die Anneliden, Bivalven und Gastropoden der

Antarktischen Kreideformation. *Wissenschaftliche Ergebnisse der Swedischen Südpolar Expedition 1901-1903, 3*(12):1-132.

1911 Die Mollusken der Antarktischen Tertiärformation. *Wissenschaftliche Ergebnisse der Swedischen Südpolar-expedition 1901-1903, 3*(13):1-62.

1924 Die Tertiäre fauna der Cockburn-Insel (Westantarktika). *Further Zoological results of the Swedish Antarctic Expedition 1901-1903, 1*(5):1-18.

Winckworth, R.
1932 The British marine Mollusca. *Journal of Conchology, 19*(7):217-252.

Wood, W., and S. Hanley
1856 *Index Testaceologicus, an illustrated catalogue of British and Foreign shells,* containing about 2800 figures accurately coloured after nature. A new and entirely revised edition, with ancient and modern appellations, synonyms, localities, etc., etc. Willis and Sotheran, London, 234 p.

Woodring, W. P.
1925 *Miocene Mollusca from Bowden Jamaica, pelecypods and scaphopods.* Carnegie Institute of Washington Publication, *366*:1-564.

Woodward, S. P.
1851-1856 *A Manual of the Mollusca: or a rudimentary treatise of recent and fossil shells.* London.

Wrenn, J. H., and G. F. Hart
1988 Paleogene dinoflagellate cyst biostratigraphy of Seymour Island, Antarctic Peninsula. *In* R. M. Feldmann and M. O. Woodburne (Eds.), *Geology and Paleontology of Seymour Island, Antarctic Peninsula,* Geological Society of America Memoir 169, p. 321-447.

Zinsmeister, W. J.
1976a A new genus and species of the family Struthiolariidae, *Antarctodarwinella ellioti,* from Seymour Island, Antarctica. *The Ohio Journal of Science, 76*(3):111-114.

1976b The Struthiolariidae (Gastropoda) fauna from Seymour Island, Antarctic Peninsula. *Actas VI Congresso Geologico Argentino,* Sept. 21-27, 1975, Bahia Blanca, Argentina, *1*:609-618.

1977 Note on a new occurrence of the Southern Hemisphere aporrhaid gastropod *Struthioptera* Finlay and Marwick on Seymour Island, Antarctica. *Journal of Paleontology, 51*(2):399-404.

1978 Eocene nautiloid fauna from the La Meseta Formation of Seymour Island, Antarctic Peninsula. *Antarctic Journal of the United States, 13*(4):24-25.

1979 Biogeographic significance of the late Mesozoic and early Tertiary molluscan faunas of Seymour Island (Antarctic Peninsula) to the final breakup of Gondwanaland. In *Historical Biogeography, Plate Tectonics, and the Changing Environment,* Oregon State University Press, Corvallis, p. 349-355.

1982 Late Cretaceous-early Tertiary molluscan biogeography of the southern circum-Pacific. *Journal of Paleontology, 56*(1):84-102.

1984 Late Eocene bivalves (Mollusca) from the La Meseta Formation, collected during the 1974-1975 joint Argentine-American Expedition to Seymour Island, Antarctic Peninsula. *Journal of Paleontology, 58*(6):1497-1527.

1988 Early geological exploration of Seymour Island, Antarctica. *In* R. M. Feldmann and M. O. Woodburne (Eds.), *Geology and Paleontology of Seymour Island, Antarctic Peninsula.* Geological Society of America Memoir *169*:1-16.

Zinsmeister, W. J., and H. H. Camacho.
1980 Late Eocene Struthiolariidae (Mollusca: Gastropoda) from Seymour Island, Antarctic Peninsula and their significance to the biogeography of early Tertiary shallow-water faunas of the Southern Hemisphere. *Journal of Paleontology, 54*:1-14

1982 Late Eocene (to possibly earliest Oligocene) molluscan fauna of the La Meseta Formation of Seymour Island, Antarctic Peninsula. *In* C. Craddock (Ed.), *Antarctic Geosciences,* pp. 299-304. University of Wisconsin Press, Madison.

Zinsmeister, W. J., and C. E. Macellari.
1988 Bivalvia (Mollusca) from Seymour Island, Antarctic Peninsula. *In* R. M. Feldmann and M. O. Woodburne (Eds.), *Geology and Paleontology of Seymour Island, Antarctic Peninsula.* Geological Society of America Memoir *169*:253-284.

Zinsmeister, W. J., and J. D. Stilwell
1990 First record of the family Ringiculidae (Gastropoda) from the middle Tertiary of Antarctica. *Journal of Paleontology, 64*(3):373-376.

Zittel, K. A.
1864 Fossile Mollusken und Echinodermen aus Neuseeland. Nebst beiträgen von den Herren Bergrath Franz Ritter v. Hauer und Professor Eduard Suess. *In* F. von Hochstetter, M. Hörnes, und F. Ritter von Hauer (Eds.), *Paläontologie von Neu-seeland, Beiträge zur Kenntnis der Fossilien Flora und Fauna der Provinzen Auckland und Nelson.* Reise der Österreichischen Fregatte Novara um die Erde, Geologischer Teil, *1*(2):15-68.

Stilwell, Jeffrey D., School of Earth Sciences, James Cook University of North Queensland, Townsville, Queensland 4811 Australia

PALEOBIOLOGY AND PALEOENVIRONMENTS OF EOCENE ROCKS, MCMURDO SOUND, EAST ANTARCTICA
ANTARCTIC RESEARCH SERIES VOLUME 76, PAGES 321-323

BRYOZOAN FRAGMENTS FROM EOCENE GLACIAL ERRATICS OF MCMURDO SOUND, EAST ANTARCTICA

Urszula Hara

Biostratigraphical Department of the Geological Museum, Polish Geological Institute, Rakowiecka 4, 00-975 Warszawa, Poland.

Remains of encrusting bryozoan colonies have been found in glacial erratics of McMurdo Sound in East Antarctica. These poorly preserved bryozoans which represent malacostegan Cheilostomatida possibly belong to genus *?Membranipora* Blainville. This is the first fossil record of this genus and the oldest occurrence of Cenozoic bryozoans from East Antarctica.

INTRODUCTION

The small encrusting bryozoan colonies which represent malacostegan cheilostomes, possibly belong to *?Membranipora* de Blainville, 1830. These colonies have been found on the glacial erratic blocks of Eocene sandstone at Minna Bluff, McMurdo, East Antarctica (Figure 1). Associated with the bryozoans are the bivalves *Eurhomalea*, *Panopea*, *?Cyclorisina*, a gastropod represented by *Struthiolarella*, as well as the plant fragments including wood and encrusting serpulid worm tubes [J. D. Stilwell, personal communication, 1996]. The fossil record of bryozoans seems to be rare in East Antarctica. A younger, well-preserved Quaternary bryozoan fauna was previously described from McMurdo Sound (Speden, 1962; Hendy, et al., 1969; Hayward and Taylor, 1984]. The fossil occurrence of the genus *Membranipora* has been reported many times from the Tertiary of the Southern Hemisphere [see Waters, 1881, 1882, 1883, 1887, 1889, 1898; MacGillivray, 1895; Canu, 1904, 1908; Brown, 1952; cf. Gordon, 1984]. The bryozoan material was collected by Dr. Jeffrey D. Stilwell during his field work at Minna Bluff, McMurdo Sound, East Antarctica in the years 1993, and 1995.

SYSTEMATIC NOTE

Order Cheilostomatida Busk, 1852
Suborder Malacostegina Levinsen, 1902
Superfamily Membraniporoidea Busk, 1852
Family Membraniporidae Busk, 1852
Genus *?Membranipora* de Blainville, 1830
Figure 2a-c

Description. Worn encrusting colony with the dimensions 8 mm (height) x 22 mm (width), forms a broad, foliaceous, slightly shredded, unilaminar sheet (Figure 2a). Zooids are elongate, have a size of 0.49 - 0.56 x 0.21 - 0.27 mm, are arranged in longitudinal rows, are variable in size and have elongate-oval, hexagonal or rectangular shapes, slightly narrowing distally about the middle. Along the margins of the zooids are very poorly visible round-shaped tiny structures which might be interpreted as the traces of the spines. The base of the colony has not been preserved. The poor state of preservation of this colony may be a result of the fact that the colonies of *Membranipora* have a lightly calcified skeleton. Other morphological structures have not been observed.

The second, tiny, fragmented, very badly preserved and possibly encrusting ?bryozoan colony, of few mm in length has been found in an erratic block at Minna Bluff locality. It is associated with mollusc fragments and plant debris. The hypothetical, encrusting fragment of this ?bryozoan colony is composed of some dozen or so oval-shaped structures of 0.3 mm width x 0.4-0.6 mm length which are similar to the the autozooids of bryozoans. Unfortunately, the zoarial wall of the colony is not preserved. The structures similar to the autozooids have a convex relief, suggesting that it is an internal mould of

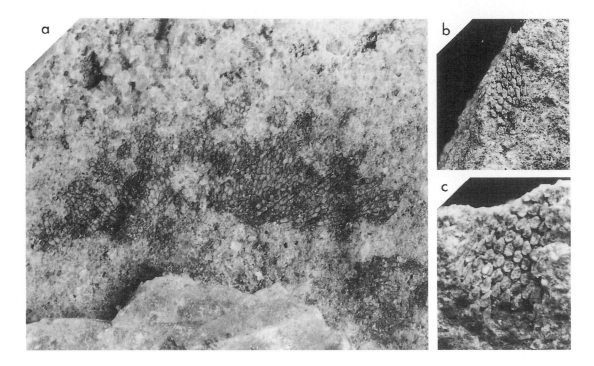

Fig 1. a. The frontal view of the bryozoan colony, USNM 498855, x 4.3, Minna Bluff; b-c. An internal mould of the possible bryozoan colony, USNM 498856, b x 5; c x 8, Minna Bluff.

the skeleton. On the basis of the comparable morphometric features it may tentatively be assumed that it is the internal mould of the same malacostegan cheilostome colony of the *?Membranipora* genus (Figure 2b-c).

Remarks. Although, the bryozoan colony examined is poorly preserved, the general pattern of arrangement of zooecia, very simple architecture of the colony, shape of zooids and their dimensions, all suggest that the colony studied belongs to *?Membranipora* genus or is very closely related. On the basis of the preserved features of the colony the systematic determination of this colony should be considered as tentative, because many species belonging formerly to the genus *Membranipora* are presently included into different genera [cf. Gordon, 1986]. Autozooids of the genus *Membranipora* are lightly calcified, the frontal surface is almost entirely membranous [Hayward, 1995].

In the Recent, the species of *Membranipora* are essentially epiphytes of kelp. As a consequence of their possessing long-lived planktotrophic larvae, most have very broad geographic distributions, although none have been reported from Antarctic waters [Hayward, 1995]. It is worth noting that in the modern estuaries which show, among other features, great fluctuations in salinity, the anascan cheilostome genera such as *Membranipora* and

Conopeum are frequent inhabitants [Ross, 1979]. Waters [1881, 1882, 1883, 1887, 1889, 1898], MacGillivray [1895], Canu [1904, 1908] and Brown [1952] recorded a fossil variety of this genus from the Miocene of the Southern Hemisphere [cf. Gordon, 1984]. The previous, oldest fossil occurrence of *Membranipora* was recorded from the Lower Miocene of South America [Canu, 1904, 1908] and New Zealand [Brown, 1952].

Occurrence. Minna Bluff, McMurdo Sound, East Antarctica.

FINAL REMARKS

Taking into consideration that this might be the first and the oldest fossil record of bryozoan fauna from East Antarctica it should be *in-situ*. Because the bryozoan colony belongs to an encrusting form on glacial erratics it is necessary to consider that the fauna studied might be younger. A list of the recognized bryozoan fauna from McMurdo Sound of East Antarctica did not include this taxon [Speden, 1962; Hendy et al., 1962; Hayward and Taylor, 1984].

Acknowledgements. I am immensely grateful to Dr. Jeffrey D. Stilwell (James Cook University, Townsville, Australia) for donation of the studied bryozoan fauna. Many thanks are due to Professor Dr. Andrzej Gazdzicki (Polish Academy of Sciences, Warszawa, Poland) for all critical comments and valuable remarks. Cordial thanks are extended to Professor Dr. Peter-Noel Webb for his final improvement of my English text (Ohio State University, Columbus, USA). I am grateful to Mrs. J. Mierzejewska (Polish Geological Institute, Warszawa, Poland) for taking the photographs.

REFERENCES

Brown, D. A.
1952   The Tertiary Cheilostomatous Polyzoa of New Zealand. British Museum (N.H.). 404 pp. London.

Canu, F.
1904   Les Bryozoaires du Patagonien. Échelle des Bryozoaires pour les terrains tertiaires. Mémoires de la Société Géologique de France, Paleontologie, 12(3): 1-30.

Canu, F.
1908   Iconographie des Bryozoaires fossiles de l'Argentine. Anales del Museo Nacional de Buenos Aires, 3: 245-341.

Gordon, D. P.,
1984   The Marine Fauna of New Zealand: Bryozoa: Gymnolaemata from the Kermadec Ridge. New Zealand Oceanographic Institute Memoir, 91: 23-24, Wellington.

Gordon, D.P.
1986   The Marine fauna of New Zealand: Bryozoa; Gymnolaemata (Ctenostomata and Cheilostomata Anasca) from the Western South Island Continental Shelf and Slope. New Zealand Oceanographic Institute Memoir, 95: 24-25.

Hayward, P. J.
1995   Antarctic Cheilostomatous Bryozoa. 355 pp, Oxford University Press. Oxford, New York, Tokyo.

Hayward, P.J., and P.D. Taylor
1984   Fossil and Recent Cheilostomata (Bryozoa) from the Ross Sea, Antarctica, 18: 71-94.

Hendy , C.H. , V.E. Neall and A.T. Wilson
1969   Recent marine deposits from Cape Barne, McMurdo Sound, Antarctica, New Zealand Journal of Geology and Geophysics,. 12: 707-712.

MacGillivray, P.H.
1895   A monograph of the Tertiary Polyzoa of Victoria. Transactions of the Royal Society of Victoria, 4: 1-166.

Ross, J.R.P.
1979   Ectoproct adaptation and ecological strategies. In G.P. Larwood (Ed.), Advances in Bryozoology. pp. 283-293. Academic Press, London-New York:

Speden, I.G.
1962   Fossiliferous Quaternary marine deposits in the McMurdo Sound region, Antarctica, New Zealand Journal of Geology and Geophysics, 5: 746-777.

Waters, A. W.
1881   On fossil cheilostomatous Bryozoa from South-west Victoria, Australia, Quarterly Journal of the Geological Society, 27: 309-347.

Waters, A.W.
1882   On cheilostomatous Bryozoa from Mount Gambier, South Australia. Quarterly Journal of the Geological Society, 38: 257-276.

Waters, A.W.
1883   Fossil cheilostomatous Bryozoa from Muddy Creek, Victoria, 39, p. 423.

Waters, A.W.
1887   Bryozoa from New South Wales, North Australia, Annual Magazine of the Natural History, scr. 4, 20(2): 181-203.

Waters, A.W.
1889   Bryozoa from New South Wales. Annual Magazine of the Natural History, ser. 6, 4 (1): 1-24.

Waters, A.W.
1898   Observation on Membraniporidae. Journal of the Linnean Society London, Zoology, 26: 654-693.

Urszula Hara, Biostratigraphical Department of the Geological Museum, Polish Geological Institute, Rakowiecka 4, 00-975 Warszawa, Poland.

PALEOBIOLOGY AND PALEOENVIRONMENTS OF EOCENE ROCKS, MCMURDO SOUND, EAST ANTARCTICA
ANTARCTIC RESEARCH SERIES VOLUME 76, PAGES 325–327

RHYNCHONELLIDE BRACHIOPODS FROM EOCENE TO EARLIEST OLIGOCENE ERRATICS IN THE MCMURDO SOUND REGION, ANTARCTICA

Daphne E. Lee

Department of Geology, University of Otago, Box 56, Dunedin, New Zealand

Jeffrey D. Stilwell

School of Earth Sciences, James Cook University, Townsville, Queensland 4811, Australia

The costellate rhynchonellide brachiopod, *Tegulorhynchia*, is recorded from crratics of calcareous sandstone and conglomerate of Middle Eocene to earliest Oligocene age near Mount Discovery, McMurdo Sound, Antarctica. This record extends the mid Cenozoic distribution of *Tegulorhynchia* from Australia, New Zealand, and Antarctic Peninsula to mainland Antarctica.

INTRODUCTION

Macroinvertebrate fossils including bivalves, gastropods, scaphopods, cirripeds, bryozoans, decapods and brachiopods were collected from erratic boulders near McMurdo Sound during four field seasons between 1991-1996 [Stilwell, this volume]. The brachiopods are the first to be described from rocks of Eocene or Oligocene age from mainland Antarctica, although about 20 brachiopod genera have been described from the Eocene to lowermost Oligocene La Meseta Formation on Seymour and Cockburn Islands, Antarctica Peninsula [Owen, 1980; Wiedman et al., 1988; Bitner, 1996].

About a dozen entire and fragmentary ribbed brachiopod fossils were collected from calcareous, well-cemented, poorly-sorted, coarse sandstone and conglomerate erratics at several sites near Mount Discovery, McMurdo Sound. A Middle Eocene to earliest Oligocene age for the brachiopod-yielding sediments is based on associated molluscs, dino flagellates [Levy, this volume], and spores/pollen) [Askin, this volume].

SYSTEMATIC PALEONTOLOGY

Order Rhynchonellida Kuhn, 1949
Superfamily Rhynchonelloidea Gray, 1848
Family Hemithyrididae Rzhonsnitskaya, 1956
Genus *Tegulorhynchia* Chapman and Crespin, 1923
Tegulorhynchia cf. *Timbricata* (Buckman, 1910)
Plate 1, figs a-g

Material. Three complete specimens (one juvenile), and several fragmentary valves. The figured specimens are housed in the collections of the Department of Paleobiology, National Museum of Natural History (USNM), Smithsonian Institution, Washington, D.C.

Dimensions. USNM 495681, E168, L=7.5 mm, W=8.5 mm; USNM 495683, E168, L=15.5 mm, W=15.8 mm, T=5.8 mm; USNM 495684, E169, L=18.9 mm, W=18.0 mm; USNM 495685 E378, L=17.7 mm, W=19.4 mm,T=8.3 mm.

Localities. E168, Mt. Discovery; E 169, Site W; E378, Mt. Discovery.

Description. Impunctate, medium-sized, subpentagonal in outline; anterior commissure rectimarginate in the juvenile specimen, broadly uniplicate in adults; surface of the better-preserved specimens finely costellate. The fine costellate ornamentation, and uniplicate folding enable the brachiopods to be placed in the genus *Tegulorhynchia*, but none of the material is sufficiently well preserved for a firm identification to species level. Three complete specimens exhibit well-preserved ribbing (Figure 1a-c), but in most examples the outer shell layers have been partly or completely abraded, and the only trace of ribbing remaining is around the anterior margin (Figure 1d-e, f-g). No internal features were observed. These specimens most closely resemble *Tegulorhynchia imbricata* (Buckman) from the Late Eocene La Meseta Formation on Seymour Island, Antarctic Peninsula.

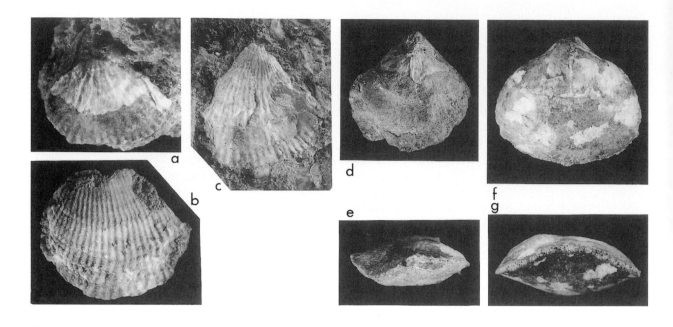

Plate 1, figs. a-g. *Tegulorhynchia cf. T.imbricata* (Buckman)

Fig. a. Exterior of dorsal valve of juvenile specimen, USNM 495681, locality E168, X4.

Fig. b. Finely ribbed, incomplete dorsal valve, USNM 495682, locality E168, X3.2

Fig. c. Exterior of ventral valve, USNM 495684, locality E169, X2.

Figs d-e. Dorsal and anterior views of entire specimen with asymmetric uniplicate fold, USNM 495683, locality E168, X2. Note that ribbing is visible only at anterior margin.

Figs. f-g. Dorsal and anterior views of entire decorticated specimen with broad symmetric uniplicate fold, USNM 495685, locality E378, X2.

Remarks. The only other records of Cenozoic rhynchonellide brachiopods from Antarctica are from James Ross, Cockburn and Seymour Islands, Antarctica Peninsula. Buckman [1910] described five species of ribbed brachiopods from Cockburn Island, which he placed in the genus *Hemithiris* D'Orbigny. Owen [1980] named a new species of the coarsely ribbed southern hemisphere genus *Notosaria*, *N. seymourensis* from Seymour Island, and figured *Hemithiris antarctica* Buckman, *Tegulorhynchia imbricata* (Buckman), and *Plicirhynchia sp.* from the "Lower Tertiary" of Cockburn Island. In the most recent study of the Seymour Island brachiopod fauna, Bitner [1996] described and figured new, well-preserved material of *H. antarctica*, *N. seymourensis*, and *T. imbricata*. She also described a new species of *Tegulorhynchia*, *T. ampullacea*, and a new genus and species of semicostate rhynchonellide,

Paraplicirhynchia gazdzickii. It is possible that further, better-preserved material from the Mount Discovery erratics might yield specimens comparable to the Seymour Island brachiopod fauna.

Paleoecology. The single extant species of *Tegulorhynchia*, *T. doederleini* (Davidson), lives in moderately deep water in the western Pacific [Lee, 1980]. The various fossil species of *Tegulorhynchia* seem to have lived on hard substrates at mid shelf depths, although the closely related genus, *Notosaria*, occurs in very shallow water around New Zealand at the present day [Lee, 1978]. None of the brachiopod specimens from the erratics was in life position, but the abraded shells, and coarse, poorly sorted nature of the matrix suggests deposition in high-energy, relatively near-shore conditions.

Biogeography. *Tegulorhynchia* is a widespread genus in the early-mid Cenozoic of the southern hemisphere. Species are recorded from Paleocene to Miocene age strata in Australia [McNamara, 1983; Lee unpublished records] and New Zealand [Lee, 1980], and from the Late Eocene of Seymour Island [Owen, 1980; Bitner, 1996]. This new record of *Tegulorhynchia* from mainland Antarctica strengthens the close relationships between shelf faunas from Australia, New Zealand and Antarctica in the early to middle Tertiary. The northward migration of the genus to the Indo-Pacific region in the Late Cenozoic [Lee, 1980] corresponds closely to the migration of numerous other invertebrate taxa including brachiopods, from southern regions into the Indo-Pacific [Fleming, 1979] as cooling of the Southern Ocean took place.

REFERENCES

Askin, this volume.

Bitner, M.A.
1996.   Brachiopods from the Eocene La Meseta Formation of Seymour Island, Antarctic Peninsula. *Palaeontologia Polonica* 55: 65-100.

Buckman, S.S.
1910.   Antarctic fossil Brachiopoda collected by the Swedish South Polar Expedition. Wiss. Ergebn. schwed. *SŸdpolarexped.*, Stockholm, 3(7): 1-40.

Fleming, C.A.
1979.   *The geological history of New Zealand and its life.* Auckland University Press. 141 pp.

Lee, D.E.
1978.   Aspects of the ecology and paleoecology of the brachiopod *Notosaria nigricans* (Sowerby). *Journal of the Royal Society of New Zealand* 8:93-113.

Lee, D.E.
1980.   Cenozoic and Recent rhynchonellide brachiopods of New Zealand: systematics and variation in the genus *Tegulorhynchia*. *Journal of the Royal Society of New Zealand* 10: 223-245.

Levy, this volume

McNamara, K.J.
1983.   The earliest *Tegulorhynchia* (Brachiopoda: Rhynchonellida) and its evolutionary significance. *Journal of Paleontology* 57 (3): 461-473.

Owen, E.F.
1980.   Tertiary and Cretaceous brachiopods from Seymour, Cockburn, and James Ross Islands, Antarctica. *Bulletin of the British Museum (Natural History), Geology,* 33:123-145.

Stilwell, this volume

Wicdman, L.A., R.M. Feldmann, D.E. Lee, and W.J. Zinsmeister
1988.   Brachiopoda from the La Meseta Formation (Eocene), Seymour Island, Antarctica. *Geological Society of America Memoir* 169: 449-457.

Daphne E. Lee, Department of Geology, University of Otago, Box 56, Dunedin, New Zealand

Jeffrey D. Stilwell, School of Earth Sciences, James Cook University, Townsville, Queensland 4811, Australia

A NEW SPECIES OF *Austrobalanus* (CIRRIPEDIA, THORACICA) FROM EOCENE ERRATICS, MOUNT DISCOVERY, MCMURDO SOUND, EAST ANTARCTICA

John St. J.S. Buckeridge

Faculty of Science and Engineering, Auckland Institute of Technology, Auckland, New Zealand.

A new species of the sessile balanomorph *Austrobalanus* is described from erratic Eocene rocks recovered from the Mount Discovery area, McMurdo Sound, Antarctica. This new species is the second austrobalanine recorded from the Antarctic region, and the third known species of *Austrobalanus*. The known distribution of *Austrobalanus* remains as Australasia - Antarctica. The preservation of *Austrobalanus antarcticus* sp. nov. is discussed, and the inference is made that the species probably lived in similar environmental conditions to other recorded *Austrobalanus*, namely shallow, warm water. The record of fossil balanomorphs in the Antarctic reflects the change to a cold water, shallow marine environment following the opening of the Drake Passage and initiation of the Circum-Antarctic Current.

INTRODUCTION

Deposits of Early Tertiary age are rare in the McMurdo Sound area. Isolated erratics of Eocene age however, have been collected from the Mt Discovery area. Lithologically, the erratics are arkosic grits that have been derived from the weathering and erosion of ancient granitic and gabbroic rocks. In some cases these are fossiliferous and have been found to contain disarticulated fragments of a new species of the balanomorph barnacle *Austrobalanus*. This paper describes the species and discusses its biogeographic significance. Specimens have been photographed and illustrated using a Wild binocular microscope and drawing tube. Type specimens are given the prefix USNM, and are held within the Department of Paleobiology, United States National Museum, Washington D.C. 20560.

SYSTEMATICS

Subclass **Cirripedia** Burmeister, 1834
Superorder **Thoracica** Darwin, 1854
Order **Sessilia** Lamarck, 1818
Suborder **Balanomorpha** Pilsbry, 1916
Superfamily **Tetraclitoidea** Newman, 1993

Family **Tetraclitidae** Gruvel, 1903
Subfamily **Austrobalaninae** Newman & Ross, 1976
Genus *Austrobalanus* Pilsbry, 1916

Diagnosis (emend.). Shell of 6 solid, calcareous compartmental plates, comprising compound rostrum, carina, and paired 1st and 2nd carinolatera; basal edges roughened with irregular points and ridges; basis, if present, thin, calcareous; radii narrow or wanting, with irregular sutural crenulations; scutum with moderately low articular and adductor ridges; tergum with end of spur rounded.

Type species. *Balanus imperator* Darwin, 1854 O.D. Recent, Australia.

Distribution. The genus is now represented by three known species, one of which is further split into two subspecies (Figure 1). These are *Austrobalanus antarcticus* sp. nov., (Eocene, McMurdo, Antarctica), *Austrobalanus macdonaldensis* Buckeridge (late Eocene to early Oligocene, South Island, New Zealand and Seymour Island, Antarctic Peninsula); *Austrobalanus imperator* (Darwin) (Pleistocene to Recent, New South Wales and Torres Strait, Australia); *Austrobalanus imperator aotea* Buckeridge (middle Oligocene to early Miocene, South Island, New Zealand).

Remarks. The age of this material has been determined as late Eocene, (Bartonian to Early Priabonian), on the basis of molluscs, dinoflagellates, and pollen (Jeff Stilwell, personal communication). As *Austrobalanus macdonaldensis* is of late Eocene age, *Austrobalanus antarcticus* sp. nov. is now seen to constitute the earliest record of the genus. The revision of compartment nomenclature [Yamaguchi and Newman, 1990, Buckeridge and Newman, 1992], has necessitated the above emendation to the diagnosis of *Austrobalanus*.

Figure 1: Distribution of ***Austrobalanus***.
Austrobalanus antarcticus sp. nov.
figures 2a-g, 3a,b.

Diagnosis. Shell wall conic, small, exterior moderately plicate; internal ribbing developed basally, forming an irregular basal margin; scutum broad basally, isoscelene triangular, interior with elevated depressor muscle pit, distal end of articular ridge depending, exterior with moderately well formed transverse growth lines, crossed near the articular angle by strong apico-basal striae.

Material examined. Field Number E151, *Drepanocheilus* gastropod block, (1 compartmental plate, collected 9 December, 1993); Field Number E155, Site V, (3 compartmental plates, collected December, 1993); Field Number E189, Site K2, (25 compartmental plates, collected 16 December, 1993); Field Number E308 (1 scutum, 5 compartmental plates, collected 15 November, 1995); Field Number E380, (2 compartmental plates, collected November 26, 1995). All collections made by Jeffrey D. Stilwell and party, University of Nebraska-Lincoln Expeditions (1993-95), from the Mount Discovery area, McMurdo Sound, Antarctica.

Lithology. A medium to light grey-brown, poorly sorted, moderately indurated, arkosic grit. The cirripede remains vary in preservation, with most showing evidence of pressure solution pitting [see Buckeridge, 1989]. As a consequence, there is difficulty in determining the detail of fine structure on compartments and opercula.

Figure 2: *Austrobalanus antarcticus* sp. nov.

Type material. Holotype: USNM 492751, scutum from E308 (occludent margin 4.7 mm); Paratypes: USNM 492752, first carinolatus (length 5.6 mm); USNM 492753, first carinolatus (length 3.4 mm); USNM 492754, rostrum (length 3.5 mm); USNM 492755, second carinolatus (length 3.8 mm); USNM 492756, first carinolatus (length 5.1 mm). Specimens USNM 492752-6 were collected from site E189.

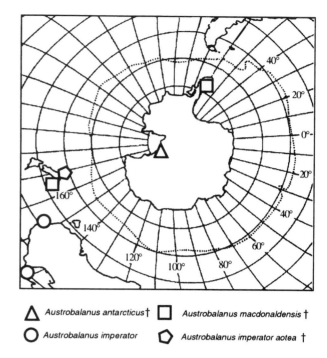

△ *Austrobalanus antarcticus*†   ▢ *Austrobalanus macdonaldensis* †

○ *Austrobalanus imperator*   ⬠ *Austrobalanus imperator aotea* †

Figure 1: Distribution of ***Austrobalanus***. The genus is currently restricted to the Southern Hemisphere, with only three species known. The earliest record is probably from the Eocene of McMurdo Sound, Antarctica. Dotted line indicates present day Antarctic Convergence; † denotes extinct taxa.

Description. Shell wall conic, small, exterior plicate, with moderately strong longitudinal ribs crossed by weaker transverse growth lines; interior of paries with moderate to weakly developed sheath, transverse growth lines weak, developed in upper two thirds; internal ribbing irregular, moderately developed basally, forming an irregular basal margin; radii very narrow or wanting, sutural edges either weakly denticulate or smooth; alae with weak growth lines parallel to basal margin; scutum (partially obscured), isoscelene triangular, broad basally, interior with elevated, moderately large, depressor muscle pit, depending articular ridge extending for two thirds length of articular margin, exterior with moderately well formed transverse growth lines, crossed near articular angle by sharp apico-basal striae, articular angle broadly obtuse, removed from articular margin by approximately one quarter width of scutum; basis and tergum unknown.

Remarks. A preliminary survey of the material suggested that more than one taxon was present, particularly with respect to features such as the sheath and internal ribbing, which ranged from poorly developed to well formed. Close observation of all specimens available confirmed, however, that there is only one species, with

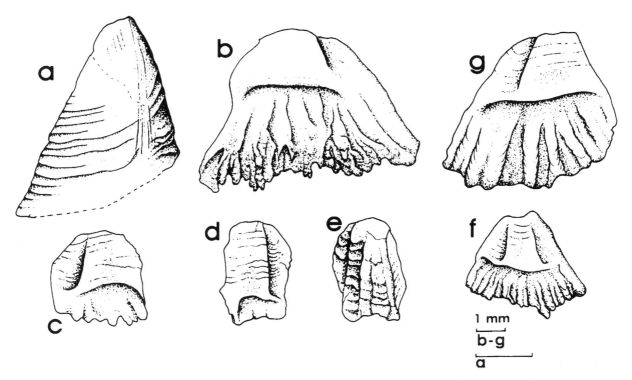

Figure 2: *Austrobalanus antarcticus* sp. nov. a: Holotype: scutum, (right, exterior), USNM 492751; b: 1st carinolatus, (right, interior), USNM 492752; c: first carinolatus, (left, interior), USNM 492753; d: second carinolatus, (right, interior), USNM 492755; e: second carinolatus, (right, exterior), USNM 492755; f: rostrum, (interior), USNM 492754; g: first carinolatus, (left, interior), USNM 492756.

many variable features, interpreted here as reflecting ontogeny. As discussed earlier, the mode of preservation prevents clear determination of delicate features. In particular, this applies to the nature of the sutural edges of compartments and the nature of the alae.

On the sole scutum, basal growth lines on the articular flange do not parallel earlier lines, but become very irregular. This is probably due however, to repair after unsuccessful predation, as some secondary thickening is noted on other parts of the plate.

Excavation of the scutum was stopped after the apical region became detached, but this did however, provide an opportunity for the interior of the plate to be observed before reattaching the fragment with polyvinyl acetate. In addition, the basal margin of the plate has not been fully exposed, primarily because of the brittleness of the plate, and the strong cementation of the sediment grains.

Austrobalanus antarcticus sp. nov. resembles the late Eocene species *Austrobalanus macdonaldensis* Buckeridge, but can be distinguished by the scutum, which is broader basally, has a higher adductor muscle scar internally, and externally possesses both well formed apico-basal striae near the articular angle and stronger transverse growth lines. In addition, the exterior of the compartments in *A. antarcticus* is generally more plicate than is the condition in specimens of *A. macdonaldensis*. *A. antarcticus* differs from *Austrobalanus imperator* in its more regular external appearance (*A. imperator* is often very rugose), and by the presence of apico-basal striae near the articular angle of the scutum.

BIOGEOGRAPHY

The opening of the Drake Passage during the Oligocene [Andrews, 1977, Barker and Burrell, 1977], and the subsequent initiation of the Circum-Antarctic Current effected significant climatic changes in the Antarctic region. This change is clearly demonstrated by sessile barnacles (Table 1). The earliest balanomorph species, such as *Austrobalanus macdonaldensis* and *Austrobalanus antarcticus* are interpreted as reflecting warm water conditions during the Eocene [Buckeridge, 1983]. Although no fossil barnacles of Oligocene or Miocene age have yet been recovered from the region, late Neogene taxa (such as *Bathylasma corolliforme*),

Figure 3: *Austrobalanus antarcticus* sp. nov. a: rostrum, (interior), USNM 492754; b: first carinolatus, (right, interior).

reflect cold temperatures, probably not significantly warmer than currently experienced [Buckeridge 1989].

CONCLUSION

Balanomorph barnacles are rare in Antarctic rocks. They do however, provide an additional opportunity to view the changes in the Antarctic environment over the last 50 million years. In particular, faunal changes reflect the warmer water shallow marine conditions that existed around the continent in the early Cainozoic. Warm ocean currents at that time swept down the coast of South America, providing the environmental conditions that are currently needed to sustain a warm, shallow marine fauna. This regime was to change significantly following the opening of the Drake Passage, and the initiation of the Circum Antarctic Current. Post Miocene faunae are distinctly cold water, reflecting the extant faunal assemblages around the continent.

Table 1: Fossil Balanomorpha of Antarctica. Whilst Palæogene taxa are interpreted as warm water species, Neogene taxa are distinctly cold water, reflecting the post Eocene climatic changes that occurred with the opening of the Drake Passage and initiation of the Circum Antarctic Current [Barker and Burrell, 1977]. Note that the Scallop Hill Formation is now accepted as Pliocene [Buckeridge, 1989]. † Indicates extinct species.

Late Eocene	**Tetraclitidae** † *Austrobalanus macdonaldensis* Buckeridge — Seymour Island, Antarctic Peninsula [Zullo, Feldmann and Wiedman, 1988]. † *Austrobalanus antarcticus* sp. nov. — Mount Discovery, McMurdo Sound (this paper). **Archaeobalanidae** † *Solidobalanus* sp. — Seymour Island, Antarctic Peninsula [Zullo, Feldmann and Wiedman, 1988].
Pliocene	**Bathylasmatidae** *Bathylasma corolliforme* (Hoek) — Scallop Hill Formation, McMurdo Sound [Speden 1962, Newman and Ross 1971, Buckeridge 1983, 1989].
Pleistocene	**Balanidae** †*Fosterella hennigi* (Newman) — Cockburn Island, Antarctic Peninsula [Newman 1979, Buckeridge 1983].
Holocene	**Bathylasmatidae** *Bathylasma corolliforme* (Hoek) — Taylor Formation, McMurdo Sound [Buckeridge 1989].

REFERENCES

Andrews, P. B.
  1977   Depositional facies and the early phase of ocean basin evolution in the Circum-Antarctic Region. *Marine Geology 25*: 1-13.

Barker, P. F. and J. Burrell
  1977   The opening of Drake Passage. *Marine Geology 25*: 15-34.

Buckeridge, J. S.
  1983   Fossil barnacles (Cirripedia: Thoracica) of New Zealand and Australia. *New Zealand Geological Survey Paleontological Bulletin 50*: 1-150 + pls. 1-13.

Buckeridge, J. S.
  1989   Marine invertebrates from late Cainozoic
              deposits in the McMurdo Sound region,
              Antarctica. *Journal of the Royal Society of New
              Zealand 19*(3): 333-342.
Buckeridge, J. S. and W. A. Newman
  1992   A re-examination of genus *Waikalasma*
              (Cirripedia : Thoracica) and its significance in
              balanomorph phylogeny. *Journal of
              Paleontology 66*(2): 341-345.
Newman, W. A.
  1979   On the biogeography of balanomorph barnacles
              of the Southern Ocean including new balanid
              taxa: a subfamily, two genera and three species.
              Proceedings of the International Symposium on
              Marine Biogeography and Evolution in the
              Southern Hemisphere. Auckland, July, 1978.
              *New Zealand Department of Scientific and
              Industrial Research Information Series 137*(1):
              279-306.
Newman, W. A. and A. Ross
  1971   *Antarctic Cirripedia.* American Geophysical
              Union Antarctic Research Series *14*: 1-257.

Speden, I. G.
  1962   Fossiliferous Quaternary marine deposits in the
              McMurdo Sound Region, Antarctica. *New
              Zealand Journal of Geology and Geophysics
              5*(5): 746-777.
Yamaguchi T. and W. A. Newman
  1990   A new and primitive barnacle (Cirripedia :
              Balanomorpha) from the North Fiji Basin
              Abyssal Hydrothermal Field, and its evolution-
              ary implications. *Pacific Science 44*: 135-55.
Zullo, V.A., R.M. Feldmann and L.A. Wiedman
  1988   Balanomorph Cirripedia from the Eocene La
              Meseta Formation. Geology and Paleontology
              of Seymour Island, Antarctic Peninsula.
              Geological Society of America Memoir 169:
              459-464.

John St. J.S. Buckeridge, Faculty of Science and
Engineering, Auckland Institute of Technology, Auckland, New
Zealand.

PALEOBIOLOGY AND PALEOENVIRONMENTS OF EOCENE ROCKS, MCMURDO SOUND, EAST ANTARCTICA
ANTARCTIC RESEARCH SERIES VOLUME 76, PAGES 335–347

CALLICHIRUS? SYMMETRICUS (DECAPODA: THALASSINOIDEA) AND ASSOCIATED BURROWS, EOCENE, ANTARCTICA

Carrie E. Schweitzer and Rodney M. Feldmann

Department of Geology, Kent State University, Kent, Ohio

Examination of 24 specimens of the ghost shrimp *Callichirus? symmetricus* (Feldmann and Zinsmeister) from erratic blocks of Eocene age in the vicinity of Mount Discovery, East Antarctica, permits refinement of the definition of the species. The species inhabited a heavily bioturbated, littoral to shallow sublittoral, medium-sized sand substratum, probably in a temperate water setting. Geometry of associated burrows, some of which contain claw elements, indicates that the burrows tend to be sparsely lined to unlined, about twice the height of the major claw and generally straight with few bifurcations. The pelleted external morphology is similar to *Ophiomorpha* Lundgren, some of which may be attributed to extant species of *Callichirus* Stimpson. Energy dispersive X-ray analysis of regions within and around the burrows reveals little elemental variation.

INTRODUCTION

Burrows assigned to the ichnogenera *Ophiomorpha* Lundgren, *Spongeliomorpha* de Saporta, and *Thalassinoides* Ehrenberg have been identified from rocks ranging throughout most of the Mesozoic and Cenozoic [Häntzschel, 1975]. Abundant *Thalassinoides* has been described in marine sediments of Ordovician age [Sheehan and Schiefelbein, 1984] and Silurian age [Watkins and Coorough, 1996]. In modern habitats, burrows constructed by callianassid and ctenochelid arthropods, the ghost shrimp, are virtually identical to some forms collected from the fossil record. As a result, members of this group of arthropods typically are cited as the probable tracemakers [Frey, Howard, and Pryor, 1978; Myrow, 1995; Bromley, 1996]. It is clear, however, that the genera of organisms responsible for construction of burrows in modern habitats did not range throughout the entire geologic range of *Ophiomorpha*. Clearly, numerous taxa, presumably all arthropods, were responsible for making very similar kinds of burrows at different times in the geological record.

On the other hand, studies of the architecture of ghost shrimp burrows in modern habitats [Griffis and Suchanek, 1991] have revealed several important morphologic characters that aid in identifying the tracemak-er as well as its preferred manner of feeding. Thus, within the ghost shrimps, several distinct burrow morphologies can be recognized and attributed to generic-level taxa, each of which has different ecological requirements. Their findings introduce the possibility that similarly detailed interpretations may be made from the fossil record, provided the tracemaker is known.

Associating tracemakers with *Thalassinoides*, *Ophiomorpha*, and *Spongeliomorpha* has been difficult owing to the paucity of body fossils associated with burrows. Furthermore, until the recent work of Manning and Felder [1991], it had been difficult or impossible to identify callianassids on the basis of chelipeds which are the only elements typically preserved. Although their work dealt solely with living callianassids, they recognized several characters of the chelipeds that are diagnostic at the generic level. Because we now have a more complete understanding of the systematics of callianassids and the variation of burrow morphologies attributable to these taxa, it is possible to begin compiling information about burrow architecture in the fossil record.

The purpose of this paper is to describe the morphology, geometry, and chemistry of burrows attributable to a callianassid tentatively assigned to *Callichirus* Stimpson. Specimens were collected from erratic blocks of Eocene age from the region of Mount Discovery, East

Antarctica. In the course of this work, some smaller burrows will also be described.

The primary tracemaker originally was identified as *Callianassa symmetrica* Feldmann and Zinsmeister [1984], based upon claw elements preserved in sandstone. The age of those specimens, also from the area of Mount Discovery, was inferred to be Eocene. Subsequently, a few additional claws and arms were collected from the same site and tentatively were assigned to *Callichirus* [Stilwell et al., 1997]. These specimens were recognized as having been preserved within small fragments of burrows. Collection of additional erratic blocks has now yielded enough material to describe the relationship of claws to burrows and the gross morphology and geometry of the burrow systems, which may be assigned to the ichnogenus *Ophiomorpha*. This material forms the basis for the current study.

LITHOLOGIC DESCRIPTION

The burrows and fossil specimens of *Callichirus? symmetricus* were collected in Eocene boulders contained in glacial till. The rock enclosing the burrows is a calcite-cemented quartz sandstone. The lithology of the rock in which the burrows are preserved is variable but always contains a large (75-85%) fraction of quartz grains. The quartz grains are typically angular to rounded and poorly sorted, ranging from fine to coarse-sized sand. This range in angularity and size of quartz grains indicates that there was mixing of rounded quartz grains with much less mature angular quartz material. Other grains include chert, mafic volcanic rocks, and possibly a small amount of magnetite. The magnetite and/or mafic volcanics appear to be the source of the red stain on the rocks, since in several cases the black grains are altered to limonite. In some specimens, there is a larger quantity of lithic and chert fragments greater than 2 mm.

The intergranular portion of the rock also appears to be finely granular in nature and includes calcite cement or matrix, a moderate quantity of mud matrix, and fine sand grains which are evident upon dissolution of the calcite cement. The calcite loosely cements the grains together. There is a large quantity of material between the sand grains, so much that the grains appear to float in the intergranular material. The calcite at times forms botryoidal crusts on the surface of the boulders and sometimes occurs in tabular vugs within the rock.

The rock surrounding most of the *Ophiomorpha* specimens is friable and breaks apart easily, although some of the rock samples are moderately well indurated. This appears to be due either to weathering of the cement or due to the rather poor cementation of the rock. No bedding planes or other structures were observable in any of the fossil-bearing rocks. Absence of bedding may be due to intense bioturbation. The traces that can be readily recognized represent only the last generation of burrowing, with previous episodes of burrowing quite possible. Several species of pelecypods and gastropods also were preserved in these rocks.

SYSTEMATIC PALEONTOLOGY

Order DECAPODA Latreille, 1803
Infraorder THALASSINIDEA Latreille, 1831
Superfamily THALASSINOIDEA Latreille, 1831
Family CALLIANASSIDAE Dana, 1852
Subfamily CALLICHIRINAE Manning and Felder, 1991
Genus *Callichirus*? Stimpson, 1866

Remarks. The genus *Callichirus* was introduced in 1866 by Stimpson who designated *Callianassa major* (Say) as the type species [Manning and Felder, 1986]. After several revisions and treatments by various workers, Manning and Felder [1986] redefined the genus and established a set of diagnostic characters for it including possession of an elongate manus and carpus of the chelipeds, a meral hook on the chelipeds, a narrow uropodal endopod, a short broad telson with a posterior emargination, and distinctive ornamentation of grooves and integumental glands on the third through fifth abdominal somites [Manning and Felder, 1986; 1991]. Manning and Felder [1991] noted that the carpus may be especially elongate in sexually mature males. Manning and Felder [1991] recognized five species as assignable to the genus, including *Callichirus major* (Say), *C. islagrande* (Schmitt), *C. seilacheri* (Bott), *C. garthi* (Retamal), and *C. adamas* (Kensley).

The form of the major cheliped may be variable within the genus *Callichirus*. The major cheliped of *Callichirus major* possesses a meral hook, and the chelipeds are unequal in adult males [Manning and Felder, 1991]. However, *C. islagrande* and *C. adamas* have an ischial hook on the major cheliped, not a meral hook [Kensley, 1974; Manning and Felder, 1986]. *Callichirus garthi* does not possess a pronounced hook on either the merus or ischium of the major cheliped [Retamal, 1975]. In *Callichirus adamas*, *C. islagrande* and *C. major*, the movable finger of the major cheliped of males appears to be strongly hooked at the tip, while the fixed finger appears to be shorter and nearly straight [Manning and

Felder, 1986, pp. 439-440]. In males of those two species, there is a pronounced emargination in the distal margin above the position of the fixed finger [Manning and Felder, 1986, pp. 439-440]. However, in males of *C. seilacheri* and *C. garthi*, the movable finger of the major cheliped is more weakly arched, and there does not appear to be any emargination in the distal margin above the position of the fixed finger [Manning and Felder, 1986, pp. 441]. Both chelipeds of *C. seilacheri* are reported to be "of minor form," perhaps indicating that the illustrated specimen is an immature male [Manning and Felder, 1986, pp. 441, Figures 3c and d]. Members of species of the genus may be variable in other aspects as well. For example, Staton and Felder [1995] reported that there are differences in overall body size, shapes of eyestalks and appendages, and habitat between two different populations of members of the species *C. major*.

Stilwell, et al. [1997] reassigned *Callianassa symmetricus* to the genus *Callichirus* but did not provide an explanation for this move. The specimens herein referred to *Callichirus? symmetricus* are tentatively placed in that genus based upon their possession of an equidimensional to elongate manus and an elongate carpus, which are diagnostic for the genus. No other thalassinoid genus possesses such an elongate carpus [see Manning and Felder, 1991]. They also possess an indentation on the distal margin of the manus above the fixed finger, a weakly to strongly arched movable finger, and a nearly straight fixed finger (Figures 1b-1d). These three characters are present in three of the five species of *Callichirus*. Additionally, *Callichirus? symmetricus* possesses a triangular tooth on the lower margin of the movable finger proximal to the midlength of the finger (Figure 1a). This character can be observed in most members of the genus *Callichirus* including *C. major*, *C. adamas*, *C. islagrande*, and *C. garthi*. Only *Callichirus seilacheri* does not possess such a tooth on the movable finger. Finally, Feldmann and Zinsmeister [1984] reported that the major and minor cheliped of *Callichirus? symmetricus* were approximately equal in size, hence the specific name. This particular character can be accommodated by the genus *Callichirus*, since only mature males are known to possess unequal chelipeds.

There are two problems with assignment of this material to *Callichirus*, accounting for the tentative placement of the material in that genus. The specimens studied herein do not appear to possess a meral hook; however, that area on each of the specimens that possess a merus is damaged or obscured. Also, note that *Callichirus islagrande*, *C. garthi*, and *C. adamas* do not possess a meral hook, indicating that possession of a meral hook

may not be diagnostic for the genus as was reported by Manning and Felder [1991]. Specimens of recent species of *Callichirus* do not appear to possess a longitudinal keel on the merus; the fossil specimens of *C?. symmetricus* do possess such a keel. However, the keel on the merus is weakly developed and is not nearly as well developed as in other thalassinoidean genera such as *Callianopsis* de Saint Laurent. As has been noted, the form of the major cheliped in the genus *Callichirus* is variable, so it is possible that the genus can accommodate the variability in form of the major cheliped of *Callichirus? symmetricus*. Therefore, the material is referred tentatively to the genus based upon the numerous similarities listed.

Other callianassid genera share some important characters with the Antarctic material but cannot accommodate it. Members of the genus *Notiax* Manning and Felder also possess an equidimensional to elongate manus, a weakly arched movable finger which is ornamented with a large tooth, a keel on the merus, and a nearly straight fixed finger [Holthuis, 1952], all of which the specimens referred to *C.? symmetricus* share. However, members of *Notiax* have a large meral spine, which the Antarctic specimens lack, and they also have a short, equidimensional carpus. The Antarctic specimens possess a markedly elongate carpus. Furthermore, *Notiax* possesses a tooth on the distal margin of the manus which the Antarctic specimens lack. Species of *Trypaea* Dana possess an elongate carpus similar to that seen in the Antarctic material, but they also have narrow fingers, an extremely large tooth on the movable finger, and a blunt, rounded tooth on the merus [Poore and Griffin, 1979], neither of which are seen in the Antarctic material.

These specimens, if legitimate members of *Callichirus*, would mark the first occurrence of the genus in the fossil record. Modern members of the genus are known from North America, Chile, and Africa in temperate to tropical environments from intertidal to sublittoral habitats [Kensley, 1974; Retamal, 1975; Manning and Felder, 1986; 1991; Staton and Felder, 1995]. This suggests that the climate in Antarctica during the Eocene was at least cool temperate, and that the rocks in which the fossils were recovered were likely shallow marine deposits.

Callichirus? symmetricus (Feldmann and Zinsmeister, 1984)

Text Figure 1

Callianassa symmetrica Feldmann and Zinsmeister, 1984, p. 1042, Figure 2.
Callichirus? symmetricus Stilwell, et al., 1997.

Fig. 1. *Callichirus? symmetricus* Feldmann and Zinsmeister. a, movable finger showing position of tooth, USNM 491765. b, cheliped with strongly arched movable finger and weakly developed tooth on movable finger, USNM 491767. c, Impression of claw showing weakly arched movable finger, USNM 491764. d, latex cast of USNM 491764. Notice very weakly developed tooth on movable finger. Scale bar = 1 cm.

Emendation to Description. Ischium of cheliped ornamented with small granules; much longer than high; nearly straight upper and lower margins; articulating with the merus near the upper margin. Merus of cheliped longer than high; upper margin nearly straight; proximal margin sinuous; lower margin not well preserved but appearing to be slightly convex; distal margin unknown; weakly developed keel paralleling upper margin; surface ornamented with fine granules, granules most dense along upper and lower margin. Carpus much longer than high (L/H = 1.5); distal margin nearly straight; upper margin nearly straight; lower margin slightly convex proximally and straightening distally; carpus-manus joint positioned approximately centrally along distal margin of carpus. Manus longer than high or nearly equidimensional (L/H ranges from 0.95-1.6); proximal, upper, and lower margins nearly straight; distal margin slightly sinuous with deep emargination just above position of fixed finger; weakly arched transversely

TABLE 1. Measurements (in mm) taken on specimens of *Callichirus? symmetricus* Feldmann and Zinsmeister. H1 = height of manus, L1 = length of hand, L2 = length of propodus, L3 = length of movable finger, H2 = height of carpus, L4 = length of carpus, H3 = height of the merus, L5 = length of merus, L6 = length of the ischium. All numbers other than six digit USNM numbers denote uncataloged specimens. Annotated six digit USNM numbers indicate specimens with more than one claw.

Specimen Number	H1	L1	L2	L3	H2	L4	H3	L5	L6
491770#1	~10.5	17.4	21.2	-	-	21.1	-	-	-
491770#2	11.7	16.2	-	~11.7	-	-	-	-	-
C1	~14.6	~14.4	-	-	-	-	-	-	-
491768	14.0	22.0	-	-	~12.8	20.7	-	-	-
C2	-	15.5	-	10.2	14.6	-	-	-	-
C3	-	16.2	-	-	9.9	18.4	-	-	-
C4	~13.7	16.4	-	-	-	-	-	-	-
C5	13.5	~12.2	21.5	10.7	14.1	~18.3	-	-	-
C6	>14.2	22.5	-	-	-	-	-	-	-
C7	-	-	-	-	13.9	21.4	9.2	12.8	15.0
C8	~11.0	13.0	-	-	-	-	-	-	-
C9	13.3	~14.6	25.3	~13.8	14.2	18.8	-	-	-
C10	16.4	23.6	-	-	-	>25.9	-	-	-
491766-E14	7.2	8.8	11.8	>5.3	-	-	3.6	-	>7.0
491766-C11	9.2	12.8	>12.4		9.8	12.3	>4.6	>8.6	>8.8
491766-C12	7.0	8.0	>10.6	>5.0	-	-	-	-	-
C13	12.1	12.0	>12.7	11.9	-	-	-	-	-
C14	11.9	14.5	>17.9	9.2	11.9	-	-	-	-
C15	12.2	16.0	-	-	-	-	-	-	-
C16	12.0	11.4	15.7	8.9	12.7	-	-	-	-
491769	13.9	~16.6	21.9	-	17.9	>11.8	-	-	-
491766-E141B	7.4	9.9	>12.9	-	7.4	11.3	-	-	-
491767	>9.5	17.9	>22.2	13.1	-	>15.4	-	-	-
491764	12.5	19.0	25.0	13.6	14.6	25.5	-	-	-

and longitudinally. Fixed finger projecting nearly straight, with very slightly convex lower margin and slightly concave upper margin, wide at base and narrowing distally, rounded in cross-section, slightly more inflated at base and along upper margin. Movable finger stouter in cross-section than fixed finger, somewhat longer than fixed finger; ranging from strongly arched to weakly arched; possessing small denticle at base and tiny serrations along lower margin.

Referred material. The holotype (USNM 353896) and other specimens, including those described by Stilwell, et al. [1997] (USNM 488273-488275) and the material herein described (USNM 491764-491768) are deposited in the United States National Museum of Natural History, Washington, D. C. All other specimens referred to this species and discussed in this paper are uncataloged and bear manuscript numbers only.

Measurements. Measurements, in mm, of specimens of *Callichirus? symmetricus* are presented in Table 1.

Localities and Stratigraphic Position. The material herein described was collected from glacial erratics of Eocene age from the Mount Discovery area. Specific locality information may be found elsewhere in this volume.

Remarks. The fossil specimens of *Callichirus? symmetricus* exhibit a range of variation in several characters. The length of the carpus is variable with the ratio of length to height ranging from 1.3 to 1.9. The ratio of the length of the manus to the height ranges from 0.95 to 1.6. The ratio of the length of the carpus to the length of the manus ranges from 0.94 to 1.5. The movable finger ranges from being only very weakly arched to strongly arched among the specimens studied herein, and the manus ranges from being nearly equidimensional to longer than high. These ranges, especially the size ratios of the manus and carpus characters, may merely represent differences in form between the major and minor cheliped, since it is not possible to distinguish between the two in these specimens. Alternatively, the range of variation could be the result of differences between mature males, immature males, and females which are known to differ from each other in recent species [Manning and Felder, 1986; 1991].

In addition to being the possible first notice of the genus *Callichirus* in the fossil record, the fossils are important because several specimens have been recovered enclosed within burrows. This has allowed a detailed description of the burrows and the relationship of the claws to the burrows, to be discussed below. Specimens preserving callianassid elements within the burrows are rare in the fossil record, and this discovery provides a unique opportunity to directly relate fossil decapod taxa with the burrows they excavated. It also permits comparison of the fossil burrows and tracemakers with modern taxa and burrow systems in order to make environmental interpretations. Undoubtedly, the provisional nature of the generic assignment of *Callichirus? symmetricus* makes direct environmental comparisons with modern taxa tentative. Modern species of the genus *Callichirus* typically inhabit burrows in normal marine conditions in intertidal to sub littoral areas, mostly in fine siliceous substrates typical of beaches and sand bars [Weimer and Hoyt, 1964; Philips, 1971; Felder, 1978; Williams, 1984; Manning and Felder, 1986; 1991; Griffis and Suchanek, 1991; Bromley, 1996]. Until a more definitive generic identification is possible, *Callichirus? symmetricus* can be inferred to have inhabited a similar environment.

Members of the genus *Trypaea* have been reported from the intertidal zone, mudflats, and estuarine areas [Manning and Felder, 1991]. Presence of articulate brachiopods in the deposits containing the thalassinids and the burrows indicates that the rocks enclosing the thalassinid fossils were clearly deposited in an environment having normal marine salinity [Lee and Stilwell, this volume]. *Notiax brachyophthalma* (Milne Edwards) is known from subtidal areas in fine sand or mud [Manning and Felder, 1991]; the burrows described here are excavated in fine to coarse sized sand. The characteristics of the burrows and the habitat of modern members of *Callichirus* are more similar to the burrows and the depositional environment of the rocks described here than are those of the other two genera. This lends further support to the interpretation that these fossils may indeed belong to the genus *Callichirus*.

ASSOCIATED BURROWS

Burrows Containing Callichirus? *Fossils*

Two major types of burrows were discovered in association with fossil specimens of *Callichirus? symmetricus*. One burrow type contained fragments of *C? symmetricus* within the burrows (Figure 2a), while the other burrow type did not possess fragments within the burrow but were associated with fossil fragments in the surrounding matrix (Figure 3a). The burrows are assignable to *Ophiomorpha* because they exhibit a weakly knobbed external surface, barely visible in the photographs (Figure 2a).

The burrows that frequently contain *Callichirus? symmetricus* fragments exhibit several distinctive characters. These burrows have a tubular shape with sharp external boundaries and a plane of weakness between the burrow and the surrounding sediment. One specimen, USNM 491769, possesses a dark red-brown rim around the edge of the burrow, perhaps indicating that the inner walls of the burrow were lined. The outer surfaces of the burrows in specimen USNM 491769 and a burrow in specimen USNM 488273 appear to have very small, poorly defined pellets which seem to be aggregates of sand grains. Modern species of *Callichirus* secrete a gelatinous substance with which they cement the sand grains of the burrow wall, which could explain the presence of a weakly developed lining in the fossil burrows [Williams, 1984].

The burrow fillings are essentially the same color, yellowish-gray (5Y 7/2) to grayish orange (10YR 7/4) [Goddard et al., 1951], as the surrounding rock matrix in each specimen, although the outside edge of several of the burrows appears to be iron stained. The preserved burrow sections of this type are straight, and the claw fragments contained within the burrows are oriented parallel to the long axis of the burrow (Figure 2a). The diameter of this type of burrow ranges from 12.0-30.4 mm, and the length of the preserved burrow segments ranges from 50-134 mm (Table 2). Five specimens of *Callichirus? symmetricus* were found within burrows, and they were found in burrows that are oriented at right angles to one another. The ratio of the height of the manus of the specimen contained within the burrow to the diameter of the burrow ranges from 0.44-0.58, suggesting that this ratio may be specific to the type of animal making the burrow (Table 2). This burrow type appears to be composed predominantly of long, presumably vertical, shafts.

Fig. 2. a, *Ophiomorpha* burrow enclosing *Callichirus? symmetricus* claw aligned parallel to long axis of burrow, indicated by arrow, USNM 492769. Note finely pelletal outer margin of burrow and distinct external boundaries. b, Burrows not associated with claw fragments. Arrow A indicates arcuate burrow, arrow B indicates straight burrow, and arrow C indicates anastomosing burrows as discussed in the text, USNM 491773.

Serial sections of an erratic block containing several burrows of this type revealed a detailed portrait of the three dimensional burrow structure (Figure 4, USNM 491772). Sections were spaced approximately 2.5 cm apart and eight sections were made. Analysis of the serial sections indicates that the burrow system also contains horizontal and oblique connecting shafts as well as vertical shafts. The burrow orientation with the most elements has arbitrarily been assigned as "vertical", and elements oriented perpendicular to those are therefore assigned as "horizontal". There are many more "vertical" shafts than "horizontal" tunnels in the serially sectioned specimen. In that specimen, some of the "vertical" shafts parallel each other so closely that they almost touch. This could

indicate either that there are several generations of burrows in the serially sectioned specimen or less likely, that two different animals created burrows that impinge upon one another. The burrows end in bluntly rounded terminations. Several of the burrows in the serially sectioned specimen vary in diameter along the length of the preserved burrow, exhibiting a maximum range of 10.6 mm in one burrow section (Table 3). This specimen has several small burrow segments, but there are no long continuous ones, as would be expected if the animal were constructing a long vertical shaft deep into the sediment. The reason for this is unknown, but it could be a result of the serial sectioning angle, of incomplete preservation of burrows, or of bioturbation subsequent to formation of

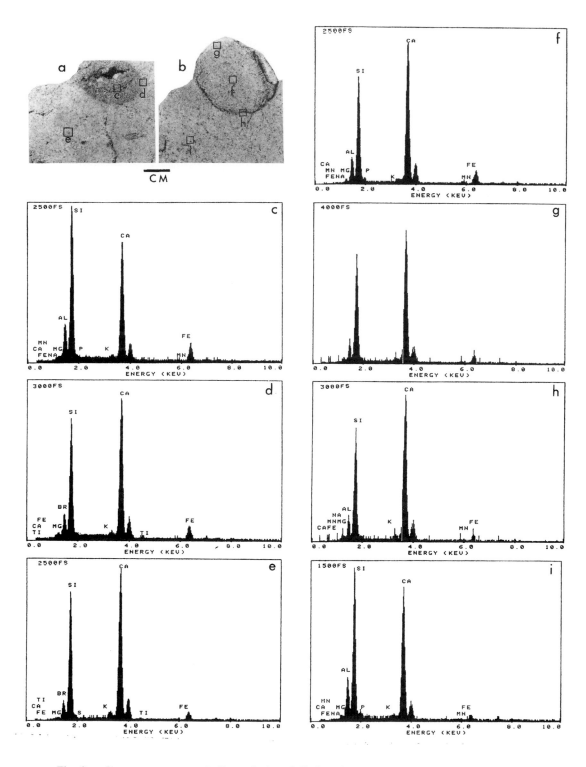

Fig. 3. a, Burrow type not typically enclosing *Callichirus? symmetricus* claws. Boxes c-e indicate portion of specimen sampled for EDX analysis. b, Burrow specimen USNM 491769 enclosing *Callichirus? symmetricus* claws. Boxes f-i indicate portions of specimen sampled for EDX analysis. f-i, Energy dispersive x-ray patterns determined within regions indicated in Figures 3.1 and 3.2.

TABLE 2. Burrow specimens and their respective diameters and lengths as well as the height of enclosed hands and the ratio of diameter to hand. Specimens other than USNM specimens are uncataloged specimens.

Specimen Number	Diameter (in mm)	Length (in mm)	Height of Hand (in mm)	Ratio of Diameter to Height of Hand
1.1	27	>130	N/A	N/A
1.2	26	N/A	15	0.58
USNM 488273	28	> 100	16	0.57
USNM 488273	20	N/A	-	-
3	30	-	-	-
4.1	28	50	-	-
4.2	30	-	16	0.53
5	21 or 28	-	-	-
USNM 491770	35	>143	-	-
302	18	-	-	-

the burrows that partially destroyed them. It is probable that the burrow system possessed bifurcating burrows, but actual junctions or bifurcations were not observed.

The second burrow type typically does not possess fragments of *Callichirus? symmetricus*. This burrow type does not have a well defined three-dimensional form and only differs from the yellowish-gray (5Y 3/2) surrounding matrix only because the burrow fill is a yellowish-brown color (10YR 6/2) with iron staining in isolated areas [Goddard et al., 1951] (Figure 3a). There is a distinct boundary between the burrow and the matrix because of the marked color change. The diameter of this burrow type ranges from 18 to 35 mm, and the length of the measured segment of USNM 491770 was >143 mm (Table 2). The dark colored burrows appear to possess long shafts as in USNM 491770. Burrow patterns in USNM 491771 indicate that there are horizontal tunnels as well.

Chemistry of Burrow Structures

The callianassid burrows that contain remains of *Callichirus* are lithologically similar to the surrounding matrix except at the wall. The wall is defined by a distinct, iron stained region that is only a few sand grains thick. The burrows that do not contain callianassid remains differ from those that do because there is no discrete wall in the burrows without callianassid fossils. Instead, the whole of the burrow interior is stained uniformly.

In order to determine the chemical variations that might produce these observed differences, specimens were prepared for energy dispersive X-ray spectroscopy (EDX) examination. An International Scientific Instruments Model SX-40A SEM with attached Princeton Gamma Tech System 4 Plus energy dispersive x-ray spectrometer was employed. The specimens were carbon coated using a BIO-RAD SEM coating system. Because the emission spectra of carbon are entirely absorbed by the beryllium window of the detector apparatus, presence of carbon will not affect the results.

During an initial trial, a single specimen of the interior of a burrow was examined. The resultant spectrum had detectable peaks for sodium, bromine, chlorine, and potassium. Presence of these elements in the erratics was attributed to evaporites resulting from exposure to seawater. In fact, the specimens tasted salty. As a result, all subsequent specimens were subjected to two periods of immersion in distilled water to minimize the effects of evaporites.

Seven samples from two burrow structures were examined. Using a small diamond saw blade on a dental drill, samples were extracted from the center of the two burrow types, from the rims of the burrows, and from the matrix approximately 1.5 cm from the wall of the burrow (Figure 3a-3b). The X-ray energy range from 0 to 10 thousand electron volts was scanned which would permit identification of most elements with atomic weights greater than the beryllium filter.

The resulting EDX spectra generally were quite similar to one another (Figures 3f-3i). Silicon, aluminum, calcium, and to a lesser extent iron, were the dominant

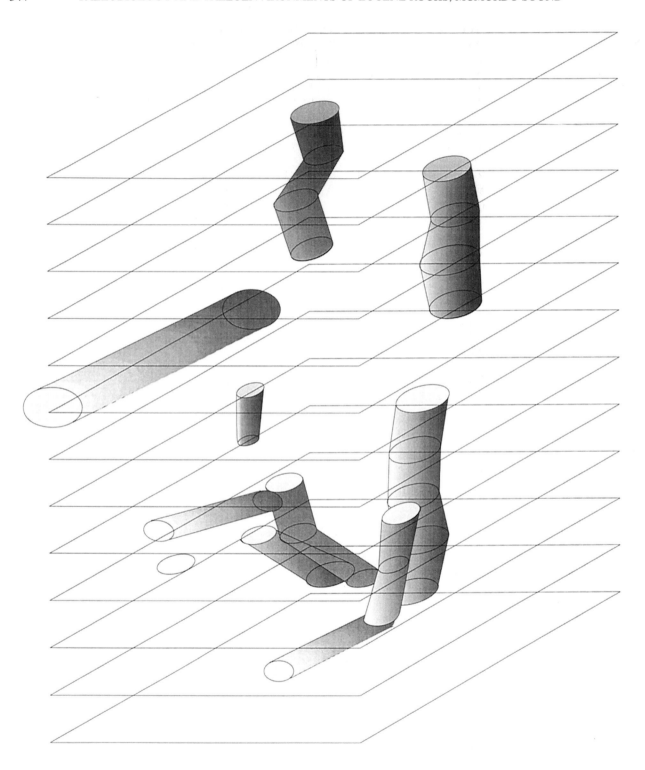

Fig. 4. Reconstruction of burrow patterns interpreted from serial sections cut through USNM 491772.

TABLE 3. Burrows preserved in the serially sectioned rock and their respective diameters, lengths, the height of enclosed hands, and the possible orientation in space of each burrow. All measurements are in millimeters.

Burrow Number	Number of Diameters Taken	Average Diameter	Range of Diameters	Length	Height of Enclosed Hand	Relative Orientation
1A	3	26.4	25.3 - 27.6	>50.0	N/A	Vertical
1B	4	24.6	21.0 - 27.3	>49.0	N/A	Vertical
1C	4	23.3	21.0 - 27.3	>134.0	N/A	Horizontal
1D	5	28.1	28.2 - 30.4	>57.0	N/A	Vertical
1E	4	17.1	12.0 - 22.6	>48.8	N/A	Horizontal
1F	1	26.2	26.2	?	N/A	Vertical
IG	1	24.1	24.1	?	N/A	Vertical
1H	3	23.4	20.8 - 25.7	?	10.3	Vertical
1J	2	20.7	20.8 - 23.8	48.7	N/A	Horizontal

components. The calcium was likely present as cement whereas the other elements could predictably be constituents of the sand grains. The chemistry of the unwalled, stained burrow was somewhat more complex than that of the matrix, however (Figures 3c-3e). Both the center and the rim of the stained burrow yielded amounts of iron, magnesium, and titanium that were higher than in the matrix. A combination of these elements, possibly as oxides, could account for the color of the stain. Because the spectra are solely qualitative, it is not possible to estimate relative percentages of the elements. However, variations in relative peak heights from one sample to another are assumed to represent an approximation of changes in relative proportions of elements.

The results were somewhat different within the samples taken from the walled burrow (Figures 3f-3i). Although the overall pattern of elemental composition was similar to that of the stained, unwalled burrow, there was a noticeable difference in abundance of iron, manganese, and magnesium in the rims. These elements, probably present as oxides, would account for the color difference between the rim and either the center of the burrow or the matrix surrounding the burrow. Within the latter two regions a minor phosphorous peak was detected. This element is of particular interest, because, if the tracemaker lined the walls of the burrow with fecal material, phosphorous might be concentrated in the burrow wall. The opposite seems to be the case.

The results of the EDX examination suggest that the chemical difference between different regions within the burrow systems and between burrow and matrix are subtle. Any significant chemical differences that may have

originally existed have been eliminated by diagenetic events. No clear evidence of a concentration of fecal material along the burrow rim was detected. Therefore there is no evidence that a thick wall or burrow lining was ever present.

OTHER ICHNOFOSSILS

Other ichnofossils were discovered in the glacial boulders in addition to the burrows associated with specimens of *Callichirus? symmetricus* (Figure 3b, USNM 491773). One type is a burrow that possesses a somewhat uneven surface, varies in diameter, is subcircular in cross section, and exhibits an arcuate morphology. This burrow diameter ranges from 5.2 to 7.3 mm, and the burrow length is at least 27.8 mm. It may connect with a series of similar burrows that occur approximately perpendicular to it and appear to be anastomosing. These anastomosing burrows range in diameter from 2.7 mm to 5.1 mm. Associated with both of these types of burrows may be a third type, which is situated immediately beside the arcuate burrow for about half its length but which is nearly straight instead of curved. The two do not appear to be connected. The straight burrow, with a diameter of 3.5 mm, is smaller in diameter than the arcuate burrow, but is about the same diameter as the anastomosing burrows. The straight burrow is at least 26.0 mm long. All of the burrows in this sample have a sediment fill that does not differ appreciably from the surrounding matrix sediment, except that the fill material is stained with limonite. No bedding or other geopetals exist in this sample to indicate which way is up.

DISCUSSION

The burrows that contain *Callichirus? symmetricus* fossils are not identical to those produced by extant species of the genus *Callichirus*, but there are some similarities between the two. Modern and the fossil burrow types commonly are produced in medium-grained sand. Modern *Callichirus* burrows typically are lined with calcium phosphate. However, the fossil burrows possess only weak indication that they may have been thinly lined, and EDX analysis indicated no chemical difference in the inferred lining. Lack of definitive geochemical evidence for a lining could be a result of diagenesis. The burrows of some (but not all) modern species of *Callichirus* burrows possess a knobby or pelleted outer surface. Some of the fossil burrow specimens have weakly pelleted outer surface. Modern species of the genus *Callichirus* are known to produce burrows with long vertical shafts leading down to deeper, branching galleries [Griffis and Suchanek, 1991]. The serial sections of burrows indicate that there are both vertical and horizontal components with the branching burrow systems. However, there are no geopetal indicators, such as bedding, in the boulders containing the burrows to indicate what the actual orientation of the burrows would have been. The modern species *Callichirus major* constructs burrows with long vertical shafts and horizontal tunnels deep in the sediment. The fossil burrows could be similar, although very few long sections of burrow were preserved, which could be a result of poor preservation, subsequent bioturbation, or diagenesis. The modern species *C. islagrande* constructs burrows that primarily consist of branching horizontal tunnels, but this does not appear to be the case with *C.? symmetricus* since actual branches were not observed. However, if the predominant burrow direction is assumed to be horizontal instead of vertical, then the burrows could fit this morphology. It is also possible that *C.? symmetricus* produces an entirely different type of burrow structure, since clearly there is a range of morphologies of burrows produced by this genus.

Weimer and Hoyt [1964, p. 762] reported occurrences of trace fossils that they believed were "exact replicas" of those made by *Callichirus major* (originally assigned to *Callianassa*) in Pleistocene rocks of Georgia. They proposed that members of the genus *Callianassa* sensu lato are responsible for constructing structures referable to the ichnogenus *Ophiomorpha* in the fossil record [Weimer and Hoyt, 1964]. Weimer and Hoyt [1964] believed that the occurrence of modern *Callianassa* burrows was an indicator of nearshore environments and that *Ophiomorpha* was therefore an important environmental indicator in ancient sediments. However, Bottjer, Droser, and Jablonski [1988] reported that the ichnogenus *Ophiomorpha* has been distributed in near shore to deep sea environments since at least the Cretaceous, indicating that this particular trace fossil is not necessarily a reliable indicator of shallow water environments. They did note that *Ophiomorpha* is most common in nearshore habitats [Bottjer, Droser, and Jablonski, 1988]. Mesozoic and Tertiary submarine canyons and inner fan deposits sometimes contain a low-diversity ichnofauna containing *Thalassinoides* and *Ophiomorpha* predominating, which have commonly been considered to be shallow-water ichnogenera [Crimes and Fedonkin, 1994]. Because the fossils are most likely referable to *Callichirus* and because modern members of *Callichirus* inhabit nearshore, shallow marine areas, these burrows suggest that these rocks were deposited in a near-shore, shallow water environment. Placement of the burrows in *Ophiomorpha*, a predominantly shallow water ichnotaxon, supports this interpretation.

Acknowledgments. Extremely useful reviews were provided by A. Ekdale and an anonymous reviewer. T. Miller, Department of Geology, Kent State University, assisted with the EDX analysis.

REFERENCES

Bottjer, D. J., M. L. Droser, and D. Jablonski.
  1988   Palaeoenvironmental trends in the history of trace fossils. *Nature, 333*(6170): 252-255.

Bromley, R. G.
  1996   *Trace Fossils,* 2nd edition. 361 pp. Chapman and Hall, London.

Dana, J. D.
  1852   Macrura. Conspectus Crustaceorum. Conspectus of the Crustacea of the Exploring Expedition under Capt. C. Wilkes, U.S.N. *Proceedings of the Academy of Natural Sciences of Philadelphia, 6*:10-28.

Crimes, T. P., and M. A. Fedonkin.
  1994   Evolution and dispersal of deep sea traces. *Palaios, 9*: 74-83.

Felder, D. L.
  1978   Osmotic and ionic regulation in several western Atlantic Callianassidae (Crustacea, Decapoda, Thalassinidea). *Biological Bulletin, 154*: 409-429.

Feldmann, R. M., and W. J. Zinsmeister.
  1984   First occurrence of fossil decapod crustaceans (Callianassidae) from the McMurdo Sound region, *Antarctica. Journal of Paleontology, 58*(4): 1041-1045.

Frey, R. W., J. D. Howard, and W. A. Pryor.
1978   *Ophiomorpha*: its morphologic, taxonomic, and environmental significance. *Palaeogeography, Palaeoclimatology, and Palaeoecology, 23*(3/4): 199-229.

Goddard, E. N., P. D. Trask, R. K. DeFord, O. N. Rove, J. T. Singewald, Jr., and R. M. Overbeck.
1951   *Rock Color Chart*. Geological Society of America, New York, New York.

Griffis, R. B., and T. H. Suchanek.
1991   A model of burrow architecture and trophic modes in thalassinidean shrimp (Decapoda: Thalassinidea). *Marine Ecology Progress Series, 79*:171-183.

Häntzschel, W.
1975   Trace fossils and problematica. *In* C. Teichert (Ed.), *Treatise on Invertebrate Paleontology*, 2nd edition. Geological Society of America and Kansas University Press, Lawrence, KS.

Holthuis, L. B.
1952   The Crustacea Decapoda Macrura of Chile. *Reports of the Lund University Chile Expedition, 1948-49, Vol. 5:* 109 pp.

Kensley, B.
1974   The genus *Callianassa* (Crustacea, Decapoda, Thalassinidea) from the west coast of South Africa with a key to the South African species. *Annals of the South African Museum, 62*: 265-278.

Latreille, P. A.
1802-1803 *Histoire naturelle, générale et particulière,des crustacés et des insectes, Volume 3*. F. Dufart, Paris, 468 p.

Latreille, P. A.
1831   Cours d'Entomologie ou de l'histoire naturelle des Crustacés, des Arachnides, des Myriapodes, et des Insectes, etc. *Annales I. Atlas, Roret*, Paris, France, 26 p.

Manning, R. B., and D. L. Felder.
1986   The status of the callianassid genus *Callichirus* Stimpson, 1866 (Crustacea: Decapoda: Thalassinidea). *Proceedings of the Biological Society of Washington, 99*(3): 437-443.
1991   Revision of the American Callianassidae (Crustacea: Decapoda: Thalassinidea). *Proceedings of the Biological Society of Washington, 104*(4): 764-792.

Myrow, P. M.
1995   *Thalassinoides* and the enigma of early Paleozoic open-framework burrow systems. *Palaios, 10*(1): 58-74.

Phillips, P. J.
1971   Observations on the biology of mudshrimps of the genus *Callianassa* (Anomura: Thalas-

sinidea) in Mississippi Sound. *Gulf Research Reports, 3*: 165-196.

Poore, G. C. B., and D. J. G. Griffin.
1979   The Thalassinidea (Crustacea: Decapoda) of Australia. *Records of the Australian Museum, 32*: 217-321.

Retamal, M. A.
1975   Descripción de una nueva especie del género *Callianassa* y clave para reconocer las especies chilenas. *Boletín de Sociedad de Biología de Concepción, 49*: 177-184.

Sheehan, P. M. and D. R. J. Schiefelbein.
1984   The trace fossil *Thalassinoides* from the Upper Ordovician of the eastern Great Basin: deep burrowing in the early Paleozoic. *Journal of Paleontology, 58*: 440-447.

Staton, J. L. and D. L. Felder.
1995   Genetic variation in populations of the ghost shrimp genus *Callichirus* (Crustacea: Decapoda: Thalassinoidea) in the western Atlantic and Gulf of Mexico. *Bulletin of Marine Science, 56*(2): 523-536.

Stilwell, J. D., R. H. Levy, R. M. Feldmann, and D. M. Harwood.
1997   On the rare occurrence of Eocene callianassid decapods (Arthropoda) preserved in their burrows, Mount Discovery, East Antarctica. *Journal of Paleontology, 71*(2): 284-287.

Stimpson, W.
1866   Descriptions of new genera and species of Macrurous Crustacea from the coasts of North America. *Proceedings of the Chicago Academy of Sciences, 1*: 46-48.

Watkins, R. and P. J. Coorough.
1996   Silurian *Thalassinoides* in an offshore marine community. *Geological Society of America Abstracts with Programs, 28*(7): A-296.

Weimer, R. J. and J. H. Hoyt.
1964   Burrows of *Callianassa major* Say, geologic indicators of littoral and shallow neritic environments. *Journal of Paleontology, 38*(4): 761-767.

Williams, A. B.
1984   *Shrimps, lobsters, and crabs of the Atlantic coast of the eastern United States, Maine to Florida*. xviii+550 p. Smithsonian Institution Press, Washington, D. C.

C. E. Schweitzer and R. M. Feldmann, Department of Geology, Kent State University, Kent, Ohio, 44242.

FISH REMAINS FROM THE EOCENE OF MOUNT DISCOVERY, EAST ANTARCTICA

Douglas J. Long

Dept. Of Ichthyology, California Academy of Sciences, Golden Gate Park, San Francisco, CA 94118, USA

Jeffrey D. Stilwell

School of Earth Sciences, James Cook University of North Queensland, Townsville Q 4811, Australia.

Recent geological and paleontological investigations into the Eocene deposits of East Antarctica yielded teeth from one teleost tentatively identified as Gadidae genus and species indeterminate, and two taxa of sharks, one *Carcharias* sp. cf. *C. macrota*, and one triakid, *Galeorhinus* sp. These fossils represent the first Cenozoic fishes from that section of the Southern Hemisphere. This small ichthyofauna suggests a relatively shallow, temperate marine climate of the Eocene of East Antarctica, similar to the better known Eocene ichthyofaunas of the Antarctic Peninsula.

INTRODUCTION

The middle to upper Eocene marine deposits of the La Meseta Formation of Seymour Island on the Antarctic Peninsula have revealed a rich ichthyofauna, and have produced the most diverse Paleogene teleost and elasmobranch fauna from anywhere in the Southern Hemisphere [Balushkin, 1994; Cione and Reguero, 1994, 1995, 1998; Cione, et al., 1995; Doktor, et al. 1996; Jerzmanska, 1988, 1991; Jerzmanska and Swidnicki, 1992; Long, 1991, 1992a-b, 1994a; Eastman and Grande, 1991; Ward and Grande, 1991; Welton and Zinsmeister, 1980]. However, virtually nothing is known about Paleogene fish faunas in other areas of Antarctica. This situation is due largely to the inaccessibility of other Paleogene deposits on the continent, and to the lack of prospecting for fossils in areas outside of the Antarctic Peninsula.

Recently, investigations into Paleogene deposits of Mount Discovery in East Antarctica have uncovered new fossil-bearing strata, preserved as glacial erratics, including Eocene marine units that have produced several specimens of teleost and shark teeth. While fragmentary in nature, this material is identifiable to at least three taxa, and represent some new and important occurrences in both Antarctica and the Southern Hemisphere. For addi-

tional information on geography, geology, stratigraphy, and age of these deposits, please see other papers in this volume.

SYSTEMATIC PALEONTOLOGY

Taxonomy for the shark and teleost taxa presented follows Compagno [1984] and Cohen, et al. [1990], and tooth morphology terminology follows Long [1992a].

Class Chondrichthyes
Order Lamniformes Compagno, 1973
Family Odontaspididae Mueller and Henle, 1838
Genus *Carcharias* Rafinesque, 1810
Carcharias sp. cf. *C. macrota* (Agassiz, 1843)
Plate 1, Figures a and b

Description. Two isolated tooth crowns; they are mesodistally narrow and apicobasally elongate, showing a weakly convex labial crown face and a smooth, moderately convex lingual crown face; the crown apex is very acute, and a sharp, non-serrated cutting edge extends from this apex to just above the crown foot on the mesial and distal sides. No root or lateral cusplets are present on these broken teeth.

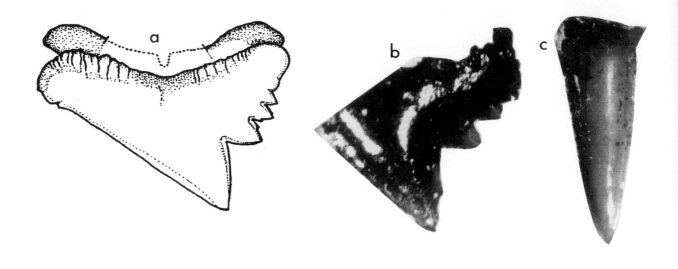

Plate 1

Figs. a and b. *Galeorhinus* sp. A. Camera-lucida drawing of the labial face of an upper lateral tooth from the Eocene of Mt Discovery, East Antarctica, USNM 494034, c.x15. B. Photograph of the same specimen, c.x15.

Fig. c. *Carcharias* sp. *cf. C. macrota* (Agassiz). Photograph of isolated tooth crown from the Eocene of Mt Discovery, East Antarctica, USNM 494032, c.x2.

Figured specimens. USNM 494032; USNM 494033.

Localities. Site T, E145; Site J, E151.

Remarks. These tooth crowns are identifiable as juvenile or subadult odontaspidid teeth, as characterized by their sigmoidal narrow crowns, the sharp but unserrated cutting edge, acute crown apex, very convex lingual crown face, and weakly convex labial crown face. However, exact specific designation is difficult because of their fragmentary nature. These specimens lack the increased sigmoidal curvature and the strongly convex lingual crown face, and narrow crown base of *Odontaspis*, but rather show characters like a moderately convex lingual crown face, weakly sigmoidal curvature, and widened crown base attributable to *Carcharias*. In comparison to smaller specimens of *C. macrota* from the Eocene of Seymour Island, they show the basic similarities in most aspects of crown morphology and are likely assignable to that species. Additionally, there have been no other species of *Carcharias* reported from the Eocene of Antarctica.

Order Carcharhiniformes Compagno, 1973
Family Triakididae Gray, 1851
Genus ***Galeorhinus*** (Linnaeus, 1758)
Galeorhinus sp.

Plate 1, Figure c

Description. This single complete upper lateral tooth is embedded in dense sandstone with only the labial crown face exposed. The tooth consists of a mesodistally expanded crown with a single large, distally inclined central cusp and a weakly convex crown face. The mesial edge is nearly straight, and a smooth but sharp cutting edge extends from the acute crown apex of the moderately triangular cusp to the upper anterior edge of the mesial root lobe. Three well developed distal cusplets are posterior to the central cusp; the cusplets are triangular and blunt with a smooth cutting edge; the cusplets decrease in size away from the cusp. The root lobes are widely divergent and rounded; the crown foot shows a moderate apical arch, and strong but short plications are present on the crown foot on both sides of the arch to near the ends of the root lobes. Little of the root is exposed, but it appears to be apicobasally narrow and slightly recessed under the crown foot; it extend slightly past the distal root lobe, but does not extend beyond the mesial root lobe.

Figured specimen. USNM 494034.

Locality. Site T, E145.

Remarks. The morphology of this tooth is consistent with basic familial characters diagnostic for Triakididae and for generic characters diagnostic for *Galeorhinus*, [see Compagno, 1970, 1988], but a specific designation is problematical. This genus contains many nominal species from the Late Cretaceous and Cenozoic; most of these have poor original diagnoses and illustrations, or are regional taxa that have not been validated by later workers [see partial reviews in Antunes and Jonet, 1970; Cappetta, 1970, 1987; Herman, 1977; Long, 1994b]. Additionally, many of these fossil forms may prove to be from other extant genera of Carcharhiniform sharks not yet identified from the fossil record [see Compagno, 1970, 1988]. *Galeorhinus* also shows a wide range of individual and ontogenetic variation that is often over-looked [see Long 1994b], potentially creating more confusion when identifying fossil taxa. For these reasons, and because only a single specimen is known from East Antarctica, it is identified to the generic level only. However, the species shows characters consistent with other fossil teeth identified as *Galeorhinus minor* and *G. minutissimus* as shown and described in Arambourg [1952], but additional specimens of these teeth, and clarification of *Galeorhinus* species-level taxonomy, are essential for a correct species assignmcnt.

Class Osteichthyes
Subclass Actinopterygii
Subdivision Teleostei
Order Gadiformes (*sensu* Cohen, 1984)
Gadidae Rafinesque, 1810
Gadidae genus and species indeterminate
not figured

Description. A single bony tooth core embedded in a piece of sandstone; narrow and triangular with a some-what blunt crown apex and sub-rounded tooth base; labial and lingual crown faces are moderately convex and devoid of enameloid but show some very weak apicobasally oriented striations.

Museum specimen. USNM 494035.

Locality. Site T, E372.

Remarks. This single broken and weathered tooth lacks diagnostic features definitely attributable to previously identified Eocene bony fishes from Antarctica [e.g. Long, 1991, 1992b; Cione, et al., 1994; Jerzmanska, 1988, 1991]. The tooth lacks the type of thick enamel consistent with any other potential marine vertebrate such as archaeocetes and crocodilians. It is also dissimilar from the thick, enameloid-covered caniform teeth of Labrid fishes, and the long, lanceolate teeth of Trichiurid fishes, both of which are known from the Eocene of Antarctica [Long 1991, 1992b]. However, this tooth

shows similarities with the teeth of an as yet unidentified teleost commonly collected from the Eocene La Meseta Formation. The La Meseta teeth have been assigned to a taxa of gadoid teleost genus informally named ìMesetaichthysî [Jerzmanska and Swindnicki, 1992]. Since this name is used tentatively in the literature and no species was formally designated for the La Meseta specimens, a specific identification of this tooth is not possible at this time. The diverse and often fragmentary nature of the fossil material attributable to Gadiform fishes suggests that there are likely several different undescribed and unidentified taxa from the Eocene of Antarctica [Doktor, et al., 1996; Eastman and Grande, 1991; Grande and Eastman, 1986; Jerzmanska, 1988; Jerzmanska and Swindnicki, 1992; Long, unpublished data]. Additionally, some of this material may eventually be identified as other non-gadoid taxa, such as nothenoid fishes [Grande and Eastman, 1986; Balushkin, 1994].

DISCUSSION

The teeth of *Carcharias* sp. *cf. C. macrota* and *Galeorhinus* sp. from E145 in the moraine deposits of Mount Discovery are associated with several macroinvertebrate taxa, including *Linucula? mcmurdoensis* n. sp., *Leionucula nova* [Wilckens], *Yoldiella?* n. sp., *Neilo beui* Stilwell and Zinsmeister, *Saxolucina sharmani* [Wilckens], *Nemocardium (Pratulum?) minutum* n. sp., *Crassatella* sp., *Hiatella harringtoni* n. sp., *Struthiolarella mcmurdoensis* n. sp., *Perissodonta* n. sp.? *cf. P. laevis* [Wilckens], *?Penion australocapax* Stilwell and Zinsmeister, *Acteon eoantarcticus* Stilwell and Zinsmeister, *Crenilabium suromaximum* Stilwcll and Zinsmeister, and Dentaliidae genus and species indeterminate [see Stilwell, this volume, for details of these taxa]. These invertebrate taxa along with the teeth, recovered from the medium-grained quartzose sandstone facies of E145, corroborate a shallow shelf environment of deposition.

Although this ichthyofauna is very limited in its taxonomic diversity, it does provide some paleoecological and biogeographical information that can be used to better interpret the Eocene marine deposits of East Antarctica. The presence of *Carcharias* is not unusual, since it is a cosmopolitan genus that lives in shallow tropical to temperate waters, and has previously been recorded from the Eocene of Antarctica [Long, 1992a and c]. This new locality record suggests *C. macrota* had a circum-Antarctic distribution in the Eocene.

Galeorhinus has not been recorded from Antarctica, and this is the first such Paleogene record of the genus from the Southern Hemisphere. The extant species *Galeorhinus galeus* is found world-wide, and its range extends well into shallow, cool temperature waters of the Southern Hemisphere [Compagno, 1984]. Fossil examples of this species in the Southern Hemisphere were previously known only from Pliocene deposits in Chile [Long, 1993]. Although the identity of this specimen of *Galeorhinus* is currently unknown, it may prove to belong to a Paleogene species known from other localities in the Northern Hemisphere.

Gadiform fishes are usually abundant in temperate to polar waters around the globe [Cohen, et al., 1990]. These suspected Eocene gadiform fossil forms apparently had a circum-Antarctic distribution as well, but since the identity of the fossil tooth remains uncertain, more pertinent biogeographical information is presently unattainable.

This new Eocene East Antarctic marine fauna includes widely distributed taxa known from other Northern Hemisphere localities. Such occurrences suggest that there was little regional endemicity of the ichthyofauna during that time, and that the Southern Hemisphere ichthyofauna was largely cosmopolitan in nature [Long, 1992a, 1994a]. Like the better known Eocene faunas from the La Meseta Formation Seymour Island, this fauna seems to consist of taxa that are associated with a temperate to cool temperate marine environment in relatively shallow waters [Long, 1992c]. Further discovery and interpretation of new fossil taxa will greatly assist in forming a more concrete paleoecological and bio- geographical framework for the Eocene marine environments of East Antarctica.

Acknowledgments. We would like to give thanks to D. Harwood, R. Levy, J. Kaser, and S. Bohaty of the University of Nebraska, Lincoln for all of their efforts in the field and enthusiastic support of this work, D. Catania of the Department of Ichthyology, California Academy of Sciences for use of scientific equipment to examine the specimens, and to A.L. Cione for comments on an early draft of the manuscript.

REFERENCES

Antunes, M.T. and S. Jonet
1970   Requins de líHelvétien supérieur et du Tortonien de Lisbonne. *Universidade de Lisboa Revista da Faculdade de Ciencias. Series 2C - Ciencias Naturais, 16*: 119-280.

Arambourg, C.
1952   Les vertébres fossiles des gisements de phosphates (Maroc, Algérie, Tunisie). *Service Géologique Maroc, Notes et Mémoires, No.92*: 1-372.

Balushkin, A.V.
1944   A fossil notothenoid, not gadiform, fish *Proeleginops grandeastmanorum* gen. et sp. nov. (Perciformes, Notothenoidea, Eleginopsidae) from late Eocene of the Seymour Island (Antarctic). *Vaprosy Ikhtiologii, 34*, 298-307.

Cappetta, H.
1970   Les sélaciens de Miocene de la region de Montpellier. *Paleovertebrata, Mémoire Extraordinaire, 1/2*, 1-139.

Cappetta, H.
1987   Handbook of Paleoichthyology, Chondrichthyes II: Mesozoic and Cenozoic Elasmobranchii. Gustav Fisher Verlag, New York, 193pp.

Cione, A.L. and M. Reguero
1994   New records of the sharks *Isurus* and *Hexanchus* from the Eocene of Seymour Island, Antarctica. *Proceedings of the Geologists Association, 105*, 1-14.

Cione, A.L. and M. Reguero
1995   Extension of the range of Hexanchid and Isurid sharks in the Eocene of Antarctica and comments on the occurrence of Hexanchids in Recent waters of Argentina. *Ameghiniana, 32*, 151-157.

Cione, A.L. and M.A. Reguero
1998   A middle Eocene basking shark (Lamniformes, Cetorhinidae) from Antarctica. *Antarctic Science, 10*, 83-88.

Cione, A.L., M. De las Mercedes Azpelicueta, and D.R. Bellwood
1994   An Oplegnathid fish from the Eocene of Antarctica. *Palaeontology, 37*, 931-940.

Cohen, D.M., T. Inada, T. Iwamoto, and N. Scialabba
1990   FAO Species Catalogue. Vol. 10. Gadiform Fishes of the World (Order Gadiformes). An annotated and illustrated catalogue of cods, hakes, grenadiers and other gadiform fishes known to date. *FAO Fisheries Synopsis (125)10*, 1-442.

Compagno, L.V.J.
1970   Systematics of the genus *Hemitriakis* (Selachii: Carcharhinidae), and related genera. *Proceedings of the California Academy of Sciences, Fourth Series, 38*, 63-98.

Compagno, L.V.J.
1984   FAO Species Catalogue. Vol. 4. Sharks of the World. An annotated and illustrated catalogue of shark species known to date. *FAO Fishers Synopsis (125), 4* (1-2), 1-655.

Compagno, L.V.J.
1988   *Sharks of the Order Carcharhiniformes.* Princeton University Press, Princeton, New Jersey, 486pp.

Doktor, M., A. Gazdzicki, A. Jerzmanska, S.J. Porebski, and E. Zastawniak

1996   A plant-and-fish assemblage from the Eocene La Meseta Formation of Seymour Island (Antarctic Peninsula) and its environmental implications. *Palaeontologia Polonica, 55,* 127-146.

Eastman, J.T. and L. Grande,

1991   Late Eocene gadiform (Teleostei) skull from Seymour Island, Antarctic Peninsula. *Antarctic Science, 3,* 87-95.

Grande, L. and J.T. Eastman

1986   A review of Antarctic ichthyofaunas in the light of new fossil discoveries. *Palaeontology, 29,* 113-137.

Herman, J.

1977   Les sélaciens des terrains néocrétaces and paleocénes de Belgique and des contrées limitrophes. Eléments díune biostratigraphie intercontinentale. *Mémoirs Pour Servir a líExplication des Cartes Géologiques et Minieres de la Belgique, Mémoire 15,* 1-450.

Jerzmanska, A.

1988   Isolated vertebrae of teleostean fishes from the Paleogene of Antarctica. *Polish Polar Research, 9,* 421-435

Jerzmanska, A.

1991   First articulated teleost fish from the Paleogene of West Antarctica. *Antarctic Science, 3,* 309-316

Jerzmanska, A. and J. Swidnicki

1992   Gadiform remains from the La Meseta Formation (Eocene) of Seymour Island, West Antarctica. *Polish Polar Research, 13,* 241-253.

Long, D.J.

1991   Fossil cutlassfish (Perciformes: Trichiuridae) teeth from the La Meseta Formation (Eocene) of Seymour Island, Antarctic Peninsula. *PaleoBios, 13*: 3-6.

Long, D.J.

1992a   Sharks from the La Meseta Formation (Eocene), Seymour Island, Antarctic Penin-sula. *Journal of Vertebrate Paleontology, 12:* 11-32.

Long, D.J.

1992b   An Eocene wrasse (Perciformes: Labridae) from Seymour Island. *Antarctic Science, 4,* 235-237.

Long, D.J.

1992c   Paleoecology of Eocene Antarctic sharks. *Antarctic Research Series, 56,* 131-139.

Long, D.J.

1993   Late Miocene and early Pliocene fish assemblages from the north central coast of Chile. *Tertiary Research, 14,* 117-126.

Long, D.J.

1994a   Quaternary colonization or Paleogene persistence?: historical biogeography of skates (Chondrichthyes: Rajidea) in the Antarctic ichthyofauna. *Paleobiology, 20:* 215-228.

Long, D.J.

1994b   Historical Biogeography of Sharks from the Northeastern Pacific Ocean. Unpublished Ph.D. Thesis, University of California, Berkeley; 371 pp.

Ward, D.J. and L. Grande

1991   Chimaeroid fish remains from Seymour Island, Antarctic Peninsula. *Antarctic Science, 3.* 323-330.

Welton, B.J. and W.J. Zinsmeister

1980   Eocene neoselachians from the La Meseta Formation, Seymour Island, Antarctic Peninsula. *Contributions in Science, Los Angeles County Museum of Natural History, 329.* 1-10.

Douglas J. Long, Dept. Of Ichthyology, California Academy of Sciences, Golden Gate Park, San Francisco, CA 94118, USA

Jeffrey D. Stilwell, School of Earth Sciences, James Cook University of North Queensland, Townsville Q 4811, Australia.

A PROBABLE PISCIVOROUS CROCODILE FROM EOCENE DEPOSITS OF MCMURDO SOUND, EAST ANTARCTICA

Paul M.A. Willis

Quinkana Pty Ltd, 3 Wanda Cres, Bewowra Hts, N.S.W. 2082, Australia

Jeffrey D. Stilwell

School of Earth Sciences, James Cook University, Townsville, Qld 4811, Australia.

A single, poorly preserved tooth from Eocene deposits in McMurdo Sound, Antarctica, is described and tentatively identified as belonging to a piscivorous crocodile. This is the first probable record of crocodiles from Antarctica. Three groups of piscivorous crocodiles are identified as candidates for the correct allocation of this crocodile; these are the thoracosaurs, gavialids, or dryosaurs. This tooth may provide the Austral high-latitude counterpart to coeval crocodiles recorded from Ellesmere Island.

INTRODUCTION

Fossil crocodilians have not previously been recorded from Antarctica. A specimen originally reported as being of part of a jaw of a sea-going crocodile [Anonymous, 1987; Case et al., 1987] from Upper Cretaceous to Eocene deposits of Seymour Island has more recently been reassessed as non-crocodilian (M. O. Woodburne, personal communication, 1996). Reassessment of the specimen and associated teeth has identified them as being from a gadid fish (Judd A. Case, personal communication, 1997).

The Seymour Island specimen consists of a single poorly preserved tooth, which may belong to a piscivorous crocodile. If this identification is correct, the tooth could most likely represent a thoracosaur, gavialid or dyrosaur, but each of these possibilities has biogeographic or temporal problems. Although both thoracosaurs and dryosaurs were contemporaneous with the deposits of McMurdo Sound [Steel, 1973], the palaeobiogeographic distribution of both these groups renders both as unlikely candidates for the McMurdo tooth although a dyrosaur has been reported from Upper Cretaceous deposits of Patagonia [Gasparini and Spalletti, 1990] and dyrosaur teeth have been recorded from the Paleocene of Bolivia [Argollo et al., 1987].

Early gavials are known from South America and offer a more viable possibility, but the earliest known gavials are considerably younger than the McMurdo deposits [Gasparini, 1983; Langston, 1965].

Of note, crocodilians have been used to deduce paleoenvironmental conditions. Markwick [1998] used environmental parameters determined from modern crocodilians to recreate paleoecological conditions for deposits where crocodilian fossils occur. Following Markwick the "thermal limits" of crocodilians indicate that temperatures in the Seymour Island region would have been 5.5° for the coldest month and 14.2° for the mean annual temperature. The tooth may represent the southern hemisphere counterpart to early Eocene crocodiles on Ellesmere Island at a paleolatitude of 71.4°C [see Markwick, 1998]. However, work in preparation by the senior author and associated colleagues raises serious questions about the suitability of crocodilians as paleoenvironmental indicators based on phylogenetic, geological and ecological factors.

AGE AND ASSOCIATED FAUNA

The tooth was derived from a glacial erratic boulder (number E312) composed of a rather mature, coarse, pebbly, fossiliferous sandstone with a high quartz component

1 cm

Plate 1

Figs. A-F. Probable crocodile tooth, crown only, USNM 494135, E312, Mount Discovery, views of tooth with corresponding fragments, all figures x3.

and only minor feldspar and opaque clastic constituents. The fossils are scattered sparsely in the matrix and are generally fragmentary. Only one species of bivalve, *"Eurhomalea" claudiae* n. sp. [see description by Stilwell, this volume] and a small fragment of a gastropod are recognized. The specimens of *"E." claudiae* n. sp. are disarticulated and are largely represented by decorticated casts. This species is closely allied with *"E." newtoni* [Wilckens] of the La Meseta Formation, Seymour Island, Antarctic Peninsula. On Seymour Island, *"E." newtoni* is long-ranging, present in Units I-VI [see Stilwell and Zinsmeister, 1992, tab. 1, figs. 40 and 42], and is not particularly age diagnostic. The age of the tooth and associated fauna and flora in the erratics is difficult to ascertain because of the strong overall endemic component of these taxa. A dinoflagellate complex derived from the erratics belongs to a group of mostly long-ranging taxa and are not amenable to precise biostratigraphical control [R. H. Levy, personal communication, 1996]. However, the overlapping ranges of the dinoflagellate taxa recovered thus far suggest an overall age range of late early to middle Eocene [Levy, this volume; Stilwell et al., 1997]. Thus, microfossils, molluscs and other associated macroinvertebrates/vertebrates indicate a middle to late Eocene age range for the taxa in the erratics.

DESCRIPTION

The specimen, USNM 494135 from E312, is in two parts which, together, form the crown of a tooth. The tooth lacks its tip and measures 13mm long and 6mm wide across the base. The tooth is slender in proportions with a gentle taper toward the tip and a gentle medial curvature. The surface is corroded and retains no significant features. Similarly the base of the tooth is encrusted with matrix obscuring any features in this region.

IDENTIFICATION

The tentative identification as crocodilian is based on the overall proportions of the tooth and the fracture pattern involved in the separation of the smaller part of the specimen from the larger. This damage reveals a multi-layered structure arranged in concentric cones within the long axis of the tooth. The fractures consist of lengthwise, straight fractures running perpendicular to the tooth surface and a transverse fracture laying at a shallow angle to the tooth surface. Similar fracture patterns have been observed in numerous other crocodilian teeth, particularly fossilized teeth, from numerous deposits around the world.

If this tooth is crocodilian, the proportions and size indicates that it probably belongs to a piscivore. Reptilian piscivores have long, slender teeth that are gently curved [Massare, 1987]. Unfortunately, the specimen appears to be corroded and any distinguishing surface features have been obliterated.

POSSIBLE AFFINITIES

Known groups of piscivorous crocodiles that are contemporaneous or near-contemporaneous with the McMurdo Sound tooth include thoracosaurs, dryosaurs and gavials. Thoracosaurs are marine and freshwater piscivores with a wide distribution [Steel, 1973]. During the Eocene, thoracosaurs were restricted to Europe, North Africa, eastern and central Asia and North America [Steel, 1973; Li, 1975; Tchernov, 1986; Yeh, 1958; Young, 1964]. They appear in the Pacific, Japan and southern Asia in the Pliocene and Pleistocene [Ishizaki, 1987; Kobatake and Kamei, 1966; Molnar, 1982]. Based on paleogeography and known distribution, it appears unlikely that there would be a thoracosaur in Antarctica during the Eocene.

During the Eocene, dryosaurs are known only from northern Africa and western Asia [Buffetaut, 1976a, 1976b, 1977; Buffetaut et al., 1982; Storrs, 1986]. However, a dyrosaur from Upper Cretaceous deposits of Patagonia [Gasparini and Spalletti, 1990] and dyrosaur teeth from the Paleocene of Bolivia [Argil et al., 1987] have been recorded.

The earliest known records of gavials are from the Oligocene of South America [Buffetaut, 1978; Gasparini, 1983; Langston, 1965; Patterson, 1936; Rovereto, 1912; Savage, 1951] . The geographic proximity of South America to Antarctica during the early Tertiary suggests that this tooth could represent the oldest-known gavial and that the group first appeared in Antarctica in the Eocene before dispersing into South America in the Oligocene.

Acknowledgments. The writers wish to thank D. Harwood, R. Levy, J. Kaser, and S. Bohaty of University of Nebraska-Lincoln for their enthusiastic support and assistance in the field. We also thank the Department of Earth Sciences staff at James Cook University for their assistance throughout the course of this investigation. This research was made possible by a NSF grant to D. Harwood and an Australian Research Council Fellowship and Grant to J.D.S.

REFERENCES

Anonymous
1987   Antarctic "Rosetta Stone" provides clues to the early Southern Hemisphere history. *Antarctic Journal of the United States*, 22: 1-2.

Argollo, J., E. Buffetaut, H. Cappetta, M. Fornari, G. Herail, G. Laubacher, B. Sige, and G. Vizcarra
1987   Decouverte de vertébrés aquatiques présumés Paléocènes dans les andes septentrionales de Bolivie (Rio suches, Syncrinorium de Putine). *Geobios, 20* 123-127.

Buffetaut, E.
1976a   Une nouvelle definition de la famille des Dyrosauridae de Stefano, 1903 (Crocodylia, Mesosuchia) et ses consequences: inclusion des genres *Hyposaurus* et *Sokotosuchus* dans les Dyrodauridae, *Geobios, 9*: 333- 336.

1976b   Sur la répartition géographique hors d'Afrique des Dyrosauridae, Crocodiliens mésosuchiens du Crétacé, terminal et du Paléogène. *C. R. Acad. Sc. Paris, 283*: 487-490.

1977   Données nouvelles sur les crocodiliens paléogènes du Pakistan et de Birmane, *C. R. Acad. Sc. Paris, 285:* 869 - 872.

1978   Sur l'historie phylogénétique et biogéographique des Gavialidae (Crocodylia, Eusuchia), *C. R. Acad. Sc. Paris, 287*: 911-914.

Buffetaut, E., V. d. Buffrénil, A. d. Ricqlès, and Z. V. Spinar
1982   Remarques anatomiques et paléohistologiques sur *Dyrosaurus phosphaticus*, crocodilien meso-suchien des phosphates *Yprésiens de Tunisie, 68*: 327 - 341.

Case, J. A., M. O. Woodburne and D. S. Chaney
1987   A gigantic phorarhacoid(?) bird from Antarctica, *J. Paleontology, 61:* 1280-1284.

Gasparini, Z.
1983   Un nuevo cocodrilo (Eusuchia) Cenozoico de America del Sur. *Ameghiniana, 20*: 51 - 53.

Gasparini Z., and L. A. Spalletti
1990   Un nuevo crocodilo en los depositos mareales Maastrichtianos de la Patagonia noroccidental. *Ameghiniana, 27*: 141 - 150.

Ishizaki, K.
1987   The crocodile *Tomistoma machikanense* and ostracodes from Southwest Japan in the Middle Pleistocene. *Fossils, 43*: 35 - 38.

Kobatake N., and T. Kamei
1966   The first discovery of fossil crocodile from Central Honshû, Japan. *Proc. Japan Acad., 42:* 264 - 269.

Langston, W.
1965   Fossil crocodilians from Colombia and the Cenozoic history of the crocodile in South America. Uni. California Pub. *Geol. Sci., 52,* 157 p.

Li, G.,
1975   A new material on *Tomistoma petrolica* of Maoming, Kwantung. *Vertebrata Palasiatica, 13:* 190 - 191.

Markwick, P. J.
1998   Fossil crocodilians as indicators of Late Cretaceous and Cenozoic climates: implications for using palaeontological data in reconstructing palaeoclimate. *Palaeogeography, Palaeoclimatology, Palaeoecology, 137:* 205-271.

Massare, J. A.
1987   Tooth morphology and prey preference of Mesozoic marine reptiles. *J. Vert. Paleont., 7:* 121 - 137.

Molnar, R. E.
1982   A longirostrine crocodilian from Murua (Woodlark) Island, Solomon Sea. *Memoirs of the Queensland Museum, 20:* 675 - 685.

Patterson, B.
1936   *Caiman latirostris* from the Pleistocene of Argentina, and a summary of South American Cenozoic Crocodilia. *Herpetologica, 1:* 43 - 54.

Rovereto, C.
1912   Los crocodilos fósiles en las Capas del Parana. *Anales del Museo Nacional de Buenos Aires, 3:* 339 - 368.

Savage, D. E.
1951   Report on fossil vertebrates from Upper Magdalena Valley, Colombia. *Science, 114:* 186 - 187.

Steel, R.
1973   Encyclopedia of paleoherpetology; Part 16, Crocodilia. Gustav Fisher Verlag, Stuttgart, 116 p.

Stilwell, J. D., R. H. Levy, R. M. Feldmann, and D. M. Harwood
1997   On the rare occurrence of Eocene callianassid decapods (Arthropoda) preserved in their burrows, Mount Discovery, East Antarctica. *Journal of Paleontology, 71*(2): 284-287.

Stilwell, J. D., and W. J. Zinsmeister
1992   Molluscan systematics and biostratigraphy. Lower Tertiary La Meseta Formation, Seymour Island, Antarctic Peninsula. *American Geophysical Union Antarctic Research Series 55*, 192 p.

Storrs, G. W.
1986   A dyrosaurid crocodile (Crocodylia: Mesosuchia) from the Paleocene of Pakistan, *Postilla, 197:* 1 - 16.

Tchernov, E.
1986   Evolution of the crocodiles in east and north Africa. *Éditions du Centre National de la Recherche Scientifique*, Paris, 65 p.

Yeh, H.
1958   A new crocodiles from Maoming Kwantung. *Vertebrata Palasiatica, 2:* 237 - 243.

Young, C. C.
1964   New fossil crocodiles from China. *Vertebrata Palasiatica, 8:* 189 - 208.

Jeffrey D. Stilwell, School of Earth Sciences, James Cook University, Townsville, Qld 4811, Australia.

Paul M. A. Willis, Quinkana Pty Ltd, 3 Wanda Cres, Bewowra Hts, N.S.W. 2082, Australia

THE FIRST RECORD OF A FOSSIL BIRD FROM EAST ANTARCTICA

Craig M. Jones

Department of Geological Sciences, University of Canterbury, Christchurch New Zealand

The first occurrence of a fossil bird from continental Antarctica is reported here. The specimen, a fragment of humerus shaft from a large volant bird was recovered from a marine sandstone erratic in the Mt Discovery area of McMurdo Sound, Ross Sea, East Antarctica. Associated dinoflagellate and molluscan taxa indicate a middle to late Eocene age. The specimen is too incomplete to be positively identified to species level but is referred to the Family Pelagornithidae (Pelecaniformes; pseudodontorns) on the basis of size and proportions. This is the second occurrence of pseudodontorns in the Antarctic region (La Meseta Formation, Seymour Island). It suggests these birds had a high latitude circumpolar distribution. In contrast to other high latitude Eocene localities no fossil penguins were found. Pseudodontorns are presumed to oceanic birds and are often associated with areas of high ocean productivity. The paleoenvironmental implications of this association are briefly discussed.

INTRODUCTION

To date the fossil bird record in Antarctica has been restricted to a few localities around the northern tip of the Antarctic Peninsula. Remains of fossil penguins (Sphenisciformes) were discovered earlier this century in the richly fossiliferous late Eocene-early Oligocene aged La Meseta Formation on Seymour Island [Zinsmeister, 1988; Wiman, 1905]. This diverse fauna comprises at least five genera and six species [Marples, 1953; Simpson, 1946, 1971b; Fordyce and Jones, 1990; Case, 1992]. More recently isolated fragmentary elements belonging to non-spheniscid fossil birds have been recorded from the La Meseta formation. Taxa recognized so far include a phorusrhacoid, a falconid, a presbyornithid, procellariids, diomedeids, pseudodontorns, and two different charadriiforms including a possible phoenicopterid [Case et.al, 1987; Noriega and Tambussi, 1996; Tonni, 1980; Tonni and Tambussi, 1985]. Several birds have also been recorded from the Late Cretaceous López de Bertodano Formation on Seymour and nearby Vega Island. They include a loon-like bird [Chatterjee, 1989;

Olson, 1992] and indeterminate presbyornithids [Noriega and Tambussi, 1995]. Footprints belonging to a large non-volant bird and a variety of smaller volant taxa have also been reported from Oligocene-Miocene age sediments on King George Island in the South Shetlands [Covacevich and Rich, 1982].

Until now no fossil bird material has been recorded from continental Antarctica. In this paper, a fragmentary long bone of a large volant bird is described. The specimen was found in an erratic boulder of middle to late Eocene age collected from moraine deposits near Mt Discovery, McMurdo Sound, in the Ross Sea region of East Antarctica. The bone, though very incomplete and indeterminate to genus level, is referable to the Family Pelagornithidae (pseudodontorns sensu Olson 1985) and is an important early high latitude southern record of pseudodontorns. The discovery in an erratic of this single fragmentary bone contrasts with the diverse and relatively abundant fossil avifauna described from similar aged sediments of the La Meseta Formation on Seymour Island. The paleobiology of pseudodontorns are discussed in light of this new fossil occurrence.

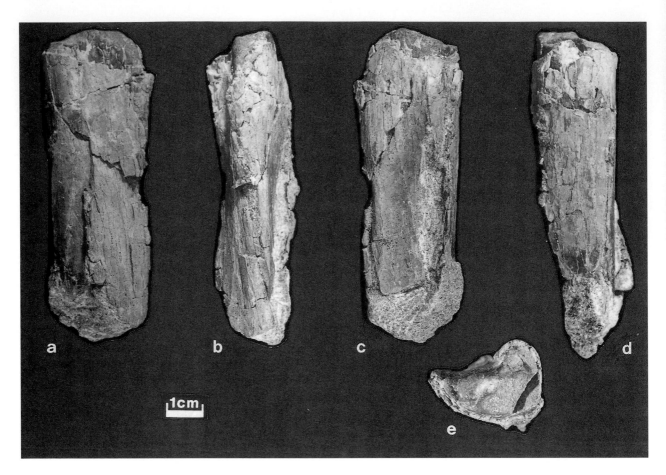

Fig. 1. Distal portion of humerus shaft of USNM 494035, an Eocene aged pseudodontorn bird (Pelagornithidae; Pelecaniformes), found in erratic A303, Mt Discovery, McMurdo Sound, Ross Sea, East Antarctica. (a) cranial view, (b) dorsal view, (c) caudal view, (d) ventral view, (e) distal view, all x1.

DESCRIPTION

Material. USNM 494035 (Fig. 1) a severely crushed fragment of long bone of a large volant bird. The specimen lacks positively identifiable osteological landmarks but comparisons with a range of extant and fossil material identify it as a portion of a shaft from a right humerus, proximal to the brachial fossa (Fossa m. brachialis). The specimen is housed in the collections of the Department of Paleobiology, National Museum of Natural History (USNM), Smithsonian Institution, Washington DC, USA. Osteological nomenclature follows Baumel and Witmer (1993)

Dimensions (in millimeters). Maximum length, 85mm; maximum dorso-ventral width (uncrushed end), 32mm; maximum cranial-caudal depth (uncrushed end), 22mm; thickness of bone wall, 1.5mm.

Locality. Found in a glacial erratic (A303) composed of fossiliferous, poorly sorted, moderately indurated, calcareous marine litharenite. The erratic was collected from moraine deposits flanking the northwestern side of Mt Discovery, McMurdo Sound, Ross Sea, Antarctica (see Introduction in this volume for locality data).

Age. The dinoflagellate taxa associated with the bone indicates a middle to late Eocene age (R. Levy, this volume), including the following species; *Deflandrea antarctica* Wilson, *Senagalinium? asymmetricum* (Wilson), *Enneadocysta partridgei* Stover and Williams, *Lejeunecysta hyalina* (Gerlach), *Hystrichosphaeridium truswelliae* Wrenn and Hart. Marine invertebrate macrofossils also recovered from the erratic include the turritellid gastropod, *Colposigma euthenia* Stilwell and Zinsmeister [Stilwell, this volume], a venerid bivalve *Eurhomalea* sp. (J. Stilwell, personal communication,

1996), fragments of the decapod *Callichirus? symmetrica* (Feldmann and Zinsmeister) [Stilwell et al., 1997], and unidentifiable mollusc and plant fragments. The gastropod *Colposigma euthenia* is known from the La Meseta Formation (TELM I to TELM VI of Sadler, 1988) on Seymour Island [Stilwell and Zinsmeister, 1992] where it ranges in age from late early to late Eocene. Therefore, on the basis of the associated microflora and macro-invertebrate fauna, the age of USNM 494035 is probably early middle to late Eocene.

Description. USNM 494035 is a straight, hollow shaft of bone, lacking any sign of a sigmoidal curve (Figure 1a-e). The distal end of the bone, originally exposed on the surface of the erratic, is relatively undistorted and is approximately triangular in cross-section (Figure 1e). Sediment infilling the bone at this end preserves impressions of cancellous bone (Figure 1a), indicating proximity to the distal articulation. The rest of the shaft is crushed. Preservation suggests the specimen was broken and infilled with sediment prior to being crushed.

The cranial surface of the bone is flat to slightly convex. A wide, shallow, flat-bottomed furrow representing the proximal end of the brachial fossa, is present on the distal third of the surface (Figure 1a, 1e). The elevated ventral side of the fossa forms a broad rounded ridge that merges into the surface proximally. The dorsal margin of the fossa is narrow and less well defined, merging abruptly with the dorsal surface of the bone (Figure 1b). At the distal end of the shaft the dorsal margin begins to diverge laterally, making the bone a little wider at this point (Figure 1a).

The caudal surface of the shaft is broken and crushed proximally (Figure 1c); distally, however, it is uncrushed and is slightly convex in distal view (Figure 1e). A thin ridge developed on the distal end of the surface may be due to post-burial compaction or it could be the proximal end of the ridge separating the scapulotriceps groove (sulcus scapulotricipitalis) from the humerotriceps groove (sulcus humerotricipitalis). Also preserved on the caudal surface are indistinct scratches that run across the shaft perpendicular to the compaction fractures and bone fabric. These features are identified as post-mortem bite marks of scavenging sharks or other fish. Similar marks have been noted on marine vertebrate material from New Zealand [McKee, 1987a] (personal observation of fossil penguin and whale bones, Otago University Geology Department collections). The ventral surface of the shaft is flat to slightly convex and approximately perpendicular to the caudal surface (Figure 1d).

Affinities. Despite the bone's large size, its hollow nature and very thin walls clearly rule out affinities with the penguins or non-volant terrestrial birds (ratites and phorusrhacids), which typically have non-pneumatic or thick-walled limb bones [Simpson, 1976; Chandler, 1994]. The bone was compared with a range of volant taxa from several orders (Procellariiformes Pelecaniformes, Anseriformes, Gruiformes and Charadriiformes). The straight, rather than sigmoid shaft and large size were most similar to the humeri of Diomedeidae (Procellariiformes; albatrosses) and the extinct Pelagornithidae (Pelecaniformes, pseudodontorns). However, the shaft diameter is significantly larger than extant or described fossil albatross humeri [Howard 1966c, 1978, 1982]. Also, the more pronounced development of the brachial fossa and the very thin walls of the bone suggest closer affinities to Pelagornithidae. The uncrushed end of the bone compares closely in diameter and bone wall thickness to the humerus shaft, proximal to the distal articulations, of the large Miocene taxa *Pelagornis miocaenus* Lartet [Harrison and Walker 1976b] and *Osteodontornis orri* Howard 1957. However the lack of further diagnostic features on the humerus makes useful comparisons with other described pseudodontorn material and exact taxonomic assignment of the bone within the Pelagornithidae impossible.

DISCUSSION

This new discovery provides an opportunity to make some taphonomic and avifaunal comparisons with the similarly aged late Eocene-early Oligocene La Meseta Formation on Seymour Island, Antarctic Peninsula Region. The La Meseta bird fauna is overwhelmingly dominated by large fossil penguins, both in numbers of species [Simpson, 1971b; Case, 1992] and amount of material, with many hundreds of specimens in various institutional collections [Wiman, 1905; Simpson, 1946, 1971b; Marples, 1953; Myrcha et.al, 1990; Fordyce and Jones, 1990;]. This is due, in part, to preservation potential of the robust non-pneumatized penguin limb bones which appear to have survived preferentially the shallow nearshore marine facies of the La Meseta Formation [Stilwell and Zinsmeister 1992; Marenssi et al., 1994]. Non-penguin fossil bird material is much rarer and more fragmentary, with taxa being represented by one or two bones. However, despite the paucity of material, a diverse fauna including indeterminate pseudodontorns have been identified [Noriega and Tambussi, 1996; Tonni, 1980; Tonni and Tambussi, 1985].

In contrast to the La Meseta Formation, the single fragmentary volant bird bone described here is the only fossil bird recovered from the hundreds of fossiliferous erratics collected in the Mt Discovery and Minna Bluff area and is clearly not a penguin. The apparent absence of fossil penguins in the Mt Discovery area may be due to differences in paleoenvironment. The La Meseta Formation is interpreted as having been deposited off a coast of moderate relief with channel, delta–barrier island and lagoonal facies represented in the formation [Stilwell and Zinsmeister 1992]. By comparison the predominantly coarse sandstones of the erratics in the Mt Discovery area poorly sorted suggesting rapid deposition close to the sediment source (the Transantarctic Mountains?). It is unlikely fossil penguins were absent from the Ross Sea region, as both New Zealand and Australia have nearly contemporaneous fossil penguin faunas [Fordyce and Jones, 1990]. It is possible that the steep coastlines of the nearby Transantarctic mountains may have been unsuitable for colonies of large fossil penguins, due to inaccessibility or breeding space [Warheit and Lindberg 1988; Warheit, 1992.

Pseudodontorns, first recognized from the early Eocene of England [Harrison and Walker, 1976b], are known from the middle-late Eocene of Antarctica, late Eocene of US Pacific coast, the early Oligocene of Nigeria, the Miocene of France, Japan, New Zealand, US Atlantic and Pacific coasts, and the Pliocene of New Zealand, US Atlantic and Pacific coasts [Olson, 1985; Goedert, 1989; McKee 1985].

Paleobiological Implications. This new occurrence of pseudodontorns is the first for the Southwest Pacific prior to the Miocene-Pliocene records in New Zealand, and corroborates an apparent early Paleogene link between East and West Antarctica suggested by invertebrate faunal elements [Stilwell et al., 1993]. Pseudodontorns were medium to very large, pelagic seabirds with large robust bills that supported rows of bony tooth-like projections. Their thin-walled wing bones were adapted for gliding, rather than flapping flight, for which they required strong steady winds, in the same way modern albatrosses do today [Olson 1985].

Pseudodontorns are often found in association with diverse fossils vertebrate faunas, including whales, other seabirds, leatherback turtles, teleost fish and sharks. Some authors [Case, 1992; Warheit, 1992] have suggested such diverse vertebrate faunas indicate highly productive oceanic conditions associated with upwelling systems. The Mt Discovery erratics contain relatively abundant, diverse, fossil phytoplankton assemblages (D. Harwood personal communication, 1999; various authors this volume) which also indicate high oceanic productivity. Such upwelling systems can often be generated by strong steady onshore winds. Such winds were probably ideal for large gliding bird such as pseudodontorns.

These birds appear to have been an important component of Eocene high-latitude avifauna and may have had a circum-polar distribution by the late Eocene [Goedert 1989]. This new occurrence also indicates the potential of glacial deposits, such as those around Mt Discovery, to provide important new information on the fossil avifauna of Antarctica, prior to its burial beneath continental ice sheets.

Acknowledgements. The author wishes to thank the following people: J. D. Stilwell for providing the opportunity to work on the specimen; D. Harwood for useful paleoenvironmental discussions; K. Swanson for help in the photography and preparation of the figures; P. Denholm for help with the initial draft; C. Tambussi and S. Emislie for reviewing the manuscript; Department of Geological Sciences, University of Canterbury and the Institute of Geological and Nuclear Sciences for allowing the time to work on this paper.

REFERENCES

Baumel, J. J. and L. M. Witmer
1993   Osteologia, In J. J. Baumel, (Ed.), *Handbook of Avian Anatomy: Nomina anatomica avium* (2nd Ed.). pp 45-133. Nuthall Ornithological Club, Cambridge Massachusetts

Case, J. A.
1992   Evidence from fossil vertebrates for a rich Eocene Antarctic marine environment, In J. P. Kennett and D. A. Warnke (Eds.), *The Antarctica paleoenvironment: A perspective on global change,* Antarctic. Res. Ser., 56: pp. 119-130. American Geophysical Union, Washington D.C.

Case, J. A., M. O. Woodburne, and D. S. Chaney
1987   A gigantic phororhacoid (?) bird from Antarctica. *J. Paleont.,* 61 (6): 1280-1284.

Chandler, R. M.
1994   The wing of Titanis walleri (Aves: Phorusrhacidae) from the Late Blancan of Florida. *Bulletin of the Florida State Museum, Biological Sciences,* 36(6): 175-180.

Chatterjee, S.
1989   The oldest Antarctic bird (abstract). *Journal of Vertebrate Paleontology* 9(3): 16A.

Covacevich, V. and P. V. Rich
1982   New bird ichnites from Fildes Peninsula, King George Island, West Antarctica, In C. Craddock, C. (Ed.), *Antarctic Geoscience.* pp 245-254. University of Wisconsin Press, Madison.

Fordyce R. E., and C. M. Jones
1990   The history of penguins and new fossil penguin material from New Zealand., In L. S. Davis and J. T. Darby (Eds.), *Penguin Biology.* pp 419-446. Academic Press, San Diego.

Goedert, J. L.
1989   Giant Late Eocene marine birds (Pelecaniformes: Pelagornithidae) from northwestern Oregon. *J. Paleont.,* 63(6): 939-944.

Harrison J. O. and C. A. Walker
1976b   A review of the Bony-toothed birds (Odontopterygiformes): with descriptions of some new species. *Tertiary Res. Special Paper 2.* 72 pp. Tertiary Research Group, London.

Howard, H.
1957a   A gigantic "toothed" marine bird from the Miocene of California. *Santa Barbara Mus. Nat His. Dept. Geol. Bull.,* 1: 1-23.

1966c   Additional avian records from the Miocene of Sharktooth Hill, California. Los Angeles Co. *Mus. Contrib. Sci.,* 114: 1-11.

1978   Late Miocene marine birds from Orange County, California. *Nat. Hist. Mus. Los Angeles Co. Contrib. Sci.,* 290: 1-26.

1982   Fossil birds from Tertiary marine beds at Oceanside, San Diego County, California, with description of two new species of the genera Uria and Cepphus (Aves: Alcidae). *Nat. Hist. Mus. Los Angeles Co. Contrib. Sci.,* 341: 1-15.

Marples, B. J.
1953   Fossil penguins from the mid-Tertiary of Seymour Island. *Falkland Islands Dependency Survey-Science Report,* 5: 1-15

Marenssi, S. A., M. A. Reguero, S. N. Santillana, S. F. Vizcaino.
1994   Eocene land mammals from Seymour Island, Antarctica: paleobiogeographical implications. *Antarctic Science* 6 (1): 3-15.

McKee, J. W. A.
1985   A pseudodontorn (Pelecaniformes: Pelagornithidae) from the middle Pliocene of Hawera, Taranaki, New Zealand. *New Zealand Journal of Zoology,* 12:181-184

1987a   The occurrence of Tereingaornis moisleyi (Sphenisciformes; Spheniscidea) at Hawera, Taranaki, New Zealand. *New Zealand Journal of Zoology,* 14: 557-561.

Myrcha, A., A. Tatur, and R. del Valle.
1990   A new species of fossil penguin from Seymour Island (West Antarctica), *Alcheringa,* 4: 195-205.

Noriega, J. I., C. P. Tambussi
1995   A Late Cretaceous Presbyornithidae (Aves: Anseriformes) from Vega Island, Antarctic Peninsula: Paleobiogeographic implications. *Ameghiniana,* 32(1): 57-61

1996   The non-penguin avifauna from Eocene (early Oligocene?) of Seymour Island, Antarctic Peninsula (abstract). Society of Avian Paleontology and Evolution, Program and Abstracts 4th International Meeting, Washington D.C., 13.

Olson S. L.
1985   The fossil record of birds, In D. S. Farner, J. R. King and K. C. Parkes (Eds.), *Avian. Biology.* pp. 79-238. Academic Press, New York.

1992   Neogaeornis wetzeli Lambrecht, a Cretaceous Loon from Chile (Aves: Gaviidae). Journal of Vertebrate Paleontology, 12 (1): 122-124.

Sadler, P. M.
1988   Geometry and stratification of uppermost Cretaceous and Paleogene units on Seymour Island, northern Antarctic Peninsula. In R. M. Feldmann and M. O. Woodburne (Eds.), *Geology and paleontology of Seymour Island , Antarctic Peninsula. Mem Geol. Soc. Am.* 169: 303-320

Simpson, G. G.
1946   Fossil penguins. *Bull. Am. Mus. Nat. Hist.,* 87: 1-99.

1971b   Review of fossil penguins from Seymour Island. *Proceedings of the Royal Society of London. Series,* 178B: 357-387

1976   *Penguins Past and Present, Here and There,* pp.150. Yale University Press, New Haven.

Stilwell, J. D., and W. J. Zinsmeister.
1992   *Molluscan systematics and biostratigraphy, Lower Tertiary La Meseta Formation, Seymour Island, Antarctic Peninsula.* American Geophysical Union Antarctic Research Series, 55: 1-192

Stilwell, J. D., R. H. Levy, R. M. Feldmann, and D. M. Harwood.
1997   On the rare occurrence of Eocene callianassid decapods (Arthropoda) preserved in their burrows, Mount Discovery, East Antarctica. *Journal of Paleontology,* 71(2): 284-287.

Tambussi C. P., and J. I. Noriega.
1996   Summary of the avian fossil record from southern South America. In G. Arratia (Ed), *Contributions of southern South America to vertebrate paleontology.* München Geowiss. Abh. (A)30: 245-264.

Tambussi, C. P., and E. P. Tonni.
1988   Un Diomedeidae (Aves, Procellariiformes) del Eoceno tardío de la Antártida, In. J. C. Quiroga and A. L. Cione (Eds.), *5th Jornadas Argentiniana de Paleontologia de Vertebrados.* 4. Universidad de La Plata, La Plata.

Tonni, E. P.
1980   Un Pseudodontornithidae (Pelecaniformes, Odontopterygia) de gran tarmaño del Terciario temprano de Antártida. *Ameghiniana,* 18(3): 273-276.

Tonni, E. P., and C. P. Tambussi.
1985   Nuevos restos de Odontopterygia (Aves: Pelecaniformes) del Terciario temprano de Antártida. *Ameghiniana,* 21(2-4): 121-124.

Warheit, K. I.
  1992   A review of the fossil seabirds from the Tertiary of the North Pacific: plate tectonics, paleoceanography and faunal change. *Paleobiology,* 18(4): 401-424.

Warheit, K. I., and D. R. Lindberg.
  1988   Interactions between seabirds and marine mammals through time: interference competition at breeding sites. In J. Burger (Ed.), *Seabirds and other marine vertebrates. Competition, predation and other interactions.* pp. 292-328. Columbia University Press, New York.

Wiman, C.
  1905   Über die Alttertiaren Vertebrata der Seymour-Insel. *Wissentschaftliche Ergebnissen der Swedische Sudpolar-Expedition,* 3(1): 1-37.

Zinsmeister W. J.
  1988   Early geologic exploration of Seymour Island, Antarctica. In R. M. Feldmann and M. O. Woodburne (Eds.), *Geology and paleontology of Seymour Island, Antarctic Peninsula. Mem. Geol. Soc. Am.* 169: 1-16.

Craig M. Jones, Department of Geological Sciences, University of Canterbury, Christchurch, New Zealand

PALEOBIOGEOGRAPHIC SYNTHESIS OF THE EOCENE MACROFAUNA FROM MCMURDO SOUND, ANTARCTICA

Jeffrey D. Stilwell

School of Earth Sciences, James Cook University, Townsville, Q4811, Australia

William J. Zinsmeister

Department of Earth and Atmospheric Sciences, Purdue University, West Lafayette, IN 47906-1397

The macrofauna from Eocene erratics of McMurdo Sound, East Antarctica, is extremely important from a paleobiogeographic perspective as, for the first time, a comparison can be made between coeval taxa of East Antarctica and the better known fauna from Seymour and Cockburn islands, Antarctic Peninsula. As many as 22 mollusc taxa, a single brachiopod, and a shark may be in common to both East and West Antarctica during the Eocene, reflecting unequivocal marine links between these regions at this time. Sea-surface waters in the McMurdo Sound region during the Eocene may have been as warm as warm temperate, based on the marked percentages of characteristic Indo-Pacific/Tethyan (~ 41%) and cosmopolitan (~ 29%) mollusc taxa in the fauna. Approximately 11% of 136 mollusc genera and all species, recorded from the Eocene of Antarctica, are endemic and indicate that the continent belonged to a distinct biotic province by this time.

INTRODUCTION

From a global viewpoint, the Eocene fauna and flora of McMurdo Sound, East Antarctica provide a great deal of information and bridge a major gap in our knowledge on the evolution and paleobiogeographic history of southern circum-Pacific biotas. Up to now, virtually no data have been available from this region of Antarctica and we have had to extrapolate from the quite diverse record of Eocene life from the Antarctic Peninsula (La Meseta Formation of Seymour and Cockburn islands), to gather a glimpse of the biotic composition of Eocene Antarctic faunas and floras. For the first time it is possible to examine the relationship between Paleogene organisms of East and West Antarctica to deduce the presence/absence of geographic links during the Eocene, and to evaluate the importance of Antarctica's role in the evolution of Austral biotas. Further, data obtained from this study provide significant insight into the degree of

isolation of the Antarctica biota during the Paleogene and the maintenance and also origin of biodiversity patterns in the southern high latitudes to ascertain long-term Antarctic taxonomic diversity gradient trends. The purpose of this paper is to explore the evolutionary and paleobiogeographic significance of the McMurdo biota and present a synthesis of the results in both an Antarctic and global context.

EOCENE PALEOGEOGRAPHY AND PALEOCEANOGRAPHY OF ANTARCTICA

Because most of Antarctica is today masked by extensive ice cover, the paleogeography and paleoceanography of the now frozen continent must be inferred from the quite limited rock exposure (~ 2%) and sediments surrounding the continent. Notwithstanding these deficiencies in the rock record, a vast amount of information has been gained through exploration and diligent research since the latter part of last century. It is now known that during the late Mesozoic the western sector of the supercontinent Gondwana was characterized by a broad and stable, low-lying Australian-East Antarctic craton with mountains of no more than 1,000 m height [Grindley, 1967; Drewry, 1975; Zinsmeister, 1987]. It is possible that the uplift in Late Cretaceous time might well have given topography greater than 1000 m, and the Ellsworth Mountains likewise may have been quite high at the time [D. H. Elliot, pers. commun., 1999]. This quiescent tectonic phase for this part of Gondwana was disrupted during the Late Jurassic or Early Cretaceous when incipient rifting between Australia and East Antarctica commenced. It is interesting to note that the age of the separation of East Antarctica from Australia has been pushed consistently further back into time over the last 25 years of research from a Paleocene separation (53 Ma) of the two continents [Weissel and Hayes, 1972], to a Late Cretaceous separation (110 to 84 Ma) [Cande and Mutter, 1982; see also Veevers, 1986], to a Neocomian separation (c. 125 Ma) [Stagg and Willcox, 1992], to a much earlier separation during the Late Jurassic-Early Cretaceous (Tithonian-Barremian) [see comments by Zinsmeister, 1987; Symonds et al., 1996]. The major phase of intracontinental extension that saw the beginnings of the end of Gondwana proper occurred along a west to east zone of rifting that split the center of the craton between Australia and East Antarctica. The spreading between Australia and Antarctica was slow during the late Mesozoic, but became much more rapid during the Paleogene, particularly in the early Eocene [see review by Stevens, 1989]. Fossil data from marine macroinvertebrates and microfossils indicate that a continuous marine seaway with deep-water circulation connecting the Indian Ocean with the Pacific Ocean did not emerge until the early Tertiary [see Kennett, 1980; Zinsmeister, 1982]. Stilwell [1997] found that few mollusc taxa at the genus-level were in common to the Western Australian and New Zealand-Chatham Islands regions during the Maastrichtian except for a few bivalves, indicating that the Miria Formation fauna of the Carnarvon Basin in Australia belonged to a different biotic province. This supports the work of Kennett and Zinsmeister that marine links between these regions were weak before Tertiary time.

Zinsmeister [1978] suggested that a seaway (Shackleton Seaway) existed between East and West Antarctica during the Late Cretaceous and Paleogene until it was closed by the formation of the West Antarctic Ice Sheet during the early Neogene. The occurrence of a number of molluscs (i.e. gastropods such as Struthiolariidae, *Taioma*, *Struthioptera*; bivalves such as *Lahillia*) in both the southwestern margin of the Pacific and the east side of the Antarctic Peninsula and their absence along western Australia support the existence of the Shackleton Seaway. Additional support for this seaway comes from the presence of latest Cretaceous microfossils in reworked glacial diamictites in the Transantarctic Mountains. Huber [1992] suggested that this microfossil occurrence indicated that some circum-Antarctic flow of shallow surface waters existed between East and West Antarctica among other possible marine corridors from the latest Cretaceous into the Paleogene. These seaways were probably ephemeral and existed at particular periods of time during the Cretaceous and early Tertiary. There were probable marine links between the Ross, Wilkes, and Pensacola basins prior to the primary mid Eocene phase of uplift of the Transantarctic Mountains [see comments by Huber, 1992], and some of these links may have persisted until at least the late Eocene. As will be shown below, some mollusc taxa that are probably mid-late Eocene in age are common to both the McMurdo Sound region and Seymour Island. Incipient uplift of the Transantarctic Mountains probably occurred much earlier than as previously proposed by the latest Cretaceous [D. H. Elliot, pers. commun., 1995].

Few paleogeographic maps of Antarctica during the middle Eocene are available and these are based on limited data [see Scotese and Denham, 1988; Lawver et al., 1992; Barrera and Huber, 1993]. Thus, new data gleaned from the Eocene erratics of McMurdo Sound are immensely important in reconstructing Antarctic paleogeography and paleoceanography. Of note, work is in progress to possibly deduce where the erratics were derived in East Antarctica. With the knowledge of many invertebrate taxa common to both West and East Antarctica during the mid to late Eocene, it appears unequivocal that marine communication did exist between these regions [Stilwell, 1995] (See Fig. 1 herein). Mollusc taxa common to both McMurdo Sound and Seymour and Cockburn islands include *Leionucula nova* [Wilckens, 1911], *Neilo beui* Stilwell and Zinsmeister, 1992, *Cucullaea* cf. *C. donaldi* Sharman and Newton, 1894, *Aulacomya* sp. cf. *A. anderssoni* Zinsmeister, 1984, *Saxolucina sharmani* [Wilckens, 1911], *?Anisodonta truncilla* Stilwell and Zinsmeister, 1992, *?Gomphina iheringi* Zinsmeister, 1984, *Cyclorismina?* n. sp. cf. "*C.*" *marwicki* Zinsmeister, 1984, *?Eumarcia (Atamarcia) robusta* Stilwell and Zinsmeister, 1992, *Panopea akerlundi* Stilwell and Zinsmeister, 1992, *Panopea* n. sp.? cf. *P. philippii* Zinsmeister, 1984,

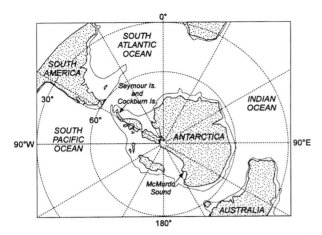

Fig. 1. Middle Eocene (43 Ma) paleogeographic map of Antarctica showing the location of McMurdo Sound, East Antarctica and Seymour and Cockburn islands, Antarctic Peninsula, West Antarctica. The dashed line around the continents denotes the position of the edge of the continental shelf. Map modified from Barrera and Huber [1993].

Periploma n. sp.? *cf. P. topei* Zinsmeister, 1984, *Cellana feldmanni* Stilwell and Zinsmeister, 1992, *Colposigma euthenia* Stilwell and Zinsmeister, 1992, *Arrhoges (Antarctohoges) diversicostata* [Wilckens, 1911], *Perissodonta* n. sp.? *cf. P. laevis* [Wilckens, 1911], *Polinices (Polinices) cf. P. (P.) subtenuis* [von Ihering, 1897], *?Penion australocapax* Stilwell and Zinsmeister, 1992, *?Eobuccinella brucei* Stilwell and Zinsmeister, 1992, *Acteon eoantarcticus* Stilwell and Zinsmeister, 1992, *Crenilabium suromaximum* Stilwell and Zinsmeister, 1992, and *Ringicula (Ringicula) cockburnensis* Zinsmeister and Stilwell, 1990. In total, up to 22 species may be common to both regions. See Stilwell [this volume, Tables 1-2], Stilwell and Zinsmeister [1992, Table 1], and Wilckens [1924] for lists of the Eocene Mollusca of Antarctica. Non- mollusc taxa common to both regions include the brachiopod *Tegulorhynchia* sp. *cf. T. imbricata* [Buckman, 1910] [see Lee and Stilwell, this volume] and the shark *Carcharias* sp. *cf. C. macrota* [Agassiz] [see Long and Stilwell, this volume]. Given the great geographic distance between Seymour Island and McMurdo Sound during the Eocene, these mollusc and non-mollusc species, with the exception of *Carcharias*, apparently had long-ranging planktotrophic (or teleplanic) larval dispersal capabilities. Some groups had more restricted dispersal capabilities and evolved distinct Antarctic lineages, such as *Solemya*, *Brachidontes*, *Crassatella*, *Hiatella*, *Struthiolarella*, possibly *Perissodonta*, *Euspira*, and

probably others. What do the fauna and flora tell us about the climate and oceanic circulation of surface waters along the shallow shelf of Antarctica during the Eocene?

The precise surface water circulation pattern of Eocene Antarctica is uncertain because of the limited nature of Upper Cretaceous and Paleogene exposures in Antarctica. As a consequence, it is not known whether the area along the Antarctic Peninsula region at this time was a series of islands separated by shallow seaways or was a continuous landmass. It seems more probable that the Peninsula was a continuous landmass that was the source of sediment for the James Ross Island basin and other basins on the east flank [D. H. Elliot, pers. commun., 1999]. Floral evidence indicates that at least part of the Peninsula was forested and above sea level during the Late Cretaceous and Paleogene [Askin and Jacobson, 1996; Askin, 1992, 1988; Francis, 1991, 1986]. Since the presence of small continental fragments in the Drake Passage region would have blocked any circumpolar deepwater circulation in Antarctica [Lawver et al., 1992], invertebrate larvae were transported during the Eocene either by: 1) surface waters through marine corridors in the Antarctic interior akin to a shallow archipelago type of circulation; or 2) by shallow and perhaps weak surface proto-circumpolar currents due to the new isolation of Antarctica during the Eocene; or 3) a combination of a circumpolar current and intracontinental marine links. At present, there is insufficient information to propose an Antarctic circulation model except that there was indeed a marine connection between the McMurdo Sound region or nearby Transantarctic Mountains and the northern tip of the Antarctic Peninsula based on invertebrate distributions.

Information on Eocene Antarctic climate is also problematic with conflicting reports of sea-surface temperatures and the absence/presence of significant ice volume. The late Paleocene to early Eocene interval was characterized by a maximum high-latitude warming event when there was a low latitudinal temperature gradient, which was followed in the middle to late Eocene by a combination of gradual cooling and increased continental ice accumulation in Antarctica [Barrera and Huber, 1993, and references therein]. The presence/absence of glaciation during the Paleogene in Antarctica is still quite controversial and as yet unresolved. It is not possible to disprove that the probable ice in the McMurdo Sound region was simply local mountain glaciation [D. H. Elliot, pers. commun., 1999]. Sea surface temperatures in the Antarctic Indian Ocean during the middle Eocene are estimated to have been between ~5° and 8°C in the late middle Eocene interval with deepwater temperatures slightly lower by about one degree [Barrera and Huber, 1993]. Shackleton and Kennett [1975] reported a progressive decline of sea surface temperatures in the southern oceans from a maximum of nearly 20°C at the start of the Eocene to 11°C at the close of the Eocene, followed by a marked decline to 7°C during the earliest Oligocene. Because no Eocene sediments have been cored in the Ross Sea, only indirect geologic evidence provides any clues on early Paleogene climate. Hambrey and Barrett [1993] reported that the earlier Cenozoic Ross Sea climate was moderately warm and probably cold to cool temperate. Floral evidence from the erratics indicate a seasonal climate [Francis, this volume; Pole, this volume] and the presence of *Nothofagus* suggests a summer temperature of 5°C [Hambrey and Barrett, 1993]. Additional information on the Eocene climate comes from the Antarctic Peninsula. Marine invertebrates from the La Meseta Formation indicate a range of cool to warm temperate conditions, and although a warm temperature scenario is preferred, the absence in the fauna of characteristic warm temperate taxa such as *Glycymeris, Limopsis, Miltha,* and *Pitar* is puzzling [Stilwell and Zinsmeister, 1992]. However, *Limopsis (Limopsista?)* and possible *Miltha* have been recognized in the McMurdo region erratics indicating temperate sea surface temperatures during the Eocene. Also, the presence of tropical taxa such as *Cardita* in the erratics implies even warmer conditions. As Stilwell and Zinsmeister [1992] suggested, the absence of these taxa may reflect high-latitude seasonality or cooler temperatures in the Seymour Island region. Conversely, recent work by Ditchfield *et al.* [1994] on the high latitude paleotemperature variation in the James Ross Basin, Antarctic Peninsula, rock record established that cold temperate or sub-polar conditions would have been established during the Eocene. The molluscan faunas of Seymour Island and McMurdo Sound do not advocate cold temperate or sub-polar conditions in Antarctica during the Eocene. Further research is required to resolve these conflicting reports of paleoclimate.

PALEOBIOGEOGRAPHY OF EOCENE FAUNA

The unique character of the southern hemisphere biota was recognized early on and has been discussed at length by various early workers such as Darwin [1859], Hutton [1872, 1896], and von Ihering [1892, 1905-07, 1925], among others. The first report of fossils in Antarctica was made by James Eights during the first American expedition to Antarctica in 1830, when a fragment of carbonized wood in a conglomerate was discovered (locality uncertain, possibly King George Island) [Eights, 1833]. Very little was known about the ancient life of Antarctica until the last decade of last century when a Norwegian whaling expedition on the barque *Jason* landed on Seymour Island in 1892 and the crew were sent to search for food. Captain Larsen of the *Jason* discovered rich deposits of fossils on Seymour Island. The significance of these fossils, now known to be Eocene in age, was portrayed by the naturalist of the *Balaena*, William S. Bruce *in* Murdoch [1894, pp. 356, 364], who wrote that "...They are probably of Tertiary in age, and indicate a warmer climate than now prevails in these high southern latitudes...". Bruce's account has generally been overlooked in the literature and more often than not Nordenskjöld *et al.* [1904] and Nordenskjöld [1905] have been given the credit for the first scientists to recognize that Antarctica has not always been locked in ice through time and that warmer climes did prevail on the southern continent in the distant past.

The paleobiogeographic significance of the Eocene fossils received little attention until the work of Zinsmeister [1979, 1982] and Stilwell and Zinsmeister [1992], although comparisons of the Seymour Island fossils with other predominantly Austral taxa were made by various authors including Wilckens [1911] who contributed to the reports of the Swedish South Polar Expedition of 1901-1903 under the command of Otto Nordenskjöld. Similarities of the Late Cretaceous and early Tertiary fauna and flora of Antarctica with those around the rim of the southern circum-Pacific led Zinsmeister [1979] to erect the Weddellian Biotic Province, based predominantly on mollusc distributions. As the final phase of fragmentation of Gondwana occurred at the end of the Cretaceous, the Austral Province of Kauffman [1973] is considered to have lost its identity and split into smaller provinces. The Weddellian Province was one of these provinces (as defined by Zinsmeister, 1979) and comprised the region south of the northeastern coast of Australia and New Zealand extending westward including the continental shelf areas along the Pacific margin of Antarctica and southern South America. The northern limits of the Weddellian Province were probably constrained by temperature [Zinsmeister, 1979] and oceanic circulation [Stilwell, 1997]. The floral/faunal assemblages of the Weddellian Province have received much attention in the literature and the concept of the province has generally been accepted. The original idea of the province has been expanded to include New Caledonia and the Chatham Islands, based on molluscan similarities [Stilwell, 1991, 1994a, b, 1997]. Zinsmeister argued that the Weddellian

Province was an entity until the early Tertiary, at least the Eocene, but Stilwell [1997] provided evidence that the Province was short-lived, extending through Campanian-Maastrichtian time. The Weddellian Province split into smaller biogeographic entities by earliest Tertiary time, much earlier than proposed by Zinsmeister.

Approximately 11% of a total of 136 mollusc genera/subgenera from the Eocene of Antarctica are endemic and unknown outside the continent. Thus, if at least 10% endemism at subgenus-/genus-level is required to erect biotic provinces/subprovinces, then Antarctica belonged to a separate province by the middle Eocene. There is further support for an Antarctic Eocene province at the species-level as all recorded mollusc species are endemic, attesting to the isolation of Antarctica's macrofauna, geographically and genetically, by at least middle Eocene time [Stilwell, 1995; this work].

The McMurdo Sound mollusc fauna can be grouped into four biogeographic categories at the genus/subgenus-level; these are endemic, paleoaustral, Indo-Pacific/Tethyan, and cosmopolitan. The endemic component of the McMurdo Sound fauna is quite weak at approximately 3% of the total number of confidently identified, well-preserved taxa (34 genera/subgenera). This percentage may be slightly greater at 5% if some questionable taxa have been accurately identified. The paleoaustral element [Fleming, 1963] refers to those taxa that have fossil records extending back into the Tertiary or Mesozoic and also groups with inferred poor dispersal capabilities whose present or known fossil distribution reflects past land connections of the southern continents. The paleoaustral element is inclusive of endemic taxa and the concept of this element originated from ideas of faunal distributions dating from Hutton [1872]. The paleoaustral element is stronger than the endemic component at about 26% (possibly s high as 42%, if poorly preserved taxa are correctly identified). The cirriped barnacle *Australobalanus* identified in the Eocene erratics [Buckeridge, this volume] also has a paleoaustral distribution in the Paleogene of East and West Antarctica and New Zealand [Zullo *et al*., 1988]. The single record of a probable crocodile, which may be related to Austral gavials, is from McMurdo Sound [Willis and Stilwell, this volume]. The Indo-Pacific/Tethyan component is largely based on the concepts of Fleming [1967] and Darragh [1985]. The latter author expanded Fleming's concept to include the northern influence in the fauna, comprising extinct and extant taxa with their predominant distribution in the tropical Indo-Pacific Realm, or part of it with their probable origin in the Indo-Pacific Basin. Groups that were largely distributed in the early

Tertiary through the Paleogene Tethyan Realm of the Mediterranean region east to Indonesia and west to the Caribbean region were also included in the category. Because of the limited fossil record of Cretaceous and Tertiary molluscs in Antarctica, it is rather difficult to establish the time of origin of many taxa. The Indo-Pacific/Tethyan grouping is necessarily broad and includes inferred warm-water taxa that were/are distributed in these areas and beyond the Austral Realm during the Cretaceous and Tertiary. The Indo-Pacific/Tethyan component of the McMurdo Sound molluscan fauna was approximately 41%, but may have been somewhat lower at nearly 30% if some poorly preserved taxa have been accurately identified. The cosmopolitan element in the fauna was about 29% and may have been slightly lower at ~24% again if some questionable taxa are accurately identified. Other cosmopolitan groups encountered in the erratics include the sharks *Carcharias* and *Galeorhinus* [Long and Stilwell, this volume] and a possible pseudodontorn bird [Jones, this volume]. The cosmopolitan element is one of the most difficult to document because establishing the geographic distribution of a particular group is more of a challenge than knowing its temporal range [*cf.* Smith, 1989, p. 263]. These percentages of biogeographic elements are interesting and important in that they support the contention that the McMurdo Sound fauna was derived predominantly by evolutionary divergence from pre-existing cosmopolitan or widespread Mesozoic and earliest Cenozoic stocks. These groups experienced gradual range restrictions resulting from the final break-up of Gondwana and concomitant continental reorganization. Many genera in the McMurdo Sound erratics are long ranging with origins in the Mesozoic when they were widespread. These taxa include *Leionucula, Saccella, Neilo, Solemya, Cucullaea, Brachidontes, Eburneopecten, Crassostrea, Thyasira (Conchocele), Crassatella, Cyclorismina?, Panopea, Teredo, Sigapatella, Euspira, Polinices s. s., Acteon,* and *Ringicula*. Further, the Antarctic Cretaceous barnacles are part of the cosmopolitan fauna that is related to western Europe, eastern North America, and Australasia forms with a marked Tethyan influence [Zullo *et al*., 1988]. There is still a moderate degree of endemism at the genus-level and a high degree at the species-level in the Antarctic faunal record, which attests to the idea that the continent was, indeed, well isolated biologically if not physically by middle Eocene time. Research on the origin and evolution of other southern circum-Pacific faunas such as those of mainland New Zealand and Chatham Islands [see Stilwell, 1994a-b, 1997] indicate a similar evolutionary history as portrayed above.

REFERENCES

Askin, R.A.
1988   Campanian to Paleocene palynological succes-
            sion of Seymour and adjacent islands, northeast-
            ern Antarctic Peninsula. *In* R. M. Feldmann and
            M. O. Woodburne (Eds.), *Geology and
            Paleontology of Seymour Island, Antarctic
            Peninsula. Mem. Geol. Soc. Am.*, *169*: 131-153.
1992   Late Cretaceous-early Tertiary Antarctic out-
            crop evidence for past vegetation and climate. *In*
            J. P. Kennett and D. A. Warnke (Eds.), *The
            Antarctic Environment: a perspective on global
            change.* American Geophysical Union,
            Antarctic Research Series, *56*: 61-73.

Askin, R. A., and S. R. Jacobson
1996   Palynological change across the Cretaceous-
            Tertiary boundary on Seymour Island,
            Antarctica: environmental and depositional fac-
            tors. *In* N. Macleod and G. Keller (Eds.),
            *Cretaceous-Tertiary Mass Extinctions: biotic
            and environmental changes.* W. W. Norton &
            Co., New York and London, pp. 7-25.

Barrera, E., and B. T. Huber
1993   Eocene to Oligocene oceanography and temper-
            atures in the Antarctic Indian Ocean. *In* J. P.
            Kennett and D. A. Warnke (Eds.), *The Antarctic
            Paleoenvironment: a perspective on global
            change.* American Geophysical Union,
            Antarctic Research Series, *60*: 49-65.

Buckman, S. S.
1910   Antarctic fossil Brachiopoda collected by the
            Swedish South Polar Expedition. *Wiss. Ergebn.
            Schwed. Südpolar-exp.*, *3*(7): 1-43.

Buckeridge, J. St. J. S.
1998   A new species of *Austrobalanus* (Cirripedia,
            Thoracica) from Eocene erratics, Mount
            Discovery, McMurdo Sound, East Antarctica. *In*
            J. D. Stilwell and R. M. Feldmann (Eds.),
            *Paleobiology and Paleoenvironments of Eocene
            Rocks, McMurdo Sound, East Antarctic*
            American Geophysical Union, Antarctic
            Research Series

Cande, S. C., and J. C. Mutter
1982   A revised identification of the oldest sea floor
            spreading anomalies between Australia and
            Antarctica. *Earth Planetary Science Letters, 58*:
            151-160.

Darragh, T. A.
1985   Molluscan biogeography and biostratigraphy of
            the Tertiary of south-eastern Australia.
            Alcheringa, 9: 83-116.

Darwin, C.
1859   *On the Origin of Species by Means of Natural
            Selection, or the preservation of favoured races*

            in the struggle for life. J. Murray, London, 502p.

Ditchfield, PW., J. D. Marshall, and D. Pirrie
1994   High latitude palaeotemperature variation: new
            data from the Tithonian to Eocene of James
            Ross Island, Antarctica. *Palaeogeography,
            Palaeoclimatology, Palaeoecology, 107*: 79-101.

Drewry, J.
1975   Initiation and growth of the East Antarctic Ice
            Sheet. *J. Geol. Soc.* London, *131*: 255-273.

Eights, J.
1833   Descriptions of a new crustaceous animal and on
            the shores of the South Shetland Islands with
            remarks on their natural history. *Trans. Albany
            Inst., 2*(1): 53-69.

Fleming, C. A.
1963   The nomenclature of biogeographic elements in
            the New Zealand biota. *Trans. R. Soc. N.Z.*
            (General), *1*: 13-22.
1967   Cenozoic history of Indo-Pacific and other
            warm-water elements in the marine Mollusca of
            New Zealand. *Venus, 25* (3-4): 105-117.

Francis, J. E.
1986   Growth rings in Cretaceous and Tertiary wood
            from Antarctica and their paleoclimatic implica-
            tions. P*alaeontology, 29*(4): 665-684.
1991   Paleoclimatic significance of Cretaceous-early
            Tertiary fossil forests of the Antarctic Peninsula.
            In M. R. A. Thomson, J. A. Crame and J. W.
            Thomson (Eds.), *Geological Evolution of
            Antarctica*, London, Cambridge University
            Press, pp. 623-627.
1998   Fossil wood from Eocene rocks, McMurdo
            Sound. *In* J. D. Stilwell and R. M. Feldmann
            (Eds.), *Paleobiology and Paleoenvironments of
            Eocene Rocks, McMurdo Sound, East
            Antarctica.* American Geophysical Union,
            Antarctic Research Series.

Grindley, G. W.
1967   The geomorphology of the Miller Range,
            Transantarctic Mountains, with notes on the
            glacial history and neotectonics of East
            Antarctica. *New Zealand Journal of Geology
            and Geophysics*, *10*(2): 557-598.

Hambrey, M. J., and P. J. Barrett
1993   Cenozoic sedimentary and climatic record, Ross
            Sea Region, Antarctica. *In* J. P. Kennett and D.
            A. Warnke (Eds.), *The Antarctic paleoenviron-
            ment: a perspective on global change,* American
            Geophysical Union, Antarctic Research Series,
            60: 91-124.

Huber, B. T.
1992   Paleobiogeography of Campanian-
            Maastrichtian foraminifera in the southern high
            latitudes. *Palaeogeography, Palaeoclimatology,
            Palaeoecology, 92*: 325-360.

Hutton, F. W.

1872   On the geographical relations of the New Zealand fauna. *Trans. N. Z. Inst., 5*: 227-256.

1896   Theoretical explanation of the distribution of southern faunas. *Proc. Linnean Soc. New S. Wales, 1*: 36-47.

Ihering, von, H.

1892   On the ancient relations between New Zealand and South America. *Trans. N. Z. Inst., 24*: 431-435.

1897   Os molluscos dos terrenos tertiarios da Patagonia. *Revista Mus. Paul., 2*: 217-328.

1905-07   Les mollusques fossiles du Tertiaire et du Crétacé Supérieur de la'Argentine. *An. Mus. Nacl. Buenos Aires*, Ser. 3 (7): 1-611. (published in 39 parts, November 21, 1905-September 13, 1907: date of each part appears on bottom page of each part).

1925   Die Kreide-Eocän Ablagerunge Antarktis. Neues Jahrb. Mineral. *Paläontol., 51*: 240-301.

Jones, THIS VOLUME

Kauffman, E.

1973   Cretaceous Bivalvia. *In* A. Hallam (Ed.), *Atlas of Palaeobiogeography*, Elsevier, pp. 353-383.

Kennett, J. P.

1980   Paloceanographic and biogeographic evolution of the Southern Ocean during the Cenozoic, and Cenozoic datums. *Palaeogeography, Palaeoclimatology, Palaeoecology, 31*: 123-152.

Lawver, L. A., L. M. Gahagan, and M. F. Coffin

1992   The development of paleoseaways around Antarctica. *In* J. P. Kennett and D. A. Warnke (Eds.), *The Antarctic Paleoenvironment: a perspective on global change*, American Geophysical Union, Antarctic Research Series, *56*: 7-30.

Lee, D. E., and J. D. Stilwell

This volume   Rhynchonellide brachiopods from Late Eocene erratics in the McMurdo Sound region, Antarctica. *In* J. D. Stilwell and R. M. Feldmann (Eds.), *Paleobiology and Paleoenvironments of Eocene Rocks, McMurdo Sound, East Antarctica*, American Geophysical Union, Antarctic Research Series.

Long, D. J., and J. D. Stilwell

This volume   Fish remains from the Eocene of Mount Discovery, East Antarctica. *In* J. D. Stilwell and R. M. Feldmann (Eds.), *Paleobiology and Paleoenvironments of Eocene Rocks, McMurdo Sound, East Antarctica*, American Geophysical Union, Antarctic Research Series.

Murdoch, W. G. Burn

1894   From Edinburgh to the Antarctic. An artist's notes and sketches during the Dundee Antarctic Expedition of 1892-93; with a chapter by W. S.

Bruce, naturalist of the barque "Balaena". Longmans, Green and Co., London, 364 p.

Nordenskjöld, O.

1905   *Antarctica, or Two Years Amongst the Ice of the South Pole* (English Edition). MacMillan Co., New York, 608 p.

Nordenskjöld, O., J. Andersson, C. A. Larsen, and C. Skottsberg

1904   Antarctic. *Två år bland syspoloens isar.* 2 vols. Albert Bonniers, Stockholm, *1*(486 p.); *2*(544 p.).

Pole, M., R. Hill, and D. M. Harwood

This volume   Plant macrofossils from erratics, McMurdo Sound. *In* J. D. Stilwell and R. M. Feldmann (Eds.), *Paleobiology and Paleoenvironments of Eocene Rocks, McMurdo Sound, East Antarctica*, American Geophysical Union, Antarctic Research Series.

Sharman, G., and E. T. Newton

1894   Notes on some fossils from Seymour Island, in the Antarctic regions, obtained by Dr. Donald. Transactions of the Royal Society of Edinburgh, *37*(3): 707-709.

Smith, P. L.

1989   Paleoscene #11. Paleobiogeography and plate tectonics. *Geoscience Canada, 15*(4): 261-279.

Scotese, C. R., and C. R. Denham

1988   *Plate tectonics for the Macintosh.* Earth in Motion Technologies, Austin, Texas.

Shackleton, N. J., and J. P. Kennett

1975   Paleotemperature history of the Cenozoic and the initiation of Antarctic glaciation: oxygen and carbon isotope analyses in DSDP sites 277, 279, and 281. *Initial Reports of the Deep Sea Drilling Project, 29*: 743-755.

Stevens, G. R.

1989   The nature and timing of biotic links between New Zealand and Antarctica in Mesozoic and early Cenozoic times. *In* J. A. Crame (Ed.), *Origins and Evolution of the Antarctic Biota. Geol. Soc. London Spec. Publ., 47*: 141-166.

Stagg, H.. J., and J. B. Willcox

1992   A case for Australia-Antarctica separation in the Neocomian (ca. 125 Ma). *Tectonophysics, 210*: 21-32.

Stilwell, J. D.

1991   Late Campanian to Eocene Mollusca of New Zealand: changes in composition as a consequence of the break-up of Gondwana. *In* 8th International Symposium on Gondwana, 24-28 June 1991, Hobart, Tasmania, Abstract. Published on the behalf of the Subcommission on Gondwana Stratigraphy (IUGS), C/-Gondwana 8, pp. 78-79.

1994a   New insights into changes in faunal composition of Gondwana Realm Mollusca across the

Cretaceous-Tertiary boundary: the New Zealand fauna. *In* 9th International Symposium on Gondwana, 10-14 January, 1994, Hyderabad, India, Abstract. Published on behalf of the Subcommission on Gondwana and Stratigraphy (IUGS), C/- Gondwana 9, pp. 44-45.

1994b   Latest Cretaceous to earliest Paleogene molluscan faunas of New Zealand: changes in composition as a consequence of the break-up of Gondwana and extinction. Ph. D. Thesis, University of Otago, Dunedin, New Zealand, 1630 p.

1995   Eocene Mollusca of high southern latitudes: evidence for marine communication between East and West Antarctica. *In* Congreso Paleogeno de America del Sur y Antartida, Santa Rosa, La Pampa, Argentina, 14-18 de Mayo de 1996, Abstracts Volume, pp. 37-38.

Tectonic and paleobiogeographic significance of the Chatham Islands, South Pacific, Late Cretaceous. *Palaeogeography, Palaeoclimatology, Palaeoecology, 136*: 97-119.

This volume   Eocene Mollusca (Bivalvia, Gastropoda and Scaphopoda) from McMurdo Sound: systematics and paleoecologic significance. *In* J. D. Stilwell and R. M. Feldmann (Eds.), *Paleobiology and Paleoenvironments of Eocene Rocks, McMurdo Sound, East Antarctica.* American Geophysical Union, Antarctic Research Series.

Stilwell, J. D., and W. J. Zinsmeister
1992   *Molluscan systematics and biostratigraphy. Lower Tertiary La Meseta Formation, Seymour Island, Antarctic Peninsula.* American Geophysical Union, Antarctic Research Series, *55*, 192 p.

Symonds, P. A., J. B. Colwell, H. I. M. Struckmeyer, J. B. Willcox, and P. J. Hill
1996   Mesozoic rift basin development off Eastern Australia. *In* Mesozoic Geology of the Eastern Australia Plate Conference, 23-26 September 1996, Queensland, Geological Society of Australia, Extended Abstracts, *43*: 528-542.

Veevers, J. J.
1986   Break-up of Australia and Antarctica estimated as mid-Cretaceous (95 ± 5 Ma) from magnetic and seismic data at the continental margin. *Earth and Planetary Science Letters, 77*: 91-99.

Weissel, J. K., and D. E. Hayes
1972   *Magnetic anomalies in the southeast Indian Ocean.* American Geophysical Union, Antarctic Research Series, *19*: 165-196.

Wilckens, O.
1911   Die Mollusken der antarktischen Tertiärformation. W*iss. Ergeb. Swed. Südpolar-*

Exped. 1901-1903, *3*(1): 1-62.
1924   Die Tertiäre fauna der Cockburn-insel (Westantarktika). *Wiss. Ergeb. Swed. Südpolar-Exped.* 1901-1903, *1*(5): 1-18.

Willis, P. M. A., and J. D. Stilwell
This volume   A possible piscivorous crocodile from Eocene deposits of McMurdo Sound, East Antarctica. *In* J. D. Stilwell and R. M. Feldmann (Eds.), *Paleobiology and Paleoenvironments of Eocene Rocks, McMurdo Sound, East Antarctica,* American Geophysical Union Antarctic Research Series.

Zinsmeister, W. J.
1978   Effect of the formation of the West Antarctic Ice Sheet on shallow-water marine faunas of Chile. *Ant. Journal of the United States, 8*(4): 25-26.

1979   Biogeographic significance of the Late Mesozoic and early Tertiary molluscan faunas of Seymour Island (Antarctic Peninsula) to the final breakup of Gondwanaland. *In* J. Gray and A. J. Boucot (Eds.), *Historical Biogeography, Plate Tectonics, and the Changing Environment,* Oregon State University Press, pp. 349-355.

1982   Late Cretaceous-early Tertiary molluscan biogeography of the southern circum-Pacific. *Journal of Paleontology, 55*: 1083-1102.

1984   Late Eocene bivalves (Mollusca) from the La Meseta Formation, collected during the 1974-75 joint Argentine-American expedition to Seymour Island, Antarctic Peninsula. *Journal of Paleontology, 58*(6): 1497-1527.

1987   Cretaceous paleogeography of Antarctica. *Palaeogeography, Palaeoclimatology, Palaeoecology, 59*: 197-206.

Zinsmeiser, W. J., and J. D. Stilwell
1990   First record of the family Ringiculidae (Gastropoda) from the middle Tertiary of Antarctica. *Journal of Paleontology, 64*(3): 373-376.

Zullo, V. A., R. M. Feldmann, and L. A. Wiedman
1988   Balanomorph Cirripedia from the Eocene La Meseta Formation, Seymour Island, Antarctica. *In* R. M. Feldmann and M. O. Woodburne (Eds.), *Geology and Paleontology of Seymour Island, Antarctic Peninsula, Memoirs of the Geological Society of America, 169*: 459-464.

Jeffrey D. Stilwell, School of Earth Sciences, James Cook University, Townsville, Q4811, Australia

William J. Zinsmeister, Department of Earth and Atmospheric Sciences, Purdue University, West Lafayette, IN 47906-1397